# Molecular Plant–Microbe Interactions

# MOLECULAR PLANT–MICROBE INTERACTIONS

Edited by

**Kamal Bouarab**

*Département de Biologie*
*Université de Sherbrooke*
*Sherbrooke*
*Québec*
*Canada*

**Normand Brisson**

*Department of Biochemistry*
*Université de Montréal*
*Montréal*
*Québec*
*Canada*

and

**Fouad Daayf**

*Department of Plant Science*
*University of Manitoba*
*Winnipeg*
*Manitoba*
*Canada*

www.cabi.org

**CABI is a trading name of CAB International**

CABI Head Office
Nosworthy Way
Wallingford
Oxon OX10 8DE
UK

Tel: +44 (0)1491 832111
Fax: +44 (0)1491 833508
E-mail: cabi@cabi.org
Website: www.cabi.org

CABI North American Office
875 Massachusetts Avenue
7th Floor
Cambridge, MA 02139
USA

Tel: +1 617 395 4056
Fax: +1 617 354 6875
E-mail: cabi-nao@cabi.org



A catalogue record for this book is available from the British Library, London, UK.

**Library of Congress Cataloging-in-Publication Data**

Molecular plant-microbe interactions / edited by Kamal Bouarab, Normand Brisson and Fouad Daayf.
p. cm.
Includes bibliographical references and index.
ISBN 978-1-84593-574-0 (alk. paper)
1. Plants--Disease and pest resistance. 2. Plant-microbe relationships. 3. Fungal diseases of plants. 4. Virus diseases of plants. I. Bouarab, Kamal. II. Brisson, Normand, 1955- III. Daayf, Fouad. IV. Title.

SB750.M67 2009
632'.3--dc22

2009007330

ISBN-13: 978 1 84593 574 0

Typeset by Columns Design, Reading.
Printed and bound in the UK by the MPG Books Group, Bodmin.

The paper used for the text pages in this book is FSC certified. The FSC (Forest Stewardship Council) is an international network to promote responsible management of the world's forests.

# Contents

# Contributors

Abd El Rahman, Taha, *Centre de Recherche en Amélioration Végétale, Département de Biologie, Université de Sherbrooke, 2500 Boulevard de l'Université, Sherbrooke, Québec, J1K 2R1, Canada.*

Asai, Shuta, *Laboratory of Defense in Plant–Pathogen Interactions, Graduate School of Bioagricultural Sciences, Nagoya University, Chikusa, Nagoya 464-8601, Japan.*

Bathily, Hamed, *Centre de Recherche en Amélioration Végétale, Département de Biologie, Université de Sherbrooke, 2500 Boulevard de l'Université, Sherbrooke, Québec, J1K 2R1, Canada.*

Beaulieu, Carole, *Centre SÈVE, Département de biologie, Université de Sherbrooke, Sherbrooke, Québec, J1K 2R1, Canada; carole.beaulieu@usherbrooke.ca*

Bouarab, Kamal, *Centre de Recherche en Amélioration Végétale, Département de Biologie, Université de Sherbrooke, 2500 Boulevard de l'Université, Sherbrooke, Québec, J1K 2R1, Canada; Kamal.Bouarab@USherbrooke.ca*

Boyle, Patrick, *Department of Biological Sciences, Brock University, 500 Glenridge Avenue, St Catharines, Ontario, L2S 3A1, Canada.*

Brisson, Normand, *Department of Biochemistry, Université de Montréal, Montréal, Québec, Canada; normand.brisson@umontreal.ca*

Cappadocia, Laurent, *Department of Biochemistry, Université de Montréal, Montréal, Québec, Canada.*

Daayf, Fouad, *Department of Plant Science, University of Manitoba, 222, Agriculture Building, Winnipeg, Manitoba, R3T 2N2, Canada; Daayff@cc.umanitoba.ca*

Després, Charles, *Department of Biological Sciences, Brock University, 500 Glenridge Avenue, St Catharines, Ontario, L2S 3A1, Canada; cdespres@brocku.ca*

Desveaux, Darrell, *Centre for the Analysis of Genome Evolution and Function & Department of Cell and Systems Biology, University of*

*Toronto, 25 Willcocks Street, Toronto, Ontario, M5S 3B2, Canada; darrell.desveaux@utoronto.ca*

Dow, J. Maxwell, *BIOMERIT Research Centre, Department of Microbiology, National University of Ireland, Cork, Ireland.*

Dunoyer, Patrice, *Institut de Biologie Moléculaire des Plantes, CNRS, 67084 Strasbourg Cedex, France; patrice.dunoyer@ibmp-ulp.u-strasbg.fr*

El Hadrami, Abdelbasset, *Department of Plant Science, University of Manitoba, 222, Agriculture Building, Winnipeg, Manitoba, R3T 2N2, Canada.*

El Hadrami, Ismail, *University Cadi Ayyad, Marrakech, Morocco.*

El Oirdi, Mohamed, *Centre de Recherche en Amélioration Végétale, Département de Biologie, Université de Sherbrooke, 2500 Boulevard de l'Université, Sherbrooke, Québec, J1K 2R1, Canada.*

Fobert, Pierre R., *National Research Council Canada, Plant Biotechnology Institute, 110 Gymnasium Place, Saskatoon, Saskatchewan, S7N 0W9, Canada.*

González-Lamothe, Rocío, *Centre de Recherche en Amélioration Végétale, Département de Biologie, Université de Sherbrooke, 2500 Boulevard de l'Université, Sherbrooke, Québec, J1K 2R1, Canada.*

Hamed, Bathily, *Centre de Recherche en Amélioration Végétale, Département de Biologie, Université de Sherbrooke, 2500 Boulevard de l'Université, Sherbrooke, Québec, J1K 2R1, Canada.*

Harris, Linda J., *Eastern Cereal and Oilseed Research Centre, Agriculture and Agri-Food Canada, Ottawa, Ontario, K1A 0C6, Canada.*

Hijri, Mohamed, *Institut de Recherche en Biologie Végétale, Université de Montréal, 4101 rue Sherbrooke Est, Montréal, Québec, H1X 2B2, Canada.*

Ichikawa, Tatsushi, *Laboratory of Defense in Plant–Pathogen Interactions, Graduate School of Bioagricultural Sciences, Nagoya University, Chikusa, Nagoya 464-8601, Japan.*

Kobayashi, Michie, *Laboratory of Defense in Plant–Pathogen Interactions, Graduate School of Bioagricultural Sciences, Nagoya University, Chikusa, Nagoya 464-8601, Japan.*

Lewis, Jennifer D., *Department of Cell and Systems Biology, University of Toronto, 25 Willcocks Street, Toronto, Ontario, M5S 3B2, Canada.*

Maréchal, Alexandre, *Department of Biochemistry, Université de Montréal, Montréal, Québec, Canada.*

Miyagawa, Noriko, *Laboratory of Defense in Plant–Pathogen Interactions, Graduate School of Bioagricultural Sciences, Nagoya University, Chikusa, Nagoya 464-8601, Japan.*

Moffett, Peter, *Département de Biologie, Université de Sherbrooke, 2500 Boulevard de l'Université, Sherbrooke, Québec, J1K 2R1, Canada; Peter.Moffett@cornell.edu*

Nasmith, Charles G., *Eastern Cereal and Oilseed Research Centre, Agriculture and Agri-Food Canada, Ottawa, Ontario, K1A 0C6, Canada.*

Ouellet, Thérèse, *Eastern Cereal and Oilseed Research Centre, Agriculture and Agri-Food Canada, Ottawa, Ontario, K1A 0C6, Canada.*

Parent, Jean-Sébastien, *Department of Biochemistry, Université de Montréal, Montréal, Québec, Canada.*

Roy, Sébastien, *Centre SÈVE, Département de biologie, Université de Sherbrooke, Sherbrooke, Québec, J1K 2R1, Canada.*

Sacco, Melanie A., *Department of Biological Science, California State University, Fullerton, 800 North State College Blvd., Fullerton, CA 92831-3599, USA.*

Sansregret, Raphaël, *Centre de Recherche en Amélioration Vegétale, Département de Biologie, Université de Sherbrooke, 2500 Boulevard de l'Université, Sherbrooke, Québec, J1K 2R1, Canada.*

Schreiber, Karl, *Department of Cell and Systems Biology, University of Toronto, 25 Willcocks Street, Toronto, Ontario, M5S 3B2, Canada.*

Simao-Beaunoir, Anne-Marie, *Centre SÈVE, Département de biologie, Université de Sherbrooke, Sherbrooke, Québec, J1K 2R1, Canada.*

Song, Fengming, *National Key Laboratory for Rice Biology, Institute of Biotechnology, Zhejiang University, Hangzhou, Zhejiang 310029, China; fmsong@zju.edu.cn*

St-Arnaud, Marc, *Institut de Recherche en Biologie Végétale, Université de Montréal, 4101 rue Sherbrooke Est, Montréal, Québec, H1X 2B2, Canada.*

Subramaniam, Rajagopal, *Eastern Cereal and Oilseed Research Centre, Agriculture and Agri-Food Canada, Ottawa, Ontario, K1A 0C6, Canada; subramaniamra@agr.gc.ca*

Vojnov, Adrián A., *Instituto de Ciencia y Tecnolgía Dr. Cesar Milstein, CONICET; avojnov@fundacioncassara.org.ar*

Wadsworth, Santiago, *Institut de Biologie Moléculaire des Plantes, CNRS, 67084 Strasbourg Cedex, France.*

Yoshioka, Hirofumi, *Laboratory of Defense in Plant–Pathogen Interactions, Graduate School of Bioagricultural Sciences, Nagoya University, Chikusa, Nagoya 464-8601, Japan; hyoshiok@agr.nagoya-u.ac.jp*

Yoshioka, Miki, *Laboratory of Defense in Plant–Pathogen Interactions, Graduate School of Bioagricultural Sciences, Nagoya University, Chikusa, Nagoya 464-8601, Japan.*

Zhang, Huijuan, *National Key Laboratory for Rice Biology, Institute of Biotechnology, Zhejiang University, Hangzhou, Zhejiang 310029, China.*

Zhang, Shuqun, *Department of Biochemistry, University of Missouri-Columbia, MO 65211, USA.*

Zimmerman, Erin, *Institut de Recherche en Biologie Végétale, Université de Montréal, 4101 rue Sherbrooke Est, Montréal, Québec, H1X 2B2, Canada; erin.zimmerman@umontreal.ca*

# Preface

Recent developments in molecular biology and in the burgeoning omics have brought about a great deal of new data in all areas of plant sciences. However, the use of these data towards their exploitation into higher performance plants has proved to be a slow process. For example, although producing genomics and proteomics data has become routine in a large number of laboratories around the world, functional genomics studies to understand the meaning of the accumulating data still lag behind. Carrying out these studies in such important areas as plant development, photosynthesis or plant responses to abiotic stress, bounces within a large window of complexity and difficulty levels. These levels escalate when more than one organism is involved, in either a mutually beneficial or an antagonistic interaction with the plant. However, navigating through such a network of interactions makes the journey more exciting, at least from the view of a plant pathologist. Studying molecular plant–microbe interactions is very stimulating indeed. There is no guarantee that a microbe, even from the same species or race within it, would act exactly the same way in its coevolution with the host plant. The same applies to the plant regarding 'upgrading' its arsenal to fight external threats. This creates a certain dynamism that fuels new discoveries, and which scientists in the molecular plant–microbe interactions field value so much. In such a dynamic discipline, it is useful to revisit the field more often than, say, every 20 years. In this volume, authors of world repute in different aspects of molecular plant–microbe interactions have agreed to contribute a chapter about their research and their views on the current developments in the fields of plant defences, pathogen counter-defences and mutually beneficial plant–microbe interactions.

This book explores recent discoveries in the area of molecular plant–microbe interactions. It focuses mainly on the mechanisms controlling plant disease resistance and the cross talk among the signalling pathways involved, and the strategies used by fungi and viruses to suppress these defences.

Furthermore, two chapters related to the role of symbionts during their interactions with plants are included.

We would like to thank all the contributors for their hard work towards meeting the deadlines for this volume.

The Editors
**Kamal Bouarab**
**Normand Brisson**
**Fouad Daayf**

# 1 Plant RNA-silencing Immunity and Viral Counter-defence Strategies

SANTIAGO WADSWORTH AND PATRICE DUNOYER

*Institut de Biologie Moléculaire des Plantes, Strasbourg, France*

## Abstract

RNA silencing is a conserved eukaryotic process mediated by small RNA molecules that inhibit gene expression at the transcriptional, mRNA-stability or translational level through sequence-specific interactions. Diverse roles have been identified for RNA silencing such as genome defence against mobile DNA elements or downregulation of specific factors during plant and animal development. In plants, RNA silencing plays a crucial role in antiviral defence by inhibiting viral accumulation and sometimes preventing systemic infection. As a counter-defence mechanism, viruses have evolved a set of anti-silencing strategies, of which the most common is the production of viral suppressors of RNA silencing (VSRs). Here we review the different strategies underlying VSRs action including prevention of viral-derived small (vs)RNAs synthesis, vsRNAs sequestration or inhibition of vsRNA-guided effector complexes. We will also underline the consequences of this molecular arms race on the evolution of both viral and host genomes.

## 1.1 Introduction

RNA silencing is an ancient eukaryotic process involved in the control of gene expression. It is triggered by double-stranded (ds)RNA and causes a sequence-specific shut down of the expression of genes with sequences identical or highly similar to the initiating dsRNA (Fire *et al.*, 1998; Wesley *et al.*, 2001). RNA silencing may act at both the RNA and the DNA levels. Mechanisms of silencing at the RNA level (called post-transcriptional gene silencing (PTGS) in plants or RNA interference (RNAi) in animals) include mRNA cleavage or translational repression. RNA silencing at the DNA level involves DNA and/or histone methylation and subsequent transcriptional gene silencing (TGS) through heterochromatin formation and maintenance (Bartel, 2004; Jones-Rhoades *et al.*, 2006). All these manifestations of RNA silencing rely on the action of small RNA (sRNA) molecules of 21–24 nucleotides (nt) in length, which originate from the processing of the dsRNA trigger (Hamilton and Baulcombe, 1999; Elbashir *et al.*, 2001). During PTGS, these sRNA molecules control stability or regulate translation of their mRNA targets by guiding endogenous effector complexes (Hammond *et al.*, 2000; Bartel, 2004; Jones-Rhoades *et al.*, 2006).

Originally, PTGS was first observed in attempts to overexpress an endogenous gene by transforming petunia plants with an extra copy of the gene of interest, which led to the extinction of both endogenous and transgenic copies (Napoli *et al.*, 1990). Subsequently, the inhibition of gene expression was confirmed in the nematode *Caenorhabditis elegans* by injecting sense or antisense copies of the target mRNA (S-PTGS and AS-PTGS, respectively) (Fire *et al.*, 1991; Guo and Kemphues, 1995). Nowadays, the strongest and most commonly used mechanism to inhibit gene expression in eukaryotic cells is PTGS triggered by an inverted-repeat construct (IR-PTGS) that produces a dsRNA molecule with a hairpin-like structure (Beclin *et al.*, 2002; Giordano *et al.*, 2002).

Besides its essential role in plant and animal development, the first biological role attributed to RNA silencing in plants was defence against viral infections (Lindbo *et al.*, 1993; Ratcliff *et al.*, 1997; Anandalakshmi *et al.*, 1998; Brigneti *et al.*, 1998; Kasschau and Carrington, 1998; Hamilton and Baulcombe, 1999). Attempts to overexpress an endogenous gene from recombinant viral vectors did not result in enhanced protein accumulation but led to the specific degradation of the corresponding mRNA (Ruiz *et al.*, 1998). This phenomenon, called virus-induced gene silencing (VIGS), probably triggered by the dsRNA intermediates formed during viral replication, was suggested to be a manifestation of an RNA silencing-based antiviral defence response (Ratcliff *et al.*, 1999). Currently, several indications suggest that RNA silencing is the primary immune system against viruses in plants: (i) sRNAs derived from a viral genome (viral-derived small RNAs or vsRNAs) are invariably detected during infection; (ii) mutant plants affected in the silencing pathways are hypersusceptible to viral infections; and (iii) as a counter-defensive strategy, plant viruses express proteins that are able to suppress the antiviral silencing response (Li and Ding, 2006; Ding and Voinnet, 2007). Indeed, back in 1998, two different viral proteins, previously identified as important pathogenicity determinants, were characterized as effective suppressors of RNA silencing (Anandalakshmi *et al.*, 1998; Brigneti *et al.*, 1998). Since then, expression of viral suppressors of RNA silencing (VSRs) has emerged as a major counter-defence mechanism against the antiviral RNA silencing response (Li and Ding, 2006; Ding and Voinnet, 2007). After a short presentation of the core mechanism of RNA silencing and its different pathways in the model plant *Arabidopsis thaliana*, we first discuss how viruses induce the antiviral silencing response in plants. We then review the different strategies underlying VSRs action as well as their effect on endogenous silencing pathways and on the evolution of both viral and host genomes.

## 1.2 Plant RNA Silencing Pathways

### The core mechanism

The initial step of RNA silencing depends on the recognition of a dsRNA molecule by an RNase III-like enzyme called Dicer (or DCL, for Dicer-like in plants) that processes the trigger into sRNA duplexes of 21–24 nt in length with 2 nt 3' overhangs (Hammond, 2005). These sRNA are divided in two main classes in both plants and animals: small interfering RNAs (siRNAs) and microRNAs (miRNAs).

siRNAs are processed from long, perfectly based-paired dsRNAs, whereas miRNAs originate from primary miRNA transcripts (pri-miRNAs) that form imperfect intramolecular hairpin-like structures. A single Dicer enzyme in worms and humans produces both miRNAs and siRNAs, whereas in *Drosophila* they are produced by Dicer-1 and Dicer-2, respectively (Hammond, 2005). In plants, Dicer-like proteins are even more diverse. For instance, *A. thaliana* encodes four Dicer-like enzymes (DCL 1–4). Whereas most mature miRNAs synthesis relies on DCL1 (Bartel, 2004), populations of 21, 22 and 24 nt-long siRNAs are synthesized from dsRNA through DCL4, DCL2 and DCL3 activity, respectively (Brodersen and Voinnet, 2006; Vazquez, 2006; Chapman and Carrington, 2007).

sRNAs are then incorporated into argonaute (AGO)-containing effector complexes termed RNA-induced silencing complexes (RISCs) and RNA-induced initiation of transcriptional silencing complexes (RITS) and, as part of these complexes, direct sequence-specific PTGS or TGS, respectively (Hammond *et al.*, 2000; Ekwall, 2004). During PTGS, the outcome of RISC activity is either cleavage and/or translational inhibition of the target mRNA and this probably relies on the degree of complementarity with the sRNA and on the different cellular factors involved. Proteins of the AGO family (ten different members in *Arabidopsis*) have been crucially implicated in the functions of both RISC and RITS complexes. AGO proteins contain at least one single strand (ss)RNA-binding PAZ domain and a PIWI domain that confers the endonucleolytic (or slicer) activity. So far, slicing activity in *Arabidopsis* has been only demonstrated for AGO1, AGO4 and AGO7 (Baumberger and Baulcombe, 2005; Qi *et al.*, 2006; Montgomery *et al.*, 2008) and the former was shown to direct both miRNA- and 21 nt-long siRNA-mediated target cleavage without requiring further protein partners (Baumberger and Baulcombe, 2005; Dunoyer *et al.*, 2007). Upon unwinding of the sRNA duplex and loading into the AGO protein, the selected strand (guide strand) guides RISC to target all RNA molecules presenting sequence complementarity to the incorporated sRNA, whereas the non-selected strand (passenger strand) is degraded. Interestingly, although the imperfect complementary strand of the miRNA, called miRNA star (miRNA*), is also normally degraded, a recent study suggests a potential biological role for miRNA*s in *Drosophila* (Okamura *et al.*, 2008). Finally, during their synthesis, a plant methyltransferase enzyme called HEN1 protects all classes of sRNAs from uridylation and subsequent degradation through addition of a methyl group on the 2' hydroxy group at their 3' end termini (Yang *et al.*, 2006).

A third class of sRNA that do not depend on Dicer but rather on AGO-like proteins for their biosynthesis are called the piwi-associated interfering RNAs (piRNAs). However, as so far evidence of these 26–30 nt-long RNAs is confined exclusively to the germline of fruit flies and vertebrates (Zamore, 2007), they will not be covered in the present chapter.

## Endogenous *Arabidopsis* silencing pathways

In plants, most miRNAs seem to function primarily like siRNAs: they are incorporated into an AGO1-containing RISC that retrieves and cleaves cellular mRNAs (Llave *et al.*, 2002), many of which encode transcription factors (TF) involved in important developmental processes (Rhoades *et al.*, 2002). Based on those observations, it has been proposed that miRNAs ensure clearance of regulatory transcripts from specific daughter cell lineages and thereby enable cell differentiation and tissue identity, an idea that has now received experimental validation in plants (Kidner and Martienssen, 2004; Parizotto *et al.*, 2004).

In addition to the four DCL and ten AGO paralogues, the *Arabidopsis* genome encodes six RNA-dependent RNA polymerases (RDR). RDR proteins participate in mechanisms that account, in plants, nematodes, fungi and other organisms for sRNA amplification through *de novo* dsRNA synthesis from ssRNA template molecules (a process also known as 'transitivity'; Himber *et al.*, 2003). Recent studies revealed that several pathways involving specific DCL/AGO/RDR combinations produce a highly diverse set of sRNAs acting in PTGS or TGS. For instance, RDR6 (also known as SDE1 or SGS2; Dalmay *et al.*, 2000; Mourrain *et al.*, 2000) has been implicated, together with DCL4 and DRB4 (a double-stranded RNA binding protein (dsRBP)), in the production of 21 nt-long *trans*-acting siRNAs (tasiRNAs) that require AGO1 or AGO7 functions to mediate post-transcriptional silencing of genes controlling heteroblasty and leaf polarity (Xie *et al.*, 2005; Adenot *et al.*, 2006; Fahlgren *et al.*, 2006; Hunter *et al.*, 2006). tasiRNAs derive from non-coding, single-stranded transcripts that are, upon miRNA-guided cleavage, converted into dsRNA by RDR6 giving rise to siRNAs produced in a specific 21 nt phase registry (Peragine *et al.*, 2004; Vazquez *et al.*, 2004; Yoshikawa *et al.*, 2005).

RDR2 is required for the production of 24 nt-long DCL3-dependent siRNAs (called repeat-associated siRNAs or rasiRNAs; Xie *et al.*, 2004). In yeast, rasiRNAs interact with AGO to form the RITS complex. In plants, a hypothetical RITS-like complex, containing AGO4 and/or AGO6 and a plant-specific RNA polymerase called PolIVb, guides RNA-directed DNA methylation (RdDM), leading to TGS of repeated DNA loci and mobile elements (Herr *et al.*, 2005; Kanno *et al.*, 2005; Matzke and Birchler, 2005; Pontier *et al.*, 2005; Zaratiegui *et al.*, 2007). Another class of endogenous 24 nt siRNAs, named natural-antisense-transcript siRNAs (nat-siRNAs), arise from dsRNA formed by two stress-induced overlapping transcripts through a complex pathway involving the action of DCL2/RDR6/PolIV (Borsani *et al.*, 2005). As yet, no AGO has been assigned to the nat-siRNA pathway.

## Spreading of RNA silencing

In plants and in some animals, an outstanding property of RNA silencing is that its effects can extend beyond the sites of its initiation, owing to the movement of signal molecules. These silencing signals must have RNA components that account for the nucleotide sequence-specificity of their effects. In plants, a first movement process involves cell-to-cell trafficking through the plasmodesmal channels connecting plant cells (Himber *et al.*, 2003). For instance, in *A. thaliana*, tissue-specific expression of an inverted repeat leads to the production of a signal molecule that normally spreads and directs the cleavage of target mRNA over ten to 15 cells. This 'short-range' cell-to-cell silencing movement is independent of both the presence of the targeted RNA in the recipient cells and the RNA-dependent RNA polymerase activity of RDR6 and is mediated by DCL4-dependent 21 nt-long siRNAs (Himber *et al.*, 2003; Dunoyer and Voinnet, 2005). This short-range silencing signal can be further amplified to give extensive cell-to-cell spread, ultimately invading the entire leaf lamina, through reiteration of short-distance signalling events. This second process is coined 'long-range' cell-to-cell silencing movement. It requires, in cells that have received the short-range signal, the production of secondary 21 nt-long siRNAs from homologous transcripts converted into new dsRNA through the activity of RDR6. This transitivity process re-amplifies the initial pool of primary 21 nt-long siRNAs (Himber *et al.*, 2003). The secondary siRNAs, produced in a DCL4-dependent manner (Moissiard *et al.*, 2007), are generated from both 5' and 3' regions of the sequence initially targeted by the primary siRNAs resulting in silencing amplification (Dalmay *et al.*, 2000).

A third silencing movement process, known as the systemic 'long-distance' movement, triggers silencing of the cognate messenger in tissues remotely located from the initiation zone. This other non-cell-autonomous silencing phenomenon was first described, through grafting-experiments, in solanaceous plant species where it follows a strict source-to-sink pattern, indicating phloem-mediated transport (Palauqui *et al.*, 1997; Voinnet *et al.*, 1998). Although the nature of the systemic signal is still unknown, previous work has suggested that it is distinct from the cell-to-cell signal (Himber *et al.*, 2003). For instance, treatment with non-toxic cadmium concentrations prevented systemic but not cell-to-cell silencing of a reporter transgene indicating that this two-stage process can be pharmacologically uncoupled (Ueki and Citovsky, 2001). Interestingly, a tight correlation between the accumulation of 24 nt-long siRNA and the onset of silencing in systemic leaves was shown even if, so far, there is no clear evidence that this molecule represents the systemic signal per se (Hamilton *et al.*, 2002; Himber *et al.*, 2003). RDR6, together with RDR2, PolIVa, DCL3 and to a lesser extent AGO4, have been critically implicated in the perception of the long-distance silencing signal whereas the genetic requirements for its production remain elusive (Schwach *et al.*, 2005; Brosnan *et al.*, 2007).

## 1.3 RNA Silencing as the Antiviral Immune System in Plants

### RNA viruses trigger sequence-specific RNA degradation

An early observation linking antiviral defence and a sequence-specific mechanism operating at the RNA level in plants came from the observation that transgenic tobacco plants expressing an untranslatable version of the Potato virus Y (PVY) or Tobacco etch virus (TEV) coat protein, became resistant to the virus from which the transgene was derived (Lindbo *et al.*, 1993; Smith *et al.*, 1994). This resistance was either effective in the inoculated leaves or was induced in emerging tissues, in which case it was called 'recovery'. Interestingly, recovered plants became resistant to later challenges with the same virus or with heterologous recombinant viruses carrying a fragment of the initial viral genome, whereas other, unrelated, viruses established normal systemic infection indicating the sequence specificity of this phenomenon (Ratcliff *et al.*, 1997).

Subsequently, some studies showed that recovery does not require homology between the viral and plant genomes. For instance, wild-type (wt) plants infected with a recombinant Potato virus X (PVX) expressing the green fluorescent protein (GFP) gene were resistant to later challenges with a recombinant Tobacco mosaic virus (TMV) carrying the same GFP sequence, but not to TMV carrying an unrelated GUS insert (Ratcliff *et al.*, 1999). This RNA-mediated resistance explained, at least partly, the principle of 'cross-protection', whereby attenuated strains of a given virus are used to immunize crops against severe strains of the same virus (Sequeira, 1984). Finally, the process known as VIGS whereby recombinant RNA viruses can trigger silencing of endogenous mRNAs in wt plants provided that they share homology with exon sequences of host nuclear genes is also a consequence of antiviral silencing (Kumagai *et al.*, 1995; Ruiz *et al.*, 1998). This leads to low levels of both viral and endogenous mRNA indicating that viruses are both trigger and targets of RNA silencing.

### In the beginning was dsRNA

The finding that viral-derived siRNAs (vsRNAs) of both positive (+) and negative (–) polarities accumulate in infected plants confirmed unambiguously the antiviral role of RNA silencing (Hamilton and Baulcombe, 1999). Since dsRNA is known to trigger RNA silencing, vsRNAs were proposed to be Dicer products resulting from the processing of dsRNA intermediates that transiently accumulate during RNA virus replication. This was confirmed by the cloning and sequencing of relative equal amounts of vsRNAs coming from both the (+) and (–) strands of Cucumber yellows closterovirus (CuYV) and Turnip mosaic potyvirus (TuMV) infected plants (Yoo *et al.*, 2004; Ho *et al.*, 2006). However, this was not the case for all virus families. Indeed, recent studies described an asymmetric accumulation of vsRNAs that preferentially derived from discrete hotspots present within the (+)-stranded RNA genome of tombusviruses and carmoviruses (Molnar *et al.*, 2005; Ho *et al.*, 2006). Structural analyses of these regions revealed that they correspond to stem loop-like structures formed by intramolecular base pairing that can be recognized and processed by plant Dicers to produce the vsRNAs (Plate 1). Moreover, as this non-uniform distribution of vsRNAs was also observed in the case of TMV and PVX, it suggests that this feature might be a general characteristic of (+)-strand RNA viruses (Molnar *et al.*, 2005). Interestingly, Itaya and colleagues have shown that during potato spindle tuber viroid (PSTVd) infection, viroid-derived siRNAs were predominantly generated from discrete regions of the (+) strand genome, suggesting that highly structured RNA is also the primary substrate for DCL activity during viroid infection (Itaya *et al.*, 2007).

### *Plate 1*

Plate 1 shows vsRNA production has different Dicer requirements regarding the nature of the viral genome. In the pathway shown in Plate 1(a) DCL4 is the primary Dicer to process RNA viruses and is replaced by DCL2 if the former is not functional. In the case of DNA viruses in the pathway shown in Plate 1(b), DCL1 may facilitate the access of viral double-stranded (ds)RNA structures to the other three Dicers. vsRNAs of 21, 22 and 24 nt in length are produced by DCL4, DCL2 and DCL3, respectively. DCL3-dependent vsRNAs may inhibit DNA virus accumulation through DNA/histone methylation of their genomic DNA. In the pathway shown in Plate 1(c) aberrant (ab) viral mRNA lacking a cap or a poly A tail serve as substrate to produce *de novo* dsRNA, through the action of host-RNA dependent RNA polymerase, that will be further processed to produced vsRNAs. After being stabilized through HEN1-dependent 2′ *O*-methylation, vsRNAs are unwound by an ATP-dependent RNA helicase and then incorporated into an AGO-containing RISC. AGO1 is presented as the major antiviral slicer but other AGO paralogues are likely to be involved. The RISC complex is then directed to the viral mRNAs sharing sequence complementarity with the incorporated-guide strand, while the non-incorporated-passenger strand is degraded. Targeted viral mRNAs are then degraded following RISC mediated cleavage. Alternatively viral mRNAs may be translationally repressed, but this possibility is yet to be demonstrated. DCL proteins are represented in association with a dsRNA-binding protein.

In the case of plant DNA viruses, vsRNAs may be produced by two different mechanisms, depending on the nature of the viral genome. The 5' end of the polycistronic transcript of the dsDNA Cauliflower mosaic pararetrovirus (CaMV), called the '35S leader' region, represents the major source of vsRNAs due to its extensive secondary structure that serves as Dicer substrate. The fact that such a structure had not been counter-selected is probably explained by its biological role during the viral life cycle (Moissiard and Voinnet, 2006). In the case of ssDNA viruses, such as geminiviruses, it has been suggested that vsRNAs may arise from dsRNA formed by pairing of sense and antisense transcripts produced during the transcription of their circular genomes (Chellappan *et al.*, 2004) (Plate 1).

Finally, another possible source of vsRNAs might rely on the activity of endogenous RDRs to produce new Dicer substrates from viral transcripts, similarly to what occurs during S-PTGS. Indeed, high replication rates of viruses can produce aborted viral transcription products that will be recognized as aberrant mRNAs, a known template for RDRs (Plate 1). The involvement of *Arabidopsis* RDRs during antiviral defence has already received experimental support. For instance, tobacco plants with reduced RDR6 activity are more susceptible than wt plants to a broad set of viruses (Qu *et al.*, 2005; Schwach *et al.*, 2005). *Arabidopsis rdr6* mutant displays hypersensitivity to Cucumber mosaic virus (CMV) infection but not to TMV or Tobacco rattle virus (TRV) (Dalmay *et al.*, 2000; Mourrain *et al.*, 2000). Similarly, Tobacco mutants with an impaired RDR1 activity were found to be hypersusceptible to potex- and tobamoviruses (Xie *et al.*, 2004). Interestingly, overexpression of RDR1 only restored resistance to the latter suggesting a potential redundancy between RDRs members during antiviral defence (Yu *et al.*, 2003; Yang *et al.*, 2004). However, the involvement of RDRs in this antiviral response has to be carefully interpreted. Indeed, at least in the case of RDR6, its activity does not seem to be required for the production of vsRNAs per se and for the initiation of silencing in infected leaves but is rather involved to limit the spread of viruses in young emerging tissue (Qu *et al.*, 2005; Schwach *et al.*, 2005), in a process probably related to its role in the perception of the systemic silencing signal (see sections 'Spreading of RNA silencing' and 'Ready to repel the invader here and there: the systemic aspects of antiviral silencing' of this chapter).

## Dicing: transforming your enemy into a weapon

Identification of the antiviral Dicer in *Arabidopsis* was not straightforward as no individual DCL mutant displayed enhanced susceptibility to virus infection, with the possible exception of Turnip crinkle virus (TCV) on *dcl2* mutant plants (Xie *et al.*, 2004). This observation strongly suggested functional redundancy among DCLs in plant antiviral immunity (Brodersen and Voinnet, 2006). Accordingly, hypersusceptibility to several RNA viruses was recently found to occur only upon inactivation of both DCL4 and DCL2 (Bouche *et al.*, 2006; Deleris *et al.*, 2006; Fusaro *et al.*, 2006; Diaz-Pendon *et al.*, 2007). DCL4 is the primary antiviral Dicer and produces 21 nt-long vsRNAs (Plate 1). However, upon DCL4 inactivation, DCL2 rescues antiviral silencing by generating 22 nt-long vsRNAs that are normally below detection limits in wt infected plants, indicating the surrogate role of DCL2 in antiviral defence.

By contrast, no significant contribution was found for DCL1 and DCL3 in immunity against RNA viruses. Indeed, in the triple *dcl2/dcl3/dcl4* mutant background, the production of vsRNAs by the miRNA-specific DCL1 was shown to be almost negligible and similar viral accumulation was observed for CMV and TCV between wt and *dcl1* mutant plants (Deleris *et al.*, 2006). DCL3-dependent 24 nt-long vsRNAs are only significantly produced when DCL4, alone or in combination with DCL2, is inactivated, further supporting the hierarchical access of the four *Arabidopsis* DCLs to the viral dsRNA substrates. Moreover, presumably because DCL3 products normally guide transcriptional silencing at the DNA level, these 24 nt-long vsRNAs are not able to mediate degradation of homologous transcripts as evidenced by lack of DCL3-directed VIGS of the endogenous phytoene desaturase (PDS) gene: infection of recombinant TRV carrying a fragment of the PDS gene (TRV-PDS) leads to photobleaching of the emerging *Arabidopsis* leaves in all single or double Dicer combination mutants, except in the *dcl2/dcl4* double mutant, where only 24 nt vsRNAs are produced (Deleris *et al.*, 2006).

Interestingly, Dicer requirements in antiviral defence change when looking at DNA viruses replicating in the nucleus. In this case, the four DCLs seem to cooperate to mediate antiviral defence (Blevins *et al.*, 2006; Moissiard and Voinnet, 2006). In wt plants infected by CaMV and Cabbage leaf curl geminivirus (CaLCuV), vsRNAs accumulated mainly as 21 and 24 nt products of DCL4 and DCL3, respectively, indicating that these two were the prevalent Dicers (Plate 1). Similarly to what occurs with RNA viruses, DCL2 activity was mostly evident following DCL4 inactivation, further supporting its subordinate role in antiviral silencing. Increased susceptibility to CaMV was only observed when DCL4, DCL2 and DCL3 were simultaneously inactivated, suggesting that 24 nt-long vsRNA may trigger transcriptional silencing of viral DNA genomes that replicate in the nucleus as minichromosomes (Moissiard and Voinnet, 2006). Thus DCL3 has an antiviral function during DNA virus infection. Finally, inactivation of DCL1 led to a general reduction of CaMV-derived 21 nt and 24 nt-long vsRNA accumulation, whereas in the triple *dcl2/dcl3/dcl4* mutant background, production of DCL1-dependent vsRNAs was almost negligible. It was proposed that DCL1 acts early in the dicing pathway by excising the 35S leader region (the major source of CaMV-derived vsRNAs) from the primary transcript to facilitate its subsequent processing by DCL4 and DCL3 (Moissiard and Voinnet, 2006) in a process reminiscent of the nuclear and DCL1-dependent pri-miRNA to precursor miRNA (pre-miRNA) conversion step (Plate 1). This DCL1 facilitating activity was also recently shown to be similarly involved in the processing of inverted repeat transcripts produced during IR-PTGS (Dunoyer *et al.*, 2007). It probably does not affect hairpin structures present within RNA viruses because they are cytoplasmic.

Beside their intrinsic affinity for various dsRNA substrates, the different Dicer requirements for production of vsRNAs from RNA or DNA viruses cannot be exclusively explained by the subcellular localization of viral dsRNA and DCLs. Indeed, reporter gene fusion experiments showed that, for instance, DCL4 accumulates exclusively in the nucleus (Hiraguri *et al.*, 2005) yet it is the main antiviral Dicer for cytoplasmically replicating RNA viruses. Therefore, either the localization data are inaccurate or, may be more interestingly, DCLs might relocalize during viral infection. This question will hopefully be addressed in the near future.

## Setting up antiviral silencing complexes

The existence of an antiviral RISC in plants was not obvious as dicing of viral dsRNA is, in principle, sufficient to inhibit virus replication. However, VIGS experiments with RNA or DNA recombinant virus carrying a fragment of endogenous gene trigger symptoms that phenocopy those of the corresponding null mutants. This observation indicated that vsRNAs are able to inhibit the expression of homologous cellular transcripts, at least *in trans*, through RISC-guided cleavage (Ruiz *et al.*, 1998; Blevins *et al.*, 2006). But the effect of vsRNA-loaded RISC directly on viral RNA (*in cis*) was still to be demonstrated.

First indications of the existence of an antiviral RISC comes from TRV-mediated VIGS experiments. Equivalent amount of 21, 22 or 24 nt-long vsRNAs were detected in *dcl2/dcl3*, *dcl3/dcl4* and *dcl2/dcl4* mutant plants, respectively. However, only the latter exhibited hypersusceptibility and elevated viral titres suggesting that dicing per se of the viral RNA is not sufficient to mediate antiviral defence (Deleris *et al.*, 2006). A more direct evidence for antiviral RISC activity came from the study of recovered plants following in-fection by an attenuated strain of Cymbidium ringspot tombusvirus (CymRSV) (Pantaleo *et al.*, 2007). Previous experiments showed that these virus-infected plants contained vsRNAs, which predominantly originated from folded regions of the (+) strand of the viral RNA (Szittya *et al.*, 2002; Molnar *et al.*, 2005). Therefore a vsRNA-guided RISC was expected to target cleavage mainly at a symmetrical position in the (–) strand. These cleavage products were indeed detected and carried non-templated U residues at the predicted vsRNA-directed cut sites, a known signature of RISC-mediated cleavage (Pantaleo *et al.*, 2007).

Several lines of evidence indicate that, among the ten *Arabidopsis* AGOs, at least AGO1 associates with the antiviral RISC (Plate 1). As stated above, AGO1 is one of the three Argonaute proteins for which slicer activity has been demonstrated (Baumberger and Baulcombe, 2005). Hypomorphic *ago1* mutant was found hypersusceptible to CMV infection and accumulates higher amounts of viral RNAs than wt plants (Morel *et al.*, 2002). Moreover, *Arabidopsis* FLAG-tagged AGO1 coimmunoprecipitates vsRNAs from plants infected with CMV and Turnip yellow mosaic virus (TYMV) (Zhang *et al.*, 2006). Finally, both CymRSV vsRNAs and cellular miRNAs cofractionate in two protein complexes that are likely to correspond to free AGO1 and partially or fully assembled RISC (Pantaleo *et al.*, 2007).

Importantly, in addition to perfect complementarity to the (–) strand, cloned vsRNAs from infected plants show also partial complementarity to the (+) strand viral RNAs from which they derived (Szittya *et al.*, 2002; Molnar *et al.*, 2005). In plants, RISC can mediate mRNA cleavage when there is perfect or near perfect base pairing between targeted mRNAs and the sRNA (Llave *et al.*, 2002) or translational repression when there is partial complementarity (Aukerman and Sakai, 2003; Chen, 2004). Therefore, it is conceivable that besides RNA degradation, vsRNAs can also mediate translational inhibition of their targets. Another argument for the possible involvement of RISC-mediated translational repression during antiviral defence comes from a recent study showing that even miRNAs and siRNAs perfectly complementary to their targets also trigger translational repression (Brodersen *et al.*, 2008). Finally, whereas AGO1 binds preferentially sRNAs with a 5' terminal uridine, AGO2, AGO4 and AGO7 recruit sRNAs with a 5' terminal adenosine, and AGO5 sRNAs that initiate with cytosine (Qi *et al.*, 2006; Mi *et al.*, 2008; Montgomery *et al.*, 2008; Takeda *et al.*, 2008). The involvement of these other Argonaute proteins during antiviral immunity will deserve careful attention.

## Ready to repel the invader here and there: the systemic aspects of antiviral silencing

Obviously, plants did not elaborate the diverse and sophisticated signalling systems described above (see section 'Spreading of RNA silencing') for the purpose of long-distance silencing of transgenes. The link with antiviral defence first became apparent from the striking similarities between the timing and pathways of systemic silencing and virus movement in plants (Cruz *et al.*, 1996, 1998; Voinnet *et al.*, 1998). Because viruses were found to be potent triggers of RNA silencing within infected cells, it was speculated that non-cell autonomous silencing could represent the systemic arm of this response, whereby transmission of a virus-induced-silencing signal ahead of the infection front would prime silencing in naïve cells that are yet to be infected. Consequently, movement of the pathogen into those cells would be delayed or precluded (Voinnet *et al.*, 1998). This hypothesis received support from elegant VIGS experiments. Recombinant PVX virus carrying a fragment of the *ribulose bisphosphate carboxylase small subunit* (*rbcs*) endogenous gene was inoculated on lower leaves of wt tobacco plants. Systemic spread of silencing in the upper leaves was monitored through the appearance of the characteristic chlorotic phenotype due to *rbcs* silencing. By using movement-deficient mutants of this recombinant virus, which are restricted to the inoculated leaves, the authors showed that a PVX-derived silencing signal was able to reach non-inoculated, systemic leaves. Moreover, systemic silencing was only apparent when replication competent PVX was used as an inoculum, supporting the idea that the observed signal, or at least part of it, has an RNA component because PVX has no DNA phase in its biology (Voinnet *et al.*, 2000).

As mentioned previously (see section 'Spreading of RNA silencing'), an intact RDR6 activity is a prerequisite for efficient perception (Schwach *et al.*, 2005; Brosnan *et al.*, 2007) of the systemic silencing signal. Moreover, RDR6 has been involved in defence against several viruses by inhibiting viral accumulation in newly emerging leaves and excluding virus from the apical growing point (Qu *et al.*, 2005; Schwach *et al.*, 2005). These observations strongly suggest that the virus-induced silencing signal primes an RDR6-mediated silencing mechanism that inhibits viral accumulation and impairs viral spread in naïve cells that have perceived this signal. How this RDR6-mediated mechanism exactly operates is still an open question, as the nature of the systemic signal remains unknown. Assuming the signal is an sRNA, it could be used by RDR6 as a primer in order to convert viral RNA *de novo* into dsRNA in cells that have just been infected. Alternatively, if the signal is a long ssRNA, it might be used directly as an aberrant template by RDR6 for synthesis of the complementary strand. Similarly to what occurs for transgene-derived production of primary and secondary siRNAs (see section 'Spreading of RNA silencing'), this new dsRNA would then be processed by DCL4 into 21 nt-long vsRNAs. Consistent with the fact that 21 nt are the prominent, if not exclusive, sRNA involved in cell-to-cell silencing movement (Dunoyer *et al.*, 2005, 2007), these vsRNAs would then move ahead of the infection front and incorporate into an antiviral RISC to ensure that a potent antiviral response is mounted in cells that are about to be infected, in a process similar to immunization.

## 1.4 Viral Suppressors of RNA Silencing

### A brand new set of arms: identification, diversity and ubiquity of VSRs

Studies of viral synergisms provided deeper insights into RNA silencing as an antiviral defence mechanism. In 1991, Vance and colleagues showed that coinfection with the potyvirus PVY promoted up to tenfold more PVX RNA accumulation compared to a single PVX infection in tobacco plants (Vance, 1991). Later on, expression of the potyviral P1/HcPro protein, either transgenically or from a recombinant virus, was shown to induce dramatic hypersusceptibility to PVX, TMV or CMV, prompting the idea that this viral protein potentially inactivates a general antiviral defence system (Pruss *et al.*, 1997). This was indeed confirmed when HcPro, simultaneously with the CMV 2b protein, was shown to inhibit RNA silencing (Anandalakshmi *et al.*, 1998; Brigneti *et al.*, 1998; Kasschau and Carrington, 1998). The two proteins were thus defined as the first virus-encoded silencing suppressors (VSRs). Since then, many more viral proteins have been shown to suppress silencing and have been identified from virtually all types of phytoviruses (Voinnet *et al.*, 1999; Voinnet, 2005; Li and Ding, 2006; Ding and Voinnet, 2007) (see Table 1.1). Interestingly, most of them had been previously defined as pathogenicity factors or virulence determinants. The ubiquity of this viral counterstrategy further reinforced the importance of RNA silencing as a general antiviral defence in plants.

Two main techniques are used to identify VSRs (Moissiard and Voinnet, 2004). The first one is based on expression of the candidate viral protein from a recombinant viral vector. This usually leads to enhanced disease symptoms if the protein displays silencing suppression activity. Alternatively or concurrently, inoculation of the recombinant virus on silenced transgenic plants can lead to silencing reversion if the candidate protein is a VSR. Initial studies involved PVX as expression vector because it appeared to be neutral on its own in this reversal assay (Brigneti *et al.*, 1998). However, another approach (see below) revealed that PVX encodes a VSR, the P25 protein (Voinnet *et al.*, 2000) raising the possibility of an additive or a synergistic effect of P25 on the originally tested suppressors. This also underlines the importance of the choice of the viral vector used in the experiment.

The second approach relies on transient expression through transferred DNA (T-DNA) of a recombinant *Agrobacterium* culture where the candidate suppressor is codelivered with a transgene construct that triggers RNA silencing (either S-PTGS or IR-PTGS) of a stably integrated reporter transgene (most commonly encoding GFP). In the absence of silencing suppression activity, the reporter mRNA is rapidly degraded whereas in the presence of a VSR its accumulation is usually stabilized (Llave *et al.*, 2000; Voinnet *et al.*, 2000; Dunoyer *et al.*, 2002; Hamilton *et al.*, 2002; Pfeffer *et al.*, 2002; Takeda *et*

**Table 1.1.** Viral suppressors of RNA silencing.

| Virus type | Viral family | Virus | VSRs | Other functions | Reference(s) |
|---|---|---|---|---|---|
| Positive-strand RNA viruses | Aureusvirus | Pothos latent virus | P14 | | Merai *et al.* (2005) |
| | Carmovirus | Turnip crinkle virus | P38 | Coat protein | Thomas *et al.* (2003) |
| | Closterovirus | Beet yellows virus | P21 | Replication enhancer | Reed *et al.* (2003), Lu *et al.* (2004), Chiba *et al.* (2006) |
| | | Citrus tristeza virus | P20 | Replication enhancer | |
| | | | P23 | Nucleic acid binding | |
| | | | CP | Coat protein | |
| | | Grapevine leaf roll-associated virus 2 | P22 | | |
| | Crinivirus | Sweet potato chlorotic mosaic virus | P22 | | Kreuze *et al.* (2005) |
| | Comovirus | Cowpea mosaic virus | S protein | Small coat protein | Liu *et al.* (2004) |
| | Cucumovirus | Cucumber mosaic virus, Tomato aspermy virus | 2b | Host specific movement | Brigneti *et al.* (1998), Zhang *et al.* (2006), Diaz-Pendon *et al.* (2007) |
| | Furovirus | Soil-borne wheat mosaic virus | 19K | | Te *et al.* (2005) |
| | Hordeivirus | Barley yellow mosaic virus | $\gamma$b | Replication enhancer, movement, seed transmission, pathogenicity determinant | Yelina *et al.* (2002) |
| | Pecluvirus | Peanut clump virus | P15 | Movement | Dunoyer *et al.* (2002) |
| | Polerovirus | Beet western yellows virus, Curcubit aphid-born yellows virus | P0 | Pathogenicity determinant | Pfeffer *et al.* (2002), Baumberger *et al.* (2007), Bortolamiol *et al.* (2007) |
| | Potexvirus | Potato virus X | P25 | Movement | Voinnet *et al.* (2000) |
| | Potyvirus | Potato virus Y, Tobacco etch virus, Turnip mosaic virus | HcPro | Movement, polyprotein processing, aphid transmission, pathogenicity determinant | Anandalakshmi *et al.* (1998), Brigneti *et al.* (1998), Kasschau and Carrington (1998) |

*Continued*

**Table 1.1.** – *Continued*

| Virus type | Viral family | Virus | VSRs | Other functions | Reference(s) |
|---|---|---|---|---|---|
| | Sobemovirus | Rice yellow mottle virus | P1 | Movement, pathogenicity determinant | Voinnet *et al.* (1999) |
| | Tobamovirus | Tobacco mosaic virus, Tomato mosaic virus | P122, P130 | Replication | Kubota *et al.* (2003), Csorba *et al.* (2007) |
| | Tobravirus | Tobacco rattle virus | 16K | | Liu *et al.* (2002) |
| | Tombusvirus | Tomato bushy stunt virus, Cymbidium ringspot virus, Carnation Italian ringspot virus | P19 | Movement, pathogenicity determinant | Voinnet *et al.* (1999), Silhavy *et al.* (2002) |
| | Tymovirus | Turnip yellow mosaic virus | P69 | Movement, pathogenicity determinant | Chen *et al.* (2004) |
| | Vitivirus | Grapevine virus A | P10 | | Chiba *et al.* (2006) |
| Negative-strand RNA viruses | Teniuvirus | Rice hoja blanca virus | NS3 | Unknown | Bucher *et al.* (2003), Hemmes *et al.* (2007) |
| | Tospovirus | Tomato spotted wilt virus | NSs | Pathogenicity determinant | Bucher *et al.* (2003) |
| Double-stranded RNA viruses | Phytoreovirus | Rice dwarf virus | Pns10 | | Cao *et al.* (2005) |
| Single-stranded DNA viruses | Begomovirus | African cassava mosaic virus (Kenyan strain) | AC2 | Transcriptional activator protein (TrAP) | Voinnet *et al.* (1999), van Wezel *et al.* (2002), Chellappan *et al.* (2005), Trinks *et al.* (2005), Zrachya *et al.* (2007), Glick *et al.* (2008) |
| | | African cassava mosaic virus (Cameroon strain)) | AC4 | | |
| | | Tomato yellow leaf curl virus | C2 V2 | | |
| | | Tomato leaf curl virus | C2 | | |
| | | Mungbean yellow mosaic virus | C2 | | |
| | Curtovirus | Beet curly top virus | L2 | Transcription factor | Wang *et al.* (2005) |
| Double-stranded DNA viruses | Caulimovirus | Cauliflower mosaic virus | P6 | Replication factor | Love *et al.* (2007) |

*al.*, 2002; Bucher *et al.*, 2003). These rapid assays allowed the identification of more than 30 VSRs from many distinct virus types (see Table 1.1). Moreover, a single type of virus might encode several distinct VSRs as was found with Citrus tristeza virus (Lu *et al.*, 2004).

Strikingly, VSRs are highly divergent in terms of sequence and structure, and represent different strategies to suppress silencing, providing a compelling example of evolutionary convergence (Plate 2). The small size of viral genomes and the low number of encoded proteins has also forced viruses to usually cumulate several functions into a single protein. In line with this observation, VSRs often display other functions during the virus life cycle. For instance, TCV- and TMV-encoded silencing suppressors are the coat protein (CP) of the virion (Qu *et al.*, 2003; Thomas *et al.*, 2003) and the p122 subunit of the viral replicase, respectively (Csorba *et al.*, 2007). Other VSRs are also often encoded by novel, overlapping genes contained within more ancient ones. These new genes are usually created by 'overprinting', whereby an existing coding sequence is translated from a different open reading frame. This evolutionary process creates isolated VSR lineages in virus phylogenies. These overlapping VSR genes include the poleroviral P0, geminiviral AC2 and AC4, tombusviral P19 and P14, cucumoviral 2b or tymoviral P69 proteins (Li and Ding, 2006). VSR have also been isolated from fungus, insect and mammalian viruses and their activity have been demonstrated to be retained in cross-kingdom analyses indicating that they are likely to be targeting key steps in the RNA silencing pathways (Delgadillo *et al.*, 2004; Dunoyer *et al.*, 2004; Li *et al.*, 2004; Reavy *et al.*, 2004; Segers *et al.*, 2006; Hemmes *et al.*, 2007; Schnettler *et al.*, 2008). The corollary of this observation is that RNA silencing also represents an antiviral defence mechanism in other eukaryotic organisms (Li *et al.*, 2002; Wang *et al.*, 2006; Segers *et al.*, 2007). However, in the case of vertebrates' VSRs, this assumption is yet to be firmly demonstrated as identification of these proteins were obtained in non-vertebrate systems such as plants or insect cells and not characterized during the context of natural infection (Bucher *et al.*, 2004; Delgadillo *et al.*, 2004; Li *et al.*, 2004).

### *Plate 2*

Plate 2 shows different strategies for suppression of RNA silencing. One strategy to suppress silencing is to avoid vsRNA production as exemplified with potyviral HcPro that inhibit dsRNA processing probably through direct binding of the dsRNA trigger, thereby blocking access to DCLs. VSRs can also directly block Dicer activity as shown in the case of the TCV P38 that inhibits DCL4 or PVX P25 that impairs production of DCL3-dependent 24 nt-long siRNA. Inhibition of DCL4 by P38 revealed the redundant antiviral function of DCL2, which generates 22 nt-long siRNA. The antiviral activity of the 22 nt vsRNA is also compromised by P38 through an unknown mechanism. Inhibition of HEN1-mediated protection of vsRNA, for instance by HcPro or TMV P122, leads to destabilized and subsequent degradation of vsRNAs. Another commonly used silencing suppression strategy is shown by beet yellows virus (BYV) P21 and CymRSV P19 that both bind small RNA duplexes. Although P21 does not show any size specificity, P19 acts as a head-to-tail homodimer that specifically measures 21 bp duplexes that are the product of DCL4. This binding impairs their subsequent incorporation into AGO1-containing RISC. Alternatively, African cassava mosaic virus (ACMV) AC4 sequesters single stranded small RNA molecules after unwinding of the duplex thereby preventing RISC-mediated cleavage of targeted RNA. AGO1 activity can also be inhibited through direct interaction with the VSR as shown with CMV 2b or through stimulation of its degradation rate through an ubiquitin-mediated proteolysis pathway as recently found for the poleroviral P0 protein. The DCL4-dependent 21 nt-long siRNA are the prominent if not exclusive cell-to-cell silencing signal that move ahead of the infection front and incorporate into an antiviral RISC to set up a potent antiviral response, similarly to immunization. Sequestration of the 21 nt siRNA duplexes by P19 precludes cell-to-cell movement of the silencing signal ahead of the infection front. Although the nature of the systemic silencing signal is still unknown, P25 and 2b have been shown to prevent the synthesis, spread or perception of this signal ahead of the infection front. Finally, how DCL3-dependent 24 nt-long vsRNA are produced during cytoplasmically replicating virus infection is an open question. Either some viral RNA enters the nucleus or DCL3 is delocalized from the nucleus to the cytoplasm during viral infection.

## Many ways of invading your enemy's territory: viral suppression strategies

The diverse genomic and evolutionary origins of VSRs, supporting the idea that they appeared independently in each virus family, is the basis of the increasing diversity found in their mode of action.

### *Suppression of siRNA accumulation*

The simplest way for a virus to avoid the RNA silencing response is to inhibit the process from the beginning, by preventing Dicer from accessing the viral dsRNA trigger(s). For instance, transgenic expression of the potyviral HcPro has been shown to inhibit the DCL-dependent processing of dsRNA (Mallory *et al.*, 2002; Dunoyer *et al.*, 2004). Indeed, when coexpressed with an inverted-repeat transgene designed to silence the *Arabidopsis* endogenous gene *chalcone synthase* (CHS, CHS-RNAi line), suppression of silencing was correlated with the accumulation of unprocessed CHS dsRNA. It is noteworthy that HcPro mainly inhibits the accumulation of the DCL4-dependent 21 nt-long siRNA and has a much reduced effect on the DCL3-dependent 24 nt-long siRNA. Given that HcPro is a cytoplasmic protein, this observation is in contradiction with the presumed nuclear localization of DCL4. Hence, the stronger effect of HcPro on 21 nt- rather than 24 nt-long siRNA accumulation, is probably explained by the fact that biogenesis of the former most likely occurs in the same subcellular compartment as the VSR, whereas biogenesis of the latter occurs in the nucleus (Li *et al.*, 2006; Pontes *et al.*, 2006). Interestingly, this inhibition of Dicer processing was only partial, as substantial levels of 21 nt-long CHS siRNAs were still detected (Dunoyer *et al.*, 2004). However, despite those residual siRNA levels, degradation of the CHS mRNA was prevented, suggesting that in addition to its effect on Dicer activity, HcPro also inhibits the activity of RISC. This proposal is consistent with the effect of HcPro on miRNA-guided cleavage of endogenous transcripts or with its sRNA binding capacity (see below).

In contrast to HcPro, P1 from Rice yellow mottle virus (RYMV) and the PVX P25 suppressor specifically impair 24 nt-long siRNA production during S-PTGS (Hamilton *et al.*, 2002; Himber *et al.*, 2003). Moreover, PVX-infected wt plants only accumulate 21 nt-long vsRNAs suggesting that this DCL3-specific inhibition by P25 also occurs during viral infection (Schwach *et al.*, 2005). In line with this observation, P25 was shown to be localized in the nucleus from early time points of the infection (Samuels *et al.*, 2007). However, the fact that 24 nt-long vsRNAs are detected in the absence of P25 still raises the question of how DCL3 gains access to the cytoplasmic viral dsRNA (Plate 2). Moreover, how HcPro, P1 and P25 interfere with the processing of dsRNA by Dicers is also unsolved. As HcPro and P25 have been shown to display RNA binding property in a non-sequence-specific manner, one possibility is that these proteins directly bind to the viral dsRNA triggers thereby blocking access to DCLs (Urcuqui-Inchima *et al.*, 2000; Kasschau and Carrington, 2001; Leshchiner *et al.*, 2006). Alternatively, these VSRs can directly inhibit Dicer activity, as recently observed with the TCV P38 protein (Plate 2). Indeed, when *dcl2* mutant plants were found more susceptible to TCV infection, the initial thought was that DCL2 was the main antiviral Dicer in plants (Xie *et al.*, 2004). Supporting this idea, DCL2-dependent 22 nt-long vsRNAs were the predominant TCV-derived sRNAs to accumulate in wt infected plants. However, the finding that wt plants yield 21 nt instead of 22 nt-long vsRNA when infected with the VSR-deficient TCV (TCV-ΔP38) indicated that DCL2 only substitutes DCL4 when its activity is compromised by P38 (Deleris *et al.*, 2006). As the 22 nt-long sRNAs were not competent to mediate viral RNA cleavage in the presence of P38, this suggests that P38 also impairs the activity of the DCL2-dependent siRNAs (Deleris *et al.*, 2006), a property probably related to its sRNA binding capacity (Merai *et al.*, 2006).

One point to keep in mind when analysing sRNA levels in transgenic or infected plants is that we are looking at steady state levels at a given time point, which takes into account rate of production and rate of degradation. Study of the Peanut clump virus (PCV) P15 protein in the CHS-RNAi system indicated that this VSR triggers a strong reduction in CHS siRNA accumulation without interfering with dsRNA processing, suggesting that this protein likely acts downstream of Dicers (Dunoyer *et al.*, 2004). Reduced siRNA accumulation levels may result from a reduced incorporation into RISC, which may cause their instability. Alternatively, as P15 has been shown to bind sRNAs and to be localize in peroxisomes (Dunoyer *et al.*, 2002; Merai *et al.*, 2006), this VSR might be responsible for the transport of the sRNAs inside this highly acidic organelle thereby increasing their degradation rate.

Finally, another way of accelerating sRNA turnover is to prevent their protection by HEN1. Resistance to beta-elimination experiments indicated that plant vsRNAs are methylated at their 3' end, just like cellular sRNAs. Moreover, *hen1* mutants accumulate less vsRNAs from RNA and DNA viruses and exhibit reduced virus-induced gene silencing (Blevins *et al.*, 2006) making HEN1 an attractive target for VSR. Accordingly, the potyviral HcPro, the tobamoviral P122 and the tombusviral P19 were all demonstrated to interfere with HEN1 activity (Ebhardt *et al.*, 2005; Yu *et al.*, 2006; Csorba *et al.*, 2007; Vogler *et al.*, 2007) (Plate 2).

### *Sequestration of vsRNAs*

Another strategy to suppress silencing relies on sequestration of sRNAs, which prevents their incorporation into effector complexes. One of the best examples of this kind is provided by the P19 protein encoded by tombusviruses. Initially, gel mobility shift assays showed that glutathione S-tranferase (GST)–P19 fusion protein specifically binds *in vitro* synthetic siRNAs with 2 nt 3' overhang, a characteristic of Dicer-dependent sRNA products. In contrast, GST–P19 exhibited poor, if any, affinity to blunt-ended siRNA duplexes, long dsRNA and ssRNA (Silhavy *et al.*, 2002). *In vivo* binding was subsequently demonstrated through coimmunoprecipitation experiments of the VSR specifically with 21 nt-long siRNA duplexes (Chapman *et al.*, 2004; Dunoyer *et al.*, 2004). The definitive insight into P19 suppression mechanism was obtained by the crystallization of P19 in direct association with a siRNA duplex (Vargason *et al.*, 2003; Ye *et al.*, 2003). The crystal structure revealed that P19 acts as a head-to-tail homodimer that binds to and specifically measures 21 nt-long siRNA duplexes (Plate 2). Length measurement is performed by a pair of tryptophan residues that interact with the last nucleotide of each strand of the duplex. Therefore, P19 works like a molecular calliper to selectively interact, in a non-sequence specific manner, with sRNA duplexes that, incidentally, have the same size as those produced by the main antiviral Dicer, DCL4.

Many additional VSRs have been shown to bind sRNA such as BYV P21, potyviral HcPro, PCV P15, Barley stripe mosaic virus (BSMV) γb, Rice hoja blanca tenuivirus (RHBV) NS3, TMV P122 or TCV P38 (Lakatos *et al.*, 2006; Merai *et al.*, 2006; Csorba *et al.*, 2007; Hemmes *et al.*, 2007). These observations led to the idea that dsRNA binding is a general silencing suppression strategy. However, in most cases, these results have to be interpreted with caution as: (i) most of the evidence comes from *in vitro* assays; (ii) binding is sometimes observed under a non-physiological amount of VSR; and (iii) RNA binding is often non-specific. For instance, closteroviral BYV P21 monomers have been shown to form an octameric ring that displays equal affinity for short, long, single- and double-stranded RNA (Ye and Patel, 2005). Finally, dsRNA binding might be unrelated to silencing suppression and only reflect additional VSR functions that require a close association with viral nucleic acids. For instance TCV P38 and TMV P122 are also involved in encapsidation and replication of the viral RNA, respectively. Therefore, whether dsRNA binding is a genuine feature of silencing suppression still remains to be addressed in most cases. Nevertheless, in some cases, it can be inferred that sRNA binding is an authentic property of VSRs. Indeed, Northern blot analyses of the small RNA fraction from P19-, HcPro- and P21-expressing transgenic lines, or TMV-infected plants, showed a remarkable stabilization of the otherwise labile strand (miRNAs*) of endogenous miRNA duplexes (Chapman *et al.*, 2004; Dunoyer *et al.*, 2004; Csorba *et al.*, 2007). As miRNAs* are usually degraded upon unwinding of the miRNA/miRNA* duplex, their stabilization by those VSRs is probably explained by sequestration of the small RNAs as duplexes, prior to their unwinding, as it was demonstrated in the case of P19. The fact that these VSRs have also been shown to inhibit HEN1-mediated methylation on both miRNA and miRNA* strands suggests that HEN1 uses duplexes as substrates and illustrates how VSRs can sometimes be informative about the mechanisms underlying endogenous silencing pathways.

The sequestration of small RNA duplexes by P19, P21 and HcPro was effective, at least *in vitro*, in preventing formation of an active RISC complex. Direct competition assays for RISC assembly in *Drosophila* embryo extracts revealed that coincubation of synthetic siRNAs with P19 impaired their incorporation into RISC and inhibited slicing of targets. In contrast, when the RISC complex was pre-assembled by incubating siRNAs in embryo extracts prior to adding P19, target cleavage occurred (Lakatos *et al.*, 2006; Csorba *et al.*, 2007). Moreover, transgenic plants expressing P19, HcPro and P21 display developmental defects that are strikingly similar and the severity of these phenotypes correlates well with inhibition of miRNA-guided cleavage (Chapman *et al.*, 2004; Dunoyer *et al.*, 2004) (Plate 2). This suggests that binding of miRNA/miRNA* duplexes impair formation of a functional RISC *in vivo* as well, even though direct evidence for this is still lacking.

Interestingly, the geminiviral AC4 protein from ACMV Cameroon strain is the only VSR reported so far that specifically binds 21 nt ssRNA (Chellappan *et al.*, 2005). Indeed, in sharp contrast to P19, P21 or HcPro transgenic plants, AC4 expression does not stabilize the miRNA* strand. Moreover, AC4 displays *in vitro* binding activity that is specific to single-stranded small RNA, but has no affinity towards double-stranded small RNA duplexes and also coimmunoprecipitates selectively with miRNA guide strands. As AC4 expression was associated with a strong increase in miRNA target accumulation, this suggests that AC4 prevents the assembly of a functional RISC by capturing single-stranded small RNAs after unwinding of the duplex (Plate 2).

### *Targeting of key factors*

Besides, or in addition to, these two first strategies, silencing suppression can be mediated by direct effects on key components of the silencing pathway, either through inhibition of host silencing effectors or through the recruitment of endogenous silencing suppressors. The existence of cellular negative regulators has been genetically identified in *C. elegans* (Kennedy *et al.*, 2004; Duchaine *et al.*, 2006). One of them, ERI-1 (for 'enhanced RNAi-1') defines a novel subfamily of DEDDh nucleases (with conserved orthologues in many organisms including *Arabidopsis*) that apparently processes siRNA into shorter, inactive forms (Kennedy *et al.*, 2004). Recently, negative regulators of RNA silencing have been identified in *Arabidopsis* as well, based on their enhanced silencing phenotype (Herr *et al.*, 2006; Gy *et al.*, 2007). These include the nuclear exoribonucleases XRN2, XRN3 and the nucleotidase/phosphatase FIERY1 (FRY1). XRN proteins have a 5'–3' exoribonuclease activity which, when disabled, leads to the accumulation of uncapped (i.e. aberrant) transgene mRNAs, which are favoured templates for RDR6 (Gazzani *et al.*, 2004). *fry1* mutant plants recapitulate developmental and molecular characteristics of *xrn* mutants by corepressing XRN2, XRN3 and XRN4. This increases the availability of aberrant RNAs to conversion into dsRNA by RDRs (Gy *et al.*, 2007).

Recruitment of endogenous negative regulators by VSRs was first exemplified with the potyviral HcPro. A yeast two-hybrid screen identified an interaction with a tobacco calmodulin-related protein, called rgsCaM whose overexpression mimics, at least macroscopically, the silencing suppression mediated by HcPro (Anandalakshmi *et al.*, 2000). Although the interaction is yet to be demonstrated *in planta*, rgsCaM expression was shown to be induced by HcPro, indicating that HcPro either directly or indirectly controls rgsCaM mRNA levels. Likewise, the geminiviral transactivator AC2 protein suppresses silencing by inducing transcription of host genes that may well encode negative regulators of RNA silencing (Trinks *et al.*, 2005).

The 2b protein of cucumoviruses was one of the first identified silencing suppressors along with HcPro (Brigneti *et al.*, 1998). However, 2b had no effect in tissues where silencing had been already established, but it was able to prevent initiation of gene silencing in *Agrobacterium*-infiltrated leaves or in newly emerging tissues receiving the systemic silencing signal (Brigneti *et al.*, 1998; Li *et al.*, 1999). Early studies demonstrated that 2b contains a single nuclear localization signal (NLS) essential for its suppressor function (Lucy *et al.*, 2000). None the less, the molecular details of its action remained elusive until recently, when it was observed that transgenic plants expressing 2b from CMV isolate 'FNY' exhibit developmental defects strikingly similar to those of *ago1 Arabidopsis* mutants. Accordingly, several miRNA targets were found to accumulate ectopically in transgenic FNY2b plants. Since miRNAs* also over-accumulate in these plants, the initial hypothesis was that 2b sequesters sRNAs duplexes, similarly to P19. However, 2b neither bound siRNA *in vitro* nor prevented their loading into AGO1 *in vivo* (Zhang *et al.*, 2006). Rather, 2b was shown to coimmunoprecipitate with AGO1 through a direct interaction of the VSR with a region of AGO1 that corresponds to one surface of the sRNA binding PAZ domain and part of the PIWI domain required for slicing. Moreover, 2b did not prevent loading of sRNAs in AGO1 but strongly inhibited target RNA cleavage by a pre-assembled RISC (Zhang *et al.*, 2006) (Plate 2).

Even more recently, a novel silencing suppression strategy has been elucidated in the case of the poleroviral P0 proteins. Following its identification as a potent silencing suppressor (Pfeffer *et al.*, 2002), it was shown that P0 contains an F-box domain that is essential for its suppressor function (Pazhouhandeh *et al.*, 2006). Through this domain, P0 interacts with *Arabidopsis* kinase-related protein 1 (SKP1) orthologues ASK1 and ASK2, which are components of the SKP1-Cullin-F box (SCF) family of E3 ubiquitin ligases involved in proteasome-mediated degradation of polyubiquitinylated proteins. Mutations abolishing P0-SKP1 interactions impair P0-mediated

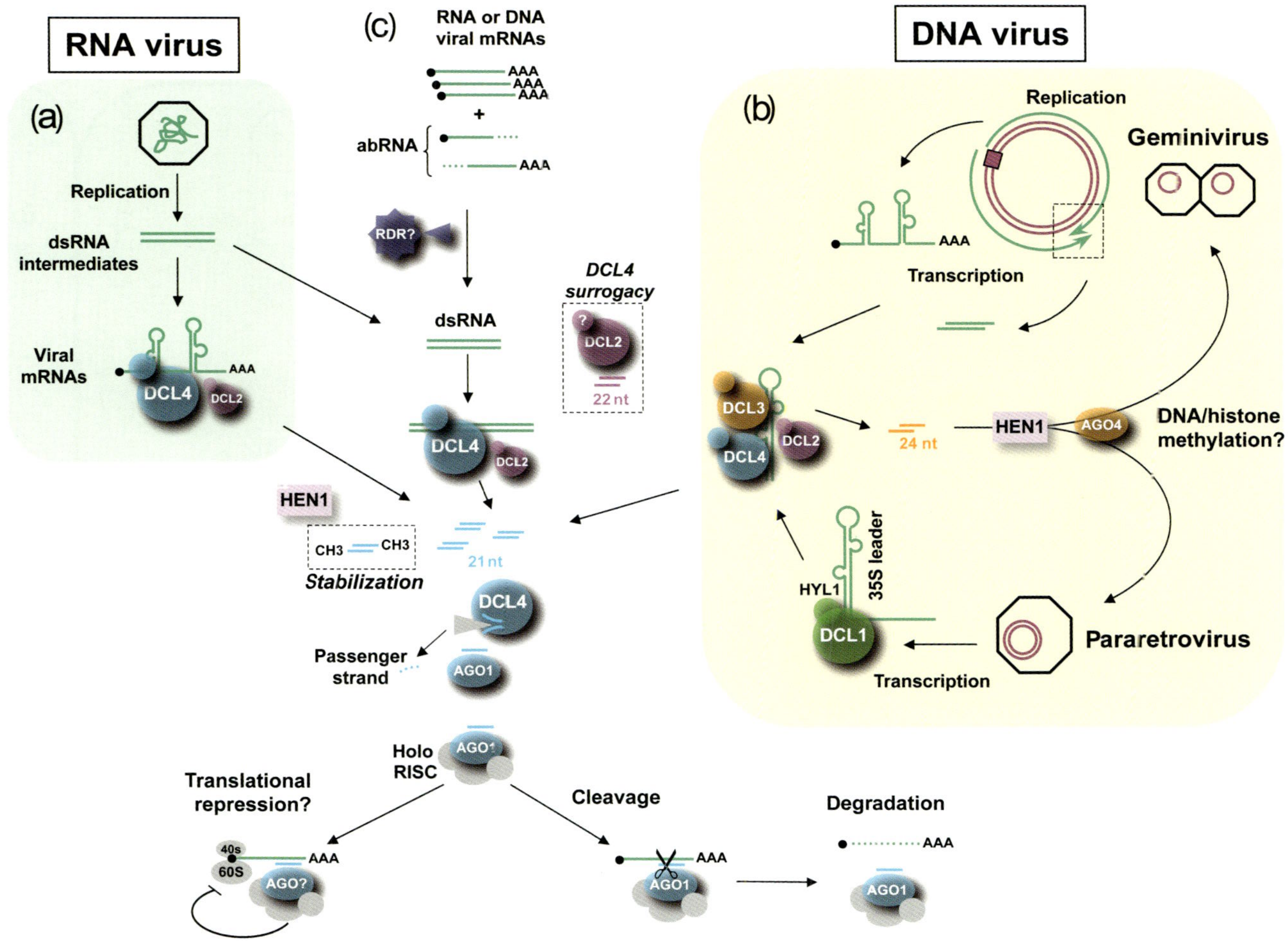


**Plate 1.** Antiviral silencing pathways in plants. (a) Antiviral Dicers and RNA viruses; (b) antiviral Dicers and DNA viruses; (c) alternative sources of viral-derived small RNA (vsRNA). abRNA, aberrant RNA; AGO, argonaute; DCL, RNase III-like enzyme called Dicer; dsRNA, double-stranded RNA; HEN1, a plant methyltransferase enzyme; HYL1, a dsRNA binding protein; RDR, RNA-dependent RNA polymerases; RISC, RNA-induced silencing complexes.

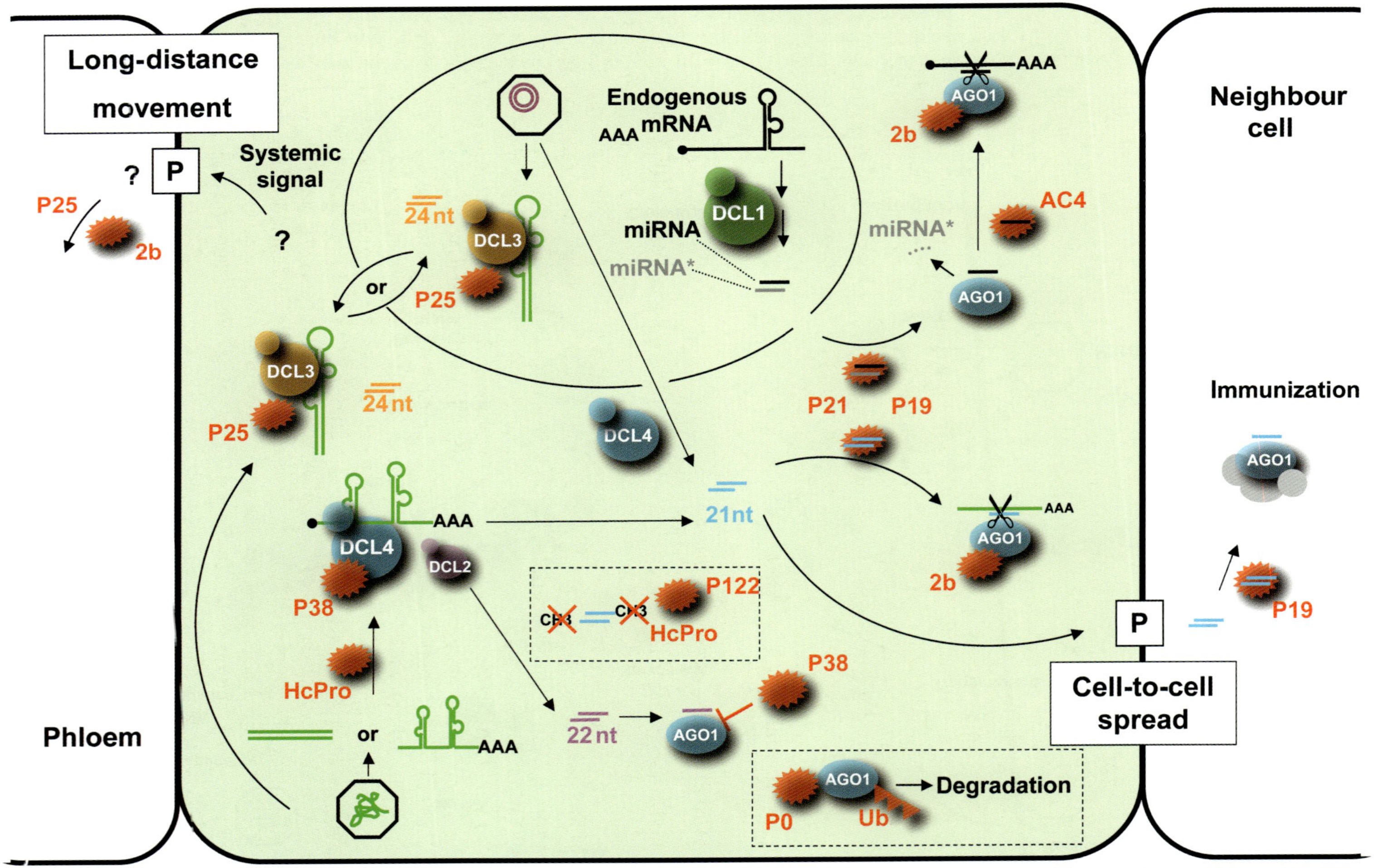


**Plate 2.** Viral strategies for suppression of RNA silencing. AC4, a geminiviral protein; AGO, argonaute; DCL, RNase III-like enzyme called Dicer; HcPro, HcPro protein; miRNA, microRNA; P (boxed), plasmodesmata; P (followed by a number and in orange), different viral suppressors of RNA silencing; Ub, ubiquitin.

silencing suppression activity and led to an increased resistance to poleroviral infection suggesting that poleroviruses use an ubiquitin-mediated proteolysis machinery to counter the antiviral silencing pathway (Pazhouhandeh *et al.*, 2006). Interestingly, P0 expression was shown to trigger degradation of the AGO1 protein and, accordingly, transgenic P0 plants displayed developmental defects and over-accumulated miRNA targets (Baumberger *et al.*, 2007; Bortolamiol *et al.*, 2007) (Plate 2). Therefore, it was postulated that, by acting as an F-box protein in an SCF complex, P0 targets AGO1 for degradation by the proteosome. However, as P0-mediated AGO1 decay was insensitive to MG132 treatment, a known inhibitor of the proteosome machinery, it was suggested that an alternative pathway might be involved in this specific degradation process (Baumberger *et al.*, 2007).

Finally, the V2 suppressor of Tomato yellow leaf curl virus (TYLCV) has been found to associate with SlSGS3, a tomato homologue of the *Arabidopsis* SGS3, an interaction required for its silencing suppression activity (Glick *et al.*, 2008). SGS3 is involved, together with RDR6, in the production of dsRNA from 'aberrant' ssRNA template (Mourrain *et al.*, 2000; Peragine *et al.*, 2004). Although the exact mechanism of V2 silencing suppression is still unknown, inactivation of SGS3 was found to promote hypersusceptibility to a virus related to TYLCV, demonstrating the significance of SGS3 and its suppression in antiviral silencing (Muangsan *et al.*, 2004).

#### *Inhibition of systemic silencing*

Further support for the antiviral role of non-cell-autonomous RNA silencing came from the finding that some VSRs are specifically directed against cell-to-cell or systemic silencing movement, as opposed to intracellular silencing. The first evidence comes from the finding that systemic silencing from recombinant, movement-deficient PVX could only be achieved if the P25 was deleted from the viral genome, whereas silencing within the inoculated region remained mostly unaffected by the presence or absence of P25 (Voinnet *et al.*, 2000). Later on, agro-infiltration of the RYMV P1 and PVX P25 was found to abolish the phloem-dependent movement of silencing without altering its short distance spread (Hamilton *et al.*, 2002; Himber *et al.*, 2003). Conversely, the ACMV AC2 suppressor eliminated cell-to-cell movement without affecting long-distance silencing. As already mentioned, P1, P25, but not AC2, inhibited the accumulation of the 24 nt-long siRNAs correlating this latter with the onset of systemic silencing. Interestingly, as observed in RDR6-deficient plants, expression of a P25 homologue also causes a loss of meristem exclusion to the virus probably resulting from the suppression of systemic silencing (Foster *et al.*, 2002). Thus P25 most likely prevents the synthesis or spread of a virus-induced silencing signal ahead of the infection front (Plate 2).

An early report indicated that CMV 2b is also implicated in long-distance movement of the virus through the phloem (Ding *et al.*, 1995). A possible explanation was provided by elegant grafting experiments in *Nicotiana benthamiana*. Indeed, the long-distance silencing signal produced in the rootstock was not able to trigger silencing of the reporter gene in 2b-expressing scions.

Moreover, the mobile signal could not induce specific silencing of a reporter gene in scions if 2b was expressed in an intergraft between the rootstock and the scion (Guo and Ding, 2002). Thus, 2b may inhibit long-distance trafficking and the onset of systemic silencing in the growing tissue by sequestering the mobile signal. This mechanism may well explain the ability of CMV 2b to promote systemic spread of normally phloem-restricted viruses (Ryang *et al.*, 2004). The recent identification of a strong nuclear component in the perception of the long-distance silencing signal, with the involvement of the nuclear RDR2/PolIVa/DCL3 factors (Brosnan *et al.*, 2007), is also consistent with previous findings that nuclear import of CMV 2b is mandatory for suppression of systemic silencing (Lucy *et al.*, 2000). It is noteworthy that P25 has also been shown to be localized in the nucleus.

Study of the CymRSV P19 protein provided detailed analysis of the contribution of silencing suppression to systemic viral infection (Havelda *et al.*, 2003). As discussed previously, P19 sequesters siRNAs thereby probably preventing their incorporation into RISC. Surprisingly, despite this effect on intracellular silencing, previous work had established that P19-defective CymRSV accumulates to wt levels in single cells, suggesting that P19 targets a non-cell autonomous step of RNA silencing (Silhavy *et al.*, 2002). By using a combination of *in situ* hybridization and immunohistochemistry, it was subsequently shown that phloem-dependent movement and replication of CymRSV in and around the vascular bundles of systemic leaves were not altered by the lack of P19 (Havelda *et al.*, 2003). Rather, lack of p19 apparently prevents further viral invasion of the leaf lamina, which, although virus free, exhibits nucleotide sequence-specific resistance to secondary infection with either CymRSV or an unrelated recombinant virus carrying sequence identity to CymRSV (Szittya *et al.*, 2002). These observations indicate that P19 most likely prevents the onset or the movement of a mobile virus-induced silencing signal that, upon phloem unloading of the pathogen, primes the destruction of viral RNAs ahead of the infection front. It is noteworthy that P19 specifically binds DCL4-dependent 21 nt-long siRNA that have been shown to be the prominent, if not exclusive, cell-to-cell silencing signal (Vargason *et al.*, 2003; Ye *et al.*, 2003; Dunoyer *et al.*, 2005, 2007) (Plate 2).

## Victory without suppressors: other pathogen responses to RNA silencing

Most viruses deploy protein-based strategies to suppress the antiviral silencing response in plants. Nevertheless, in a few cases, other types of counter-defence responses can be found. One of these alternatives was first demonstrated in human cells infected by adenoviruses. These viruses, in addition to viral proteins, encode a highly structured RNA of approximately 160 nt called VA1. VA1 is expressed at extraordinarily high levels during adenoviral replication (Mathews and Shenk, 1991) and was shown to inhibit RNA silencing at a physiological level. This suppression apparently occurs through binding of Dicer as a competitive substrate resulting in its reduced processing activity (Lu and Cullen, 2004; Andersson *et al.*, 2005). In plants, this strategy has been exemplified in the case of Red clover necrotic mosaic virus (RCNMV) infection. RNA silencing suppression by this virus was not attributed to any of the viral-encoded proteins but rather required multiple viral components including the viral replicase and the viral RNAs. Moreover, *cis*-elements required for (-) strand viral RNA synthesis were found to be mandatory for silencing suppression, indicating a strong link between the viral RNA replication process and inhibition of antiviral silencing. Interestingly, RCNMV interferes with miRNA biogenesis and, in a transient assay, with the accumulation of siRNA, suggesting that viral replicative intermediate dsRNA or highly structured RNA akin to VA1 may outcompete Dicer processing (Takeda *et al.*, 2005).

Another strategy to avoid degradation can rely on the intrinsic property of the pathogen's genome itself, as observed with viroids. Viroids do not encode any protein but replicate autonomously and spread in plants by recruiting host proteins. Their circular ssRNA genome has evolved strong secondary structures that make them very good targets for Dicer processing (Papaefthimiou *et al.*, 2001; Martinez de Alba *et al.*, 2002; Landry and Perreault, 2005). However, structured viroid RNAs are strongly resistant to RISC-mediated degradation (Itaya *et al.*, 2007), despite the fact that these viroid-derived siRNA are otherwise biologically functional as shown by the efficient cleavage of exogenous sensor transcripts. Furthermore, expression of a hairpin RNA derived from potato spindle tuber viroid (PSTVd) triggers the appearance of symptoms similar to those observed during PSTVd infection, suggesting that these pathogens cause disease symptoms by directing RNA silencing against physiologically important host genes (Wang *et al.*, 2004). Viroid RNAs are resistant to RISC-mediated cleavage because RISC is unable to unfold target secondary structures and, therefore, the efficiency of cleavage correlates directly with the accessibility of the target site (Ameres *et al.*, 2007).

## Using your enemy's force against him: viral subversion of host silencing pathways

As mentioned earlier, besides their demonstrated role in antiviral defence immunity, several silencing effectors are also involved in regulating host gene expression that controls important developmental processes. Accordingly, plants expressing VSRs that either bind sRNAs or inhibit AGO1-mediated slicing display developmental defects that phenocopy those found in mutants affected in miRNA biogenesis and/or function (Chapman *et al.*, 2004; Dunoyer *et al.*, 2004; Zhang *et al.*, 2006). However, this was thought to be a secondary consequence of the siRNA pathway inhibition of some steps shared with the miRNA pathway, rather than a deliberate viral strategy to reprogramme or alter the host genome expression (Chapman *et al.*, 2004; Dunoyer *et al.*, 2004).

Interestingly, endogenous siRNAs and miRNAs have been recently shown to play a key role in *Arabidopsis* basal and race-specific resistance against bacterial pathogens (Katiyar-Agarwal *et al.*, 2006; Navarro *et al.*, 2006). For instance, expression of an endogenous nat-siRNA is specifically induced upon infection by the bacterial pathogen *Pseudomonas syringae*. Biogenesis of this nat-siRNA requires DCL1, HYL1, HEN1, RDR6, PolIVa and SGS3 and only occurs in the presence of both *P. syringae* effector avrRpt2 and the cognate host disease resistance gene *RPS2*. Interestingly, this siRNA targets and represses pentatricopeptide repeats protein-like (PPRL), a putative negative regulator of the RPS2 resistance pathway thereby contributing to RPS2-mediated race-specific disease resistance (Katiyar-Agarwal *et al.*, 2006). As these protein-based processes of disease resistance are also involved in protecting plants against viruses, inhibition of endogenous sRNA pathways by VSRs might reflect a deliberate viral strategy to inhibit such immune systems.

Alternatively, vsRNAs can potentially target host transcripts that impact on virus fitness. For instance, several vsRNAs derived from the 35S leader region of CaMV exhibit near-perfect complementarity to *Arabidopsis* transcripts and trigger their sequence-specific degradation during infection (Moissiard and Voinnet, 2006). If some of these transcripts encode host defence factors, they may be under positive selection in the virus. Similarly, the cell-to-cell movement of vsRNAs, normally deployed by the plant to immunize naïve cells, can also potentially trigger silencing of defence factors, creating an optimal environment for virus multiplication in tissues about to be infected. Strikingly, plant viruses have been shown to elicit drastic modifications of mRNA expression at or ahead of infection fronts (Escaler *et al.*, 2000; Havelda and Maule, 2000).

On the other hand, plants also may have evolved strategies to counteract the action of VSRs. This counter-counter defence may potentially explain the contrasted susceptibility of different plant species or even ecotypes to viruses (Li *et al.*, 1999). An illustration of this concept is provided by the contrasted effect of the 2b proteins from CMV Q and FNY strains (Zhang *et al.*, 2006). Whereas both 2b alleles efficiently inhibit AGO1-mediated cleavage *in vitro*, analysis of their accumulation in transgenic plants revealed that Q2b did not accumulate. In fact, this specific suppressor allele appeared to be truncated *in planta*, possibly as a result of proteolysis that did not affect the FNY2b allele. This allelic difference is significant because it provides an explanation for the mild and severe virulence of the Q and FNY strains, respectively.

Resistance (*R*) genes, involve in the protein-based innate immune response, sense changes in the integrity of specific host defence components called 'guardees' and, in response, trigger host defence reaction. These guardees are primary targets of a set of pathogen's proteins called the virulence factors. Given that VSRs are, by essence, virulence factors, one can imagine that some of these guardees can be key silencing components and that *R* genes have evolved to specifically sense changes in one of the RNA silencing pathways. Therefore, upon expression of a VSR, plants will have the potential to trigger a host defence response ultimately through the appearance of a hypersensitive response (HR). This concept has been illustrated at least once when mutations of a VSR that affect its silencing suppression activity was also shown to compromise induction of an HR (Li *et al.*, 1999). This plant counter counter-defensive strategy would constitute a potent driving force for the evolution of silencing suppressors in order to avoid this *R* gene recognition, providing yet another illustration of the never ending molecular arms race between hosts and pathogens.

## Acknowledgements

This work was supported by a PhD grant from the French Ministry of Research to S.W. and by grant 'MiDASS' from l'Agence National de la Recherche (ANR-08-JCJC-0063-01) to P.D.

## References

Adenot, X., Elmayan, T., Lauressergues, D., Boutet, S., Bouche, N., Gasciolli, V. and Vaucheret, H. (2006) DRB4-dependent TAS3 trans-acting siRNAs control leaf morphology through AGO7. *Current Biology* 16, 927–932.

Ameres, S.L., Martinez, J. and Schroeder, R. (2007) Molecular basis for target RNA recognition and cleavage by human RISC. *Cell* 130, 101–112.

Anandalakshmi, R., Pruss, G.J., Ge, X., Marathe, R., Mallory, A.C., Smith, T.H. and Vance, V.B. (1998) A viral suppressor of gene silencing in plants. *Proceedings of the National Academy of Sciences, USA* 95, 13079–13084.

Anandalakshmi, R., Marathe, R., Ge, X., Herr, J.M., Jr, Mau, C., Mallory, A., Pruss, G., Bowman, L. and Vance, V.B. (2000) A calmodulin-related protein that suppresses posttranscriptional gene silencing in plants. *Science* 290, 142–144.

Andersson, M.G., Haasnoot, P.C., Xu, N., Berenjian, S., Berkhout, B. and Akusjarvi, G. (2005) Suppression of RNA interference by adenovirus virus-associated RNA. *Journal of Virology* 79, 9556–9565.

Aukerman, M.J. and Sakai, H. (2003) Regulation of flowering time and floral organ identity by a MicroRNA and its APETALA2-like target genes. *The Plant Cell* 15, 2730–2741.

Bartel, D.P. (2004) MicroRNAs: genomics, biogenesis, mechanism and function. *Cell* 116, 281–297.

Baumberger, N. and Baulcombe, D.C. (2005) *Arabidopsis* ARGONAUTE1 is an RNA slicer that selectively recruits microRNAs and short interfering RNAs. *Proceedings of the National Academy of Sciences, USA* 102, 11928–11933.

Baumberger, N., Tsai, C.H., Lie, M., Havecker, E. and Baulcombe, D.C. (2007) The polerovirus silencing suppressor P0 targets ARGONAUTE proteins for degradation. *Current Biology* 17, 1609–1614.

Beclin, C., Boutet, S., Waterhouse, P. and Vaucheret, H. (2002) A branched pathway for transgene-induced RNA silencing in plants. *Current Biology* 12, 684–688.

Blevins, T., Rajeswaran, R., Shivaprasad, P.V., Beknazariants, D., Si-Ammour, A., Park, H.S., Vazquez, F., Robertson, D., Meins, F., Jr, Hohn, T. and Pooggin, M.M. (2006) Four plant Dicers mediate viral small RNA biogenesis and DNA virus induced silencing. *Nucleic Acids Research* 34, 6233–6246.

Borsani, O., Zhu, J., Verslues, P.E., Sunkar, R. and Zhu, J.K. (2005) Endogenous siRNAs derived from a pair of natural *cis*-antisense transcripts regulate salt tolerance in *Arabidopsis*. *Cell* 123, 1279–1291.

Bortolamiol, D., Pazhouhandeh, M., Marrocco, K., Genschik, P. and Ziegler-Graff, V. (2007) The polerovirus F box protein P0 targets ARGONAUTE1 to suppress RNA silencing. *Current Biology* 17, 1615–1621.

Bouche, N., Lauressergues, D., Gasciolli, V. and Vaucheret, H. (2006) An antagonistic function for *Arabidopsis* DCL2 in development and a new function for DCL4 in generating viral siRNAs. *EMBO Journal* 25, 3347–3356.

Brigneti, G., Voinnet, O., Li, W.X., Ji, L.H., Ding, S.W. and Baulcombe, D.C. (1998) Viral pathogenicity determinants are suppressors of transgene silencing in *Nicotiana benthamiana*. *EMBO Journal* 17, 6739–6746.

Brodersen, P. and Voinnet, O. (2006) The diversity of RNA silencing pathways in plants. *Trends in Genetics* 22, 268–280.

Brodersen, P., Sakvarelidze-Achard, L., Bruun-Rasmussen, M., Dunoyer, P., Yamamoto, Y.Y., Sieburth, L. and Voinnet, O. (2008) Widespread translational inhibition by plant miRNAs and siRNAs. *Science* 320, 1185–1190.

Brosnan, C.A., Mitter, N., Christie, M., Smith, N.A., Waterhouse, P.M. and Carroll, B.J. (2007) Nuclear gene silencing directs reception of long-distance mRNA silencing in *Arabidopsis*. *Proceedings of the National Academy of Sciences, USA* 104, 14741–14746.

Bucher, E., Sijen, T., De Haan, P., Goldbach, R. and Prins, M. (2003) Negative-strand tospoviruses and tenuiviruses carry a gene for a suppressor of gene silencing at analogous genomic positions. *Journal of Virology* 77, 1329–1336.

Bucher, E., Hemmes, H., de Haan, P., Goldbach, R. and Prins, M. (2004) The influenza A virus NS1 protein binds small interfering RNAs and suppresses RNA silencing in plants. *Journal of General Virology* 85, 983–991.

Cao, X., Zhou, P., Zhang, X., Zhu, S., Zhong, X., Xiao, Q., Ding, B. and Li, Y. (2005) Identification of an RNA silencing suppressor from a plant double-stranded RNA virus. *Journal of Virology* 79, 13018–13027.

Chapman, E.J. and Carrington, J.C. (2007) Specialization and evolution of endogenous small RNA pathways. *Nature Reviews Genetics* 8, 884–896.

Chapman, E.J., Prokhnevsky, A.I., Gopinath, K., Dolja, V.V. and Carrington, J.C. (2004) Viral RNA silencing suppressors inhibit the microRNA pathway at an intermediate step. *Genes & Development* 18, 1179–1186.

Chellappan, P., Vanitharani, R., Pita, J. and Fauquet, C.M. (2004) Short interfering RNA accumulation correlates with host recovery in DNA virus-infected hosts, and gene silencing targets specific viral sequences. *Journal of Virology* 78, 7465–7477.

Chellappan, P., Vanitharani, R. and Fauquet, C.M. (2005) MicroRNA-binding viral protein interferes with *Arabidopsis* development. *Proceedings of the National Academy of Sciences, USA* 102, 10381–10386.

Chen, J., Li, W.X., Xie, D., Peng, J.R. and Ding, S.W. (2004) Viral virulence protein suppresses RNA silencing-mediated defence but upregulates the role of microrna in host gene expression. *The Plant Cell* 16, 1302–1313.

Chen, X. (2004) A microRNA as a translational repressor of APETALA2 in *Arabidopsis* flower development. *Science* 303, 2022–2025.

Chiba, M., Reed, J.C., Prokhnevsky, A.I., Chapman, E.J., Mawassi, M., Koonin, E.V., Carrington, J.C. and Dolja, V.V. (2006) Diverse suppressors of RNA silencing enhance agroinfection by a viral replicon. *Virology* 346, 7–14.

Cruz, S.S., Chapman, S., Roberts, A.G., Roberts, I.M., Prior, D.A. and Oparka, K.J. (1996) Assembly and movement of a plant virus carrying a green fluorescent protein overcoat. *Proceedings of the National Academy of Sciences, USA* 93, 6286–6290.

Cruz, S.S., Roberts, A.G., Prior, D.A., Chapman, S. and Oparka, K.J. (1998) Cell-to-cell and phloem-mediated transport of potato virus X. The role of virions. *The Plant Cell* 10, 495–510.

Csorba, T., Bovi, A., Dalmay, T. and Burgyan, J. (2007) The p122 subunit of tobacco mosaic virus replicase is a potent silencing suppressor and compromises both small interfering RNA- and microRNA-mediated pathways. *Journal of Virology* 81, 11768–11780.

Dalmay, T., Hamilton, A., Rudd, S., Angell, S. and Baulcombe, D.C. (2000) An RNA-dependent RNA polymerase gene in *Arabidopsis* is required for posttranscriptional gene silencing mediated by a transgene but not by a virus. *Cell* 101, 543–553.

Deleris, A., Gallego-Bartolome, J., Bao, J., Kasschau, K.D., Carrington, J.C. and Voinnet, O. (2006) Hierarchical action and inhibition of plant Dicer-like proteins in antiviral defence. *Science* 313, 68–71.

Delgadillo, M.O., Saenz, P., Salvador, B., Garcia, J.A. and Simon-Mateo, C. (2004) Human influenza virus NS1 protein enhances viral pathogenicity and acts as an RNA silencing suppressor in plants. *Journal of General Virology* 85, 993–999.

Diaz-Pendon, J.A., Li, F., Li, W.X. and Ding, S.W. (2007) Suppression of antiviral silencing by cucumber mosaic virus 2b protein in *Arabidopsis* is associated with drastically reduced accumulation of three classes of viral small interfering RNAs. *The Plant Cell* 19, 2053–2063.

Ding, S.W. and Voinnet, O. (2007) Antiviral immunity directed by small RNAs. *Cell* 130, 413–426.

Ding, S.W., Li, W.X. and Symons, R.H. (1995) A novel naturally occurring hybrid gene encoded by a plant RNA virus facilitates long distance virus movement. *EMBO Journal* 14, 5762–5772.

Duchaine, T.F., Wohlschlegel, J.A., Kennedy, S., Bei, Y., Conte, D., Jr, Pang, K., Brownell, D.R., Harding, S., Mitani, S., Ruvkun, G., Yates, J.R. and Mello, C.C. (2006) Functional proteomics reveals the biochemical niche of *C. elegans* DCR-1 in multiple small-RNA-mediated pathways. *Cell* 124, 343–354.

Dunoyer, P. and Voinnet, O. (2005) The complex interplay between plant viruses and host RNA-silencing pathways. *Current Opinion in Plant Biology* 8, 415–423.

Dunoyer, P., Pfeffer, S., Fritsch, C., Hemmer, O., Voinnet, O. and Richards, K.E. (2002) Identification, subcellular localization and some properties of a cysteine-rich suppressor of gene silencing encoded by peanut clump virus. *The Plant Journal* 29, 555–567.

Dunoyer, P., Lecellier, C.H., Parizotto, E.A., Himber, C. and Voinnet, O. (2004) Probing the microRNA and small interfering RNA pathways with virus-encoded suppressors of RNA silencing. *The Plant Cell* 16, 1235–1250.

Dunoyer, P., Himber, C. and Voinnet, O. (2005) DICER-LIKE 4 is required for RNA interference and produces the 21-nucleotide small interfering RNA component of the plant cell-to-cell silencing signal. *Nature Genetics* 37, 1356–1360.

Dunoyer, P., Himber, C., Ruiz-Ferrer, V., Alioua, A. and Voinnet, O. (2007) Intra- and intercellular RNA interference in *Arabidopsis thaliana* requires components of the microRNA and heterochromatic silencing pathways. *Nature Genetics* 39, 848–856.

Ebhardt, H.A., Thi, E.P., Wang, M.B. and Unrau, P.J. (2005) Extensive 3′ modification of plant small RNAs is modulated by helper component-proteinase expression. *Proceedings of the National Academy of Sciences, USA* 102, 13398–13403.

Ekwall, K. (2004) The RITS complex-A direct link between small RNA and heterochromatin. *Molecular Cell* 13, 304–305.

Elbashir, S.M., Lendeckel, W. and Tuschl, T. (2001) RNA interference is mediated by 21- and 22-nucleotide RNAs. *Genes & Development* 15, 188–200.

Escaler, M., Aranda, M.A., Thomas, C.L. and Maule, A.J. (2000) Pea embryonic tissues show common responses to the replication of a wide range of viruses. *Virology* 267, 318–325.

Fahlgren, N., Montgomery, T.A., Howell, M.D., Allen, E., Dvorak, S.K., Alexander, A.L. and Carrington, J.C. (2006) Regulation of AUXIN RESPONSE FACTOR3 by TAS3 ta-siRNA affects developmental timing and patterning in *Arabidopsis*. *Current Biology* 16, 939–944.

Fire, A., Albertson, D., Harrison, S.W. and Moerman, D.G. (1991) Production of antisense RNA leads to effective and specific inhibition of gene expression in *C. elegans* muscle. *Development* 113, 503–514.

Fire, A., Xu, S., Montgomery, M.K., Kostas, S.A., Driver, S.E. and Mello, C.C. (1998) Potent and specific genetic interference by double-stranded RNA in *Caenorhabditis elegans*. *Nature* 391, 806–811.

Foster, T.M., Lough, T.J., Emerson, S.J., Lee, R.H., Bowman, J.L., Forster, R.L. and Lucas, W.J. (2002) A surveillance system regulates selective entry of RNA into the shoot apex. *The Plant Cell* 14, 1497–1508.

Fusaro, A.F., Matthew, L., Smith, N.A., Curtin, S.J., Dedic-Hagan, J., Ellacott, G.A., Watson, J.M., Wang, M.B., Brosnan, C., Carroll, B.J. and Waterhouse, P.M. (2006) RNA interference-inducing hairpin RNAs in plants act through the viral defence pathway. *EMBO Reports* 7, 1168–1175.

Gazzani, S., Lawrenson, T., Woodward, C., Headon, D. and Sablowski, R. (2004) A link between mRNA turnover and RNA interference in *Arabidopsis*. *Science* 306, 1046–1048.

Giordano, E., Rendina, R., Peluso, I. and Furia, M. (2002) RNAi triggered by symmetrically transcribed transgenes in *Drosophila melanogaster*. *Genetics* 160, 637–648.

Glick, E., Zrachya, A., Levy, Y., Mett, A., Gidoni, D., Belausov, E., Citovsky, V. and Gafni, Y. (2008) Interaction with host SGS3 is required for suppression of RNA silencing by tomato yellow leaf curl virus V2 protein. *Proceedings of the National Academy of Sciences, USA* 105, 157–161.

Guo, H.S. and Ding, S.W. (2002) A viral protein inhibits the long range signaling activity of the gene silencing signal. *EMBO Journal* 21, 398–407.

Guo, S. and Kemphues, K.J. (1995) par-1, a gene required for establishing polarity in *C. elegans* embryos, encodes a putative Ser/Thr kinase that is asymmetrically distributed. *Cell* 81, 611–620.

Gy, I., Gasciolli, V., Lauressergues, D., Morel, J.B., Gombert, J., Proux, F., Proux, C., Vaucheret, H. and Mallory, A.C. (2007) *Arabidopsis* FIERY1, XRN2, and XRN3 are endogenous RNA silencing suppressors. *The Plant Cell* 19, 3451–3461.

Hamilton, A.J. and Baulcombe, D.C. (1999) A species of small antisense RNA in posttranscriptional gene silencing in plants. *Science* 286, 950–952.

Hamilton, A., Voinnet, O., Chappell, L. and Baulcombe, D. (2002) Two classes of short interfering RNA in RNA silencing. *EMBO Journal* 21, 4671–4679.

Hammond, S.M. (2005) Dicing and slicing: the core machinery of the RNA interference pathway. *FEBS Letters* 579, 5822–5829.

Hammond, S.M., Bernstein, E., Beach, D. and Hannon, G.J. (2000) An RNA-directed nuclease mediates post-transcriptional gene silencing in *Drosophila* cells. *Nature* 404, 293–296.

Havelda, Z. and Maule, A.J. (2000) Complex spatial responses to cucumber mosaic virus infection in susceptible *Cucurbita pepo* cotyledons. *The Plant Cell* 12, 1975–1986.

Havelda, Z., Hornyik, C., Crescenzi, A. and Burgyan, J. (2003) In situ characterization of cymbidium ringspot tombusvirus infection-induced posttranscriptional gene silencing in *Nicotiana benthamiana*. *Journal of Virology* 77, 6082–6086.

Hemmes, H., Lakatos, L., Goldbach, R., Burgyan, J. and Prins, M. (2007) The NS3 protein of rice hoja blanca tenuivirus suppresses RNA silencing in plant and insect hosts by efficiently binding both siRNAs and miRNAs. *RNA* 13, 1079–1089.

Herr, A.J., Jensen, M.B., Dalmay, T. and Baulcombe, D.C. (2005) RNA polymerase IV directs silencing of endogenous DNA. *Science* 308, 118–120.

Herr, A.J., Molnar, A., Jones, A. and Baulcombe, D.C. (2006) Defective RNA processing enhances RNA silencing and influences flowering of *Arabidopsis*. *Proceedings of the National Academy of Sciences, USA* 103, 14994–15001.

Himber, C., Dunoyer, P., Moissiard, G., Ritzenthaler, C. and Voinnet, O. (2003) Transitivity-dependent and -independent cell-to-cell movement of RNA silencing. *EMBO Journal* 22, 4523–4533.

Hiraguri, A., Itoh, R., Kondo, N., Nomura, Y., Aizawa, D., Murai, Y., Koiwa, H., Seki, M., Shinozaki, K. and Fukuhara, T. (2005) Specific interactions between Dicer-like proteins and HYL1/DRB-family dsRNA-binding proteins in *Arabidopsis thaliana*. *Plant Molecular Biology* 57, 173–188.

Ho, T., Pallett, D., Rusholme, R., Dalmay, T. and Wang, H. (2006) A simplified method for cloning of short interfering RNAs from *Brassica juncea* infected with turnip mosaic potyvirus and turnip crinkle carmovirus. *Journal of Virology Methods* 136, 217–223.

Hunter, C., Willmann, M.R., Wu, G., Yoshikawa, M., de la Luz Gutierrez-Nava, M. and Poethig, S.R. (2006) Trans-acting siRNA-mediated repression of ETTIN and ARF4 regulates heteroblasty in *Arabidopsis*. *Development* 133, 2973–2981.

Itaya, A., Zhong, X., Bundschuh, R., Qi, Y., Wang, Y., Takeda, R., Harris, A.R., Molina, C., Nelson, R.S. and Ding, B. (2007) A structured viroid RNA serves as a substrate for dicer-like cleavage to produce biologically active small RNAs but is resistant to RNA-induced silencing complex-mediated degradation. *Journal of Virology* 81, 2980–2994.

Jones-Rhoades, M.W., Bartel, D.P. and Bartel, B. (2006) MicroRNAS and their regulatory roles in plants. *Annual Review in Plant Biology* 57, 19–53.

Kanno, T., Huettel, B., Mette, M.F., Aufsatz, W., Jaligot, E., Daxinger, L., Kreil, D.P., Matzke, M. and Matzke, A.J. (2005) Atypical RNA polymerase subunits required for RNA-directed DNA methylation. *Nature Genetics* 37, 761–765.

Kasschau, K.D. and Carrington, J.C. (1998) A counterdefensive strategy of plant viruses: suppression of posttranscriptional gene silencing. *Cell* 95, 461–470.

Kasschau, K.D. and Carrington, J.C. (2001) Long-distance movement and replication maintenance functions correlate with silencing suppression activity of potyviral HC-Pro. *Virology* 285, 71–81.

Katiyar-Agarwal, S., Morgan, R., Dahlbeck, D., Borsani, O., Villegas, A., Jr, Zhu, J.K., Staskawicz, B.J. and Jin, H. (2006) A pathogen-inducible endogenous siRNA in plant immunity. *Proceedings of the National Academy of Sciences, USA* 103, 18002–18007.

Kennedy, S., Wang, D. and Ruvkun, G. (2004) A conserved siRNA-degrading RNase negatively regulates RNA interference in *C. elegans*. *Nature* 427, 645–649.

Kidner, C.A. and Martienssen, R.A. (2004) Spatially restricted microRNA directs leaf polarity through ARGONAUTE1. *Nature* 428, 81–84.

Kreuze, J.F., Savenkov, E.I., Cuellar, W., Li, X. and Valkonen, J.P. (2005) Viral class 1 RNase III involved in suppression of RNA silencing. *Journal of Virology* 79, 7227–7238.

Kubota, K., Tsuda, S., Tamai, A. and Meshi, T. (2003) Tomato mosaic virus replication protein suppresses virus-targeted posttranscriptional gene silencing. *Journal of Virology* 77, 11016–11026.

Kumagai, M.H., Donson, J., della-Cioppa, G., Harvey, D., Hanley, K. and Grill, L.K. (1995) Cytoplasmic inhibition of carotenoid biosynthesis with virus-derived RNA. *Proceedings of the National Academy of Sciences, USA* 92, 1679–1683.

Lakatos, L., Csorba, T., Pantaleo, V., Chapman, E.J., Carrington, J.C., Liu, Y.P., Dolja, V.V., Calvino, L.F., Lopez-Moya, J.J. and Burgyan, J. (2006) Small RNA binding is a common strategy to suppress RNA silencing by several viral suppressors. *EMBO Journal* 25, 2768–2780.

Landry, P. and Perreault, J.P. (2005) Identification of a peach latent mosaic viroid hairpin able to act as a Dicer-like substrate. *Journal of Virology* 79, 6540–6543.

Leshchiner, A.D., Solovyev, A.G., Morozov, S.Y. and Kalinina, N.O. (2006) A minimal region in the NTPase/helicase domain of the TGBp1 plant virus movement protein is responsible for ATPase activity and cooperative RNA binding. *Journal of General Virology* 87, 3087–3095.

Li, C.F., Pontes, O., El-Shami, M., Henderson, I.R., Bernatavichute, Y.V., Chan, S.W., Lagrange, T., Pikaard, C.S. and Jacobsen, S.E. (2006) An ARGONAUTE4-containing nuclear processing center colocalized with Cajal bodies in *Arabidopsis thaliana*. *Cell* 126, 93–106.

Li, F. and Ding, S.W. (2006) Virus counterdefence: diverse strategies for evading the RNA-silencing immunity. *Annual Review of Microbiology* 60, 503–531.

Li, H.W., Lucy, A.P., Guo, H.S., Li, W.X., Ji, L.H., Wong, S.M. and Ding, S.W. (1999) Strong host resistance targeted against a viral suppressor of the plant gene silencing defence mechanism. *EMBO Journal* 18, 2683–2691.

Li, H.W., Li, W.X. and Ding, S.W. (2002) Induction and suppression of RNA silencing by an animal virus. *Science* 296, 1319–1321.

Li, W.X., Li, H., Lu, R., Li, F., Dus, M., Atkinson, P., Brydon, E.W., Johnson, K.L., Garcia-Sastre, A., Ball, L.A., Palese, P. and Ding, S.W. (2004) Interferon antagonist proteins of influenza and vaccinia viruses are suppressors of RNA silencing. *Proceedings of the National Academy of Sciences, USA* 101, 1350–1355.

Lindbo, J.A., Silva-Rosales, L., Proebsting, W.M. and Dougherty, W.G. (1993) Induction of a highly specific antiviral state in transgenic plants: implications for regulation of gene expression and virus resistance. *The Plant Cell* 5, 1749–1759.

Liu, H., Reavy, B., Swanson, M. and MacFarlane, S.A. (2002) Functional replacement of the tobacco rattle virus cysteine-rich protein by pathogenicity proteins from unrelated plant viruses. *Virology* 298, 232–239.

Liu, L., Grainger, J., Canizares, M.C., Angell, S.M. and Lomonossoff, G.P. (2004) Cowpea mosaic virus RNA-1 acts as an amplicon whose effects can be counteracted by a RNA-2-encoded suppressor of silencing. *Virology* 323, 37–48.

Llave, C., Kasschau, K.D. and Carrington, J.C. (2000) Virus-encoded suppressor of posttranscriptional gene silencing targets a maintenance step in the silencing pathway. *Proceedings of the National Academy of Sciences, USA* 97, 13401–13406.

Llave, C., Xie, Z., Kasschau, K.D. and Carrington, J.C. (2002) Cleavage of Scarecrow-like mRNA targets directed by a class of *Arabidopsis* miRNA. *Science* 297, 2053–2056.

Love, A.J., Laird, J., Holt, J., Hamilton, A.J., Sadanandom, A. and Milner, J.J. (2007) Cauliflower mosaic virus protein P6 is a suppressor of RNA silencing. *Journal of General Virology* 88, 3439–3444.

Lu, R., Folimonov, A., Shintaku, M., Li, W.X., Falk, B. W., Dawson, W.O. and Ding, S.W. (2004) Three distinct suppressors of RNA silencing encoded by a 20-kb viral RNA genome. *Proceedings of the National Academy of Sciences, USA* 101, 15742–15747.

Lu, S. and Cullen, B.R. (2004) Adenovirus VA1 noncoding RNA can inhibit small interfering RNA and microRNA biogenesis. *Journal of Virology* 78, 12868–12876.

Lucy, A.P., Guo, H.S., Li, W.X. and Ding, S.W. (2000) Suppression of post-transcriptional gene silencing by a plant viral protein localized in the nucleus. *EMBO Journal* 19, 1672–1680.

Mallory, A.C., Reinhart, B.J., Bartel, D., Vance, V.B. and Bowman, L.H. (2002) A viral suppressor of RNA silencing differentially regulates the accumulation of short interfering RNAs and micro-RNAs in tobacco. *Proceedings of the National Academy of Sciences, USA* 99, 15228–15233.

Martinez de Alba, A.E., Flores, R. and Hernandez, C. (2002) Two chloroplastic viroids induce the accumulation of small RNAs associated with posttranscriptional gene silencing. *Journal of Virology* 76, 13094–13096.

Mathews, M.B. and Shenk, T. (1991) Adenovirus virus-associated RNA and translation control. *Journal of Virology* 65, 5657–5662.

Matzke, M.A. and Birchler, J.A. (2005) RNAi-mediated pathways in the nucleus. *Nature Reviews Genetics* 6, 24–35.

Merai, Z., Kerenyi, Z., Molnar, A., Barta, E., Valoczi, A., Bisztray, G., Havelda, Z., Burgyan, J. and Silhavy, D. (2005) Aureusvirus P14 is an efficient RNA silencing suppressor that binds double-stranded RNAs without size specificity. *Journal of Virology* 79, 7217–7226.

Merai, Z., Kerenyi, Z., Kertesz, S., Magna, M., Lakatos, L. and Silhavy, D. (2006) Double-stranded RNA binding may be a general plant RNA viral strategy to suppress RNA silencing. *Journal of Virology* 80, 5747–5756.

Mi, S., Cai, T., Hu, Y., Chen, Y., Hodges, E., Ni, F., Wu, L., Li, S., Zhou, H., Long, C., Chen, S., Hannon, G.J. and Qi, Y. (2008) Sorting of small RNAs into *Arabidopsis* argonaute complexes is directed by the 5' terminal nucleotide. *Cell* 133, 116–127.

Moissiard, G. and Voinnet, O. (2004) Viral suppression of RNA silencing in plants. *Molecular Plant Pathology* 5, 71–82.

Moissiard, G. and Voinnet, O. (2006) RNA silencing of host transcripts by cauliflower mosaic virus requires coordinated action of the four *Arabidopsis* Dicer-like proteins. *Proceedings of the National Academy of Sciences, USA* 103, 19593–19598.

Moissiard, G., Parizotto, E.A., Himber, C. and Voinnet, O. (2007) Transitivity in *Arabidopsis* can be primed, requires the redundant action of the antiviral Dicer-like 4 and Dicer-like 2, and is compromised by viral-encoded suppressor proteins. *RNA* 13, 1268–1278.

Molnar, A., Csorba, T., Lakatos, L., Varallyay, E., Lacomme, C. and Burgyan, J. (2005) Plant virus-derived small interfering RNAs originate predominantly from highly structured single-stranded viral RNAs. *Journal of Virology* 79, 7812–7818.

Montgomery, T.A., Howell, M.D., Cuperus, J.T., Li, D., Hansen, J.E., Alexander, A.L., Chapman, E.J., Fahlgren, N., Allen, E. and Carrington, J.C. (2008) Specificity of ARGONAUTE7-miR390 interaction and dual functionality in TAS3 trans-acting siRNA formation. *Cell* 133, 128–141.

Morel, J.B., Godon, C., Mourrain, P., Beclin, C., Boutet, S., Feuerbach, F., Proux, F. and Vaucheret, H. (2002) Fertile hypomorphic ARGONAUTE (ago1) mutants impaired in post-transcriptional gene silencing and virus resistance. *The Plant Cell* 14, 629–639.

Mourrain, P., Beclin, C., Elmayan, T., Feuerbach, F., Godon, C., Morel, J.B., Jouette, D., Lacombe, A.M., Nikic, S., Picault, N., Rémoué, K., Sanial, M., Vo, T.A. and Vaucheret, H. (2000) *Arabidopsis* SGS2 and SGS3 genes are required for posttranscriptional gene silencing and natural virus resistance. *Cell* 101, 533–542.

Muangsan, N., Beclin, C., Vaucheret, H. and Robertson, D. (2004) Geminivirus VIGS of endogenous genes requires SGS2/SDE1 and SGS3 and defines a new branch in the genetic pathway for silencing in plants. *The Plant Journal* 38, 1004–1014.

Napoli, C., Lemieux, C. and Jorgensen, R. (1990) Introduction of a chimeric chalcone synthase gene into petunia results in reversible co-suppression of homologous genes in *trans*. *The Plant Cell* 2, 279–289.

Navarro, L., Dunoyer, P., Jay, F., Arnold, B., Dharmasiri, N., Estelle, M., Voinnet, O. and Jones, J.D. (2006) A plant miRNA contributes to antibacterial resistance by repressing auxin signaling. *Science* 312, 436–439.

Okamura, K., Phillips, M.D., Tyler, D.M., Duan, H., Chou, Y.T. and Lai, E.C. (2008) The regulatory activity of microRNA* species has substantial influence on microRNA and 3' UTR evolution. *Nature Structural and Molecular Biology* 15, 354–363.

Palauqui, J.C., Elmayan, T., Pollien, J.M. and Vaucheret, H. (1997) Systemic acquired silencing: transgene-specific post-transcriptional silencing is transmitted by grafting from silenced stocks to non-silenced scions. *EMBO Journal* 16, 4738–4745.

Pantaleo, V., Szittya, G. and Burgyan, J. (2007) Molecular bases of viral RNA targeting by viral small interfering RNA-programmed RISC. *Journal of Virology* 81, 3797–3806.

Papaefthimiou, I., Hamilton, A., Denti, M., Baulcombe, D., Tsagris, M. and Tabler, M. (2001) Replicating potato spindle tuber viroid RNA is accompanied by short RNA fragments that are characteristic of post-transcriptional gene silencing. *Nucleic Acids Research* 29, 2395–2400.

Parizotto, E.A., Dunoyer, P., Rahm, N., Himber, C. and Voinnet, O. (2004) *In vivo* investigation of the transcription, processing, endonucleolytic activity, and functional relevance of the spatial distribution of a plant miRNA. *Genes & Development* 18, 2237–2242.

Pazhouhandeh, M., Dieterle, M., Marrocco, K., Lechner, E., Berry, B., Brault, V., Hemmer, O., Kretsch, T., Richards, K.E., Genschik, P. and Ziegler-Graff, V. (2006) F-box-like domain in the polerovirus protein P0 is required for silencing suppressor function. *Proceedings of the National Academy of Sciences, USA* 103, 1994–1999.

Peragine, A., Yoshikawa, M., Wu, G., Albrecht, H.L. and Poethig, R.S. (2004) SGS3 and

SGS2/SDE1/RDR6 are required for juvenile development and the production of trans-acting siRNAs in *Arabidopsis*. *Genes & Development* 18, 2368–2379.

Pfeffer, S., Dunoyer, P., Heim, F., Richards, K.E., Jonard, G. and Ziegler-Graff, V. (2002) P0 of beet western yellows virus is a suppressor of posttranscriptional gene silencing. *Journal of Virology* 76, 6815–6824.

Pontes, O., Li, C.F., Nunes, P.C., Haag, J., Ream, T., Vitins, A., Jacobsen, S.E. and Pikaard, C.S. (2006) The *Arabidopsis* chromatin-modifying nuclear siRNA pathway involves a nucleolar RNA processing center. *Cell* 126, 79–92.

Pontier, D., Yahubyan, G., Vega, D., Bulski, A., Saez-Vasquez, J., Hakimi, M.A., Lerbs-Mache, S., Colot, V. and Lagrange, T. (2005) Reinforcement of silencing at transposons and highly repeated sequences requires the concerted action of two distinct RNA polymerases IV in *Arabidopsis*. *Genes & Development* 19, 2030–2040.

Pruss, G., Ge, X., Shi, X.M., Carrington, J.C. and Bowman Vance, V. (1997) Plant viral synergism: the potyviral genome encodes a broad-range pathogenicity enhancer that transactivates replication of heterologous viruses. *The Plant Cell* 9, 859–868.

Qi, Y., He, X., Wang, X.J., Kohany, O., Jurka, J. and Hannon, G.J. (2006) Distinct catalytic and non-catalytic roles of ARGONAUTE4 in RNA-directed DNA methylation. *Nature* 443, 1008–1012.

Qu, F., Ren, T. and Morris, T.J. (2003) The coat protein of turnip crinkle virus suppresses posttranscriptional gene silencing at an early initiation step. *Journal of Virology* 77, 511–522.

Qu, F., Ye, X., Hou, G., Sato, S., Clemente, T.E. and Morris, T.J. (2005) RDR6 has a broad-spectrum but temperature-dependent antiviral defence role in *Nicotiana benthamiana*. *Journal of Virology* 79, 15209–15217.

Ratcliff, F., Harrison, B.D. and Baulcombe, D.C. (1997) A similarity between viral defence and gene silencing in plants. *Science* 276, 1558–1560.

Ratcliff, F.G., MacFarlane, S.A. and Baulcombe, D.C. (1999) Gene silencing without DNA. RNA-mediated cross-protection between viruses. *The Plant Cell* 11, 1207–1216.

Reavy, B., Dawson, S., Canto, T. and MacFarlane, S.A. (2004) Heterologous expression of plant virus genes that suppress posttranscriptional gene silencing results in suppression of RNA interference in *Drosophila* cells. *BMC Biotechnology* 4, 18.

Reed, J.C., Kasschau, K.D., Prokhnevsky, A.I., Gopinath, K., Pogue, G.P., Carrington, J.C. and Dolja, V.V. (2003) Suppressor of RNA silencing encoded by beet yellows virus. *Virology* 306, 203–209.

Rhoades, M.W., Reinhart, B.J., Lim, L.P., Burge, C.B., Bartel, B. and Bartel, D.P. (2002) Prediction of plant microRNA targets. *Cell* 110, 513–520.

Ruiz, M.T., Voinnet, O. and Baulcombe, D.C. (1998) Initiation and maintenance of virus-induced gene silencing. *The Plant Cell* 10, 937–946.

Ryang, B.S., Kobori, T., Matsumoto, T., Kosaka, Y. and Ohki, S.T. (2004) Cucumber mosaic virus 2b protein compensates for restricted systemic spread of potato virus Y in doubly infected tobacco. *Journal of General Virology* 85, 3405–3414.

Samuels, T.D., Ju, H.J., Ye, C.M., Motes, C.M., Blancaflor, E.B. and Verchot-Lubicz, J. (2007) Subcellular targeting and interactions among the potato virus X TGB proteins. *Virology* 367, 375–389.

Schnettler, E., Hemmes, H., Goldbach, R. and Prins, M. (2008) The NS3 protein of rice hoja blanca virus suppresses RNA silencing in mammalian cells. *Journal of General Virology* 89, 336–340.

Schwach, F., Vaistij, F.E., Jones, L. and Baulcombe, D.C. (2005) An RNA-dependent RNA polymerase prevents meristem invasion by potato virus X and is required for the activity but not the production of a systemic silencing signal. *Plant Physiology* 138, 1842–1852.

Segers, G.C., van Wezel, R., Zhang, X., Hong, Y. and Nuss, D.L. (2006) Hypovirus papain-like protease p29 suppresses RNA silencing in the natural fungal host and in a heterologous plant system. *Eukaryotic Cell* 5, 896–904.

Segers, G.C., Zhang, X., Deng, F., Sun, Q. and Nuss, D.L. (2007) Evidence that RNA silencing functions as an antiviral defence mechanism in fungi. *Proceedings of the National Academy of Sciences, USA* 104, 12902–12906.

Sequeira, L. (1984) Cross protection and induced resistance: their potential for plant disease control. *Trends in Biochemical Sciences* 2, 26–30.

Silhavy, D., Molnar, A., Lucioli, A., Szittya, G., Hornyik, C., Tavazza, M. and Burgyan, J. (2002) A viral protein suppresses RNA silencing and binds silencing-generated, 21- to 25-nucleotide double-stranded RNAs. *EMBO Journal* 21, 3070–3080.

Smith, H.A., Swaney, S.L., Parks, T.D., Wernsman, E.A. and Dougherty, W.G. (1994) Transgenic plant virus resistance mediated by untranslatable sense RNAs: expression, regulation and fate of nonessential RNAs. *The Plant Cell* 6, 1441–1453.

Szittya, G., Molnar, A., Silhavy, D., Hornyik, C. and Burgyan, J. (2002) Short defective interfering RNAs of tombusviruses are not targeted but trigger post-transcriptional gene silencing against their helper virus. *The Plant Cell* 14, 359–372.

Takeda, A., Sugiyama, K., Nagano, H., Mori, M., Kaido, M., Mise, K., Tsuda, S. and Okuno, T. (2002) Identification of a novel RNA silencing suppressor, NSs protein of tomato spotted wilt virus. *FEBS Letters* 532, 75–79.

Takeda, A., Tsukuda, M., Mizumoto, H., Okamoto, K., Kaido, M., Mise, K. and Okuno, T. (2005) A plant RNA virus suppresses RNA silencing through viral RNA replication. *EMBO Journal* 24, 3147–3157.

Takeda, A., Iwasaki, S., Watanabe, T., Utsumi, M. and Watanabe, Y. (2008) The mechanism selecting the guide strand from small RNA duplexes is different among argonaute proteins. *Plant Cell Physiology* 49, 493–500.

Te, J., Melcher, U., Howard, A. and Verchot-Lubicz, J. (2005) Soilborne wheat mosaic virus (SBWMV) 19K protein belongs to a class of cysteine rich proteins that suppress RNA silencing. *Journal of Virology* 2, 18.

Thomas, C.L., Leh, V., Lederer, C. and Maule, A.J. (2003) Turnip crinkle virus coat protein mediates suppression of RNA silencing in *Nicotiana benthamiana*. *Virology* 306, 33–41.

Trinks, D., Rajeswaran, R., Shivaprasad, P.V., Akbergenov, R., Oakeley, E.J., Veluthambi, K., Hohn, T. and Pooggin, M.M. (2005) Suppression of RNA silencing by a geminivirus nuclear protein, AC2, correlates with transactivation of host genes. *Journal of Virology* 79, 2517–2527.

Ueki, S. and Citovsky, V. (2001) Inhibition of systemic onset of post-transcriptional gene silencing by non-toxic concentrations of cadmium. *The Plant Journal* 28, 283–291.

Urcuqui-Inchima, S., Maia, I.G., Arruda, P., Haenni, A.L. and Bernardi, F. (2000) Deletion mapping of the potyviral helper component-proteinase reveals two regions involved in RNA binding. *Virology* 268, 104–111.

van Wezel, R., Dong, X., Liu, H., Tien, P., Stanley, J. and Hong, Y. (2002) Mutation of three cysteine residues in tomato yellow leaf curl virus-China C2 protein causes dysfunction in pathogenesis and posttranscriptional gene-silencing suppression. *Molecular Plant–Microbe Interactions* 15, 203–208.

Vance, V.B. (1991) Replication of potato virus X RNA is altered in coinfections with potato virus Y. *Virology* 182, 486–494.

Vargason, J.M., Szittya, G., Burgyan, J. and Hall, T.M. (2003) Size selective recognition of siRNA by an RNA silencing suppressor. *Cell* 115, 799–811.

Vazquez, F. (2006) *Arabidopsis* endogenous small RNAs: highways and byways. *Trends in Plant Science* 11, 460–468.

Vazquez, F., Vaucheret, H., Rajagopalan, R., Lepers, C., Gasciolli, V., Mallory, A.C., Hilbert, J.L., Bartel, D.P. and Crete, P. (2004) Endogenous trans-acting siRNAs regulate the accumulation of *Arabidopsis* mRNAs. *Molecular Cell* 16, 69–79.

Vogler, H., Akbergenov, R., Shivaprasad, P.V., Dang, V., Fasler, M., Kwon, M.O., Zhanybekova, S., Hohn, T. and Heinlein, M. (2007) Modification of small RNAs associated with suppression of RNA silencing by tobamovirus replicase protein. *Journal of Virology* 81, 10379–10388.

Voinnet, O. (2005) Induction and suppression of RNA silencing: insights from viral infections. *Nature Reviews Genetics* 6, 206–220.

Voinnet, O., Vain, P., Angell, S. and Baulcombe, D. C. (1998) Systemic spread of sequence-specific transgene RNA degradation in plants is initiated by localized introduction of ectopic promoterless DNA. *Cell* 95, 177–187.

Voinnet, O., Pinto, Y.M. and Baulcombe, D.C. (1999) Suppression of gene silencing: a general strategy used by diverse DNA and RNA viruses of plants. *Proceedings of the National Academy of Sciences, USA* 96, 14147–14152.

Voinnet, O., Lederer, C. and Baulcombe, D.C. (2000) A viral movement protein prevents spread of the gene silencing signal in *Nicotiana benthamiana*. *Cell* 103, 157–167.

Wang, H., Buckley, K.J., Yang, X., Buchmann, R.C. and Bisaro, D.M. (2005) Adenosine kinase inhibition and suppression of RNA silencing by geminivirus AL2 and L2 proteins. *Journal of Virology* 79, 7410–7418.

Wang, M.B., Bian, X.Y., Wu, L.M., Liu, L.X., Smith, N.A., Isenegger, D., Wu, R.M., Masuta, C., Vance, V.B., Watson, J.M., Rezaian, A.M., Dennis, E.S. and Waterhouse, P.M. (2004) On the role of RNA silencing in the pathogenicity and evolution of viroids and viral satellites. *Proceedings of the National Academy of Sciences, USA* 101, 3275–3280.

Wang, X.H., Aliyari, R., Li, W.X., Li, H.W., Kim, K., Carthew, R., Atkinson, P. and Ding, S.W. (2006) RNA interference directs innate immunity against viruses in adult *Drosophila*. *Science* 312, 452–454.

Wesley, S.V., Helliwell, C.A., Smith, N.A., Wang, M.B., Rouse, D.T., Liu, Q., Gooding, P.S., Singh, S.P., Abbott, D., Stoutjesdijk, P.A., Robinsons, S.P., Gleave, A.P., Green, A.G. and Waterhouse, P.M. (2001) Construct design for efficient, effective and high-throughput gene silencing in plants. *The Plant Journal* 27, 581–590.

Xie, Z., Johansen, L.K., Gustafson, A.M., Kasschau, K.D., Lellis, A.D., Zilberman, D., Jacobsen, S.E. and Carrington, J.C. (2004) Genetic and functional diversification of small RNA pathways in plants. *PLoS Biology* 2, E104.

Xie, Z., Allen, E., Wilken, A. and Carrington, J.C. (2005) DICER-LIKE 4 functions in trans-acting small interfering RNA biogenesis and vegetative phase change in *Arabidopsis thaliana*. *Proceedings of the National Academy of Sciences, USA* 102, 12984–12989.

Yang, S.J., Carter, S.A., Cole, A.B., Cheng, N.H. and Nelson, R.S. (2004) A natural variant of a host RNA-dependent RNA polymerase is associated with increased susceptibility to viruses by *Nicotiana benthamiana*. *Proceedings of the National Academy of Sciences, USA* 101, 6297–6302.

Yang, Z., Ebright, Y.W., Yu, B. and Chen, X. (2006) HEN1 recognizes 21–24 nt small RNA duplexes and deposits a methyl group onto the 2' OH of the 3' terminal nucleotide. *Nucleic Acids Research* 34, 667–675.

Ye, K. and Patel, D.J. (2005) RNA silencing suppressor p21 of beet yellows virus forms an RNA binding octameric ring structure. *Structure* 13, 1375–1384.

Ye, K., Malinina, L. and Patel, D.J. (2003) Recognition of small interfering RNA by a viral suppressor of RNA silencing. *Nature* 426, 874–878.

Yelina, N.E., Savenkov, E.I., Solovyev, A.G., Morozov, S.Y. and Valkonen, J.P. (2002) Long-distance movement, virulence, and RNA silencing suppression controlled by a single protein in Hordei- and potyviruses: complementary functions between virus families. *Journal of Virology* 76, 12981–12991.

Yoo, B.C., Kragler, F., Varkonyi-Gasic, E., Haywood, V., Archer-Evans, S., Lee, Y.M., Lough, T.J. and Lucas, W.J. (2004) A systemic small RNA signaling system in plants. *The Plant Cell* 16, 1979–2000.

Yoshikawa, M., Peragine, A., Park, M.Y. and Poethig, R.S. (2005) A pathway for the biogenesis of trans-acting siRNAs in *Arabidopsis. Genes & Development* 19, 2164–2175.

Yu, B., Chapman, E.J., Yang, Z., Carrington, J.C. and Chen, X. (2006) Transgenically expressed viral RNA silencing suppressors interfere with microRNA methylation in *Arabidopsis. FEBS Letters* 580, 3117–3120.

Yu, D., Fan, B., MacFarlane, S.A. and Chen, Z. (2003) Analysis of the involvement of an inducible *Arabidopsis* RNA-dependent RNA polymerase in antiviral defence. *Molecular Plant–Microbe Interactions* 16, 206–216.

Zamore, P.D. (2007) RNA silencing: genomic defence with a slice of pi. *Nature* 446, 864–865.

Zaratiegui, M., Irvine, D.V. and Martienssen, R.A. (2007) Noncoding RNAs and gene silencing. *Cell* 128, 763–776.

Zhang, X., Yuan, Y.R., Pei, Y., Lin, S.S., Tuschl, T., Patel, D.J. and Chua, N.H. (2006) Cucumber mosaic virus-encoded 2b suppressor inhibits *Arabidopsis* Argonaute1 cleavage activity to counter plant defence. *Genes & Development* 20, 3255–3268.

Zrachya, A., Glick, E., Levy, Y., Arazi, T., Citovsky, V. and Gafni, Y. (2007) Suppressor of RNA silencing encoded by tomato yellow leaf curl virus-Israel. *Virology* 358, 159–165.

# 2 Mitogen-activated Protein Kinase Cascades in Plant Defence Responses

FENGMING SONG,[1] HUIJUAN ZHANG[1] AND SHUQUN ZHANG[2]

[1]*Institute of Biotechnology, Zhejiang University, China;* [2]*University of Missouri-Columbia, Missouri, USA*

## Abstract

Mitogen-activated protein kinase (MAPK) cascades are highly conserved universal signalling modules in eukaryotes. A typical MAPK cascade consists of three interconnected protein kinases, a MAPK, a MAPK kinase (MAPKK, MKK or MEK), and a MAPKK kinase (MEKK or MAPKKK). MAPK cascades mediate the transmission of extracellular signals to downstream effector proteins by a mechanism of sequential phosphorylation. Recent biochemical, genetic and functional genomics analyses demonstrated that plant MAPK cascades play important roles in plant growth, development, and response to biotic and abiotic stresses. In this chapter, we summarize the recent progress on the functions, mechanisms and integration of MAPK cascades in the signalling networks involved in plant innate immunity, resistance gene-mediated defence responses and induced immunity. Future research directions regarding upstream receptors/sensors and downstream substrates/effectors of MAPK cascades in plant disease resistance signalling pathways are also discussed.

## 2.1 Introduction

Plants employ highly sophisticated signalling networks to regulate their response and adaptation to the ever-changing environment and to achieve maximal growth and normal development. Environmental cues are perceived by plant receptors/sensors, which trigger an array of signalling events leading to gene expression reprogramming and cellular responses (Glazebrook, 2005; Grant and Lamb, 2006; Jones and Dangl, 2006; Bent and Mackey, 2007). It has been shown that several types of post-translational modifications of proteins play critical roles in the signalling networks. Among them, protein phosphorylation/dephosphorylation by specific protein kinases/phosphatases is one of the major mechanisms for controlling cellular functions in response to external signals.

Mitogen-activated protein kinase (MAPK) cascades are highly conserved signalling modules in eukaryotes. The core of a MAPK cascade consists of three interconnected kinases. MAPK, the last kinase in the cascade, is activated by dual phosphorylation of the Thr and Tyr residues in a Thr-Xaa-Tyr motif located in the activation loop between subdomains VII and VIII of the kinase catalytic domain. This phosphorylation is mediated by a MAPK kinase (MAPKK or MEK), which is activated, in turn, by a MAPKK kinase (MAPKKK or MEKK) through phosphorylation. There are multiple members in each of the three tiers of kinases, which contribute to the specificity of the transmitted signal. Similar to animal and yeast systems, plant MAPKs are encoded by multi-gene families (Hamel *et al.*, 2006). In the fully sequenced *Arabidopsis* genome, 20 genes encoding MAPKs, ten genes encoding MAPKKs, and about 60 genes encoding MAPKKKs were identified (MAPK Group, 2002; Hamel *et al.*, 2006). Other plant species including rice have similar numbers of MAPK genes (Hamel *et al.*, 2006).

Significant progress has been made in our understanding of the biological processes regulated by MAPK cascades in plants, revealing the importance and complexity of MAPK signalling in growth, development, and responses to environmental cues. During the last decade, extensive biochemical and genetic studies have revealed that MAPK cascades play critical roles in regulating disease resistance in plants. In this chapter, we focus on the involvement and functions of MAPK cascades in signalling networks regulating innate immunity, resistance gene-mediated defence response and induced immunity in plants including *Arabidopsis*, tobacco, tomato and rice.

## 2.2. MAPK Cascades in *Arabidopsis* Defence Responses

Forward genetic studies by screening for mutants with altered phenotypes of defence response in *Arabidopsis* identified several genes in MAPK cascades, including EDR1 (ENHANCED DISEASE RESISTANCE 1). With the availability of various collections of insertional mutants and the use of reverse genetic approaches, novel functions have been assigned to *Arabidopsis* MAPKs, resulting in identification of a complete MAPK cascade MEKK1-MKK4/MKK5-MPK3/MPK6, which is involved in regulating innate immunity.

### *Arabidopsis* MAPK cascades in innate immunity

The active defence responses of plants against invading pathogens are regulated by a complex signalling network initiated by pathogen recognition, which is mediated either by gene-for-gene interactions between host resistance (*R*) genes and pathogen avirulence (*Avr*) genes, or by the binding of non-host specific pathogen-associated molecular patterns (PAMPs) to their receptors (Dangl and Jones, 2001; Martin *et al.*, 2003; Boller, 2005).

Biochemical studies demonstrated that the *Arabidopsis* mitogen-activated protein kinase 6 (MPK6) and MPK3, orthologues of tobacco salicylic acid (SA)-induced protein kinase (SIPK) and wound-induced protein kinase (WIPK), respectively, are activated after treatment with PAMPs or after pathogen infection (Nuhse *et al.*, 2000; Desikan *et al.*, 2001; Yuasa *et al.*, 2001; Asai *et al.*, 2002). Further genetic and biochemical studies have identified two MAPKKs, MKK4 and MKK5, that function upstream of MPK3 and MPK6 (Asai *et al.*, 2002; Ren *et al.*, 2002). The MAPKKK upstream of MKK4/MKK5 could be MEKK1 and/or MAPKKKα in *Arabidopsis* (Asai *et al.*, 2002; del Pozo *et al.*, 2004; Ren *et al.*, 2008). This complete MAPK cascade MEKK1-MKK4/MKK5-MPK3/MPK6 was initially established using a protoplast-based system and is activated in *Arabidopsis* cells treated with flg22, a peptide PAMP derived from bacterial flagellin (Asai *et al.*, 2002). However, recent genetic results showed that MEKK1 functions upstream of MPK4 in the flg22 signalling pathway because activation of MPK3/MPK6 after flg22 treatment is not altered in *mekk1* mutants (Ichimura *et al.*, 2006; Nakagami *et al.*, 2006; Meszaros *et al.*, 2006; Suarez-Rodriguez *et al.*, 2007).

Loss- and gain-of-function studies provided genetic evidence supporting a positive role of this MAPK cascade in signalling plant disease resistance. Activation of MPK3/MPK6 by PAMPs occurs within 1–5 min, representing one of the earliest detectable defence responses (Nuhse *et al.*, 2000; Desikan *et al.*, 2001; Yuasa *et al.*, 2001; Asai *et al.*, 2002). The rapid activation of these two MAPKs potentially allows them to regulate a variety of other early, intermediate and late defence responses. RNA interference (RNAi)-mediated silencing of endogenous MPK6 resulted in enhanced susceptibility to an avirulent strain of *Hyaloperonospora parasitica* and avirulent and virulent strains of *Pseudomonas syringae*, but did not affect the ability to develop induced systemic resistance, demonstrating that MPK6 plays a role in both *R* gene-mediated resistance and basal resistance (Menke *et al.*, 2004). Transient and stable expression of constitutively activated MKK4 or MKK5 in *Arabidopsis* leaves or transgenic plants results in enhanced resistance to bacterial and fungal pathogens and upregulates defence responses including activation of defence gene expression, generation of reactive oxygen species (ROS) and hypersensitive response (HR)-like cell death (Asai *et al.*, 2002; Ren *et al.*, 2002). The appearance of cell death in constitutively active MKK4- or MKK5-overexpressing transgenic plants is preceded by the generation of $H_2O_2$, suggesting that this MAPK cascade-induced HR-like cell death might be mediated by the $H_2O_2$ generation (Ren *et al.*, 2002). More recently, it was found that the MPK3/MPK6 cascade plays an important role in regulating synthesis of camalexin, the major phytoalexin in *Arabidopsis* (Ren *et al.*, 2008). Induction of camalexin by *Botrytis cinerea* is preceded by MPK3/MPK6 activation, but is compromised in *mpk3* and *mpk6* mutants. Genetic analysis revealed that the MPK3/MPK6 cascade acts upstream of PHYTOALEXIN DEFICIENT 2 (PAD2) and PAD3 in regulating synthesis of camalexin. Moreover, camalexin induction after MPK3/MPK6 activation is preceded by rapid and coordinated upregulation of multiple genes encoding enzymes involved in the camalexin biosynthetic pathway.

Thus, the MPK3/MPK6 cascade regulates camalexin synthesis through transcriptional regulation of the biosynthetic genes after pathogen infection (Ren *et al.*, 2008). These findings indicate that the MPK3/MPK6 cascade regulates multiple defence responses, either in parallel or sequentially.

Although the MPK3/MPK6 cascade regulates multiple defence responses, the underlying mechanism(s) remain to be determined. More recently, the identification of two *in vivo* substrates of the MPK3/MPK6 cascade provided new insights in the regulatory mechanisms of this MAPK cascade in the disease resistance signalling network. Biochemical and genetic analyses placed the MPK6 cascade upstream of the ethylene biosynthetic pathway (Liu and Zhang, 2004). 1-aminocyclopropane-1-carboxylic acid synthase 2 (ACS2) and ACS6, two members of the Type-1 ACS family, are substrates of MPK6. Phosphorylation of ACS2/ACS6 by MPK6 leads to the accumulation of ACS protein, the rate-limiting enzyme of ethylene biosynthesis, and thus elevates cellular ACS activity and ethylene production (Liu and Zhang, 2004). The identification of the first plant MAPK substrate revealed that MPK3/MPK6 regulates ethylene production by stabilizing a subset of ACS isoforms after direct phosphorylation (Kim, C.Y. *et al.*, 2003; Liu and Zhang, 2004; Joo *et al.*, 2008). Ethylene induction is associated with plant defence responses, and positively regulates defence gene expression, although its function in plant disease resistance is more complex.

Another substrate of the MPK3/MPK6 cascade came from a study on the identification of proteins that are differentially phosphorylated upon treatment of *Arabidopsis* suspension cultures with flg22 (Merkouropoulos *et al.*, 2008). Two highly related proteins, AtPHOS32 and AtPHOS34, became phosphorylated in *Arabidopsis* suspension-cultured cells after flg22 elicitation. Using *in vitro* kinase assays, it was found that AtPHOS32 is a substrate of both AtMPK3 and AtMPK6. The target phosphorylation site in AtPHOS32 is conserved in AtPHOS34 and among orthologues from many plant species, indicating that phosphorylation of these proteins by AtMPK3 and AtMPK6 orthologues has been conserved throughout evolution (Merkouropoulos *et al.*, 2008). However, the biological function of AtPHO32 and AtPHO34 in the flg22-mediated signalling pathway and thereby the disease resistance response is unclear.

Recently, it was shown that *MKK3*, a group B MAPK kinase, is strongly induced by *P. syringae* pv. *tomato* strain DC3000. The *mkk3* knockout plants are more susceptible than wild-type plants to the *P. syringae* pv. *tomato* DC3000 virulent strain. The *MKK3*-overexpressing plants not only have increased disease resistance but also show increased expression of several defence genes. By yeast two-hybrid analysis, coimmunoprecipitation and protein kinase assays, MKK3 was revealed to be an upstream activator of the group C MAPKs including MPK1, MPK2, MPK7 and MPK14 (Doczi *et al.*, 2007). However, the involvement and the role of this yet-to-be identified MKK3 cascade in plant defence responses need further investigation because contradictory results were observed in the *mkk3* mutants (Doczi *et al.*, 2007; Takahashi *et al.*, 2007).

EDR1 is a CTR1-like MAPKK kinase belonging to B3 subgroup of plant Raf-like kinases and functions as a negative regulator of disease resistance and ethylene-induced senescence (Frye *et al*., 2001; Tang and Innes, 2002; Tang *et al*., 2005). It was recently found that EDR1 exerts negative regulation on HR cell death and powdery mildew resistance by limiting the transcriptional amplification of RPW8.1 and RPW8.2, two related R proteins that confer broad-spectrum resistance to powdery mildew (Xiao *et al*., 2005). This suggests that EDR1 may negatively regulate a conserved basal defence pathway that is required for broad-spectrum resistance against powdery mildew.

## *Arabidopsis* MAPK cascades in induced immunity

Upon activation of a local defence response, immunity may be promoted in uninfected tissues by the induction of systemic acquired resistance (SAR), which is manifested as enhanced resistance to a subsequent challenge by pathogens (Durrant and Dong, 2004). Salicylic acid (SA) is a key signal molecule and NPR1 (NONEXPRESSOR OF PATHOGENESIS-RELATED GENES) is a critical mediator in the SAR signalling pathway. Transposon inactivation of *Arabidopsis MPK4* results in a constitutive SAR phenotype including elevated SA levels, increased resistance to virulent pathogens, and constitutive expression of SA-dependent defence genes (Petersen *et al*., 2000). Endogenous *MPK4* was found to be activated by some biotic factors, such as the bacterial elicitors flg22 and harpin (Desikan *et al*., 2001; Suarez-Rodriguez *et al*., 2007). Therefore, it was concluded that *Arabidopsis* MPK4 functions as a negative regulator of SAR (Petersen *et al*., 2000). Analysis of *mpk4/NahG* and *mpk4/npr1* double mutants indicated that SAR in the *mpk4* mutant is dependent on elevated SA levels but is independent of NPR1 (Petersen *et al*., 2000).

Like other MAP kinases, MPK4 function is mediated by its phosphorylation and activation by upstream kinases in response to specific stimuli, and by interactions with other downstream proteins as substrates. Interaction screens in yeast have indicated that MEKK1 can interact with MKK1 and MKK2, which in turn can interact with MPK4 (Ichimura *et al*., 1998; Mizoguchi *et al*., 1998). However, MKK1 and MKK2 appear to have distinct functions in interactions with MPK4. MPK4 is specifically phosphorylated and activated *in vitro* by MKK1 in response to various stress treatments, suggesting *in vivo* connection between these two kinases (Teige *et al*., 2004). Treatment of *Arabidopsis* cells with flg22 can activate the endogenous MMK1, which in turn phosphorylates and activates MPK4 *in vitro* and *in vivo*. Activation of *MPK4* by flg22 and resistance to both virulent and avirulent *P. syringae* pv. *tomato* are impaired in the *mkk1* mutant plants (Meszaros *et al*., 2006). Transgenic plants expressing constitutively active *MKK2* show enhanced levels of MPK4 activation and are more resistant to infection by *P. syringae* DC3000 and *Erwinia carotovora* subsp. *carotovora* (Brader *et al*., 2007). MEKK1, which has been implicated in the activation of MPK3 and MPK6 in response to flg22, is also required for flg22-induced activation of MPK4, although the kinase activity of MEKK1 appears not necessary for flg22-induced MPK4 activation (Ichimura *et al.*, 2006; Suarez-Rodriguez *et al.*, 2007). This indicates that MEKK1 acts upstream of MPK4 as a negative regulator of pathogen response pathways, a function that may not require MEKK1's full kinase activity. However, the MEKK1- and MPK4-mediated signalling pathways appear to be more complex in nature. Direct evidence defining the MAPK cascade involved in MEKK1, MKK1/MKK2 and MPK4 is still lacking.

Yeast two-hybrid complementary DNA (cDNA) library screening has identified a novel MPK4 substrate, MAP kinase 4 substrate (MKS1) (Andreasson *et al.*, 2005). MKS1 has potential MAPK phosphorylation sites and is phosphorylated on multiple sites by MPK4. Biochemical analysis showed that activated MPK4 immunoprecipitated from *Arabidopsis* seedlings can phosphorylate recombinant MKS1 *in vitro*. Overexpression of *MKS1* results in the dwarf phenotype similar to *mpk4* null mutants, indicating the functional link between these two proteins. Like the *mpk4* mutant, pathogenesis-related (PR) proteins that are normally induced in SAR are upregulated in *MKS1*-overexpressing transgenic plants, which are more resistant to pathogen attack. Furthermore, the *mpk4* mutants can be partially rescued by reducing *MKS1* expression, indicating that MPK4 negatively regulates MKS1 activity after phosphorylation (Andreasson *et al.*, 2005). MKS1 interacts with the WRKY transcription factors WRKY25 and WRKY33, suggesting that MKS1 may contribute to MPK4-regulated defence activation by coupling the kinase to specific WRKY transcription factors (Andreasson *et al.*, 2005; Qiu *et al.*, 2008).

Another putative MAPK cascade that regulates SAR response involves MKK7, which has previously been shown to negatively regulate polar auxin transport (Dai *et al.*, 2006). The activation-tagged *bud1/mkk7* mutant, in which the expression of *MKK7* is increased, accumulates elevated SA levels, exhibits constitutive expression of defence genes, and displays enhanced resistance to both *P. syringae* pv. *maculicola* ES4326 and *H. parasitica* Noco2 (Zhang *et al.*, 2007). Both defence gene expression and disease resistance in the *bud1/mkk7* plants depend on SA, partially depend on NPR1, and require the kinase activity of the MKK7 protein. Knockdown of MKK7 mRNA levels by antisense RNA expression not only blocks basal resistance, but also inhibits the induction of SAR. Moreover, ectopic expression of *MKK7* in local tissues induces defence gene expression and disease resistance in systemic tissues, indicating that activation of MKK7 is sufficient for generating a yet unknown mobile signal of SAR (Zhang *et al.*, 2007). Thus, unlike MPK4, MKK7 positively regulates the SAR signalling pathway.

## MAPK cascade regulating JA signalling

Jasmonic acid (JA) plays key roles in the environmental stress responses and developmental processes of plants. *Arabidopsis* MPK4 has been implicated in plant defence regulation because the *mpk4* knockout plants exhibit constitutive activation of SA-dependent defences, but fail to induce JA defence marker genes (e.g. JA-responsive *PDF1.2* and *THI2.1*) in response to JA (Petersen *et al.*, 2000). Recently, it was shown that the *mpk4* mutants are defective in defence gene induction in response to ethylene (ET) and are more susceptible than wild-type plants to *Alternaria brassicicola* (Brodersen *et al.*, 2006). Moreover, MKK2 seems to act upstream of MPK4. It was found that plants expressing constitutively active MKK2 show enhanced levels of MPK4 activation, enhanced expression of genes encoding enzymes of ET/JA synthesis and increased susceptibility to *A. brassicicola* (Brader *et al.*, 2007). Together, these findings suggest that MPK4 functions as a positive regulator of the JA/ET signalling pathways that are required for defence responses against necrotrophic fungal pathogens.

The MKK3-MPK6 module was recently shown to be an important part of the JA signalling pathway. JA activates MPK6 and full MPK6 activation by JA requires CORONATINE INSENSITIVE 1 (COI1), a key component in the JA signalling pathway. MKK3 not only activates MPK6 but also induces the activation of MPK6 by JA. ATMYC2/JASMONATE-INSENSITIVE1 (JIN1) is a positive regulator of JA-inducible gene expression and is essential for JA-dependent developmental processes in *Arabidopsis* (Lorenzo *et al.*, 2004). The MKK3-MPK6 module acts as a negative regulator in JA-dependent expression of *ATMYC2/JIN1* and affects expression of the JA-regulated genes. Both positive and negative regulation by JA may be used to fine-tune *ATMYC2/JIN1* expression to control JA signalling (Takahashi *et al.*, 2007).

Based on these analyses, MPK4 and MPK6 are two MAPKs that regulate JA signalling pathways in *Arabidopsis*. The functions of MPK4 and MPK6 in defence responses are in turn regulated by a Ser/Thr phosphatase 2C AP2C1, which inactivates MPK4 and MPK6. Plants with increased AP2C1 levels display lower activation of MPK4 and MPK6, and inhibit JA-responsive *PDF1.2* gene expression and innate immunity against the necrotrophic pathogen *B. cinerea* (Schweighofer *et al.*, 2007). Together, this finding reflects a more complicated mechanism that regulates JA signalling and defence response by MPK4 and MPK6.

## MAPK cascades as targets of pathogen suppressors

Pathogens have evolved mechanisms to suppress or mask basal plant defence and redirect plant cell functions for their benefit. This process commonly involves secretion of race-specific virulence factors to the host cells. Some of these virulence factors are potent suppressors of receptor function and MAPK activation. More recently, it has been demonstrated that phytopathogenic bacteria suppress host innate immunity response using their type III effectors that inactivate plant MAPK cascades. Applying a cell-based genetic screen, AvrPto and AvrPtoB from *P. syringae* have been identified as potent suppressors of microbe-associated molecular pattern (MAMP)-triggered early defence gene transcription and MAPK activation. Epistasis analysis with constitutively active MAPKKs and MAPKKK suggests that AvrPto and AvrPtoB suppress MAMP-triggered signalling upstream of MAPKKK in the MAPK cascade (He *et al.*,

2006). Shortly after the above finding, it was found that AvrPto binds MAMP flg22 receptor kinase FLS2 to block immune responses in the plant cells (Xiang *et al.*, 2008). HopAI1, a conserved effector from *P. syringae* strains, acts as a phosphothreonine lyase that removes the phosphate group from phosphothreonine to inactive MAPKs. MPK3 and MPK6 are direct targets of HopAI1 because they physically interact with HopAI1 and because endogenous MPK3 and MPK6 activation by flg22 is suppressed in HopAI1-overexpressing transgenic plants or in mesophyll cells. The inhibition of MPK3 and MPK6 by HopA1 suppresses the innate immunity responses including the reinforcement of cell wall defence and transcriptional activation of defence genes (Zhang *et al.*, 2007). The fact that bacterial type III effectors target the MAPK cascades to suppress innate immunity for their own benefit not only uncovers a novel mechanism by which bacteria overcome host innate immunity to promote pathogenesis, but also further demonstrates the importance of MAPK cascades in regulating plant innate immunity.

## 2.3 MAPK Cascades in Tobacco Defence Responses

One of the early pieces of evidence implicating the involvement of plant MAPKs in stress signalling is the purification and identification of SIPK from tobacco (Zhang and Klessig, 1997). In-gel kinase assays revealed a second smaller MAPK that was activated by tobacco mosaic virus (TMV) infection and elicitin treatment. This MAPK was determined to be WIPK using a member-specific antibody (Seo *et al.*, 1995; Zhang *et al.*, 1998; Zhang and Klessig, 1998a). Recently, a third MAPK, Ntf4 (where Nt denotes *Nicotiana tabacum*), which shares high homology with SIPK, was identified from tobacco (Ren *et al.*, 2006).

### MEK2-SIPK/Ntf4/WIPK module

SIPK and WIPK can be induced not only by various pathogenic signals, but also by wounding and various abiotic stresses, indicating that these MAPKs integrate different abiotic and biotic stress responses. SIPK activation by stress/PAMPs precedes the induction of SA (Zhang and Klessig, 1997, 1998a, b; Zhang *et al.*, 1998). Inhibition of SIPK and WIPK activation by staurosporine and K-252a suppresses HR-like cell death in tobacco suspension cells treated with elicitin from oomycetic pathogens (Zhang *et al.*, 2000). The activation of SIPK and WIPK by TMV is gene-for-gene specific, implying a role(s) in disease resistance and possibly HR (Zhang and Klessig, 1998a; Romeis *et al.*, 1999). Potato virus X (PVX)-mediated silencing of endogenous *SIPK/WIPK* attenuates *N* gene-mediated resistance against TMV (Jin *et al.*, 2003; Liu *et al.*, 2003). The kinetics of the WIPK activation upon elicitor treatment coincides with the onset of HR-like cell death (Zhang *et al.*, 2000). This was further confirmed by the observations that transient expression of *SIPK/WIPK* and transgenic *Ntf4* plants with elevated levels of Ntf4 protein accelerate HR-like cell death after treatment with elicitors (Zhang and Liu, 2001; Liu *et al.*, 2003; Samuel *et al.*, 2005; Ren *et al.*, 2006). Generally, the activation of SIPK/Ntf4/WIPK is very rapid, which potentially allows them to control multiple defence responses either directly or indirectly. Therefore, it is likely that SIPK, Ntf4 and WIPK are convergent points in the signalling pathway of defence responses.

Based on both the *in vitro* and the *in vivo* evidence, it was concluded that NtMEK2 is a shared common upstream MAPKK of SIPK, Ntf4 and WIPK (Yang *et al.*, 2001; Ren *et al.*, 2006). More direct evidence for the roles of SIPK, Ntf4 and WIPK in regulating defence response and HR-like cell death came from gain-of-function studies using a constitutively active mutant of NtMEK2 (Yang *et al.*, 2001; Ren *et al.*, 2006). Expression in tobacco of $NtMEK2^{DD}$ under the control of a steroid-inducible promoter activates endogenous SIPK, Ntf4 and WIPK, which leads to HR-like cell death in the absence of pathogen (Yang *et al.*, 2001; Jin *et al.*, 2003; Yoshioka *et al.*, 2003; Ren *et al.*, 2006). The magnitude and the kinetics of SIPK/Ntf4/WIPK activation by $NtMEK2^{DD}$ are similar to those induced by pathogens or pathogen-derived elicitors that induce HR-like cell death (Zhang and Klessig, 1998b; Zhang *et al.*, 2000). Using a similar approach, activation of SIPK or Ntf4 alone was shown to be sufficient to induce HR-like cell death (Zhang and Liu, 2001; Ren *et al.*, 2006). However, the HR-like phenotype is delayed in the absence of WIPK activity, consistent with a role for WIPK as a positive feed-forward regulator in the MAPK cascade, accelerating the cell death process (Liu *et al.*, 2003).

Based on these analyses, the NtMEK2-SIPK/Ntf4/WIPK module plays important roles in regulating tobacco innate immunity. Potential MAPKKKs upstream of NtMEK2-SIPK/Ntf4/WIPK include orthologues of *Arabidopsis* MEKK1 and tomato MAPKKKα (Asai *et al.*, 2002; del Pozo *et al.*, 2004). However, the identity of the MAPKKK that acts upstream of the NtMEK2-SIPK/Ntf4/WIPK module still needs to be identified. Another MAPKKK that functions in tobacco innate immunity is *Nicotiana* protein kinase 1 (NPK1), which was previously shown to play a role in cell-plate formation during cytokinesis. It was found that silencing *NPK1* expression interferes with the function of the disease resistance genes *N*, *Bs2* and *Rx*, but does not affect *Pto*- and *Cf4*-mediated resistance (Jin *et al.*, 2002). Furthermore, tobacco rattle virus (TRV)-mediated silencing of *NQK1* and *NRK1* in the cascade NPK1-NQK1-NRK1, a MAPK cascade that is required in cytokinesis, also attenuates *N*-mediated resistance to TMV (Liu *et al.*, 2004), indicating that the NPK1-NQK1-NRK1 cascade might constitute a complete MAPK cascade playing an important role in *N*-mediated resistance to TMV. However, NPK1 is not an upstream kinase of the NtMEK2-SIPK/Ntf4/WIPK module. Therefore, it appears that at least two different MAPK cascades function in *N*-mediated resistance. It is also possible that their function in TMV resistance is secondary. The lack of complete cell plate formation when *NQK1* or *NRK1* is silenced will allow the virus to spread more readily.

## Downstream events of the NtMEK2-SIPK/Ntf4/WIPK module

### *Biosynthesis of stress ethylene*

ET is involved in regulating plant responses to both biotic and abiotic stresses, in addition to its functions in plant growth and development. Increase in ethylene biosynthesis occurs in plants under a wide variety of stresses. The involvement of MAPK cascade in biosynthesis of stress ET was established by the findings from studies on the NtMEK2-SIPK/WIPK module. In a conditional gain-of-function transgenic system, the activation of SIPK by $NtMEK2^{DD}$ results in a dramatic increase in ET production. The increase in ET after the activation of SIPK coincides with a dramatic increase in ACS activity, which is followed by the activation of a subgroup of genes encoding key enzymes in the ET biosynthetic pathway. After ET production in $NtMEK2^{DD}$ plants, expression of ETHYLENE-RESPONSE FACTOR genes is strongly activated, similar to the effect in tobacco plants with the genotype NN infected with TMV (Kim, C.Y. *et al.*, 2003). Thus, the induction of ET biosynthesis is involved in defence responses mediated by the NtMEK2-SIPK/WIPK module. This finding led to the identification of the first plant MAPK substrate *Arabidopsis* ACS6, which is phosphorylated by MPK3 and MPK6, and to the discovery of a new mechanism that modulates the biosynthesis of ET by MAPK cascade (Liu and Zhang, 2004).

### *Involvement of ROS*

Several studies have shown that generation of ROS is related to the activation of stress-responsive MAPKs in plants under stresses. High concentrations of $H_2O_2$, when exogenously applied, activate SIPK/WIPK. It was also reported that NADPH oxidase is a downstream target of SIPK/WIPK in the induction of $H_2O_2$ generation. Activation of SIPK/WIPK by the active $NtMEK2^{DD}$ induces *NbrbohB* expression (Yoshioka *et al.*, 2003). However, Avr9-induced SIPK/WIPK activation is not dependent on a burst of ROS from membrane-associated NADPH oxidases (Romeis *et al.*, 1999). The rapid oxidative burst induced by elicitin is also not required for SIPK activation by elicitin either. No rapid $H_2O_2$ burst in $NtMEK2^{DD}$ plants after dexamethasone (DEX) treatment was detected (Yang *et al.*, 2001; Ren *et al.*, 2002). These results suggest that SIPK/Ntf4/WIPK activation is not involved in the early ROS burst from NADPH oxidase in pathogen-infected plants. Recently, it was found that chloroplast-generated ROS are involved in HR-like cell death after SIPK/Ntf4/WIPK activation (Liu *et al.* 2007). Therefore, the MAPK activation after the perception of pathogens/elicitors is independent of ROS burst. However, ROS generation is associated with cell death induced by activation of the NtMEK2-SIPK/WIPK module, similar to the pathogen-induced HR cell death (Ren *et al.*, 2002). Furthermore, ROS-mediated mitochondrial dysfunction precedes HR cell death induced by the activation of the NtMEK2-SIPK/WIPK module (Takahashi *et al.*, 2003; Liu *et al.* 2007; Takabatake *et al.*, 2007). It is therefore most likely that the activation of the NtMEK2-SIPK/WIPK module and ROS generation represent two parallel, interconnected downstream events after the sensing step, in which the ROS burst may positively feed into MAPK activation.

*WRKY transcription factors*

In the conditional gain-of-function NtMEK2$^{DD}$ transgenic tobacco plants, the activation of endogenous SIPK and WIPK by NtMEK2$^{DD}$ induces expression of several groups of defence genes. Gel-mobility shift assays revealed a strong increase in the binding activity to the W box in nuclear extracts from the NtMEK2$^{DD}$ plants, suggesting a direct phosphorylation regulation of WRKY transcription factors by the NtMEK2-SIPK/WIPK module. A rapid increase in the expression of several WRKY genes in the NtMEK2$^{DD}$ plants was also observed. These results placed the NtMEK2-SIPK/WIPK module upstream of the WRKY family proteins (Kim and Zhang, 2004). Yeast two-hybrid screens for direct downstream components have identified two WRKY transcription factors that can be phosphorylated by SIPK or WIPK. SIPK phosphorylates NtWRKY1, resulting in enhanced DNA-binding activity of WRKY1 to a W box sequence from *CHN50*, a defence gene encoding a chitinase. Coexpression of *SIPK* and *NtWRKY1* in *Nicotiana benthamiana* led to more rapid cell death than expression of *SIPK* alone, suggesting that WRKY1 is a putative substrate of SIPK (Menke *et al.*, 2005). Another WRKY transcription factor that acts downstream of WIPK is NtWIPK-interacting factor (NtWIF), which directly interacts with WIPK. *In vitro* phosphorylation assays demonstrated that WIPK efficiently phosphorylates NtWIF. Coexpression of *NtWIF* with *WIPK* in tobacco cells increases its transcriptional activity (Yap *et al.*, 2005). Over-expression of *NtWIF* in tobacco plants enhances HR upon TMV infection and cryptogein treatment, while its silencing by RNAi suppresses such HR (Chung and Sano, 2007). Transgenic tobacco plants overexpressing *NtWIF* exhibit constitutive accumulation of transcripts for defence genes including *PR-1a* and *PR-2* and elevated levels of SA (Waller *et al.*, 2006). Therefore, NtWIF is a transcription factor that is directly phosphorylated by WIPK and regulates defence response through influencing SA biosynthesis.

## 2.4 MAPKs in Tomato Defence Responses

Searching tomato genome databases identified at least 16 putative SlMPKs (where Sl denotes *Solanum lycopersicum*), among which SlMPK1, SlMPK2 and SlMPK3 are clustered with biotic stress-related MAP kinase orthologues from *Arabidopsis* and tobacco. Furthermore, four SlMKKs and one SlMEKK have also been identified from tomato. Functional analyses using virus-induced gene silencing revealed that MAPK cascades play different but overlapping roles in defence responses against pathogens and herbivorous insects.

## MAPKs in defence response against pathogens

Interactions of tomato–*P. syringae* pv. *tomato* and tomato–*Cladosporium fulvum* are two well-characterized 'gene-for-gene' systems and have been extensively used in studies aimed at elucidating the molecular basis of disease resistance responses and signal transduction pathways. In the tomato–*P. syringae* pv. *tomato* interaction, the Pto kinase mediates race-specific resistance by activating host defences upon recognition of the pathogen effector proteins AvrPto or AvrPtoB. The Pto-mediated resistance requires multiple signal transduction pathways and has been shown to activate many defence responses including expression of a set of defence genes and localized cell death. In-gel kinase assays revealed that SlMPK2 and SlMPK3, orthologues of tobacco SIPK and WIPK, respectively, are activated in a Pto-specific manner upon expression of AvrPto and AvrPtoB in tomato leaves (Pedley and Martin, 2004). *SlMPK3* is specifically induced during elicitation of the HR in resistant plants infected by *P. syringae* pv. *tomato* and silencing of *SlMPK3* blocks resistance to avirulent strains of *P. syringae* pv. *tomato* (Ekengren *et al.*, 2003; Mayrose *et al.*, 2004). Transient overexpression of two phylogenetically unrelated *SlMKK2* and *SlMKK4* in leaves elicits cell death (Pedley and Martin, 2004). Silencing of *SlMKK1* and *SlMKK2* by tobacco MEK1 and MEK2 fragments reduces the Pto-mediated resistance (Ekengren *et al.*, 2003). *In vitro* experiments revealed that both SlMKK2 and SlMKK4 are able to phosphorylate SlMPK1, SlMPK2 and SlMPK3 (Pedley and Martin, 2004). Phylogenetic analyses revealed that SlMKK4 is an orthologue of tobacco NtMEK2 and *Arabidopsis* AtMKK4 (where At denotes *Arabidopsis thaliana*) and AtMKK5, further supporting its link to SlMPK2 and SlMPK3. Searches for functional components in the signalling pathway downstream of the Pto–AvrPto interaction, by silencing a large number of random genes originating from a cDNA library prepared from leaf tissues exposed to various pathogens and elicitors, identified MAPKKKα that is required for the Pto-mediated HR and resistance against *P. syringae* pv. *tomato* (del Pozo *et al.*, 2004). Overexpression of *MAPKKKα* in leaves activates MAPKs including SlMPK2 and SlMPK3, and causes pathogen-independent cell death (del Pozo *et al.*, 2004; Pedley and Martin, 2004). Results from analysis of cell death caused by overexpression of *MAPKKKα* and suppression of various *MAPKK* and *MAPK* genes indicate that there are two distinct MAPK cascades that act downstream of MAPKKKα (del Pozo *et al.*, 2004). Based on these analyses, the MAPKKKα-SlMKK2-SlMPK2/SlMPK3 may represent a putative complete MAPK cascade that acts downstream of the Pto–AvrPto interaction, but the tomato MAPKKKα and its biochemical link to SlMKK2 need to be identified.

Tomato plants with the *Cf* genes recognize strains of *C. fulvum* that secrete the corresponding Avr proteins. In tobacco suspension cells and transgenic plants expressing the *Cf9* gene, SIPK and WIPK are activated upon treatment with Avr9 (Romeis *et al.*, 2000). SlMAPKKKα was also shown to be a positive regulator of the *Cf-9*-mediated HR (del Pozo *et al.*, 2004). Transgenic tomato seedlings coexpressing *Cf-4* and *Avr4* mount an HR at 20°C, which is suppressed at 33°C. Within 2 h after a shift from 33°C to 20°C, SlMPK1,

SlMPK2 and SlMPK3 are simultaneously activated in Cf-4/Avr4 seedlings. Silencing of *SlMPK2* or *SlMPK3* specifically suppresses MAPK activity and the Cf-4/Avr4-induced HR, whereas silencing of *SlMPK1* does not affect occurrence of the HR (Stulemeijer *et al.*, 2007). Interestingly, full resistance against *C. fulvum* is blocked in *SlMPK1*- and *SlMPK3*-silenced tomato plants, whereas *SlMPK2*-silenced plants still show full resistance to *C. fulvum* (Stulemeijer *et al.*, 2007). The results suggest that the SlMPKs have different but also overlapping roles in regulating the Cf-4/Avr4-induced HR and full resistance against *C. fulvum*.

### MAPKs in the defence response against herbivorous insects

The involvement of MAPKs in the wounding response is well documented. For example, wounding causes rapid activation of SIPK and Ntf4, two highly homologous MAPKs in tobacco (Seo *et al.*, 1995, 1999; Zhang and Klessig, 1998b; Ren *et al.*, 2006). JA is highly induced by wounding or insect chewing. Recently, several studies have shown that MAPKs also play important roles in the defence response against herbivorous insects. Systemin is a wound-signalling peptide that mediates defences of tomato plants against herbivorous insects. In tomato suspension cells, systemin activates SlMPK1 and SlMPK2 (Holley *et al.*, 2003; Higgins *et al.*, 2007). The activation of SlMPK1 and SlMPK2 is not reduced in the *def1* JA-biosynthesis mutant, indicating that these MAPKs function either upstream or in parallel to JA biosynthesis (Stratmann and Ryan, 1997). The transgenic 35S::prosys plants that overexpress prosystemin, the systemin precursor, accumulate high levels of defence proteins and exhibit increased resistance to herbivorous insects. Silencing of *SlMPK1*, *SlMPK2* and *SlMPK3* in single or in combination reduces MAPK kinase activity, JA biosynthesis, expression of JA-dependent defence genes and prosystemin-mediated resistance to *Manduca sexta* herbivory. Because methyl JA restored the full defence response, it is most likely that SlMPK1, SlMPK2 and SlMPK3 function upstream of JA biosynthesis and they are required for successful defences against herbivorous insects (Kandoth *et al.*, 2007). Furthermore, loss-of-function studies revealed that SlMPK1, SlMPK2 and SlMPK3 are also required for aphid resistance mediated by *Mi-1*, a tomato gene that confers resistance to potato aphids and whiteflies (Li *et al.*, 2006). Therefore, these MAPKs represent convergence points for different signalling pathways inducing the activation of defence responses against pathogens and herbivorous insects in tomato.

## 2.5 MAPKs in Rice Disease Resistance

The first MAPK identified from rice is the OsBWMK1 (where Os denotes *Oryza sativa*), whose expression is induced upon infection by blast fungus *Magnaporthe grisea* and mechanical wounding (He *et al.*, 1999). This study provides the first evidence for the existence of a MAPK cascade component in rice and the possible involvement in rice defence mechanism(s). With the completion of the rice genome sequencing project, extensive bioinformatics analyses identified 17 MPKs and eight MKKs in rice (Hamel *et al.*, 2006; Reyna and Yang, 2006). Compared with the significant progress on the biological function of the MAPK cascades in *Arabidopsis*, little is known about the function and regulation of MAPKs in rice. Up to date, only a few of them have been characterized in detail for their biological function in rice growth, development and response to environmental cues.

## Identification of MAPKs and expression patterns in defence responses

Phylogenetic analysis and pairwise comparison of *Arabidopsis* and rice MAPKs revealed that 11 rice MAPKs contain the TDY activation motif and six MAPKs contains the TEY motif. This is different from those in *Arabidopsis*, which contains more MAPKs with the TEY motif than the MAPKs with the TDY motif. Genome-wide expression analyses by qRT-PCR indicate that, upon inoculation with *M. grisea*, nine of 17 *OsMPK* genes are induced at the mRNA level during either early, late, or both stages of infection. Four of the *M. grisea*-induced *OsMPK* genes are associated with host-cell death in the lesion-mimic rice mutant, and eight of them are differentially induced in response to defence signal molecules such as JA, SA and ET (Reyna and Yang, 2006).

*OsMPK12* (*OsBWMK1*) is induced by both virulent and avirulent isolates of *M. grisea* and by treatments with SA, JA and 1-aminocyclopropane-1-carboxylic acid (ACC) within 24 h after inoculation/treatment (He *et al.*, 1999; Cheong *et al.*, 2003; Reyna and Yang, 2006). The OsMPK12 protein levels do not change in rice leaf tissues after treatments with different defence signalling molecules, indicating that the OsMPK12 activation is primarily achieved by post-translational modification (Cheong *et al.*, 2003). The *OsMPK12* gene has three alternatively spliced transcripts, *OsBWMK1L*, *OsBWMK1M* and *OsBWMK1S*, but only the *OsBWMK1* transcripts are differentially expressed in tissues and under various stress conditions. Alternative splicing of *OsBWMK1* generates three different transcript variants that produce proteins with different subcellular localizations (Koo *et al.*, 2007). *OsMPK5* (also known as *OsMSRMK2*, *OsMAPK2*, *OsBIMK1* or *OsMAP1*) is induced by various biotic and abiotic stresses, its induction by *M. grisea* and some defence signalling molecules including SA, JA and ACC is well documented by several research groups (Agrawal *et al.*, 2002; Song and Goodman, 2002; Xiong and Yang, 2003; Reyna and Yang, 2006). The *OsMPK5* gene generates at least two alternatively spliced transcripts, *OsMPK5a* and *OsMPK5b*. Kinase activity assays revealed that only OsMAPK5a exhibits autophosphorylation activity *in vitro* (Xiong and Yang, 2003). OsMAPK5 kinase activity is induced significantly by blast fungus infection and the early transient activation of OsMAPK5 activity probably is related to the resistance response to avirulent blast isolates (Xiong and Yang, 2003). *OsMPK13* (also known as *OsBIMK2*) is induced within 48 h during an incompatible interaction between rice and *M.*

*grisea*, and by treatment with SA, benzothiadiazole (BTH) and ACC, but not by treatment with JA (Reyna and Yang, 2006; Song *et al.*, 2006). *In vitro* kinase assay revealed that OsBIMK2 has an autophosphorylation activity (Song *et al.*, 2006). *OsMPK1* (also known as *OsMAPK6*) was reported to be post-translationally activated by a sphingolipid elicitor and regulated by the small GTPase OsRac1 and heterotrimeric G-protein (Lieberherr *et al.*, 2005). Genome-wide analyses also revealed that six other rice MAPK genes, *OsMPK2*, *OsMPK4*, *OsMPK7*, *OsMPK8*, *OsMPK15* and *OsMPK17*, are induced by *M. grisea* and defence signalling molecules, suggesting that they may also play roles in defence signal transduction (Reyna and Yang, 2006).

Very little is known about MKKs and MEKKs in biotic defence response in rice. *OsMKK1* is induced in rice leaves after infection by *M. grisea*, feeding by insect (*Nilaparvata lugens*) and treatment with SA, JA and ET (You *et al.*, 2007). *OsEDR1*, an orthologue of *Arabidopsis AtEDR1*, has a constitutive expression in seedling leaves and is further upregulated by treatment with JA, SA and ET (Kim, J.A. *et al.*, 2003). These preliminary observations may provide starting points to investigate defence-related MAPK cascades in rice disease resistance.

However, the involvement of most of the rice MAPKs in disease resistance was established mainly based on their inducible expression patterns in response to infection by *M. grisea* and to treatments with defence signalling molecules. Further biochemical, genetics and functional analyses are required to clarify the biological functions and mechanisms of MAPK cascades in rice disease resistance.

## Functions of MAPKs in rice disease resistance

Although a number of MAPKs have been implicated in rice defence responses, only a few of them have been characterized for their biological functions in rice disease resistance through functional genomics approaches. Suppression of *OsMPK5* expression and its kinase activity results in the constitutive expression of defence genes in dsRNAi transgenic plants and significantly enhances resistance to fungal (*M. grisea*) and bacterial (*Burkholderia glumae*) pathogens. However, these same dsRNAi lines have significant reductions in drought, salt and cold tolerance. By contrast, *OsMPK5*-overexpressing plants exhibit increased kinase activity and increased tolerance to drought, salt and cold stresses. Based on these findings, it was proposed that OsMAPK5 negatively modulates defence gene expression and broad-spectrum disease resistance and positively regulates abiotic stress tolerance (Xiong and Yang, 2003). It is likely that suppressing or knocking out of *OsMPK6*, an orthologue of *Arabidopsis AtMPK4*, enhances resistance to different races of *Xanthomonas oryzae* pv. *oryzae*, causing bacterial blight disease. The *OsMPK6*-suppressed plants show increased expression of a subset of defence-responsive genes functioning in the NH1 (an *Arabidopsis* NPR1 orthologue)-involved defence signal transduction pathway. These findings support that OsMPK6 functions as a repressor to regulate rice defence responses upon bacterial infection (Yuan *et al.*, 2007).

Ectopic expression in tobacco was also used to study the function of some rice MAPK genes in disease resistance responses. Overexpression of the *OSBIMK2* gene in transgenic tobacco results in enhanced disease resistance against tomato mosaic virus and *Alternaria alternata* (Song *et al.*, 2006). OsMPK12 (OsBWMK1) interacts with and phosphorylates OsEREBP1, a transcription factor belonging to the ethylene-responsive transcriptional factor family. Phosphorylation of OsEREBP1 by OsMPK12 enhances its *in vitro* DNA-binding activity to the GCC box element (AGCCGCC) in promoters of several defence genes. Transgenic tobacco plants overexpressing *OsMPK12* display increased expression of many defence genes, induced HR-like cell death and enhanced disease resistance against *H. parasitica* var. *nicotianae* and *P. syringae* pv. *tabacci*. Therefore, it is likely that OsMPK12 contributes to plant defence signal transduction by phosphorylating one or more transcription factors (Cheong *et al.*, 2003). This is similar to the observations that, in tobacco and *Arabidopsis*, MAPK cascades regulate defence responses by phosphorylating downstream WRKY transcription factors. Together, these findings support the idea that MAPK cascades regulate transcription of a variety of stress responsive genes through modulation of corresponding transcription factors, although the target transcription factors might be unique to different plant species.

## 2.6 Concluding Remarks

In the past decade, great progress has been made in our understanding of the biological functions of plant MAPK cascades. Although a number of MAPKs and/or putative MAPK cascades are implicated in the signalling pathways of plant defence responses, the underlying molecular mechanisms are largely unknown. At this stage, only a few MAPK cascades or modules including *Arabidopsis* MEKK1/MAPKKKα-MKK4/MKK5-MPK3/MPK6 and tobacco MKK2-SIPK/WIPK have been established that play important roles in regulating plant defence responses. An interesting phenomenon is that some of the components in different MAPK cascades can be shared. For instance, *Arabidopsis* pathogen-responsive MPK3/MPK6 and their upstream MAPKKs also play critical roles in several developmental pathways including stomatal formation, embryogenesis and ovule formation (Wang *et al.*, 2007, 2008; Bush and Krysan, 2008). How specificity is maintained when distinct functional pathways share common components is central to our understanding of MAPK cascades in plant defence signalling pathways. At this stage, it is important for us to identify all components of the signalling pathways including receptors/sensors upstream of the MAPK cascades, kinases at different tiers in the MAPK cascades, and downstream substrates/effectors of MAPK cascades using a combination of genetic, biochemical and functional genomics approaches. This will eventually allow the establishment of the MAPK signalling network and the understanding of underlying molecular mechanisms of MAPK functions in plant disease resistance.

Because of the multi-functionality of MAPKs and the fact that the disruption of plant adaptive response can compromise plant hormonal responses, growth and development, some of the phenotypic alterations observed in MAPK mutants could be secondary effects. However, the specific/direct function(s) of a MAPK cascade will be eventually backed up by the identification of specific substrates. Recently, several MAPK substrates were reported including ACS and MKS1 (Liu and Zhang, 2004; Andreasson *et al.*, 2005). Additional proteins including tobacco NtWIP and NtWRKY1 were identified as potential MAPK substrates (Menke *et al.*, 2005; Yap *et al.*, 2005). Searching for additional MAPK substrates is likely to become the focal point in the next phase of plant MAPK research. A number of approaches including classical biochemical purification, yeast two-hybrid interaction screening, high through-put protein array and phosphoproteomics will lead to the identification of new MAPK substrates. On the other hand, it is also critical to identify the specific receptors/sensors that function upstream of a MAPK cascade and understand how the recognition of stimuli/ligands activates the MAPK cascade. After the identification of each individual component in the MAPK signalling networks, *in vivo* functional analyses will allow us to piece together the linear functional pathways and eventually the signalling networks illustrating the roles of MAPK cascades in plant defence responses.

## Acknowledgements

Studies in the Song and Zhang laboratories are supported by grants from the National Natural Science Foundation of China (No. 30571209 and No. 30771399 to Song, F. and No. 30728013 to Zhang, S. and Song, F.), by a grant from the PhD Programs Foundation of Ministry of Education of China (No. 20070335111 to Song, F.) and by grants from the National Science Foundation of USA (Zhang, S.).

## References

Agrawal, G.K., Rakwal, R. and Iwahashi, H. (2002) Isolation of novel rice (*Oryza sativa* L.) multiple stress responsive MAP kinase gene, *OsMSRMK2*, whose mRNA accumulates rapidly in response to environmental cues. *Biochemical and Biophysical Research Communications* 294, 1009–1016.

Andreasson, E., Jenkins, T., Brodersen, P., Thorgrimsen, S., Petersen, N.H.T., Zhu, S., Qiu, J.-L., Micheelsen, P., Rocher, A., Petersen, M., Newman, M.-A., Nielsen, H.B., Hirt, H., Somssich, I., Mattsson, O. and Mundy, J. (2005) The MAP kinase substrate MKS1 is a regulator of plant defense responses. *EMBO Journal* 24, 2579–2589.

Asai, T., Tena, G., Plotnikova, J., Willmann, M.R., Chiu, W.L., Gomez-Gomez, L., Boller, T., Ausubel, F.M. and Sheen, J. (2002) MAP kinase signalling cascade in *Arabidopsis* innate immunity. *Nature* 415, 977–983.

Bent, A.F. and Mackey, D. (2007) Elicitors, effectors, and R genes: the new paradigm and a lifetime supply of questions. *Annual Review of Phytopathology* 45, 399–436.

Boller, T. (2005) Peptide signalling in plant development and self/non-self perception. *Current Opinion in Cell Biology* 17, 116–122.

Brader, G., Djamei, A., Teige, M., Palva, E.T. and Hirt, H. (2007) The MAP kinase kinase MKK2 affects disease resistance in *Arabidopsis*. *Molecular Plant–Microbe Interactions* 20, 589–596.

Brodersen, P., Petersen, M., Bjorn Nielsen, H., Zhu, S., Newman, M.-A., Shokat, K.M., Rietz, S., Parker, J. and Mundy, J. (2006) *Arabidopsis* MAP kinase 4 regulates salicylic acid- and jasmonic acid/ethylene-dependent responses via EDS1 and PAD4. *The Plant Journal* 47, 532–546.

Bush, S.M. and Krysan, P.J. (2008) Mutational evidence that the *Arabidopsis* MAP kinase MPK6 is involved in anther, inflorescence, and embryo development. *Journal of Experimental Botany* 58, 2181–2191.

Cheong, Y.H., Moon, B.C., Kim, J.K., Kim, C.Y., Kim, M.C., Kim, I.H., Park, C.Y., Kim, J.C., Park, B.O., Koo, S.C., Yoon, H.W., Chung, W.S., Lim, C.O., Lee, S.Y. and Cho, M.J. (2003) BWMK1, a rice mitogen-activated protein kinase, locates in the nucleus and mediates pathogenesis-related gene expression by activation of a transcription factor. *Plant Physiology* 132, 1961–1972.

Chung, K.M. and Sano, H. (2007) Transactivation of wound-responsive genes containing the core sequence of the auxin-responsive element by a wound-induced protein kinase-activated transcription factor in tobacco plants. *Plant Molecular Biology* 65, 763–773.

Dai, Y., Wang, H., Li, B., Huang, J., Liu, X., Zhou, Y., Mou, Z. and Li, J. (2006) Increased expression of *MAP KINASE KINASE7* causes deficiency in polar auxin transport and leads to plant architectural abnormality in *Arabidopsis*. *The Plant Cell* 18, 308–320.

Dangl, J.L. and Jones, J.D.G. (2001) Plant pathogens and integrated defence responses to infection. *Nature* 411, 826–833.

del Pozo, O., Pedley, K.F. and Martin, G.B. (2004) MAPKKKα is a positive regulator of cell death associated with both plant immunity and disease. *EMBO Journal* 23, 3072–3082.

Desikan, R., Hancock, J.T., Ichimura, K., Shinozaki, K. and Neill, S.J. (2001) Harpin induces activation of the *Arabidopsis* mitogen-activated protein kinases AtMPK4 and AtMPK6. *Plant Physiology* 126, 1579–1587.

Doczi, R., Brader, G., Pettko-Szandtner, A., Rajh, I., Djamei, A., Pitzschke, A., Teige, M. and Hirt, H. (2007) The *Arabidopsis* mitogen-activated protein kinase kinase MKK3 is upstream of group C mitogen-activated protein kinases and participates in pathogen signaling. *The Plant Cell* 19, 3266–3279.

Durrant, W.E. and Dong, X. (2004) Systemic acquired resistance. *Annual Review of Phytopathology* 42, 185–209.

Ekengren, S.K., Liu, Y., Schiff, M., Dinesh-Kumar, S.P. and Martin, G.B. (2003) Two MAPK cascades, NPR1, and TGA transcription factors play a role in Pto-mediated disease resistance in tomato. *The Plant Journal* 36, 905–917.

Frye, C.A., Tang, D. and Innes, R.W. (2001) Negative regulation of defense responses in plants by a conserved MAPKK kinase. *Proceedings of the National Academy of Sciences, USA* 98, 373–378.

Glazebrook, J. (2005) Contrasting mechanisms of defense against biotrophic and necrotrophic pathogens. *Annual Review of Phytopathology* 43, 205–227.

Grant, M. and Lamb, C. (2006) Systemic immunity. *Current Opinion in Plant Biology* 9, 414–420.

Hamel, L.-P., Nicole, M.-C., Sritubtim, S., Morency, M.-J., Ellis, M., Ehlting, J., Beaudoin, N., Barbazuk, B., Klessig, D., Lee, J., Martin, G., Mundy, J., Ohashi, Y., Scheel, D., Sheen, J., Xing, T., Zhang, S., Seguin, A. and Ellis, B.E. (2006) Ancient signals: comparative genomics of plant MAPK and MAPKK gene families. *Trends in Plant Science* 11, 192–198.

He, C., Fong, S.H., Yang, D. and Wang, G.-L. (1999) BWMK1, a novel MAP kinase induced by fungal infection and mechanical wounding in rice. *Molecular Plant–Microbe Interactions* 12, 1064–1073.

He, P., Shan, L., Lin, N.C., Martin, G.B., Kemmerling, B., Nürnberger, T. and Sheen, J. (2006) Specific bacterial suppressors of MAMP signaling upstream of MAPKKK in *Arabidopsis* innate immunity. *Cell* 125, 563–575.

Higgins, R., Lockwood, T., Holley. S., Yalamanchili, R. and Stratmann, J.W. (2007) Changes in extracellular pH are neither required nor sufficient for activation of mitogen-activated protein kinases (MAPKs) in response to systemin and fusicoccin in tomato. *Planta* 225, 1535–1546.

Holley, S.R., Yalamanchili, R.D., Moura, D.S., Ryan, C.A. and Stratmann, J.W. (2003) Convergence of signaling pathways induced by systemin, oligosaccharide elicitors, and ultraviolet-B radiation at the level of mitogen-activated protein kinases in *Lycopersicon peruvianum* suspension-cultured cells. *Plant Physiology* 132, 1728–1738.

Ichimura, K., Mizoguchi, T., Irie, K., Morris, P., Giraudat, J., Matsumoto, K. and Shinozaki, K. (1998) Isolation of ATMEKK1 (a MAP kinase kinase kinase)-interacting proteins and analysis of a MAP kinase cascade in *Arabidopsis*. *Biochemical and Biophysical Research Communications* 253, 532–543.

Ichimura, K., Casais, C., Peck, S.C., Shinozaki, K. and Shirasu, K. (2006) MEKK1 is required for MPK4 activation and regulates tissue-specific and temperature-dependent cell death in *Arabidopsis*. *Journal of Biological Chemistry* 281, 36969–36976.

Jin, H., Axtell, M.J., Dahlbeck, D., Ekwenna, O., Zhang, S., Staskawicz, B. and Baker, B. (2002) NPK1, an MEKK1-like mitogen-activated protein kinase kinase kinase, regulates innate immunity and development in plants. *Developmental Cell* 3, 291–297.

Jin, H., Liu, Y., Yang, K.-Y., Kim, C.Y., Baker, B. and Zhang, S. (2003) Function of a mitogen-activated protein kinase pathway in *N* gene-mediated resistance in tobacco. *The Plant Journal* 33, 719–731.

Jones, J.D. and Dangl, J.L. (2006) The plant immune system. *Nature* 444, 323–329.

Joo, S., Liu, Y., Lueth, A. and Zhang, S. (2008) MAPK phosphorylation-induced stabilization of ACS6 protein is mediated by the non-catalytic C-terminal domain, which also contains the *cis*-determinant for rapid degradation by the 26S proteasome pathway. *The Plant Journal* 54, 129–140.

Kandoth, P.K., Ranf, S., Pancholi, S.S., Jayanty, S., Walla, M.D., Miller, W., Howe, G.A., Lincoln, D.E. and Stratmann, J.W. (2007) Tomato MAPKs LeMPK1, LeMPK2, and LeMPK3 function in the systemin-mediated defense response against herbivorous insects. *Proceedings of the National Academy of Sciences, USA* 104, 12205–12210.

Kim, C.Y. and Zhang, S. (2004) Activation of a mitogen-activated protein kinase cascade induces WRKY family of transcription factors and defense genes in tobacco. *The Plant Journal* 38, 142–151.

Kim, C.Y., Liu, Y., Thorne, E.T., Yang, H., Fukushig, H., Gassmann, W., Hildebrand, D., Sharp, R.E. and Zhang, S. (2003) Activation of a stress-responsive mitogen-activated protein kinase cascade induces the biosynthesis of ethylene in plants. *The Plant Cell* 15, 2707–2718.

Kim, J.A., Agrawal, G.K., Rakwal, R., Han, K.S., Kim, K.N., Yun, C.H., Heu, S., Park, S.Y., Lee, Y.H. and Jwa, N.S. (2003) Molecular cloning and mRNA expression analysis of a novel rice (*Oryza sativa* L.) MAPK kinase kinase, *OsEDR1*, an ortholog of *Arabidopsis AtEDR1*, reveal its role in defense/stress signalling pathways and development. *Biochemical and Biophysical Research Communications* 300, 868–876.

Koo, S.C., Yoon, H.W., Kim C.Y., Moon, B.C., Cheong, Y.H., Han, H.J., Lee, S.M., Kang, K.Y., Kim, M.C., Lee, S.Y., Chung, W.S. and Cho, M.J. (2007) Alternative splicing of the *OsBWMK1* gene generates three transcript variants showing differential subcellular localizations. *Biochemical and Biophysical Research Communications* 360, 188–193.

Li, Q., Xie, Q.J., Smith-Becker, J., Navarre, D.A. and Kaloshian, I. (2006) *Mi-1*-Mediated aphid resistance involves salicylic acid and mitogen-activated protein kinase signaling cascades. *Molecular Plant–Microbe Interactions* 19, 655–664.

Lieberherr, D., Thao, N.P., Nakashima, A., Umemura, K., Kawasaki, T. and Shimamoto, K. (2005) A sphingolipid elicitor-inducible mitogen-activated protein kinase is regulated by the small GTPase OsRac1 and heterotrimeric G-protein in rice 1. *Plant Physiology* 138, 1644–1652.

Liu, Y. and Zhang, S. (2004) Phosphorylation of 1-aminocyclopropane-1-carboxylic acid synthase by MPK6, a stress-responsive mitogen-activated protein kinase, induces ethylene biosynthesis in *Arabidopsis*. *The Plant Cell* 16, 3386–3399.

Liu, Y., Jin, H., Yang, K.-Y., Kim, C.Y., Baker, B. and Zhang, S. (2003) Interaction between two mitogen-activated protein kinases during tobacco defense signaling. *The Plant Journal* 34, 149–160.

Liu, Y., Schiff, M. and Dinesh-Kumar, S.P. (2004) Involvement of MEK1 MAPKK, NTF6 MAPK, WRKY/MYB transcription factors, *COI1* and *CTR1* in *N*-mediated resistance to tobacco mosaic virus. *The Plant Journal* 38, 800–809.

Liu, Y., Ren, D., Pike, S., Pallardy, S., Gassmann, W. and Zhang, S. (2007) Chloroplast-generated reactive oxygen species are involved in hypersensitive response-like cell death mediated by a mitogen-activated protein kinase cascade. *The Plant Journal* 51, 941–954.

Lorenzo, O., Chico, J.M., Sanchez-Serrano, J.J. and Solano, R. (2004) *JASMONATE-INSENSITIVE1* encodes a MYC transcription factor essential to discriminate between different jasmonate-regulated defense responses in *Arabidopsis*. *The Plant Cell* 16, 1938–1950.

MAPK Group (2002) Mitogen-activated protein kinase cascades in plants: a new nomenclature. *Trends in Plant Science* 7, 301–308.

Martin, G.B., Bogdanove, A.J. and Sessa, G. (2003) Understanding the functions of plant disease resistance proteins. *Annual Review of Plant Biology* 54, 23–61.

Mayrose, M., Bonshtien, A. and Sessa, G. (2004) LeMPK3 is a mitogen-activated protein kinase with dual specificity induced during tomato defense and wounding responses. *Journal of Biological Chemistry* 279, 14819–14827.

Menke, F.L., van Pelt, J.A., Pieterse, C.M. and Klessig, D.F. (2004) Silencing of the mitogen-activated protein kinase MPK6 compromises disease resistance in *Arabidopsis*. *The Plant Cell* 16, 897–907.

Menke, F.L., Kang, H.-G., Chen, Z., Park, J.M., Kumar, D. and Klessig, D.F. (2005) Tobacco transcription factor WRKY1 is phosphorylated by the MAP kinase SIPK and mediates HR-like cell death in tobacco. *Molecular Plant–Microbe Interactions* 18, 1027–1034.

Merkouropoulos, G., Andreasson, E., Hess, D., Boller, T. and Peck, S.C. (2008) An *Arabidopsis* protein phosphorylated in response to microbial elicitation, AtPHOS32, is a substrate of MAP kinases 3 and 6. *Journal of Biological Chemistry* 283, 10493–10499.

Meszaros, T., Helfer, A., Hatzimasoura, E., Magyar, Z., Serazetdinova, L., Rios, G., Bardoczy, V., Teige, M., Koncz, C., Peck, S. and Bogre, L. (2006) The *Arabidopsis* MAP kinase kinase MKK1 participates in defence responses to the bacterial elicitor flagellin. *The Plant Journal* 48, 485–498.

Mizoguchi, T., Ichimura, K., Irie, K., Morris, P., Giraudat, J., Matsumoto, K. and Shinozaki, K. (1998) Identification of a possible MAP kinase cascade in *Arabidopsis thaliana* based on pairwise yeast two-hybrid analysis and functional complementation tests of yeast mutants. *FEBS Letters* 437, 56–60.

Nakagami, H., Soukupova, H., Schikora, A., Zarsky, V. and Hirt, H. (2006) A mitogen-activated protein kinase kinase kinase mediates reactive oxygen species homeostasis in *Arabidopsis*. *Journal of Biological Chemistry* 281, 38697–38704.

Nuhse, T., Peck, S.C., Hirt, H. and Boller, T. (2000) Microbial elicitors induce activation and dual phosphorylation of the *Arabidopsis thaliana* MAPK 6. *Journal of Biological Chemistry* 275, 7521–7526.

Pedley, K.F. and Martin, G.B. (2004) Identification of MAPKs and their possible MAPK kinase activators involved in the Pto-mediated defense response of tomato. *Journal of Biological Chemistry* 279, 49229–49235.

Petersen, M., Brodersen, P., Naested, H., Andreasson, E., Lindhart, U., Johansen, B., Nielsen, H.B., Lacy, M., Austin, M.J., Parker, J.E., Sharma, S.B., Klessig, D.F., Martienssen, R., Mattsson, O., Jensen, A.B. and Mundy, J. (2000) *Arabidopsis* map kinase 4 negatively regulates systemic acquired resistance. *Cell* 103, 1111–1120.

Qiu, J.L., Fiil, B.K., Petersen, K., Nielsen, H.B., Botanga, C.J., Thorgrimsen, S., Palma, K., Suarez-Rodriguez, M.C., Sandbech-Clausen, S., Lichota, J., Brodersen, P., Grasser, K.D., Mattsson, O., Glazebrook, J., Mundy, J. and Petersen, M. (2008) *Arabidopsis* MAP kinase 4 regulates gene expression through transcription factor release in the nucleus. *EMBO Journal* 27, 2214–2221.

Ren, D., Yang, H. and Zhang, S. (2002) Cell death mediated by MAPK is associated with hydrogen peroxide production in *Arabidopsis*. *Journal of Biological Chemistry* 277, 559–565.

Ren, D., Yang, K.-Y., Li, G.-J., Liu, Y. and Zhang, S. (2006) Activation of Ntf4, a tobacco mitogen-activated protein kinase, during plant defense response and its involvement in hypersensitive response-like cell death. *Plant Physiology* 141, 1482–1493.

Ren, D., Liu, Y., Yang, K.Y., Han, L., Mao, G., Glazebrook, J. and Zhang, S. (2008) A fungal-responsive MAPK cascade regulates phytoalexin biosynthesis in *Arabidopsis*. *Proceedings of the National Academy of Sciences, USA* 105, 5638–5643.

Reyna, N.S. and Yang, Y. (2006) Molecular analysis of the rice MAP kinase gene family in relation to *Magnaporthe grisea* infection. *Molecular Plant–Microbe Interactions* 19, 530–540.

Romeis, T., Piedras, P., Zhang, S., Klessig, D.F., Hirt, H. and Jones, J. (1999) Rapid Avr9- and Cf-9-dependent activation of MAP kinases in tobacco cell cultures and leaves: convergence of resistance gene, elicitor, wound, and salicylate responses. *The Plant Cell* 11, 273–287.

Romeis, T., Piedras, P. and Jones, J.D. (2000) Resistance gene-dependent activation of a calcium-dependent protein kinase in the plant defense response. *The Plant Cell* 12, 803–816.

Samuel, M.A., Walia, A., Mansfield, S.D. and Ellis, B.E. (2005) Overexpression of SIPK in tobacco enhances ozone-induced ethylene formation and blocks ozone-induced SA accumulation. *Journal of Experimental Botany* 56, 2195–2201.

Schweighofer, A., Kazanaviciute, V., Scheikl, E., Teige, M., Doczi, R., Hirt, H., Schwanninger, M., Kant, M., Schuurink, R., Mauch, F., Buchala, A., Cardinale, F. and Meskiene, I. (2007) The PP2C-type phosphatase AP2C1, which negatively regulates MPK4 and MPK6, modulates innate immunity, jasmonic acid, and ethylene levels in *Arabidopsis*. *The Plant Cell* 19, 2213–2224.

Seo, S., Okamoto, M., Seto, H., Ishizuka, K., Sano, H. and Ohashi, Y. (1995) Tobacco MAP kinase: a possible mediator in wound signal transduction pathways. *Science* 270, 1988–1992.

Seo, S., Sano, H. and Ohashi, Y. (1999) Jasmonate-based wound signal transduction requires activation of WIPK, a tobacco mitogen-activated protein kinase. *The Plant Cell* 11, 289–298.

Song, D., Chen, J., Song, F. and Zheng, Z. (2006) A novel rice MAPK gene, *OsBIMK2*, is involved in disease-resistance responses. *Plant Biology* 8, 587–596.

Song, F. and Goodman, R.M. (2002) *OsBIMK1*, a rice MAP kinase gene involved in disease resistance responses. *Planta* 215, 997–1005.

Stratmann, J.W. and Ryan, C.A. (1997) Myelin basic protein kinase activity in tomato leaves is induced systemically by wounding and increases in response to systemin and oligosaccharide elicitors. *Proceedings of the National Academy of Sciences, USA* 94, 11085–11089.

Stulemeijer, I.J., Stratmann, J.W. and Joosten, M.H. (2007) Tomato mitogen-activated protein kinases LeMPK1, LeMPK2, and LeMPK3 are activated during the Cf-4/Avr4-induced hypersensitive response and have distinct phosphorylation specificities. *Plant Physiology* 144, 1481–1494.

Suarez-Rodriguez, M.C., Adams-Phillips, L., Liu, Y., Wang, H., Su, S.-H., Jester, P.J., Zhang, S., Bent, A.F. and Krysan, P.J. (2007) MEKK1 is required for flg22-induced MPK4 activation in *Arabidopsis* plants. *Plant Physiology* 143, 661–669.

Takabatake, R., Ando, Y., Seo, S., Katou, S., Tsuda, S., Ohashi, Y. and Mitsuhara, I. (2007) MAP kinases function downstream of HSP90 and upstream of mitochondria in TMV resistance gene *N*-mediated hypersensitive cell death. *Plant and Cell Physiology* 48, 498–510.

Takahashi, Y., Berberich, T., Miyazaki, A., Seo, S., Ohashi, Y. and Kusano, T. (2003) Spermine signalling in tobacco: activation of mitogen-activated protein kinases by spermine is mediated through mitochondrial dysfunction. *The Plant Journal* 36, 820–829.

Takahashi, F., Yoshida, R., Ichimura, K., Mizoguchi, T., Seo, S., Yonezawa, M., Maruyama, K., Yamaguchi-Shinozaki, K. and Shinozaki, K. (2007) The mitogen-activated protein kinase cascade MKK3–MPK6 is an important part of the jasmonate signal transduction pathway in *Arabidopsis*. *The Plant Cell* 19, 805–818.

Tang, D. and Innes, R.W. (2002) Overexpression of a kinase-deficient form of the *EDR1* gene enhances powdery mildew resistance and ethylene-induced senescence in *Arabidopsis*. *The Plant Journal* 32, 975–983.

Tang, D., Christiansen, K.M. and Innes, R.W. (2005) Regulation of plant disease resistance, stress responses, cell death, and ethylene signaling in *Arabidopsis* by the EDR1 protein kinase. *Plant Physiology* 138, 1018–1026.

Teige, M., Scheikl, E., Eulgem, T., Doczi, R., Ichimura, K., Shinozaki, K., Dangl, J.L. and Hirt, H. (2004) The MKK2 pathway mediates cold and salt stress signaling in *Arabidopsis*. *Molecular Cell* 15, 141–152.

Waller, F., Muller, A., Chung, K.M., Yap, Y.K., Nakamura, K., Weiler, E. and Sano, H. (2006) Expression of a WIPK-activated transcription factor results in increase of endogenous salicylic acid and pathogen resistance in tobacco plants. *Plant and Cell Physiology* 47, 1169–1174.

Wang, H., Ngwenyama, N., Liu, Y., Walker, J.C. and Zhang, S. (2007) Stomatal development and patterning are regulated by environmentally responsive mitogen-activated protein kinases in *Arabidopsis*. *The Plant Cell* 19, 63–73.

Wang, H., Liu, Y., Bruffett, K., Lee, J., Hause, G., Walker, J.C. and Zhang, S. (2008) Haplo-insufficiency of MPK3 in MPK6 mutant background uncovers a novel function of these two MAPKs in *Arabidopsis* ovule development. *The Plant Cell* 20, 602–613.

Xiang, T., Zong, N., Zou, Y., Wu, Y., Zhang, J., Xing, W., Li, Y., Tang, X., Zhu, L., Chai, J. and Zhou, J.M. (2008) *Pseudomonas syringae* effector AvrPto blocks innate immunity by targeting receptor kinases. *Current Biology* 18, 74–80.

Xiao, S., Calis, O., Patrick, E., Zhang, G., Charoenwattana, P., Muskett, P., Parker, J.E. and Turner, J.G. (2005) The atypical resistance gene, *RPW8*, recruits components of basal defence for powdery mildew resistance in *Arabidopsis*. *The Plant Journal* 42, 95–110.

Xiong, L. and Yang, Y. (2003) Disease resistance and abiotic stress tolerance in rice are inversely modulated by an abscisic acid-inducible mitogen-activated protein kinase. *The Plant Cell* 15, 745–759.

Yang, K.-Y., Liu, Y. and Zhang, S. (2001) Activation of a mitogen-activated protein kinase pathway is involved in disease resistance in tobacco. *Proceedings of the National Academy of Sciences, USA* 98, 741–746.

Yap, Y.-K., Kodama, Y., Waller, F., Chung, K.M., Ueda, H., Nakamura, K., Oldsen, M., Yoda, H., Yamaguchi, Y. and Sano, H. (2005) Activation of a novel transcription factor through phosphorylation by WIPK, a wound-induced mitogen-activated protein kinase in tobacco plants. *Plant Physiology* 139, 127–137.

Yoshioka, H., Numata, N., Nakajima, K., Katou, S., Kawakita, K., Rowland, O., Jones, J.D.G. and Doke, N. (2003) *Nicotiana benthamiana* gp91$^{phox}$ homologs *NbrbohA* and *NbrbohB* participate in $H_2O_2$ accumulation and resistance to *Phytophthora infestans*. *The Plant Cell* 15, 706–718.

You, M.K., Oh, S.I., Ok, S.H., Cho, S.K., Shin, H.Y., Jeung, J.U. and Shin, J.S. (2007) Identification of putative MAPK kinases in *Oryza minuta* and *O. sativa* responsive to biotic stresses. *Molecules and Cells* 23, 108–114.

Yuan, B., Shen, X., Li, X., Xu, C. and Wang, S. (2007) Mitogen-activated protein kinase OsMPK6 negatively regulates rice disease resistance to bacterial pathogens. *Planta* 226, 953–960.

Yuasa, T., Ichimura, K., Mizoguchi, T. and Shinozaki, K. (2001) Oxidative stress activates ATMPK6, an *Arabidopsis* homologue of MAP kinase. *Plant and Cell Physiology* 42, 1012–1016.

Zhang, J., Shao, F., Li, Y., Cui, H., Chen, L., Li, H., Zou, Y., Long, C., Lan, L., Chai, J., Chen, S., Tang, X. and Zhou, J.M. (2007) A *Pseudomonas syringae* effector inactivates MAPKs to suppress PAMP-induced immunity in plants. *Cell Host & Microbe* 1, 175–185.

Zhang, S. and Klessig, D.F. (1997) Salicylic acid activates a 48-kD MAP kinase in tobacco. *The Plant Cell* 9, 809–824.

Zhang, S. and Klessig, D.F. (1998a) Resistance gene *N*-mediated de novo synthesis and activation of a tobacco mitogen-activated protein kinase by tobacco mosaic virus infection. *Proceedings of the National Academy of Sciences, USA* 95, 7433–7438.

Zhang, S. and Klessig, D.F. (1998b) The tobacco wounding-activated mitogen-activated protein kinase is encoded by SIPK. *Proceedings of the National Academy of Sciences, USA* 95, 7225–7230.

Zhang, S. and Liu, Y. (2001) Activation of salicylic acid-induced protein kinase, a mitogen-activated protein kinase, induces multiple defense responses in tobacco. *The Plant Cell* 13, 1877–1889.

Zhang, S., Du, H. and Klessig, D.F. (1998) Activation of the tobacco SIP kinase by both a cell wall-derived carbohydrate elicitor and purified proteinaceous elicitins from *Phytophthora* spp. *The Plant Cell* 10, 435–450.

Zhang, S., Liu, Y. and Klessig, D.F. (2000) Multiple levels of tobacco WIPK activation during the induction of cell death by fungal elicitins. *The Plant Journal* 23, 339–347.

Zhang, X., Dai, Y., Xiong, Y., DeFraia, C., Li, J., Dong, X. and Mou, Z. (2007) Overexpression of *Arabidopsis MAP kinase kinase 7* leads to activation of plant basal and systemic acquired resistance. *The Plant Journal* 52, 1066–1079.

# 3 Molecular Mechanisms of the Radical Burst in Plant Immunity

HIROFUMI YOSHIOKA, SHUTA ASAI, NORIKO MIYAGAWA, TATSUSHI ICHIKAWA, MIKI YOSHIOKA AND MICHIE KOBAYASHI

*Nagoya University, Chikusa, Nagoya, Japan*

## Abstract

Rapid production of nitric oxide (NO) and reactive oxygen species (ROS) has been implicated in the regulation of innate immunity in plants. There are many reports about complementary, synergistic and overlapping functions of NO and ROS in the defence responses. Recent advances provide the molecular mechanisms of NO and oxidative bursts in the defence responses. NOA1 (NO ASSOCIATED1; formerly named NOS1) and membrane-bound NADPH oxidase are believed to participate in the radical burst. Here we describe that two mitogen-activated protein kinase (MAPK) cascades, MEK2-SIPK and cytokinesis-related MEK1-NTF6, are involved in the induction of NADPH oxidase at the transcriptional level in *Nicotiana benthamiana*. On the other hand, NOA1-mediated NO burst is regulated by the MEK2-SIPK cascade. Furthermore, we discuss the activation of NADPH oxidase by the calcium-dependent protein kinase (CDPK) through the direct phosphorylation of its N-terminal region.

## 3.1 Introduction

Rapid production of nitric oxide (NO) and reactive oxygen species (ROS), called NO burst and oxidative burst, respectively, have been implicated in diverse physiological processes, such as resistance to biotic and abiotic stress, hormonal signalling and development (Doke, 1983; Kwak *et al.*, 2003; Bright *et al.*, 2006; Grun *et al.*, 2006; Takeda *et al.*, 2008). Recently, NO has attracted attention as a radical that participates in innate immunity in plants. NO induces phytoalexin accumulation (Noritake *et al.*, 1996), activates the mitogen-activated protein kinase (MAPK) cascade (Clarke *et al.*, 2000) and increases the expression of defence genes, such as those coding for phenylalanine ammonia-lyase (PAL) and pathogenesis-related proteins (Durner *et al.*, 1998). In animals,

 

NO is produced by NO synthase (NOS). The sources of NO synthesis in plants are many and include reduction in nitrite by nitrate reductase (NR), oxidation of arginine to citrulline by NOS, and a non-enzymatic NO generation system (Bethke *et al.*, 2004). Although evidence for arginine-dependent NO synthesis in plants has accumulated, no gene or protein that has a sequence similar to known mammalian-type NOS has been found in plants (Butt *et al.*, 2003; Garcia-Mata and Lamattina, 2003). Guo *et al.* (2003) identified a NOS-like enzyme from *Arabidopsis thaliana* (AtNOS1) with a sequence similar to a protein that has been implicated in NO synthesis in the snail *Helix pomatia*. The AtNOS1 protein has no NOS activity (Zemojtel *et al.*, 2006), and therefore the AtNOS1 was renamed AtNOA1 for NO ASSOCIATED1 (Crawford *et al.*, 2006). However, the *A. thaliana* mutant *noa1* is still useful for its phenotype, which shows reduced levels of NO in plant growth, fertility, hormonal signalling, salt tolerance and plant–pathogen responses (Guo *et al.*, 2003; He *et al.*, 2004; Zeidler *et al.*, 2004; Zhao *et al.*, 2007). Knocking out or down the *NOA1* provides a powerful tool to analyse the NO function.

The phagocyte enzymatic complex of NADPH oxidase consists of two plasma membrane proteins, gp91$^{phox}$ (phox for phagocyte oxidase) and p22$^{phox}$. Cytosolic regulatory proteins, p47$^{phox}$, p67$^{phox}$, p40$^{phox}$ and Rac2 translocate to the plasma membrane to form the active complex after stimulation (Lambeth, 2004). However, no homologues of the p22$^{phox}$, p67$^{phox}$, p47$^{phox}$ and p40$^{phox}$ regulators of the phagocyte NADPH oxidase were found in the *Arabidopsis* genome. On the other hand, a homologue of human Rac GTPase has been isolated from rice (*Oryza sativa*), and the constitutively active mutant of Rac activates ROS production (Wong *et al.*, 2007). Plant NADPH oxidases designated as RBOH (respiratory burst oxidase homologue) have been identified as genes related to mammalian gp91$^{phox}$ in various plants (Groom *et al.*, 1996; Keller *et al.*, 1998; Torres *et al.*, 1998; Yoshioka *et al.*, 2001, 2003; Sagi *et al.*, 2004) and carry an N-terminal extension comprising two EF-hand motifs, suggesting that $Ca^{2+}$ regulates its activity. RBOHs are localized on the plasma membrane (Kobayashi *et al.*, 2006). *Arabidopsis rbohD rbohF* double mutant greatly decreases ROS production against infection of avirulent *Pseudomonas syringae* pv. *tomato* DC3000 and *Hyaloperonospora parasitica* (Torres *et al.*, 2002) and in response to abscisic acid (Kwak *et al.*, 2003). Analysis of the loss of function of *NtRBOHD* in tobacco cells (*Nicotiana tabacum*) showed loss of ROS production following elicitor treatment (Simon-Plas *et al.*, 2002). We also showed that silencing *NbRBOHA* and *NbRBOHB* in *Nicotiana benthamiana* plants leads to less ROS production and reduced resistance to infection by the potato pathogen *Phytophthora infestans* (Yoshioka *et al.*, 2003). These reports suggest that RBOH is a key regulator of ROS production and has pleiotropic functions in plants.

The mitogen-activated protein kinase (MAPK) cascade is a major evolutionarily conserved signalling pathway used to transduce extracellular stimuli into intracellular responses among eukaryotes (MAPK Group, 2002; Nakagami *et al.*, 2005). In the MAPK signal transduction cascade, a MAPK is activated by a MAPK kinase (MAPKK), which itself is activated by a MAPKK kinase

(MAPKKK). Many studies have extensively characterized plant MAPKs, including tobacco wound-induced protein kinase (WIPK) (Seo *et al.*, 1999) and salicylic-acid-induced protein kinase (SIPK) (Zhang and Klessig, 1997) and their orthologues in other plant species (Lee *et al.*, 2004; Katou *et al.*, 2005; Pedley and Martin, 2005). WIPK and SIPK participate either in *N* gene-dependent resistance to tobacco mosaic virus (TMV) (Zhang and Klessig, 1998; Jin *et al.*, 2003) or in *Cf9*-dependent resistance to *Cladosporium fulvum*-derived elicitor *Avr9* (Romeis *et al.*, 1999) in a gene-for-gene-specific way and in response to plant species-specific elicitors, such as elicitins (Zhang *et al.*, 2000). NtMEK2, a tobacco MAPKK, is upstream of both WIPK and SIPK (Yang *et al.*, 2001). Expression of NtMEK2$^{DD}$, a constitutively active mutant of NtMEK2, induces hypersensitive response (HR)-like cell death, defence gene expression, and generation of ROS, all of which are preceded by activation of endogenous WIPK and SIPK (Yang *et al.*, 2001; Ren *et al.*, 2002). The potato (*Solanum tuberosum*) orthologue of tobacco NtMEK2, StMEK2 (formerly described as StMEK1) was also cloned (Katou *et al.*, 2003). Heterologous expression of StMEK2$^{DD}$ in *N. benthamiana* induces NbWIPK and NbSIPK activities (Katou *et al.*, 2003) and then an oxidative burst accompanied by an induction of *NbRBOHB* expression (Yoshioka *et al.*, 2003). NbMAPKKK$\alpha$ was identified as an upstream activator of NbMEK2 and NbSIPK in *N. benthamiana* (del Pozo *et al.*, 2004). Silencing the *NbMAPKKK*$\alpha$ gene eliminated HR-like cell death caused by the interaction between the *P. syringae* avirulence gene *AvrPto* and its cognate resistant gene *Pto*. Interestingly, HR-like cell death induced by NbMAPKKK$\alpha$ overexpression is compromised not only by silencing *NbMEK2* or *NbSIPK*, but also by silencing *NbMEK1* or *NbNTF6*. NPK1-MEK1/NQK1-NTF6/NRK1 is a pivotal MAPK cascade in the regulation of cytokinesis (Soyano *et al.*, 2003; Sasabe *et al.*, 2006). Like WIPK or SIPK, silencing *NbNPK1*, *NbMEK1* or *NbNTF6* attenuates *N*- and *Pto*-mediated resistance against TMV (Jin *et al.*, 2002; Liu *et al.*, 2004) and *P. syringae AvrPto*, respectively (Ekengren *et al.*, 2003). These studies indicated that MAPK cascades MEK2-WIPK/SIPK and NPK1-MEK1-NTF6 participate in disease resistance in plants. In addition, recent work has shown that tobacco NTF4 that shares 93.6% identity with SIPK is also activated by NtMEK2, and that ectopic expression of *NTF4* induces HR-like cell death (Ren *et al.*, 2006). However, the relationship between these MAPK cascades and the radical burst in response to pathogen signals is not clear.

We recently described the roles of MEK2-WIPK/SIPK/NTF4 and MEK1-NTF6 cascades in the regulation of NO and oxidative bursts in *N. benthamiana* (Asai *et al.*, 2008). Gain-of-function and loss-of-function analyses showed that the MEK2-SIPK/NTF4 cascade controls the NOA1-mediated NO burst, and that the MEK2-SIPK/NTF4 and MEK1-NTF6 cascades regulate the NADPH oxidase-dependent oxidative burst. Furthermore, we also demonstrated that the calcium-dependent protein kinase (CDPK) activates NADPH oxidase by the direct phosphorylation of its N-terminal region (Kobayashi *et al.*, 2007).

## 3.2 MAPKs Regulate NO Production

ROS and NO are believed to play important roles independently or acting in coordination in plant innate immunity. ROS generated on the plasma membrane are released to the apoplast where they induce oxidative crosslinking of glycoproteins and strengthen the cell wall against secondary infection (Bradley *et al.*, 1992), simultaneously activating the $Ca^{2+}$ channel to increase the level of cytosolic $Ca^{2+}$ (Lecourieux *et al.*, 2002). $Ca^{2+}$ may function not only as an inducer of the oxidative burst, but also as a signalling molecule downstream of the oxidative burst and causes various cellular responses, including defence (Torres and Dangl, 2005). However, NO signalling includes various messenger molecules, such as cyclic guanosine monophosphate (cGMP), cADP ribose and $Ca^{2+}$ (Durner *et al.*, 1998; Wendehenne *et al.*, 2001; Lamotte *et al.*, 2004; Romero-Puertas *et al.*, 2004), which both directly and indirectly modulate the expression of specific genes (Polverari *et al.*, 2003; Parani *et al.*, 2004). NO signalling pathways often include post-translational modification of target proteins, such as NO-dependent cysteine *S*-nitrosylation that can modulate the activity and function of different proteins (Sokolovski and Blatt, 2004; Feechan *et al.*, 2005; Lindermayr *et al.*, 2005; Romero-Puertas *et al.*, 2007). NO can also react with $O_2^-$ to form the reactive molecule peroxynitrite ($ONOO^-$). $ONOO^-$ can lead to formation of $NO_2$ and the effective oxidant hydroxyl radical ($OH^•$). $OH^•$ is a very strong oxidizing species that can rapidly attack biological membranes and all types of biomolecules, such as DNA and proteins, leading to irreparable damage, metabolic dysfunction and cell death (del Rio *et al.*, 2003). $ONOO^-$ is also responsible for tyrosine nitration (Lamattina *et al.*, 2003), which is the major toxic reactive nitrogen species in animal cells (Stamler *et al.*, 1992). $ONOO^-$ is relevant to HR and defence gene expression (Alamillo and García-Olmedo, 2001). One study emphasized that the combination of NO and hydrogen peroxide ($H_2O_2$), but not $ONOO^-$, takes part in the induction of defence responses (Delledonne *et al.*, 2001). NO or ROS are not essential for HR in plants but induce apoptosis in adjacent cells during the defence response (Tada *et al.*, 2004). HR cell death may require a fine balance between NO and ROS (Delledonne *et al.*, 2001).

Many studies have shown that NOS plays a pivotal role in NO synthesis in plants, as well as in animals, and is responsible for various stress responses, development and disease resistance (Guo *et al.*, 2003; He *et al.*, 2004; Zeidler *et al.*, 2004; Zhao *et al.*, 2007). Although controversy exists on the nature of NOA1 (Crawford *et al.*, 2006; Zemojtel *et al.*, 2006), the knockout mutant *Arabidopsis noa1* impairs arginine-dependent NOS activity (Guo *et al.*, 2003; Zhao *et al.*, 2007), increases disease susceptibility to the pathogen *P. syringae* pv. *tomato* DC3000 (Zeidler *et al.*, 2004), and is highly vulnerable to salt and oxidative stress (Zhao *et al.*, 2007).

Recently, we found that $StMEK2^{DD}$ provokes NO burst, and virus-induced gene silencing (VIGS) of *SIPK/NTF4* as well as *NbNOA1* compromised the infestin 1 (INF1)-induced NO burst in *N. benthamiana*, whereas VIGS of *WIPK* did not compromise the NO burst (Asai *et al.*, 2008) (Fig. 3.1). NTF4 is a tobacco MAPK that reveals high homology with SIPK (93.6% in identity) and

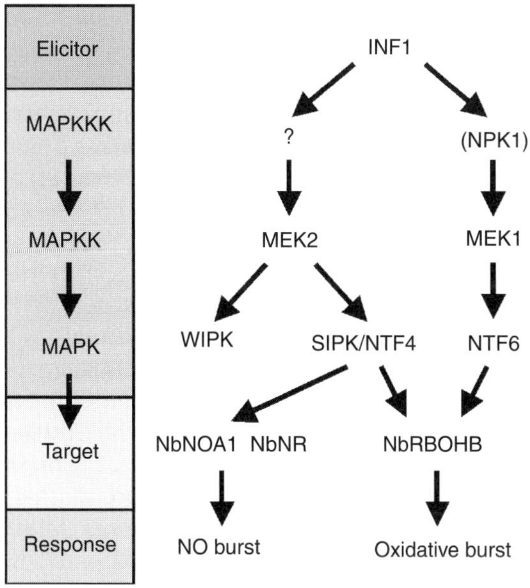


**Fig. 3.1.** Scheme of the proposed model for the MAPK cascade signalling pathway leading to radical burst. INF1 induces NO burst by means of MEK2-SIPK/NTF4 cascade and oxidative burst by means of MEK2-SIPK/NTF4 and (NPK1)-MEK1-NTF6 cascades. Based on loss-of-function and gain-of-function analyses, SIPK and NTF4 play important roles in NbNOA1- and NbNR-mediated NO burst and NbRBOHB-dependent oxidative burst. NTF6 is also responsible for NbRBOHB-dependent oxidative burst. The question mark indicates unidentified MAPKKK(s).

a similar expression pattern and activation profile of SIPK (Ren *et al.*, 2006). The conditional expression of *NTF4* induces autophosphorylation and HR-like cell death (Ren *et al.*, 2006). It was shown that NbNTF4, like NbSIPK, also regulates the radical bursts. NTF4 may play a similar role to SIPK in the signalling of basal defence (Asai *et al.*, 2008). Together, these findings show that these gene products can participate in the process of NO production by NOS activity. *NbNOA1* is not induced in *N. benthamiana* leaves by INF1-treatment (Kato *et al.*, 2008), suggesting that post-transcriptional control of NOA1-influenced NO production is effected through the MEK2-SIPK/NTF4 cascade. It has been shown that NR, a key enzyme of nitrogen assimilation, is another enzyme capable of producing NO in plants (Lamattina *et al.*, 2003; Yamamoto *et al.*, 2003). Silencing *NbNOA1* did not completely inhibit the NO burst induced by INF1 and StMEK2$^{DD}$. Furthermore, the NO burst was significantly suppressed by tungstate, an NR inhibitor, in both tobacco rattle virus (TRV) control- and *NbNOA1*-silenced plants (Asai *et al.*, 2008). These results suggest that another factor, NR, also participates in the NO burst induced by INF1 and StMEK2$^{DD}$ in agreement with the previous finding that NbNR1 partially contributes to the INF1-induced NO burst in *N. benthamiana* (Yamamoto-Katou *et al.*, 2006). However, the *NbNOA1*-silencing concomitant treatment with tungstate did not result in complete inhibition of the NO generation, suggesting that a non-enzymatic NO generation system exists as reported by Bethke *et al.* (2004).

## 3.3 MAPKs Regulate NADPH Oxidase Genes

Protein kinases are activated during plant defence responses (Peck, 2003). The MAPK cascade transduces extracellular stimuli to intracellular signals by phosphorylation cascades (MAPK Group, 2002). Transient expression of StMEK2$^{DD}$ (StMEK2 is a potato orthologue of NtMEK2) also induced HR-like cell death and ROS production in *N. benthamiana* leaves (Katou *et al.*, 2003, 2005). The transient expression of *StMEK2$^{DD}$* increased accumulation of *NbRBOHB* mRNA (Yoshioka *et al.*, 2003), suggesting that the transcriptional activation of *RBOH* is one of the functions of the MAPK cascade in ROS production. We also showed that silencing *NbRBOHA* and *NbRBOHB* in *N. benthamiana* plants leads to less ROS production and reduced resistance to *P. infestans* (Yoshioka *et al.*, 2003). We found that INF1-induced expression of *NbRBOHB* was compromised in *SIPK/NTF4/NTF6*-silenced leaves, but not in *WIPK*-silenced leaves. Furthermore conditional expression of *NbMEK1$^{DD}$* induced expression of *NbRBOHB*. These results indicated that INF1 regulates NbRBOHB-dependent ROS generation through MEK2-SIPK/NTF4 and MEK1-NTF6 cascades (Fig. 3.1). It was reported previously that treatment of BY-2 tobacco cells with INF1 induces $ONOO^-$ generation which results in tyrosine nitration (Saito *et al.*, 2006). This study supports the idea that NO and $O_2^-$ are produced simultaneously through a convergent signalling pathway, MAPK cascade.

Defence-related *RBOHs*, which play a pivotal role in ROS production in response to pathogen signals, seem to be regulated at the transcriptional and post-translational levels. Recently, we reported that a calcium-dependent protein kinase (StCDPK5) activates *N. benthamiana* NbRBOHs as well as potato StRBOHs by direct phosphorylation of their N-terminal regions (Kobayashi *et al.*, 2007) (Fig. 3.2). CDPK appears to regulate a rapid and transient accumulation of $H_2O_2$ (phase I burst) and a massive oxidative burst at 6–9 h after elicitor-treatment (phase II burst) (Yoshioka *et al.*, 2001; Kobayashi *et al.*, 2007). Liu *et al.* (2007) reported that conditional gain-of-function by NtMEK2$^{DD}$ causes rapid shutdown of carbon fixation reaction in chloroplasts, which could lead to the generation of ROS in chloroplasts under illumination. They concluded that the chloroplast burst occurs earlier than the NADPH oxidase-dependent oxidative burst by MAPK (phase II burst), and that the chloroplast-generated ROS are only a facilitator that accelerates cell death because plant cells without mature chloroplasts die eventually. They suggested that mitochondria-generated ROS might be essential in accelerating the cell-death process. Communication between chloroplasts and mitochondria exists in cells undergoing HR cell death (Yao *et al.*, 2004). Mitochondrial dysfunction appears to be caused by NtMEK2$^{DD}$. NtMEK2$^{DD}$-mediated dysfunction is prevented by Bcl-xL, which is a mammalian anti-apoptotic factor that prevents mitochondrial dysfunction in plants as it does in animals (Takabatake *et al.*, 2007). We routinely use the chemiluminescence probe L-012 to detect RBOH-generated ROS in *N. benthamiana* leaves. The probe has been used for the analysis of ROS generated by membrane-bound NADPH oxidase in neutrophils (Imada *et al.*, 1999). However, we failed to detect rapid chloroplast- and mitochondria-

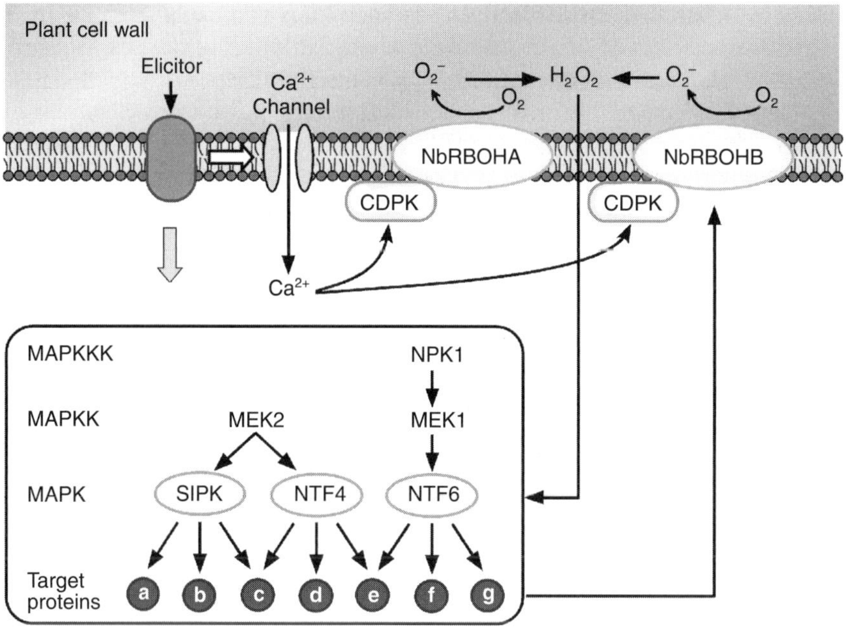


**Fig. 3.2.** Model for RBOH regulation by CDPK. The elicitor induces $Ca^{2+}$ influx. Increase of intracellular $Ca^{2+}$ concentration provokes $Ca^{2+}$ binding to EF-hand motifs of CDPK and RBOH N-terminal region. Phosphorylation of RBOH by the CDPK results in the phase I and II bursts.

generated ROS in elicited plants after treatment with INF1, $StMEK2^{DD}$, $NbMEK1^{DD}$ and fungal infection, suggesting that L-012 is suitable to detect apoplastic ROS generated by membrane-bound RBOH in plant cells.

The early chloroplast-generated ROS caused by $NtMEK2^{DD}$ and phase I burst by elicitors might contribute to the influx of $Ca^{2+}$ into cytoplasm, and the increased level of $Ca^{2+}$ might result in activation of CDPK as an inducer of the phase II burst from NADPH oxidases localized in the plasma membrane.

## 3.4 Cross Talk between Two MAPK Cascades and Radical Bursts

The expression and activation of the *Arabidopsis* gene *OXIDATIVE SIGNAL-INDUCIBLE1* (*OXI1*) encoding Ser/Thr kinase is induced in response to $H_2O_2$ (Rentel *et al.*, 2004). OXI1 is required for activation of MAPKs (AtMPK3 and AtMPK6, orthologues of WIPK and SIPK, respectively, in *A. thaliana*) after treatment with $H_2O_2$ or an elicitor. We reported that the MEK2-SIPK cascade regulates the oxidative burst resulting from the induction of *NbRBOHB* expression (Asai *et al.*, 2008). Taking these results together, it can be assumed that there is a positive feedback circuit between SIPK and NbRBOHB. Similarly, MAPK cascade MEK2-SIPK regulates the NO burst. NO also activates potential MAPKs as shown by in-gel kinase assays using myelin basic protein (MBP) as a substrate (Clarke *et al.*, 2000). SIPK may give a positive feedback between NO burst signals, as well as oxidative burst signals.

A novel MAPKK (NbMKK1) has been identified as an effective inducer of HR-like cell death in *N. benthamiana*. NbMKK1 activates NbSIPK, and NbMKK1-mediated cell death is compromised by silencing *NbSIPK* (Takahashi *et al*., 2007). The transient expression of *SIPK* is sufficient to induce HR-like cell death and the expression of *3-hydroxy-3-methylglutaryl CoA reductase*, a gene encoding a key enzyme in the phytoalexin biosynthesis pathway (Zhang and Liu, 2001). We showed that SIPK regulates the expression of *NbRBOHB* and the NO burst (Asai *et al*., 2008). These results strongly indicate that SIPK plays a central role in multiple defence responses. WIPK can function in accelerating SIPK-mediated cell death or initiating a new pathway (Liu *et al*., 2003). WIPK might function as a factor to commit SIPK-mediated NO and oxidative bursts (Asai *et al*., 2008).

The NPK1-MEK1-NTF6 cascade, a pivotal MAPK cascade in the regulation of cytokinesis (Soyano *et al*., 2003), participates in *N*- and *Pto*-mediated resistance to TMV and *P. syringae AvrPto*, respectively (Jin *et al*., 2002; Ekengren *et al*., 2003; Liu *et al*., 2004). How the NTF6 can bifurcate two distinct cellular functions, cytokinesis and innate immunity, is unclear. These puzzles require further investigation.

## 3.5 CDPK Activates NADPH Oxidase by Direct Phosphorylation of its N-Terminal Region

The regulatory mechanisms of RBOH also remain unknown while several lines of evidence indicate the importance of $Ca^{2+}$ and protein kinases in ROS production. Because overexpression of *AtRBOHD* does not result in constitutive ROS production, RBOH may require post-transcriptional regulation for its activation (Torres and Dangl, 2005). $Ca^{2+}$ influx into the cytoplasm (Chandra and Low, 1997; Piedras *et al*., 1998; Grant *et al*., 2000) and changes in protein phosphorylation (Kauss and Jeblick, 1995; Miura *et al*., 1995) are implicated in the activation process of the RBOH. Recently, *Arabidopsis* AtRBOHD was shown to be phosphorylated by a certain protein kinase activity (Benschop *et al*., 2007; Nühse *et al*., 2007), and seems to be activated synergistically by $Ca^{2+}$ binding to the EF-hand motifs at its N-terminal region (Ogasawara *et al*., 2008). Interestingly, NOX5, the human homologue of RBOH that contains four EF-hands, also seems to be activated by protein kinase C (Serrander *et al*., 2007). It was reported that RBOH-like proteins in the plasma membrane fractions of tomato (*Solanum lycopersicum*) and tobacco show NADPH oxidase activity without any cytosolic components in the renatured gel after SDS-PAGE, and the activity increases upon addition of $Ca^{2+}$ (Sagi and Fluhr, 2001). On the other hand, a homologue of human Rac GTPase has been isolated from rice, and the constitutively active mutant of Rac activates ROS production (Kawasaki *et al*., 1999). These reports suggest that the extended N-terminal region may play a key role in the regulation of the RBOH enzyme.

We previously isolated *StRBOHA* to *D* from potato plants (Yoshioka *et al*., 2001; Yamamizo *et al*., 2006). RNA gel blot analyses indicate that *StRBOHA* is constitutively expressed at a low level, whereas *StRBOHB* and *StRBOHC* are upregulated during the phase II burst. The promoter analysis of *StRBOHC* demonstrated that MEK2 regulates the gene activation at the transcriptional level (unpublished results). NADPH oxidase inhibitor diphenylene iodonium (DPI) blocked phase I and phase II bursts, while pretreatment of tubers with the protein synthesis inhibitor cycloheximide abolished only the second burst. These data suggest that StRBOHA and StRBOHB plus C contribute to phase I and phase II bursts, respectively (Yoshioka *et al.*, 2001; Yamamizo *et al.*, 2006). We found that both bursts are also inhibited by a protein kinase inhibitor or a calcium inhibitor. These findings led us to investigate the direct phosphorylation of the N-terminal region of the StRBOH protein by certain protein kinases for the activation of the enzymes. We identified Ser82 and Ser97 in the N terminus of potato StRBOHB as potential phosphorylation sites by in-gel kinase assays using the mutated N-terminal proteins of StRBOHB and mass spectrometry analysis. Moreover, an anti-phosphopeptide (pSer82) antibody indicated that the Ser82 was phosphorylated by pathogen signals *in planta*. We cloned StCDPK5 by complementary DNA (cDNA) expression screening using the anti-pSer82 antibody and cells expressing a substrate polypeptide, and mass spectrometry analyses showed that the CDPK phosphorylated only Ser82 and Ser97 in the N terminus of StRBOHB in a calcium-dependent manner. Ectopic expression of the constitutively active mutant of StCDPK5, StCDPK5VK, provoked ROS production in *N. benthamiana* leaves. The CDPK-mediated ROS production was disrupted by knockdown of *NbRBOHB* in tobacco. The loss of function was complemented by heterologous expression of wild-type potato StRBOHB, but not by a mutant (S82A/S97A). Furthermore, the heterologous expression of *StCDPK5VK* phosphorylated Ser82 of StRBOHB in tobacco. In fact, StCDPK5 phosphorylated N-terminal regions of *N. benthamiana* NbRBOHA and B as well as potato StRBOHA to D (Kobayashi *et al.*, 2007). Furthermore, analyses by the BiFC (bimolecular fluorescence complementation) method indicated that StRBOHB and StCDPK5 interact on the plasma membrane and mutations of *N*-myristoylation and palmitoylaytion sites of the StCDPK5, which are responsible for localization on the membrane, eliminate these interactions (unpublished results). These lines of evidence suggest that the StCDPK5 induces phosphorylation of RBOHs and regulates the oxidative burst (Fig. 3.2).

## 3.6 CDPK-triggered Oxidative Burst Confers Resistance to Late Blight but Enhances Susceptibility to Early Blight Pathogen in Potato

In regulating gene expression, NO can induce the expression of *PAL* and *chalcone synthase* independently of ROS, and induction of defence-related genes by NO, such as glutathione *S*-transferase, depends on $H_2O_2$ (Grun *et al.*, 2006). The cooperative and individual roles of NO and ROS during disease resistance are unclear. Transgenic potato plants containing *StMEK2*$^{DD}$ fused to a pathogen-inducible promoter, which was regulated by SIPK and WIPK, confers resistance to both near-obligate pathogen late blight pathogen *P. infestans* and necrotrophic pathogen *Alternaria solani* that causes early blight on potato plants (Yamamizo *et al.*, 2006). These results suggest that cogeneration of NO and ROS by StMEK2$^{DD}$ confers resistance to both biotrophic and necrotrophic pathogens. We generated transgenic potato plants harbouring *StCDPK5VK* under the control of the same pathogen-inducible promoter (unpublished results). Virulent isolates of *P. infestans* and *A. solani* induced HR-like cell death accompanied by ROS production at the infection sites in the transgenic plants. The transgenic potato plants conferred resistance to the biotrophic pathogen *P. infestans* but, in contrast, enhanced susceptibility to the necrotrophic pathogen *A. solani* (Fig. 3.3). The gene expression profile of the transgenic plants in response to *P. infestans* indicated that expression of salicylic acid (SA)-inducible *PR-1* was enhanced. On the other hand, infection of *A. solani* caused suppression of *PR-1* gene expression compared with wild-type potato. These results suggest that StCDPK5VK-mediated ROS production provides resistance to the biotrophic pathogen but enhancement of susceptibility to the necrotrophic pathogen by induction of SA-dependent signalling. These

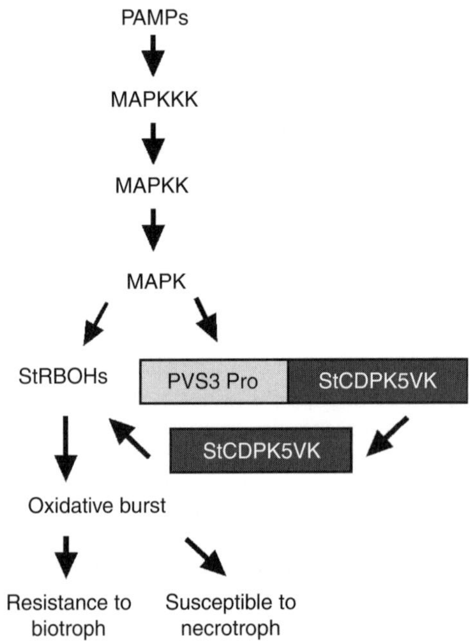


**Fig. 3.3.** Schematic representation of induction mechanism of oxidative burst in transgenic potato plants attacked by pathogens. Pathogen-associated molecular patterns (PAMPs) activate the endogenous MAPK cascade. Following the activation of MAPKs, *StCDPK5VK* driven by the potato vetispiradiene synthase promoter (PVS3 Pro) is expressed. The direct phosphorylation of StRBOHs by StCDPK5VK produces ROS. The oxidative burst confers resistance to the biotrophic pathogen, but enhances susceptibility to the necrotrophic pathogen.

results support the idea that during evolution plants may have obtained the signalling pathway which regulates both NO and ROS production to adapt to wide-spectrum pathogens.

## References

Alamillo, J.M. and García-Olmedo, F. (2001) Effects of urate, a natural scavenger of peroxynitrite-mediated toxicity, in the response of *Arabidopsis thaliana* to the bacterial pathogen *Pseudomonas syringae*. *The Plant Journal* 25, 529–540.

Asai, S., Ohta, K. and Yoshioka, H. (2008) MAPKs signaling regulates nitric oxide and NADPH oxidase-dependent oxidative bursts in *Nicotiana benthamiana*. *The Plant Cell* 20, 1390–1406.

Benschop, J.J., Mohammed, S., O'Flaherty, M., Heck, A.J.R., Slijper, M. and Menke, F.L.H. (2007) Quantitative phosphoproteomics of early elicitor signaling in *Arabidopsis*. *Molecular and Cellular Proteomics* 6, 1198–1214.

Bethke, P.C., Badger, M.R. and Jones, R.L. (2004) Apoplastic synthesis of nitric oxide by plant tissues. *The Plant Cell* 16, 332–341.

Bradley, D.J., Kjellbom, P. and Lamb, C.J. (1992) Elicitor- and wound-induced oxidative cross-linking of a proline-rich plant cell wall protein: a novel, rapid defense response. *Cell* 70, 21–30.

Bright, J., Desikan, R., Hancock, J.T., Weir, I. and Neill, S.J. (2006) ABA-induced NO generation and stomatal closure in *Arabidopsis* are dependent on $H_2O_2$ synthesis. *The Plant Journal* 45, 113–122.

Butt, Y.K.-C., Lum, J.H.-K. and Lo, S.C.-L. (2003) Proteomic identification of plant proteins probed by mammalian nitric oxide synthase antibodies. *Planta* 216, 762–771.

Chandra, S. and Low, P.S. (1997) Measurement of $Ca^{2+}$ fluxes during elicitation of the oxidative burst in aequorin-transformed tobacco cells. *Journal of Biological Chemistry* 272, 28274–28280.

Clarke, A., Desikan, R., Hurst, R.D., Hancock, J.T. and Neill, S.J. (2000) NO way back: nitric oxide and programmed cell death in *Arabidopsis thaliana* suspension cultures. *The Plant Journal* 24, 667–677.

Crawford, N.M., Galli, M., Tischner, R., Heimer, Y.M., Okamoto, M. and Mack, A. (2006) Response to Zemojtel *et al*: plant nitric oxide synthase: back to square one. *Trends in Plant Science* 11, 526–527.

Delledonne, M., Zeier, J., Marocco, A. and Lamb, C. (2001) Signal interactions between nitric oxide and reactive oxygen intermediates in the plant hypersensitive disease resistance response. *Proceedings of the National Academy of Sciences, USA* 98, 13454–13459.

del Pozo, O., Pedley, K.F. and Martin, G.B. (2004) MAPKKK is a positive regulator of cell death associated with both plant immunity and disease. *EMBO Journal* 23, 3072–3082.

del Rio, L.A., Corpas, F.J., Sandalio, L.M., Palma, J.M. and Barroso, J.B. (2003) Plant peroxisomes, reactive oxygen metabolism and nitric oxide. *International Union of Biochemistry and Molecular Biology (IUBMB) Life* 55, 71–81.

Doke, N. (1983) Involvement of superoxide anion generation in the hypersensitive response of potato tuber tissues to infection with an incompatible race of *Phytophthora infestans* and to the hyphal wall components. *Physiological Plant Pathology* 23, 345–357.

Durner, J., Wendehenne, D. and Klessig, D.F. (1998) Defense gene induction in tobacco by nitric oxide, cyclic GMP and cyclic ADP-ribose. *Proceedings of the National Academy of Sciences, USA* 95, 10328–10333.

Ekengren, S.K., Liu, Y., Schiff, M., Dinesh-Kumar, S.P. and Martin, G.B. (2003) Two MAPK cascades, NPR1, and TGA transcription factors play a role in Pto-mediated disease resistance in tomato. *The Plant Journal* 36, 905–917.

Feechan, A., Kwon, E., Yun, B.W., Wang, Y., Pallas, J.A. and Loake, G.J. (2005) A central role for *S*-nitrosothiols in plant disease resistance. *Proceedings of the National Academy of Sciences, USA* 22, 8054–8059.

Garcia-Mata, C. and Lamattina, L. (2003) Abscisic acid, nitric oxide and stomatal closure – is nitrate reductase one of the missing links. *Trends in Plant Science* 8, 20–26.

Grant, M., Brown, I., Adams, S., Knight, M., Ainslie, A. and Mansfield, J. (2000) The *RPM1* plant disease resistance gene facilitates a rapid and sustained increase in cytosolic calcium that is necessary for the oxidative burst and hypersensitive cell death. *The Plant Journal* 23, 441–450.

Groom, Q.J., Torres, M.A., Fordham-Skelton, A.P., Hammond-Kosack, K.E., Robinson, N.J. and Jones, J.D.G. (1996) *rbohA*, a rice homologue of the mammalian *gp91phox* respiratory burst oxidase gene. *The Plant Journal* 10, 515–522.

Grun, S., Lindermayr, C. and Durner, J. (2006) Nitric oxide and gene regulation in plants. *Journal of Experimental Botany* 57, 507–516.

Guo, F.-Q., Okamoto, M. and Crawford, N.M. (2003) Identification of a plant nitric oxide synthase gene involved in hormonal signaling. *Science* 302, 100–103.

He, Y., Tang, R.-H., Hao, Y., Stevens, R.D., Cook, C.W., Ahn, S.M., Jing, L., Yang, Z., Chen, L., Guo, F., Fiorani, F., Jackson, R.B., Crawford, N.M. and Pei, Z.-M. (2004) Nitric oxide represses the *Arabidopsis* floral transition. *Science* 305, 1968–1971.

Imada, I., Sato, E.F., Miyamoto, M., Ichimori, Y., Minamiyama, Y., Konaka, R. and Inoue, M. (1999) Analysis of reactive oxygen species generated by neutrophils using a chemiluminescence probe L-012. *Analytical Biochemistry* 271, 53–58.

Jin, H., Axtell, M.J., Dahlbeck, D., Ekwenna, O., Zhang, S., Staskawicz, B. and Baker, B. (2002) NPK1, an MEKK1-like mitogen-activated protein kinase kinase kinase, regulates innate immunity and development in plants. *Developmental Cell* 3, 291–297.

Jin, H., Liu, Y., Yang, K.-Y., Kim, C.Y., Baker, B. and Zhang, S. (2003) Function of a mitogen-activated protein kinase pathway in *N* gene mediated resistance in tobacco. *The Plant Journal* 33, 719–731.

Kato, H., Asai, S., Yamamoto-Katou, A., Yoshioka, H., Doke, N. and Kawakita, K. (2008) Involvement of *NbNOA1* in NO production and defense responses in INF1-treated *Nicotiana benthamiana*. *Journal of General Plant Pathology* 74, 15–23.

Katou, S., Yamamoto, A., Yoshioka, H., Kawakita, K. and Doke, N. (2003) Functional analysis of potato mitogen-activated protein kinase kinase, StMEK1. *Journal of General Plant Pathology* 69, 161–168.

Katou, S., Yoshioka, H., Kawakita, K., Rowland, O., Jones, J.D.G., Mori, H. and Doke, N. (2005) Involvement of PPS3 phosphorylated by elicitor-responsive mitogen-activated protein kinases in the regulation of plant cell death. *Plant Physiology* 139, 1914–1926.

Kauss, H. and Jeblick, W. (1995) Pretreatment of parsley suspension cultures with salicylic acid enhances spontaneous and elicited production of $H_2O_2$. *Plant Physiology* 108, 1171–1178.

Kawasaki, T., Henmi, K., Ono, E., Hatakeyama, S., Iwano, M., Satoh, H. and Shimamoto, K. (1999) The small GTP-binding protein Rac is a regulator of cell death in plants. *Proceedings of the National Academy of Sciences, USA* 96, 10922–10926.

Keller, T., Damude, H.G., Werner, D., Doerner, P., Dixon, R.A. and Lamb, C. (1998) A plant homolog of the neutrophil NADPH oxidase gp91$^{phox}$ subunit gene encodes a plasma membrane protein with $Ca^{2+}$ binding motifs. *The Plant Cell* 10, 255–266.

Kobayashi, M., Kawakita, K., Maeshima, M., Doke, N. and Yoshioka, H. (2006) Subcellular localization of Strboh proteins and NADPH-dependent $O_2^-$-generating activity in potato tuber tissues. *Journal of Experimental Botany* 57, 1373–1379.

Kobayashi, M., Ohura, I., Kawakita, K., Yokota, N., Fujiwara, M., Shimamoto, K., Doke, N. and Yoshioka, H. (2007) Calcium-dependent protein kinases regulate the production of reactive oxygen species by potato NADPH oxidase. *The Plant Cell* 19, 1065–1080.

Kwak, J.M., Mori, I.C., Pei, Z.-M., Leonhardt, N., Torres, M.A., Dangl, J.L., Bloom, R.E., Bodde, S., Jones, J.D.G. and Schroeder, J.I. (2003) NADPH oxidase *AtrbohD* and *AtrbohF* genes function in ROS-dependent ABA signaling in *Arabidopsis*. *EMBO Journal* 22, 2623–2633.

Lamattina, L., Garcia-Mata, C., Graziano, M. and Pagnussat, G. (2003) Nitric oxide: the versatility of an extensive signal molecule. *Annual Review of Plant Biology* 54, 109–136.

Lambeth, J.D. (2004) NOX enzymes and the biology of reactive oxygen. *Nature Reviews Immunology* 4, 181–189.

Lamotte, O., Gould, K., Lecourieux, D., Sequeira-Legrand, A., Lebrun-Garcia, A., Durner, J., Pugin, A. and Wendehenne, D. (2004) Analysis of nitric oxide signaling functions in tobacco cells challenged by the elicitor cryptogein. *Plant Physiology* 135, 516–529.

Lecourieux, D., Mazars, C., Pauly, N., Ranjeva, R. and Pugin, A. (2002) Analysis and effects of cytosolic free calcium increases in response to elicitors in *Nicotiana plumbaginifolia* cells. *The Plant Cell* 14, 2627–2641.

Lee, J., Rudd, J.J., Macioszek, V.K. and Scheel, D. (2004) Dynamic changes in the localization of MAP kinase cascade components controlling pathogenesis-related (PR) gene expression during innate immunity in parsley. *Journal of Biological Chemistry* 279, 22440–22448.

Lindermayr, C., Saalbach, G. and Durner, J. (2005) Proteomic identification of *S*-nitrosylated proteins in *Arabidopsis*. *Plant Physiology* 137, 921–930.

Liu, Y., Jin, H., Yang, K.-Y., Kim, C.Y., Baker, B. and Zhang, S. (2003) Interaction between two mitogen-activated protein kinases during tobacco defense signaling. *The Plant Journal* 34, 149–160.

Liu, Y., Schiff, M. and Dinesh-Kumar, S.P. (2004) Involvement of MEK1 MAPKK, NTF6 MAPK, WRKY/MYB transcription factors, *COI1* and *CTR1* in *N*-mediated resistance to tobacco mosaic virus. *The Plant Journal* 38, 800–809.

Liu, Y., Ren, D., Pike, S., Pallardy, S., Gassmann, W. and Zhang, S. (2007) Chloroplast-generated reactive oxygen species are involved in hypersensitive response-like cell death mediated by a mitogen-activated protein kinase cascade. *The Plant Journal* 51, 941–954.

MAPK Group (2002) Mitogen-activated protein kinase cascades in plants: a new nomenclature. *Trends in Plant Science* 7, 301–308.

Miura, Y., Yoshioka, H. and Doke, N. (1995) An autophotographic determination of the active oxygen generation in potato tuber discs during hypersensitive response to fungal infection or elicitor. *Plant Science* 105, 45–52.

Nakagami, H., Pitzschke, A. and Hirt, H. (2005) Emerging MAP kinase pathways in plant stress signalling. *Trends in Plant Science* 10, 339–346.

Noritake, T., Kawakita, K. and Doke, N. (1996) Nitric oxide induces phytoalexin accumulation in potato tuber tissues. *Plant and Cell Physiology* 37, 113–116.

Nühse, T.S., Bottrill, A.R., Jones, A.M. and Peck, S.C. (2007) Quantitative phosphoproteomic analysis of plasma membrane proteins reveals regulatory mechanisms of plant innate immune responses. *The Plant Journal* 51, 931–940.

Ogasawara, Y., Kaya, H., Hiraoka, G., Yumoto, F., Kimura, S., Kadota, Y., Hishinuma, H., Senzaki, E., Yamagoe, S., Nagata, K., Nara, M., Suzuki, K., Tanokura, M. and Kuchitsu, K. (2008) Synergistic activation of the *Arabidopsis* NADPH oxidase AtrbohD by $Ca^{2+}$ and phosphorylation *Journal of Biological Chemistry* 283, 8885–8892.

Parani, M., Rudrabhatla, S., Myers, R., Weirich, H., Smith, B., Leaman, D.W. and Goldman, S.L. (2004) Microarray analysis of nitric oxide responsive transcripts in *Arabidopsis*. *Plant Biotechnology Journal* 2, 359–366.

Peck, S.C. (2003) Early phosphorylation events in biotic stress. *Current Opinion in Plant Biology* 6, 334–338.

Pedley, K.F. and Martin, G.B. (2005) Role of mitogen-activated protein kinases in plant immunity. *Current Opinion in Plant Biology* 8, 541–547.

Piedras, P., Hammond-Kosack, K.E., Harrison, K. and Jones, J.D.G. (1998) Rapid, Cf-9 and Avr9-dependent, production of active oxygen species in tobacco suspension cultures. *Molecular Plant–Microbe Interactions* 11, 1155–1166.

Polverari, A., Molesini, B., Pezzotti, M., Buonaurio, R., Marte, M. and Delledonne, M. (2003) Nitric oxide-mediated transcriptional changes in *Arabidopsis thaliana*. *Molecular Plant–Microbe Interactions* 16, 1094–1105.

Ren, D., Yang, H. and Zhang, S. (2002) Cell death mediated by mitogen-activated protein kinase pathway is associated with the generation of hydrogen peroxide in *Arabidopsis*. *Journal of Biological Chemistry* 277, 559–565.

Ren, D., Yang, K.-Y., Li, G., Liu, Y. and Zhang, S. (2006) Activation of Ntf4, a tobacco MAPK, during plant defense response and its involvement in hypersensitive response-like cell death. *Plant Physiology* 141, 1482–1493.

Rentel, M.C., Lecourieux, D., Ouaked, F., Usher, S., Petersen, L., Okamoto, H., Knight, H., Peck, S.C., Grierson, C.S., Hirt, H. and Knight, M.R. (2004) OXI1 kinase is necessary for oxidative burst-mediated signalling in *Arabidopsis*. *Nature* 427, 858–861.

Romeis, T., Piedras, P., Zhang, S., Klessig, D.F., Hirt, H. and Jones, J.D.G. (1999) Rapid Avr9- and Cf9-dependent activation of MAP kinases in tobacco cell cultures and leaves: convergence of resistance gene, elicitor, wound, and salicylate responses. *The Plant Cell* 11, 273–287.

Romero-Puertas, M.C., Perazzolli, M., Zago, E.D. and Delledonne, M. (2004) Nitric oxide signalling functions in plant–pathogen interactions. *Cellular Microbiology* 9, 795–803.

Romero-Puertas, M.C., Laxa, M., Matte, A., Zaninotto, F., Finkemeier, I., Jones, A.M.E., Perazzolli, M., Vandelle, E., Dietz, K.-J. and Delledonne, M. (2007) S-nitrosylation of peroxiredoxin II E promotes peroxynitrite-mediated tyrosine nitration. *The Plant Cell* 19, 4120–4130.

Sagi, M. and Fluhr, R. (2001) Superoxide production by plant homologues of the $gp91^{phox}$ NADPH oxidase. Modulation of activity by calcium and by tobacco mosaic virus infection. *Plant Physiology* 126, 1281–1290.

Sagi, M., Davydov, O., Orazova, S., Yesbergenova, Z., Ophir, R., Stratmann, J.W. and Fluhr, R. (2004) Plant respiratory burst oxidase homologs impinge on wound responsiveness and development in *Lycopersicon esculentum*. *The Plant Cell* 16, 616–628.

Saito, S., Yamamoto-Katou, A., Yoshioka, H., Doke, N. and Kawakita, K. (2006) Peroxynitrite generation and tyrosine nitration in defense responses in tobacco BY-2 cells. *Plant and Cell Physiology* 47, 689–697.

Sasabe, M., Soyano, T., Takahashi, Y., Sonobe, S., Igarashi, H., Itoh T.J., Hidaka, M. and Machida, Y. (2006) Phosphorylation of NtMAP65-1 by a MAP kinase down-regulates its activity of microtubule bundling and stimulates progression of cytokinesis of tobacco cells. *Genes & Development* 20, 1004–1014.

Seo, S., Sano, H. and Ohashi, Y. (1999) Jasmonate-based wound signal transduction requires activation of WIPK, a tobacco mitogen- activated protein kinase. *The Plant Cell* 11, 289–298.

Serrander, L., Jaquet, V., Bedard, K., Plastre, O., Hartley, O., Arnaudeau, S., Demaurex, N., Schlegel, W. and Krause, K.H. (2007) NOX5 is expressed at the plasma membrane and generates superoxide in response to protein kinase C activation. *Biochimie* 89, 1159–1167.

Simon-Plas, F., Elmayan, T. and Blein, J.-P. (2002) The plasma membrane oxidase NtrbohD is responsible for AOS production in elicited tobacco cells. *The Plant Journal* 31, 137–147.

Sokolovski, S. and Blatt, M.R. (2004) Nitric oxide block of outward-rectifying $K^+$ channels indicates direct control by protein nitrosylation in guard cells. *Plant Physiology* 136, 4275–4284.

Soyano, T., Nishihama, R., Morikiyo, K., Ishikawa, M. and Machida, Y. (2003) NQK1/NtMEK1 is a MAPKK that acts in the NPK1 MAPKKK-mediated MAPK cascade and is required for plant cytokinesis. *Genes & Development* 17, 1055–1067.

Stamler, J.S., Singel, D.J. and Loscalzo, J. (1992) Biochemistry of nitric oxide and its redox-activated forms. *Science* 258, 1898–1902.

Tada, Y., Mori, T., Shinogi, T., Yao, N., Takahashi, S., Betsuyaku, S., Sakamoto, M., Park, P., Nakayashiki, H., Tosa, Y. and Mayama, S. (2004) Nitric oxide and reactive oxygen species do not elicit hypersensitive cell death but induce apoptosis in the adjacent cells during the defense response of oat. *Molecular Plant–Microbe Interactions* 17, 245–253.

Takabatake, R., Ando, Y., Seo, S., Katou, S., Tsuda, S., Ohashi, Y. and Mitsuhara, I. (2007) MAP kinases function downstream of HSP90 and upstream of mitochondria in TMV resistance gene *N*-mediated hypersensitive cell death. *Plant and Cell Physiology* 48, 498–510.

Takahashi, Y., Nasir, K.H.B., Ito, A., Kanzaki, H., Matsumura, H., Saitoh H., Fujisawa, S., Kamoun, S. and Terauchi, R. (2007) A high-throughput screen of cell-death-inducing factors in *Nicotiana benthamiana* identifies a novel MAPKK that mediates INF1-induced cell death signaling and non-host resistance to *Pseudomonas cichorii*. *The Plant Journal* 49, 1030–1040.

Takeda, S., Gapper, C., Kaya, H., Bell, E., Kuchitsu, K. and Dolan, L. (2008) Local positive feedback regulation determines cell shape in root hair cells. *Science* 319, 1241–1244.

Torres, M.A. and Dangl, J.L. (2005) Functions of the respiratory burst oxidase in biotic interactions, abiotic stress and development. *Current Opinion in Plant Biology* 8, 397–403.

Torres, M.A., Onouchi, H., Hamada, S., Machida, C., Hammond-Kosack, K.E. and Jones, J.D.G. (1998) Six *Arabidopsis thaliana* homologues of the human respiratory burst oxidase (*gp91$^{phox}$*). *The Plant Journal* 14, 365–370.

Torres, M.A., Dangl, J.L. and Jones, J.D.G. (2002) *Arabidopsis* gp91$^{phox}$ homologues *AtrbohD* and *AtrbohF* are required for accumulation of reactive oxygen intermediates in the plant defense response. *Proceedings of the National Academy of Sciences, USA* 99, 517–522.

Wendehenne, D., Pugin, A., Klessig, D.F. and Durner, J. (2001) Nitric oxide: comparative synthesis and signaling in animal and plant cells. *Trends in Plant Science* 6, 177–183.

Wong, H.L., Pinontoan, R., Hayashi, K., Tabata, R., Yaeno, T., Hasegawa, K., Kojima, C., Yoshioka, H., Iba, K., Kawasaki, T. and Shimamoto, K. (2007) Regulation of rice NADPH oxidase by binding of Rac GTPase to its N-terminal extension. *The Plant Cell* 19, 4022–4034.

Yamamizo, C., Kuchimura, K., Kobayashi, A., Katou, S., Kawakita, K., Jones, J.D.G., Doke, N. and Yoshioka, H. (2006) Rewiring mitogen-activated protein kinase cascade by positive feedback confers potato blight resistance. *Plant Physiology* 140, 681–692.

Yamamoto, A., Katou, S., Yoshioka, H., Doke, N. and Kawakita, K. (2003) Nitrate reductase, a nitric oxide-producing enzyme: induction by pathogen signals. *Journal of General Plant Pathology* 69, 218–229.

Yamamoto-Katou, A., Katou, S., Yoshioka, H., Doke, N. and Kawakita, K. (2006) Nitrate reductase is responsible for elicitin-induced nitric oxide production in *Nicotiana benthamiana*. *Plant and Cell Physiology* 47, 726–735.

Yang, K.-Y., Liu, Y. and Zhang, S. (2001) Activation of a mitogen-activated protein kinase pathway is involved in disease resistance in tobacco. *Proceedings of the National Academy of Sciences, USA* 98, 741–746.

Yao, N., Eisfelder, B.J., Marvin, J. and Greenberg, J.T. (2004) The mitochondrion – an organelle commonly involved in programmed cell death in *Arabidopsis thaliana*. *The Plant Journal* 40, 596–610.

Yoshioka, H., Sugie, K., Park, H.-J., Maeda, H., Tsuda, N., Kawakita, K. and Doke, N. (2001) Induction of plant gp91 *phox* homolog by fungal cell wall, arachidonic acid, and salicylic acid in potato. *Molecular Plant–Microbe Interactions* 14, 725–736.

Yoshioka, H., Numata, N., Nakajima, K., Katou, S., Kawakita, K., Rowland, O., Jones, J.D.G. and Doke, N. (2003) *Nicotiana benthamiana* gp91$^{phox}$ homologs *NbrbohA* and *NbrbohB* participate in $H_2O_2$ accumulation and resistance to *Phytophthora infestans*. *The Plant Cell* 15, 706–718.

Zeidler, D., Zahringer, U., Gerber, I., Dubery, I., Hartung, T., Bors, W., Hutzler, P. and Durner, J. (2004) Innate immunity in *Arabidopsis thaliana*: lipopolysaccharides activate nitric oxide synthase (NOS) and induce defense genes. *Proceedings of the National Academy of Sciences, USA* 101, 15811–15816.

Zemojtel, T., Frohlich, A., Palmieri, M.C., Kolanczyk, M., Mikula, I., Wyrwicz, L.S., Wanker, E.E., Mundlos, S., Vingron, M., Martasek, P. and Durner, J. (2006) Plant nitric oxide synthase: a never-ending story? *Trends in Plant Science* 11, 524–525.

Zhang, S. and Klessig, D.F. (1997) Salicylic acid activates a 48 kD MAP kinase in tobacco. *The Plant Cell* 9, 809–824.

Zhang, S. and Klessig, D.F. (1998) *N* resistance gene-mediated *de novo* synthesis and activation of a tobacco MAP kinase by TMV infection. *Proceedings of the National Academy of Sciences, USA* 95, 7433–7438.

Zhang, S. and Liu, Y. (2001) Activation of salicylic acid-induced protein kinase, a mitogen-activated protein kinase, induces multiple defense responses in tobacco. *The Plant Cell* 13, 1877–1889.

Zhang, S., Liu, Y. and Klessig, D.F. (2000) Multiple levels of tobacco WIPK activation during the induction of cell death by fungal elicitins. *The Plant Journal* 23, 339–347.

Zhao, M.-G., Tian, Q.-Y. and Zhang, W.-H. (2007) Nitric oxide synthase-dependent nitric oxide production is associated with salt tolerance in *Arabidopsis*. *Plant Physiology* 144, 206–217.

# 4 Disease Resistance in *Arabidopsis*, Starring TGA2 and also Featuring NPR1

Patrick Boyle,[1] Pierre R. Fobert[2] and Charles Després[1]

[1]*Brock University, St Catharines, Ontario, Canada;* [2]*Plant Biotechnology Institute, Saskatoon, Saskatchewan, Canada*

## Abstract

Systemic acquired resistance (SAR) is a long-lasting, broad-spectrum disease resistance that arises throughout a plant, including non-infected tissue, upon localized exposure to a microbe that causes necrosis. Induction of SAR is accompanied by the accumulation of salicylic acid (SA) and is ultimately characterized by the upregulation of *Pathogenesis-Related* (*PR*) genes, including the SAR marker gene *PR-1*. Direct transcriptional control of *PR-1* is governed by the TGA2 clade of transcription factors. TGA2, the archetypical member of this clade, demonstrates a unique dichotomy in that it is essential for mediating the repression of *PR-1* in resting tissues, yet is also a requisite for activating this gene in SA-stimulated cells. TGA2 is at all times positioned on the *PR-1* promoter and constitutes a point of integration for the genetic regulatory information, including that transmitted by transcriptional cofactors, most notably Nonexpresser of Pathogenesis-Related (PR) genes 1 (NPR1). Although the coactivator NPR1 is recognized as THE key regulator of *PR-1* gene expression, activation of *PR-1* is contingent upon its incorporation into a transactivating enhanceosome complex nucleated by TGA2.

## 4.1 Introduction

Plants employ a variety of defences in order to combat colonization by microbial pathogens. The means of defence include constitutive physical and chemical barriers, such as those offered by way of the waxy cuticle and a variety of preformed peptides and non-proteinaceous secondary metabolites. Plants also possess inducible defensive mechanisms that are deployed in response to pathogen attacks and which require the specific recognition of elicitor molecules. Elicitors are typically pathogen products, or plant-derived molecules released by the degradation of the cell wall. The perception of an elicitor triggers a signal transduction cascade that usually involves protein kinases and phosphatases (Pedley and Martin, 2005). Through the concerted actions of these entities, the signal is transmitted to the nucleus where it effects a transcriptional reprogramming that ultimately results in the coordinated activation of a battery of defence genes. A common consequence of induced plant defence is rapid cell death at the site of infection, known as the hypersensitive response (HR), which functions to limit and confine pathogen infection (Mur *et al.*, 2008).

In addition to local defensive reactions, the plant can also mount a systemic response to microbial infections in distal uninfected tissues. One of the most studied examples of induced global resistance in *Arabidopsis* is known as systemic acquired resistance (SAR). Attack by a microbe that causes necrosis, including an HR, results in the activation of a signal transduction pathway that produces a global transcriptional reprogramming, ultimately yielding a systemic, long-lasting and broad-spectrum disease resistance state known as SAR (Pieterse and Van Loon, 2004). Among the suite of genes upregulated in this global defence programme are the *Pathogenesis-Related* (*PR*) genes. A necessary prerequisite for the establishment of SAR and *PR*-gene activation is the accumulation of the mandatory molecule salicylic acid (SA). Exogenous application of SA or its chemical analogues, 2,6-dichloroisonicotinic acid (INA) and benzothiadiazole are also sufficient for *PR*-gene activation and the deployment of SAR, in a process referred to as chemical SAR (Oostendorp *et al.*, 2001).

*PR-1* gene expression is orchestrated through the concerted efforts of *cis*-regulatory elements residing in the *PR-1* promoter region, the TGA2-containing clade of transcription factors, and most notably the coactivator Nonexpresser of Pathogenesis-Related (PR) genes 1 (NPR1) (Rochon *et al.*, 2006). The contributions of TGA transcription factors in regulating the *PR-1* locus have been overshadowed by NPR1 due to functional redundancy. However, genetic and molecular approaches have now established that these factors are essential to governing the expression of *PR-1* under both resting conditions and after SA stimulation (Zhang *et al.*, 2003; Rochon *et al.*, 2006). TGA2 is recruited directly to the *PR-1* promoter in an SA- and NPR1-independent manner (Rochon *et al.*, 2006). Under resting conditions, TGA2 functions to maintain the *PR-1* gene in a repressed state (Zhang *et al.*, 2003). TGA2 retains its capacity to repress gene activation following SA stimulation (Rochon *et al.*, 2006). However, in this situation, the NPR1 coactivator is recruited to the TGA2 transcription factor resulting in the formation of an NPR1–TGA2 transactivating complex, which is directly responsible for the activation of the *PR-1* gene.

TGA2 is THE key constituent at the *PR-1* promoter because it is required for the recruitment of both the repressive and the activating apparatus. An increasing number of studies indicate that the function of a transcription factor is greatly influenced by its environment. The ability of transcription factors to mediate gene activation and repression events is not entirely determined by instrinic properties of the factor, but is rather contingent upon the cofactors (both coactivators and corepressors) available to them and the signal input they receive from diverse sources, such as DNA regulatory elements and those signals transduced through the cellular milieu. The emergence of transcription factors as processing centres, integrating the function of both molecular sensor and transcriptional switch box, is becoming evident. The current review will address some of the recent advances in transcriptional regulation and how these views apply to the regulation of *PR-1*.

## 4.2 Opening the Locked Door to Gene Activation: the Separate States of Transcription

Central to gene regulation is the ability to manifest, maintain and modulate distinct transcriptional states. The eukaryotic promoter serves as a doorway for the basal transcription machinery (BTM), enabling this apparatus to access the transcription start site, which is a prerequisite for gene expression. Thus the state of gene activation is defined largely by the status of this doorway (Fig. 4.1).

All promoters, when present as a naked DNA template in the company of the BTM, demonstrate an inherent level of gene activity referred to as the basal level of transcription (Roeder, 2005). The doorway in this case is open and therefore permissive to transcription. The extent to which the door is open will vary considerably as a function of the DNA sequence present at the promoter (Struhl, 1999). However, it should be understood that this basal level of activation is generally not observed *in vivo* in eukaryotic systems because chromatin structures impose a non-permissive transcriptional ground state (Struhl, 1999; Roeder, 2005; Heintzman and Ren, 2007; Li *et al.*, 2007). Chromatin effectively slams the door shut on the basal transcription apparatus by rendering *cis*-elements such as the TATA box, required for the recruitment of this machinery, inaccessible.

The chromatinized promoter can be viewed as a closed door, fastened shut with a bolt, which defines the ground state of eukaryotic gene activation. Just as an already closed door cannot be closed any further, there is considerable difficulty in demonstrating that promoter transcriptional output can exist below this ground state using *in vitro* transcription systems. However, this does not preclude the possibility of further negative gene regulation or repression. A closed door cannot be further closed but it can be fortified in this closed position through the introduction of various locks. Further states of repression are achieved through the recruitment of sequence-specific DNA-binding transcription factors, known as repressors, to the promoter through *cis*-regulatory elements.

It should be noted that while even the simplest of chromatin templates are sufficient to occlude the recruitment of the BTM, nucleosomes present a relatively modest barrier to the DNA-binding activity of transcription factors (Struhl, 1999; Li *et al.*, 2007). Upon binding to regulatory elements in the proximal promoter region, repressors are able to recruit corepressor complexes that possess a multitude of chromatin-modifying activities. Some of these

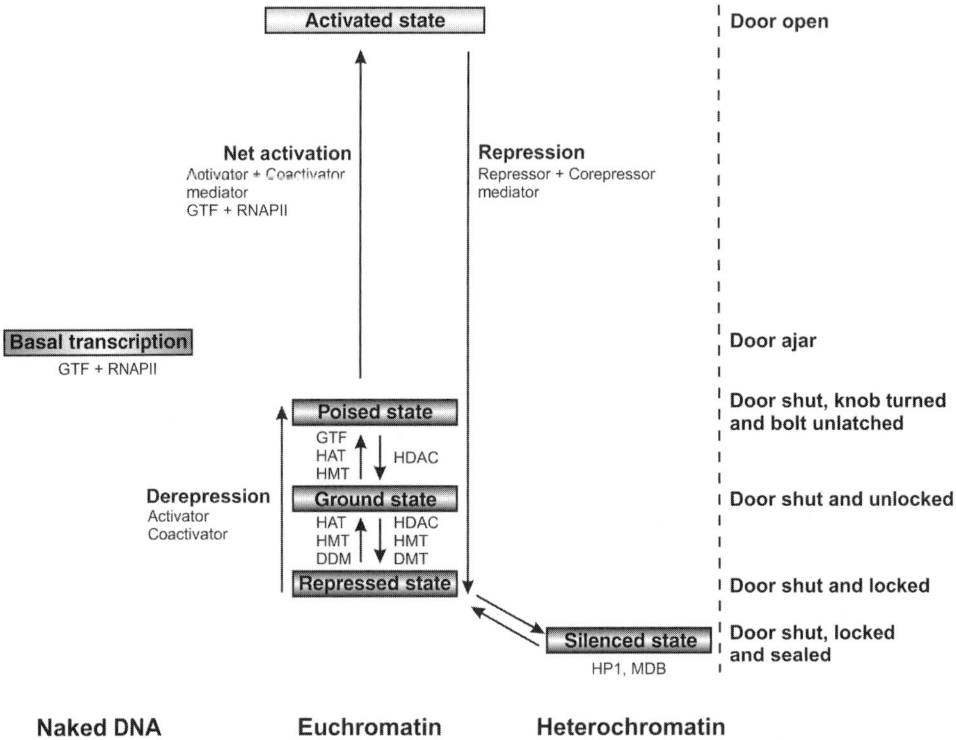


**Fig. 4.1.** States of gene regulation: the door analogy (adapted from Roeder, 2005).The basal transcription machinery (BTM) consists of the general transcription factors (GTF) and the RNA polymerase II (RNAPII). Basal transcription: A naked *in vitro* DNA template in the presence of the BTM demonstrates an intrinsic level of activation known as basal transcription. This state of gene activation is represented by a door ajar. The ground state: In eukaryotes the door is effectively maintained in a closed position by way of chromatin structures, which prevent gene activation through the occlusion of the BTM. This is the ground state. The repressed state: Just as a shut door can be locked, the chromatin barrier can be further fortified through actions of repressors that enable the recruitment of corepressors displaying histone deacetylases (HDAC), histone methyltransferase (HMT) and DNA methyltransferase (DMT) activities. The chromatin modifications mediated by these corepressors render the promoter in a repressed state. The silenced state: In addition to the occlusion of the BTM and transcriptional activators, the above chromatin modifications can also serve to recruit additional repressive entities including heterochromatin protein 1 (HP1) and methylated DNA binding proteins (MDB). The presence of these entities renders chromatin in a highly compacted and transcriptionally silent state known as heterochromatin. In this state, the door is sealed shut. Derepression: Gene activation, much like opening a locked door, is a multi-step process. Unlocking the door requires clearance of repressors and the repressive chromatin modifications from the promoter. Activators serve to recruit coactivators that boast histone acetyltransferase (HAT) and HMT and DNA demethylase (DDM) activities, which collectively contribute to the establishment of euchromatin, an open chromatin conformation permissive to the transcriptional machinery. Elements of the BTM can also be recruited prior to RNAPII creating a poised state for activation. Net activation: A fully activated state of gene expression is reached in response to the recruitment of the mediator and the complete complement of the BTM, most notably the RNAPII, to the promoter.

activities are aimed at specifically antagonizing histone modifications associated with gene activation while others work to recruit further repressive entities to the locus (Rosenfeld *et al.*, 2006). Corepressor complexes also include a family of ATP-dependent nucleosome-remodelling factors that function to further constrain chromatin structures (Rosenfeld *et al.*, 2006). It is important to note that these multi-subunit corepressor complexes can be recruited in a parallel and/or sequential manner (Rosenfeld *et al.*, 2006). The histones present in the promoters of poised and active genes are typically acetylated and phosphorylated at key residues. Corepressors commonly boast histone deacetylase (HDAC) and phosphatase activities that remove these activating marks (Roeder, 2005; Rosenfeld *et al.*, 2006; Heintzman and Ren, 2007; Li *et al.*, 2007). Other chromatin modification activities include those mediated by corepressors such as histone methyltransferases (HMTs) and histone demethylases (HDMs), which conjugate and remove methyl moieties from histone tails, respectively (Rosenfeld *et al.*, 2006; Garcia-Bassets *et al.*, 2007). Methylation of histone H3 lysines at positions 9 (H3K9) and 27 (H3K27) as well as histone H4 lysine 20 (H4K20) are associated with repression (Rosenfeld *et al.*, 2006; Li *et al.*, 2007), while methylation at H3K4 is commonly observed at the promoters of activated genes (Rosenfeld *et al.*, 2006; Garcia-Bassets *et al.*, 2007; Li *et al.*, 2007). Activities that methylate H3K9, H3K27 and H4K20 and those which demethylate H3K4 are featured among those present in corepressor complexes (Rosenfeld *et al.*, 2006; Garcia-Bassets *et al.*, 2007). Ultimately, these events occlude the recruitment of any coactivators, securing the promoter in a non-permissive state. Furthermore, they can also serve to direct the recruitment of entities that can manifest the most severely constrained and repressed chromatin structure, known as facultative heterochromatin (Rosenfeld *et al.*, 2006).

A heterochromatinized promoter is a doorway sealed shut. In this state the gene is no longer competent for activation and it is deemed transcriptionally silent. The term transcriptional silencing is rather ambiguous because it is currently employed to define a number of related yet different phenomena. When used in a strictly transcriptional context, a silent gene refers to a repressed or inactive gene. The terms 'silenced' and 'repressed' are essentially interchangeable. In the field of epigenetics, gene silencing carries a distinct connotation in that it refers to a maintained and heritable state of gene repression, which is effected through the facultative heterochromatinization of the loci (Hsieh and Fischer, 2005). In this chapter, we will be using the term 'silencing' in the epigenetic sense because it appreciates the greater state of repression that is imposed by the heterochromatin structure. From the perspective of the RNA polymerase, a heterochromatinized promoter presents a far greater barrier than that of an actively repressed promoter despite the fact that both are transcriptionally inactive, much like a doorway sealed shut is considerably more difficult to open than a locked door, even though both doorways are equally closed.

The heterochromatinization of a gene is manifested through a number of characteristic modifications at the promoter, most notably DNA cytosine methylation, primarily but not exclusively in the context of CpG dinucleotides, and histone methylation at position H3K9 (Mutskov and Felsenfeld, 2004;

Naumann *et al.*, 2005; Stancheva, 2005). These modifications demonstrate a puzzling interdependence; however, they clearly both contribute to the establishment of transcriptionally silent heterochromatinized loci (Stancheva, 2005). Beyond the interdependence of DNA methylation and H3K9 methylation, these modifications also serve in the recruitment of distinct protein entities. DNA methylation enables the recruitment of methylated DNA binding proteins (MBPs), while the H3K9 methylation mark is responsible for directing the heterochromatin protein-1 (HP1; in plants the HP1 homologue is known as like-HP1 or LHP1; Gaudin *et al.*, 2001) to the locus. These entities essentially function to constrict and compact the chromatin into the conformation known as heterochromatin.

Opening a locked door requires three separate steps: (i) unlocking the door; (ii) turning the knob to release the bolt from the latch; and (iii) finally opening the door. The activation of an actively repressed gene proceeds through a similar three-step procedure. In order to unlock a gene from a repressed state, it is necessary to alleviate the repressive chromatin modifications and structures at the promoter. Accomplishing this feat typically entails the dismissal of repressive transcription factors, allowing for the subsequent recruitment of activator(s), which are other sequence-specific DNA-binding factors that bind *cis*-elements present in the proximal promoter. There are also a number of cases in which the repressor is converted into an activator through the binding of a ligand or via a post-translational modification (Rosenfeld *et al.*, 2006).

Activators function to recruit coactivator complexes to the promoter, and much like their antagonists, the corepressors, these complexes can be placed into two distinct classes: those which serve to recruit and stabilize the transcriptional apparatus and those that effect the remodelling and modification of the chromatin (Roeder, 2005; Rosenfeld *et al.*, 2006). Members of the first class of coactivators are often referred to as adaptors. These entities form a direct bridge between the activator and the BTM. The most notable example of an adaptor is the multi-subunit mediator complex (Roeder, 2005). The mediator is conserved among most eukaryotic organisms and is a necessary component for activator-driven transcription (Roeder, 2005). Not only do the adaptor coactivators, such as the mediator, direct the recruitment of the general transcription factors and RNA polymerase II (RNAPII) to the promoter, but they also provide a means to communicate regulatory information from the activator and *cis*-regulatory elements to the transcription machinery (Roeder, 2005).

The second class of coactivator complexes, which target their activities to the chromatin, are typically grouped into two subclasses; the histone modifiers and the remodellers. The histone modifier class boasts the ability to perform a myriad of post-translational modifications, including acetylation, methylation, demethylation, phosphorylation, ubiquitylation, sumoylation and poly(ADP-ribosyl)ation, most of which are targeted to the N-terminal tails of histones H3 and H4 in the nucleosome (Rosenfeld *et al.*, 2006). Histone acetyltransferases (HATs) constitute a major component of coactivator complexes. Promoter histone hyperacetylation is a common feature of active genes (Rosenfeld *et al.*,

2006; Rando and Ahmad, 2007). This modification is proposed to facilitate gene activation by three different mechanisms. First of all, the introduction of acetyl moieties alters the net charge of nucleosome, attenuating DNA–histone interactions and ultimately rendering nucleosomes easier to displace (Li *et al.*, 2007). Secondly, acetylation of the H4K16 position has also been shown to prevent the formation of compact higher-order chromatin structures (Li *et al.*, 2007). Finally, histone acetylation provides distinct marks, or tags, that permit the recruitment of other proteins to the locus, which can facilitate various aspects of derepression and gene activation (Rosenfeld *et al.*, 2006; Heintzman and Ren, 2007; Li *et al.*, 2007). The activities associated with these coactivators are also responsible for the modifications of components of the transcriptional machinery, and such modifications control critical events in transcriptional regulation (Rosenfeld *et al.*, 2006). The second subclass of chromatin-directed coactivators, the histone remodellers, employs components of the ATP-dependent chromatin-remodelling machinery. This group includes entities such as the mating type SWItching (SWI)/sucrose non-fermentation (SNF) complex, which can compromise histone–DNA interactions in the nucleosome, enabling nucleosome sliding and eviction (Rosenfeld *et al.*, 2006; Li *et al.*, 2007; Rando and Ahmad, 2007). Such activities are essential to the displacement of nucleosomes from the TATA box, freeing this important *cis*-element for binding by the general transcription machinery (Li *et al.*, 2007; Rando and Ahmad, 2007). The remodelling machinery is also involved in histone replacement and the installment of histone variants such as H3.3 and H2A.Z, both of which are enriched in promoter regions (Li *et al.*, 2007; Rando and Ahmad, 2007). The presence of these variants is believed to primarily influence local chromatin architecture, rather than affecting the histone code-driven recruitment of ancillary factors, because the variants differ very little with respect to the sites of modification in canonical histones (Li *et al.*, 2007). It should be noted that various coactivator complexes can be recruited in parallel. However, some of these chromatin-modifying activities are required to take place first, before the recruitment of subsequent coactivators and adaptors (Struhl, 1999; Rosenfeld *et al.*, 2006). The collective efforts of the various coactivators provide a means to unlock a repressed promoter from its restricted state.

In order to open an unlocked door, you need only to turn the knob and open it. However, turning the knob is a mechanistically and possibly temporally distinct step from opening the door. The restructuring at the promoter, mediated by the coactivators, permits binding of the general transcription factors and RNAPII, giving rise to what is known as the pre-initiation complex (PIC) (Roeder, 2005; Heintzman and Ren, 2007). The assembly of the PIC renders a gene poised for activation. The mediator complex, which makes direct contact with aspects of the general transcription factors and RNAPII, plays a key role in regulating the initiation of transcription from the PIC, poised at the promoter (Roeder, 2005; Heintzman and Ren, 2007). This poised state is comparable to standing in front of a door with the knob turned and the bolt completely removed from the latch.

The final act of opening the door to gene activation begins with melting of the DNA around the transcription start site, allowing RNAPII access to the template strand, and from this point, transcription proceeds. It should be noted that activated transcription far exceeds the level of gene activity produced from the naked template and the BTM (Roeder, 2005). The collective efforts of the activators and coactivators not only alleviate the restrictive state imposed by the chromatin and repressors, but also serve to establish an environment for the optimal performance of the transcription apparatus (Roeder, 2005).

Passing through the doorway to gene activation is a complicated matter in eukaryotes because of the locked door imposed by chromatin and repressors. However, the system boasts an array of activities that can perform the separate acts of unlocking the door, releasing the latch, and opening it up wide.

## 4.3 Distinguishing Duality among Treasonous Transcription Factors

According to the conventional wisdom on transcription factors, activators recruit coactivators, resulting in gene activation, while repressors recruit corepressors, resulting in the repression of transcription. However, there are a great number of cases in which a transcription factor that activates and recruits coactivators in one instance can recruit corepressors in another (Latchman, 2001; Ma, 2005; Rosenfeld *et al.*, 2006). The term 'dual function' is assigned to many transcription factors based entirely upon their ability to mediate both gene activation and repression events. However, upon investigating the conditions under which this duality is demonstrated, it becomes apparent that there are different classes of dual-function factors. The treasonous behaviour of these transcription factors is typically demonstrated in a context- or signal-dependent manner (Latchman, 2001; Ma, 2005). The *Arabidopsis* TGA2 is one such treasonous transcription factor, as it is required for the repression of *PR-1* gene expression before stimulation with SA, but it is also necessary for *PR-1* gene induction following SA treatment (Zhang *et al.*, 2003; Rochon *et al.*, 2006).

### Context-dependent duality

The ability of a transcription factor to selectively recruit a coactivator or corepressor is not a purely intrinsic property, and is often shaped by the DNA sequence to which the factor is bound, the structure of the surrounding chromatin, and the type of molecules available in the nuclear milieu.

The dual nature of many transcription factors is promoter dependent. In these cases, a factor acts as an activator in the context of one promoter but represses in the context of another. The basis for this differential recruitment is attributed to differences in the DNA sequence of the *cis*-regulatory elements occupied by the factor (Latchman, 2001; Natoli, 2004; Ma, 2005; Rosenfeld *et al.*, 2006; Heintzman and Ren, 2007). Transcription factors tend to tolerate some amount of sequence variation in their cognate binding elements, as evidenced by their general ability to bind several degenerate sequences with a high affinity (Latchman, 2001; Natoli, 2004; Heintzman and Ren, 2007). The ability of these factors to recognize degenerate target sequences is central to their capacity to recruit different cofactors. One often neglects to consider the contributions of *cis*-regulatory elements in gene regulation. The DNA sequences in regulatory elements are much more than simply an address in the genome that is to be recognized by a specific transcription factor. DNA binding can produce drastic changes in transcription factor conformation (Natoli, 2004). The transcription factor DNA-binding domains will adopt different conformations in order to optimize interactions with a *cis*-element, and therefore different DNA sequences will have different conformational consequences (Natoli, 2004). The conformation adopted by the factor in response to DNA binding will ultimately influence the positioning and accessibility of cofactor interaction motifs in the transcription factor complex (Latchman, 2001; Natoli, 2004; Ma, 2005; Rosenfeld *et al.*, 2006; Heintzman and Ren, 2007). In essence, *cis*-elements aid in sculpting the structures and surfaces being broadcasted by DNA-bound transcription factors into the cellular milieu, directly influencing which cofactors will be recruited to the locus. This phenomenon is evidenced by the work of Leung *et al.* (2004) in which it was demonstrated that a single nucleotide mutation in the binding site for the NF-κB transcription factor results in the recruitment of a coactivator complex different from the one normally recruited when NF-κB is bound to the unmodified promoter element. A true example of promoter-dependent transcription factor duality is demonstrated by the glucocorticoid receptor (GR). This Nuclear Receptor (NR) transcription factor only binds its *cis*-regulatory elements in response to treatment with the corresponding hormone (Latchman, 2001). The steroid-bound transcription factor binds two different *cis*-elements termed GRE (glucocorticoid response element) and nGRE (negative GRE). With the former, the factor binds the element as a dimer, which results in gene activation. However, GR binds the latter as a trimer and this entity represses gene expression.

A number of factors are reported to demonstrate cell- or tissue-dependent dual activator/repressor function. However, these opposing behaviours are often manifested on different *cis*-elements and therefore, technically, constitute examples of promoter-dependent duality. That being said, there are also instances in which the capacity of a transcription factor to activate or repress a given promoter is dictated in an entirely cell-or tissue-dependent manner. The mammalian HES-1 (Hairy and Enhancer of Split 1) factor demonstrates cell-type-dependent dual function. This basic helix-loop-helix (bHLH) transcription factor acts as a repressor of the human acid α-glucosidase (GAA) gene through a 25-bp silencer element in Hep G2 cells. However, this same promoter element was found to function as an enhancer in human fibroblast cells. The level of gene activation was increased as a result of overexpressing the HES-1 factor, while deletion of the HES-1 binding site in the GAA 25-bp promoter element abrogated gene activation (Yan *et al.*, 2002). The Pit-1 (Pituitary-1) is a tissue-specific transcription factor, which demonstrates both promoter- and cell-dependent dual activator/repressor functions. The factor is required to activate the expression of *growth hormone 1* (*GH1*) in one somatotrope cell type, yet acts to repress *GH1* expression in lactotrope cells (Scully *et al.*, 2000).

Regulation of cell- or tissue-specific genes is often governed by cell-type-specific transcription factors and cofactors (Ren and Liao, 2001; Hochheimer and Tjian, 2003; Taatjes *et al.*, 2004). The ability of a transcription factor to function as an activator or repressor can be entirely the consequence of the unique complement of factors and cofactors expressed in a particular cell type (Ren and Liao, 2001; Ma, 2005). The tissue specificity of transcription factors can be manifested through competitions among these factors for certain *cis*-regulatory elements and cofactors in the target tissue type. The AP-2 (Activator Protein-2) is so-named for its ability to activate transcription. However, this factor is necessary for the repression of the *Serum Amyloid A1* (*SAA1*) gene in non-hepatic cells. The activation of the *SAA1* gene requires the transcription factor NF-κB. In this case, the NF-κB-binding site overlaps with that of the AP-2 in the *SAA1* promoter. Protein binding experiments demonstrated that the interaction of AP-2 or NF-κB with this overlapping binding site is mutually exclusive (Ren and Liao, 2001). It was also shown that the ability to repress the *SAA1* promoter activation in HeLa cells was contingent upon the presence of the AP-2-binding element (Ren and Liao, 2001). In this situation, a tissue-specific transcription factor, AP-2, serves to prevent the aberrant expression of a liver-specific gene in non-hepatic cells by displacing the activator NF-κB from its enhancer element. The prototypical dual-function transcription factor YY1 (Yin Yang 1) is proposed to mediate the repression of some genes by way of a very similar mechanism. However, this is only one of many means by which this factor can negatively regulate gene expression (Ma, 2005; Gordon *et al.*, 2006).

The competition among transcription factors extends to cofactors. The availability of these cofactors can be a key determinant of transcription factor behaviour (Ma, 2005). This concept of limiting concentrations of coactivators affecting gene regulation programmes is based largely on what is observed in the Rubenstein–Taybi syndrome (Rosenfeld *et al.*, 2006). This disorder, which is characterized by severe development abnormalities, arises as a result of haplo-insufficiency of the ubiquitous coactivator CBP (CREB binding protein, also known as p300), meaning that only half of the normal amount of this HAT-containing coactivator is present in the cells. Further supporting the notion that cofactor concentration can dictate transcription factor function can be found in the Wnt signalling pathway. Typically, following activation of the canonical Wnt pathway, the β-catenin coactivator is translocated from the cytosol to the nucleus, where it forms, along with the Leucocyte Enhancer Factor (LEF)/T Cell Factor (TCF) transcription factor, a transactivating complex that activates the expression of a number of genes (Kikuchi *et al.*, 2006). However, when non-TCF/LEF transcription factors are present at high concentrations, they can compete for interaction with β-catenin, yielding a very different transcriptional programme (Rosenfeld *et al.*, 2006).

Another example of how cofactor availability can dictate the function of a dual-acting transcription factor can be seen in the regulation of the adeno-associated virus (AAV) P5 promoter by the YY1 factor. As previously mentioned, YY1 is the prototypical dual-function transcription factor, and in this case, it mediates repression of AAV P5. However, coinfection with adenovirus results in the production of the adenovirus coactivator Early 1A (E1A) (Chang *et al.*, 1989). The E1A coactivator is recruited to the AAV P5 promoter in an YY1-dependent manner. The E1A and YY1 collectively recruit the p300 HAT coactivator complex, resulting in the activation of the AAV P5 locus. YY1 is known to exert transcriptional activation and repression through a number of different mechanisms and to mediate interactions with both HAT and HDAC cofactors (reviewed in Thomas and Seto, 1999; Gordon *et al.*, 2006). The means by which the E1A is able to convert YY1 from a repressor to an activator is unclear. However, it has been proposed that the interaction with E1A elicits a conformational change in the transcription factor that masks the repression motif while unveiling concealed activation domains (Gordon *et al.*, 2006).

The concentration of a transcription factor itself can also govern if the factor will function as an activator or a repressor at a given promoter. The Kruppel (Kr) zinc finger protein is an example of such a transcription factor. At low concentrations, Kr binds DNA as a monomer which activates transcription. However, at high concentrations, the transcription factor forms a homodimer, which binds the same DNA sequence as the monomeric species, but functions exclusively as a repressor (Sauer and Jackle, 1993; Sauer *et al.*, 1995).

Many transcription factors boast dual functions. However, the ability of a transcription factor to affect gene activation or repression is rarely inherent to the factor and is most often owed to its environment, as defined by the regulatory elements upon which it sits, the other DNA-binding factors that surround it, and the constellation of cofactors available to it.

## Signal-dependent duality

The ability of a dual-acting transcription factor to switch from a repressor to an activator or vice versa can be regulated in a signal-dependent manner. This behaviour is clearly demonstrated by the NR family of transcription factors. The ability of these factors to recruit HAT coactivators is typically contingent upon their binding of a ligand (Ma, 2005; Rosenfeld *et al.*, 2006). The ligands include a number of steroids and hormone species (Ma, 2005). In the absence of their cognate ligands, the NR transcription factors mediate the recruitment of HDAC corepressor complexes through interactions with the Nuclear Receptor-coRepressor (N-coR) and Silencing Mediator for Retinoid and Thyroid Receptors (SMRT) components (Ma, 2005; Rosenfeld *et al.*, 2006). The differential recruitment of cofactors mediated by the ligand-bound and unbound species is attributed to conformational changes in the NR-cofactor interaction interface induced by ligand binding (Ma, 2005).

Plants do not possess NR transcription factors. However, the duality of many other classes of eukaryotic transcription factors is also regulated in a signal-dependent manner, but not as directly as that observed with NR factors. Signal transduction pathways often result in the post-translational modification of transcription factors. Modifications, such as phosphorylation and sumoylation, can effect the conversion of repressor to activator and activator to repressor, respectively (Ma, 2005). For example, the CCAAT/Enhancer Binding Protein β (C/EBPβ), a basic leucine-zipper (bZIP) transcription factor, is a component of the Ras signal transduction pathway. C/EBPβ is converted from a transcriptional repressor to activator following Ras-dependent phosphorylation (Mo *et al.*, 2004). It is important to note that both the repressor and the activator functions of this factor are exerted at the same locus through the same binding site in the promoter (Mo *et al.*, 2004). Both the repressive and activating forms of C/EBPβ recruit the Mediator adaptor complex. However, following Ras-dependent phosphorylation of the transcription factor, a component of the Mediator, the Trap230/Trap240/CDK8/cyclinC subcomplex, known to be recruited to repressed genes, was absent from the complex (Conaway *et al.*, 2005). The phosphorylated and unphosphorylated forms of C/EBPβ interact with different subunits of the Mediator (Mo *et al.*, 2004). It is proposed that the conformation adopted by the Mediator complex, in the presence of the phosphorylated C/EBPβ, destabilizes the interaction between the Mediator core subunits and the Trap230/Trap240/CDK8/cyclinC subcomplex, resulting in the detachment of this repressive component (Mo *et al.*, 2004).

The ability of the Sp3 (Specificity Protein 3) zinc finger transcription factor to act as either a repressor or an activator is contingent upon an interplay between sumoylation and acetylation (Valin and Gill, 2007). Sp3 must be sumoylated in order to function as a repressor, while acetylation is required for strong activation (Valin and Gill, 2007). SUMO (Small Ubiquitin-like Modifier) is a 101-amino acid peptide that is conjugated to a lysine residue in the target protein through a process similar to ubiquitylation (Verger *et al.*, 2003). Interestingly, the expression of SUMO as a translational fusion with the GAL4 DNA-binding domain is sufficient to repress transcription in reporter gene assays (Verger *et al.*, 2003). This modification is proposed to serve as a platform that aids in the recruitment of HDAC-containing corepressors. However, there is also evidence for HDAC-independent SUMO-mediated repression (Valin and Gill, 2007).

The signal-dependent class of dual-acting transcription factors functions as molecular sensors, enabling the modulation of transcription programmes in response to various stimuli. Essential to performing this role is the conformational diversity that these factors boast, a potential that is bolstered by their ability to accommodate various types and combinations of post-translational modifications. These modifications serve to further diversify the interaction and recruitment motifs offered by the transcription factors.

Contrary to conventional beliefs, not all transcription factors behave as agents that mechanically bind a DNA sequence and recruit coactivators or corepressors based simply on their exclusive nature as either activator or repressor. While there are some examples of transcription factors that go about ignorantly imposing their function upon a gene, there are also factors that formulate their function as a result of their environment as well as others that serve as molecular sensors that can switch functions in response to a single signal. It is the collective action of these various classes of factors that coordinate the diverse yet precise transcriptional programmes in response to complex stimuli.

## 4.4 The Treasonous Nature of TGA2 is Critical to *PR-1* Gene Regulation

*PR-1* gene activation is a molecular marker for the induction of SAR. The expression of this locus is controlled through the concerted efforts of the coactivator NPR1 and the functionally redundant TGA2-containing clade (TGA2, TGA5 and TGA6) of transcription factors. In the present paradigm for *PR*-gene expression, NPR1 is recognized as the key positive regulator of *PR-1* induction. However, the TGA2 transcription factor plays an essential role in both the activation of this locus following stimulation with SA and its basal repression in resting cells. The duality demonstrated by TGA2 is critical to the regulation of *PR-1* under both resting and inducing conditions.

In resting cells, the *PR-1* gene is maintained in a repressed state by the TGA2-clade of transcription factors (Zhang *et al.*, 2003; Rochon *et al.*, 2006). A role in *PR-1* gene repression for the TGA2 transcription factor was first indicated when elevated levels of *PR-1* expression were observed in the *tga2/5/6* triple-knockout *Arabidopsis* mutant under non-inducing conditions (Zhang *et al.*, 2003). The repression function of TGA2 was definitively demonstrated through the use of an *in planta* transcription assay (Rochon *et al.*, 2006). This system showed that TGA2 could repress an activated reporter gene through the heterologous GAL4 DNA-binding domain. This system was further used to demonstrate that the native TGA2 factor could also repress reporter gene expression in the context of the *PR-1* promoter (Rochon *et al.*, 2006). These data suggest that the conformation adopted by the factor upon binding its cognate *cis*-element through its endogenous DNA-binding (DB) domain or that produced upon recruitment to the GAL4 upstream activating sequence (UAS) through the heterologous GAL4 DB domain are both sufficient to mediate the active repression of the reporter gene. However, it is not known if this repression is conducted by way of a conserved mechanism.

Linker scanning (LS) mutagenesis of the *PR-1* promoter identified the presence of both positive and negative *cis*-regulatory elements (Lebel *et al.*, 1998). TGA2 has been shown to bind the LS5 and LS7 promoter elements *in vitro* (Després *et al.*, 2000). The LS5 appears to contribute to the negative regulation of *PR-1* expression both in the absence and in the presence of SA, whereas LS7 is required for SA-mediated induction of *PR-1* (Lebel *et al.*, 1998). Chromatin immunoprecipitation (ChIP) studies have demonstrated that TGA2 is recruited to the *PR-1* promoter in resting cells and notably this recruitment does not require NPR1 (Rochon *et al.*, 2006). Due to limitations in resolution in the ChIP technique, these studies provided no indication as to which of the *PR-1* promoter *cis*-regulatory elements is occupied by the transcription factor.

In resting cells, the NPR1 protein is localized to both the nucleus and the cytosol (Després *et al.*, 2000). Through the use of the ChIP technique, it was revealed that the NPR1 coactivator is specifically present in the regulatory region of the *PR-1* gene under non-inducing conditions, and it further demonstrated that the recruitment of the coactivator to this locus is independent of the TGA2 clade of transcription factors (Rochon *et al.*, 2006). NPR1, like most coactivators, lacks a known DNA-binding domain and is therefore likely to be maintained at the repressed *PR-1* promoter by way of another protein. However, there is currently no information as to what the NPR1-anchoring entity might be, nor is there any indication of the function of NPR1 in this situation. The presence of other coactivators at unactivated or repressed promoters, although uncommon, has also been reported in *Drosophila* using the ChIP technique (Martinez and Arnosti, 2008). The presence of coactivators such as NPR1 at repressed promoters does not conform to the existing paradigm for gene regulation. However, it could be reasoned that the proximity of these latent coactivators to *cis*-regulatory elements renders them perfectly poised to activate gene expression in response to the appropriate cue. Despite the presence of NPR1 at the *PR-1* promoter in resting cells, *npr1* mutations do not affect the basal repression of the locus. Only the *tga2/5/6* triple-knockout mutant demonstrates derepression of the *PR-1* gene under non-inducing conditions (Zhang *et al.*, 2003).

The requirement of NPR1 for the induction of SAR and the activation of *PR-1* in response to SA is well documented by numerous different genetic screens (Cao *et al.*, 1994; Delaney *et al.*, 1995; Zhang *et al.*, 2003). However, both the deployment of SAR and the expression of *PR-1* also require the TGA2 clade of transcription factors. The necessity of these transcription factors in the activation of *PR-1* and the deployment of SAR was not identified in the genetic screens due to the functional redundancy between TGA2, TGA5 and TGA6. The critical role of these factors in *PR-1* regulation is in many ways undersold by their redundancy. The ability of the TGA transcription factors to interact with NPR1, both in the nucleus and *in vitro*, suggested a role in *PR-1* activation (Després *et al.*, 2000; Fan and Dong, 2002). However, the specific requirement of the TGA2 clade of factors in the activation of *PR-1* was not appreciated until the development of the *tga2/5/6* triple-knockout mutant (Zhang *et al.*, 2003).

In SA-stimulated cells, like in resting cells, TGA2 is recruited to the *PR-1* promoter in an NPR1-independent manner. The condition-invariant binding of the *PR-1* regulatory region demonstrated by TGA2 is reminiscent of a behaviour that is also exhibited by another bZIP transcription factor, HY5 (Lee *et al.*, 2007). ChIP studies have shown that HY5, the key positive regulator of photomorphogenesis, is constitutively bound to a multitude of light-induced loci and its recruitment is not affected by light conditions or light-to-dark transitions (Lee *et al.*, 2007).

Although TGA2 and NPR1 are both present in the nucleus at the *PR-1* promoter before SA treatment, the plant two-hybrid assay demonstrated that these factors do not interact until after stimulation with SA (Rochon *et al.*, 2006). In the current model for *PR-1* activation following stimulation with SA,

NPR1 is incorporated into a transactivating complex with the TGA2 transcription factor, which nucleates the formation of an enhanceosome at the *PR-1* promoter. It is important to note that the capacity for TGA2 to mediate repression is not directly affected by treatment with SA. This property of TGA2 was demonstrated through the use of an *in planta* transcription assay. In the absence of a functional NPR1, TGA2 continued to repress in the context of both the heterologous GAL4 UAS and *PR-1* promoters in SA-stimulated cells (Rochon *et al.*, 2006). Such an observation casts some doubts on the possibility that the ability of TGA2 to activate or repress transcription is modulated by a simple switch-type mechanism mediated by a post-translational modification stimulated by SA treatment. Further supporting this viewpoint is a previous study that demonstrated that the TGA2 transcription factor is phosphorylated by a casein kinase (CK)2-type kinase activity, which emerges following SA treatment. However, mutation of the phosphorylated residues in TGA2 did not affect the ability of the factor to activate *PR-1* expression in response to SA stimulation (Kang and Klessig, 2005). The activator function of TGA2 is only realized when complexed with NPR1. However, the TGA2 activator function is not confined solely to the *PR-1* promoter, since TGA2 can also activate gene expression in a heterologous context through the GAL4 DB in an SA- and NPR1-dependent manner (Rochon *et al.*, 2006). The ability of TGA2 to manifest repressor/activator duality in these two unrelated contexts might suggest that the function of TGA2 is modulated minimally through DNA-binding allosteric effects. It is quite possible that the DNA binding mediated through its endogenous DNA-binding domain or by way of the GAL4 DB domain produces equivalent changes in the NPR1–TGA2 complex conformation, resulting in a common means of activation. However, it is not possible to rule out the possibility that the complex adopts different conformations in these two contexts and that activation proceeds through different mechanisms. In this case, despite the dramatic difference in conformation changes imposed by the binding of different *cis*-elements, TGA2 would still be able to maintain the surfaces required to mediate repression and those necessary to recruit NPR1, ultimately effecting activation (Latchman, 2001; Natoli, 2004; Ma, 2005). The ability of TGA2 to maintain these interfaces in different contexts would enable the factor to retain its transcriptional duality.

The NPR1 coactivator is constitutively present at the *PR-1* promoter. However, it appears to only activate *PR-1* gene expression in SA-stimulated cells and requires the TGA2 clade of transcription factors. Recruitment of NPR1 to a heterologous promoter by way of the GAL4 DB domain is able to activate expression of a reporter gene *in planta*, but only in response to SA treatment (Rochon *et al.*, 2006). Based on these observations, it would appear that the ability of NPR1 to function as a coactivator *in planta* is controlled by its recruitment to an appropriate *cis*-regulatory element by way of a DNA-binding entity and the SA-dependent stimulation of NPR1 coactivator activity. The requirement for SA to awaken the coactivator capacity of NPR1 was first indicated by the fact that transgenic lines overexpressing NPR1 do not demonstrate constitutive *PR-1* expression (Cao *et al.*, 1998). These over-expressing lines, despite the abundance of NPR1, still require SA treatment to effect *PR-1* activation (Cao *et al.*, 1998). The agent which directly delivers the SA-dependent signal to the latent NPR1 and the modification or switch it produces in the coactivator have not yet been identified. These data suggest that NPR1 could serve as a sensor in the SA signalling pathway, creating a situation in which, upon perception of SA by the cell, a signal is transduced that results in the stimulation of NPR1, rendering it accessible and competent for recruitment to the repressor TGA2. This cascade of events culminates in the formation of the TGA2–NPR1 transactivating entity which activates *PR-1* transcription. The presence of multiple sensor elements among the transcription factor–cofactor apparatus, which is able to readily manifest dramatic yet highly directed changes in transcriptional output, is an increasingly popular theme in eukaryotic gene regulation paradigms and one that appears to apply to the regulation of *PR-1* in *Arabidopsis*. Our current understanding of *PR-1* regulation is summarized and depicted in Fig. 4.2.

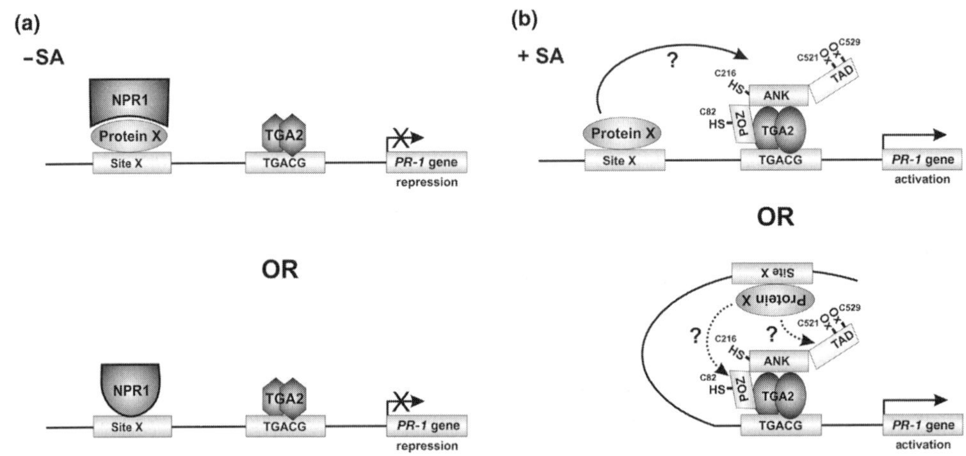


**Fig. 4.2.** The dual function of TGA2 at the *PR-1* promoter. (a) In resting cells TGA2 is recruited to a TGACG motif in the *PR-1* promoter in an NPR1-independent manner. TGA2 is required to maintain the *PR-1* gene in a repressed state. The coactivator NPR1 is also recruited to the unactivated *PR-1* promoter in a TGA2-independent manner at a site yet to be identified (Site X). However, it is not known if this recruitment is mediated by way of an unknown protein (Protein X) or via an uncharacterized DNA binding domain in the NPR1 protein. (b) In salicylic acid (SA)-stimulated cells, TGA2 recruits the coactivator NPR1 and collectively these factors contribute to the formation of an enhanceosome responsible for the activated expression of the *PR-1* gene. It is not understood if NPR1 maintains contacts with the agent or site of initial recruitment (Protein X or Site X) following SA stimulation and incorporation into the NPR1–TGA2 enhanceosome. Transactivation by the enhanceosome requires the oxidation of cysteines (C) 521 and 529 located in the transactivation domain (TAD) of NPR1 (adapted from Rochon *et al.*, 2006).

## Acknowledgements

We thank Ms Jee Yan Chu for editorial assistance. Research in our labs is supported by the NRC PBI core funding (P.R.F.), the National Science and Engineering Research Council (NSERC) discovery grant programme (C.D., P.R.F.), the Canada Foundation for Innovation (C.D.), the Ontario Innovation Trust (C.D.) and the NSERC graduate scholarship programme (P.B.).

## References

Cao, H., Bowling, S.A., Gordon, A.S. and Dong, X. (1994) Characterizaton of an *Arabidopsis* mutant that is nonresponsive to inducers of systemic acquired resistance. *The Plant Cell* 6, 1583–1592.

Cao, H., Li, X. and Dong, X. (1998) Generation of broad-spectrum disease resistance by overexpression of an essential regulatory gene in systemic acquired resistance. *Proceedings of the National Academy of Sciences, USA* 95, 6531–6536.

Chang, L.S., Shi, Y. and Shenk, T. (1989) Adeno-associated virus P5 promoter contains an adenovirus E1A-inducible element and a binding site for the major late transcription factor. *Journal of Virology* 63, 3479–3488.

Conaway, R.C., Sato, S., Tomomori-Sato, C., Yao, T. and Conaway, J.W. (2005) The mammalian Mediator complex and its role in transcriptional regulation. *Trends in Biochemical Sciences* 30, 250–255.

Delaney, T.P., Friedrich, L. and Ryals, J.A. (1995) *Arabidopsis* signal transduction mutant defective in chemically and biologically induced disease resistance. *Proceedings of the National Academy of Sciences, USA* 92, 6602–6606.

Després, C., DeLong, C., Glaze, S., Liu, E. and Fobert, P.R. (2000) The *Arabidopsis* NPR1/NIM1 protein enhances the DNA binding activity of a subgroup of the TGA family of bZIP transcription factors. *The Plant Cell* 12, 279–290.

Fan, W. and Dong, X. (2002) *In vivo* interaction between NPR1 and transcription factor TGA2 leads to salicylic acid-mediated gene activation in *Arabidopsis*. *The Plant Cell* 14, 1377–1389.

Garcia-Bassets, I., Kwon, Y.S., Telese, F., Prefontaine, G.G., Hutt, K.R., Cheng, C.S., Ju, B.G., Ohgi, K.A., Wang, J., Escoubet-Lozach, L., Rose, D.W., Glass, C.K., Fu, X.D. and Rosenfeld, M.G. (2007) Histone methylation-dependent mechanisms impose ligand dependency for gene activation by nuclear receptors. *Cell* 128, 505–518.

Gaudin, V., Libault, M., Pouteau, S., Juul, T., Zhao, G., Lefebvre, D. and Grandjean, O. (2001) Mutations in *LIKE HETEROCHROMATIN PROTEIN 1* affect flowering time and plant architecture in *Arabidopsis*. *Development* 128, 4847–4858.

Gordon, S., Akopyan, G., Garban, H. and Bonavida, B. (2006) Transcription factor YY1: structure, function, and therapeutic implications in cancer biology. *Oncogene* 25, 1125–1142.

Heintzman, N.D. and Ren, B. (2007) The gateway to transcription: identifying, characterizing and understanding promoters in the eukaryotic genome. *Cellular and Molecular Life Sciences* 64, 386–400.

Hochheimer, A. and Tjian, R. (2003) Diversified transcription initiation complexes expand promoter selectivity and tissue-specific gene expression. *Genes & Development* 17, 1309–1320.

Hsieh, T.F. and Fischer, R.L. (2005) Biology of chromatin dynamics. *Annual Review of Plant Biology* 56, 327–351.

Kang, H.G. and Klessig, D.F. (2005) Salicylic acid-inducible *Arabidopsis* CK2-like activity phosphorylates TGA2. *Plant Molecular Biology* 57, 541–557.

Kikuchi, A., Kishida, S. and Yamamoto, H. (2006) Regulation of Wnt signaling by protein–protein interaction and post-translational modifications. *Experimental and Molecular Medicine* 38, 1–10.

Latchman, D.S. (2001) Transcription factors: bound to activate or repress. *Trends in Biochemical Sciences* 26, 211–213.

Lebel, E., Heifetz, P., Thorne, L., Uknes, S., Ryals, J. and Ward, E. (1998) Functional analysis of regulatory sequences controlling *PR-1* gene expression in *Arabidopsis*. *The Plant Journal* 16, 223–233.

Lee, J., He, K., Stolc, V., Lee, H., Figueroa, P., Gao, Y., Tongprasit, W., Zhao, H., Lee, I. and Deng, X.W. (2007) Analysis of transcription factor HY5 genomic binding sites revealed its hierarchical role in light regulation of development. *The Plant Cell* 19, 731–749.

Leung, T.H., Hoffmann, A. and Baltimore, D. (2004) One nucleotide in a kappaB site can determine cofactor specificity for NF-kappaB dimers. *Cell* 118, 453–464.

Li, B., Carey, M. and Workman, J.L. (2007) The role of chromatin during transcription. *Cell* 128, 707–719.

Ma, J. (2005) Crossing the line between activation and repression. *Trends in Genetics* 21, 54–59.

Martinez, C.A. and Arnosti, D.N. (2008) Spreading of a corepressor linked to action of long-range repressor hairy. *Molecular and Cellular Biology* 28, 2792–2802.

Mo, X., Kowenz-Leutz, E., Xu, H. and Leutz, A. (2004) Ras induces mediator complex exchange on C/EBP beta. *Molecular Cell* 13, 241–250.

Mur, L.A., Kenton, P., Lloyd, A.J., Ougham, H. and Prats, E. (2008) The hypersensitive response; the centenary is upon us but how much do we know? *Journal of Experimental Botany* 59, 501–520.

Mutskov, V. and Felsenfeld, G. (2004) Silencing of transgene transcription precedes methylation of promoter DNA and histone H3 lysine 9. *EMBO Journal* 23, 138–149.

Natoli, G. (2004) Little things that count in transcriptional regulation. *Cell* 118, 406–408.

Naumann, K., Fischer, A., Hofmann, I., Krauss, V., Phalke, S., Irmler, K., Hause, G., Aurich, A.C., Dorn, R., Jenuwein, T. and Reuter, G. (2005) Pivotal role of AtSUVH2 in heterochromatic histone methylation and gene silencing in *Arabidopsis*. *EMBO Journal* 24, 1418–1429.

Oostendorp, M., Kunz, W., Dietrich, B. and Staub, T. (2001) Induced disease resistance in plants by chemicals. *European Journal of Plant Pathology* 107, 19–28.

Pedley, K.F. and Martin, G.B. (2005) Role of mitogen-activated protein kinases in plant immunity. *Current Opinion in Plant Biology* 8, 541–547.

Pieterse, C.M. and Van Loon, L. (2004) NPR1: the spider in the web of induced resistance signaling pathways. *Current Opinion in Plant Biology* 7, 456–464.

Rando, O.J. and Ahmad, K. (2007) Rules and regulation in the primary structure of chromatin. *Current Opinion in Cell Biology* 19, 250–256.

Ren, Y. and Liao, W.S. (2001) Transcription factor AP-2 functions as a repressor that contributes to the liver-specific expression of serum amyloid A1 gene. *Journal of Biological Chemistry* 276, 17770–17778.

Rochon, A., Boyle, P., Wignes, T., Fobert, P.R. and Després, C. (2006) The coactivator function of *Arabidopsis* NPR1 requires the core of its BTB/POZ domain and the oxidation of C-terminal cysteines. *The Plant Cell* 18, 3670–3685.

Roeder, R.G. (2005) Transcriptional regulation and the role of diverse coactivators in animal cells. *FEBS Letters* 579, 909–915.

Rosenfeld, M.G., Lunyak, V.V. and Glass, C.K. (2006) Sensors and signals: a coactivator/corepressor/epigenetic code for integrating signal-dependent programs of transcriptional response. *Genes & Development* 20, 1405–1428.

Sauer, F. and Jackle, H. (1993) Dimerization and the control of transcription by Kruppel. *Nature* 364, 454–457.
Sauer, F., Fondell, J.D., Ohkuma, Y., Roeder, R.G. and Jackle, H. (1995) Control of transcription by Kruppel through interactions with TFIIB and TFIIE beta. *Nature* 375, 162–164.
Scully, K.M., Jacobson, E.M., Jepsen, K., Lunyak, V., Viadiu, H., Carriere, C., Rose, D.W., Hooshmand, F., Aggarwal, A.K. and Rosenfeld, M.G. (2000) Allosteric effects of Pit-1 DNA sites on long-term repression in cell type specification. *Science* 290, 1127–1131.
Stancheva, I. (2005) Caught in conspiracy: cooperation between DNA methylation and histone H3K9 methylation in the establishment and maintenance of heterochromatin. *Biochemistry and Cell Biology* 83, 385–395.
Struhl, K. (1999) Fundamentally different logic of gene regulation in eukaryotes and prokaryotes. *Cell* 98, 1–4.
Taatjes, D.J., Marr, M.T. and Tjian, R. (2004) Regulatory diversity among metazoan co-activator complexes. *Nature Reviews, Molecular Cell Biology* 5, 403–410.
Thomas, M.J. and Seto, E. (1999) Unlocking the mechanisms of transcription factor YY1: are chromatin modifying enzymes the key? *Gene* 236, 197–208.
Valin, A. and Gill, G. (2007) Regulation of the dual-function transcription factor Sp3 by SUMO. *Biochemical Society Transactions* 35, 1393–1396.
Verger, A., Perdomo, J. and Crossley, M. (2003) Modification with SUMO. A role in transcriptional regulation. *EMBO Reports* 4, 137–142.
Yan, B., Raben, N. and Plotz, P.H. (2002) Hes-1, a known transcriptional repressor, acts as a transcriptional activator for the human acid alpha-glucosidase gene in human fibroblast cells. *Biochemical and Biophysical Research Communications* 291, 582–587.
Zhang, Y., Tessaro, M.J., Lassner, M. and Li, X. (2003) Knockout analysis of *Arabidopsis* transcription factors *TGA2*, *TGA5*, and *TGA6* reveals their redundant and essential roles in systemic acquired resistance. *The Plant Cell* 15, 2647–2653.

# 5 Disease Resistance Genes: Form and Function

MELANIE A. SACCO[1] AND PETER MOFFETT[2]

[1]*California State University, Fullerton, California, USA* [2] *Université de Sherbrooke, Sherbrooke, Québec, Canada*

## Abstract

Plants present numerous barriers to potential pathogens including structural hindrances such as waxy cuticles. Furthermore, plants, like all multicellular organisms, have evolved multiple layers of defences based on the recognition of pathogens via germline encoded receptor-like proteins. The genes encoding these receptor-like molecules confer recognition of specific pathogens or pathogen isolates and are often highly variable, resulting in cultivar or ecotype-specific differences in resistance to pathogens. In addition, many pathogens need to utilize host cellular mechanisms for their own purposes, and produce proteins that interact with host proteins to do so. However, plants also possess diversity in these factors and incompatibility between host and pathogen proteins can lead to a lack of susceptibility. Resistance to pathogens is often determined by variation at single genetic loci encoding either factors mediating active recognition of, or susceptibility to the pathogen. The genes encoded at these variable loci are broadly known as disease resistance (*R*) genes and encode multiple classes of R proteins. Recent advances in molecular genetics have lead to insights into the mechanisms by which plants either prevent pathogens from infecting them in the first place, or actively recognize and eliminate pathogens.

## 5.1 Introduction

Plants are hosts to a broad spectrum of parasitic organisms including viruses, prokaryotes (bacteria), eukaryotes (fungi and oomycetes), and highly complex multicellular parasites. In order to effectively colonize a plant, potential pathogens must overcome a number of hurdles, including physical barriers such as waxy cuticles and cell walls. For biotrophic pathogens – those that require living host tissue – pathogens must be able to interact with the host cellular machinery in order to manipulate it to their advantage. Once past these barriers, the pathogen must contend with the various layers of cell-autonomous defence mechanisms that plants have evolved in the absence of specialized circulating immune cells. Most plants are resistant to most pathogens by mechanisms that traditionally have been generally referred to as basal resistance. Basal resistance is very often mediated via the detection of pathogen-associated molecular patterns (PAMPs). PAMPs include molecules that are associated with large classes of pathogens, such as lipopolysaccharide and chitin, the cell wall components of bacteria and fungi, as well as bacterial flagellin or virus-derived double-stranded RNA (Nurnberger *et al.*, 2004). This broad-spectrum recognition induces relatively low-intensity responses, referred to as PAMP-triggered immunity (PTI), that none the less provide protection against the majority of pathogens (Chisholm *et al.*, 2006). Host-adapted pathogens, however, are able to suppress PTI, often by delivering so-called effector proteins into the host cytoplasm that interfere with PTI signalling. In turn, plants have evolved mechanisms to recognize specific pathogen effector proteins. This recognition induces much stronger responses than PTI and is known as effector-triggered immunity (ETI) (Chisholm *et al.*, 2006). Although the components and recognition capacities of PTI receptors appear to be homogeneous in a given species, the recognition specificities of ETI are often highly variable both within and between populations of the same species. This chapter discusses the natural variation that occurs in host compatibility factors and ETI components that collectively form the source of plant defences manifested genetically as gene-for-gene resistance.

## 5.2 Recessive Resistance Genes: Real Resistance or Real Incompatibility?

Upon the popularization of Mendel's laws of inheritance around the turn of the 20th century, these principles were applied to plant breeding to enhance agronomic qualities, including disease resistance. The first report of Mendelian inheritance for disease resistance in plants was by Biffen who, studying the inheritance of resistance to yellow rust in wheat, discovered 'fair proof that susceptibility and immunity are definite Mendelian characters, the former being the dominant one' (Biffen, 1905). The recessive nature of certain *R* genes suggests that they may, in some cases, encode variants that are incompatible with the pathogen components that have evolved to engage them.

Biotrophic pathogens must alter host cell processes by interacting with cellular proteins. This is particularly acute for viruses, which rely on the host machinery for the translation of their genetic material into proteins. Cellular mRNAs undergo a series of modifications during their transit from the site of polymerization in the nucleus to the cytoplasm, such as the addition of a 5'-7-methylguanosine cap and a poly(A) tail, which allow them to interact with the translational machinery. However, since the majority of plant viruses are

RNA viruses that replicate in the cytoplasm, their protein-coding transcripts are not processed in the same way, and thus viruses have evolved a number of strategies for engaging the host translational machinery (Thivierge *et al.*, 2005). Many viruses produce transcripts with a 5' cap and poly(A) tail, but others have evolved alternate mechanisms to replace these structures. The potyvirus VPg protein is covalently attached to the 5' end of the RNA genome and functionally replaces the 5' cap (Thivierge *et al.*, 2005; Kneller *et al.*, 2006). Like the 5' cap structure, potyviral VPg proteins interact with the eukaryotic initiation factor eIF-4E or its isomer eIF-4E(iso) to promote initiation of viral protein translation (Leonard *et al.*, 2000, 2004; Khan *et al.*, 2006; Beauchemin *et al.*, 2007; Roudet-Tavert *et al.*, 2007; Yeam *et al.*, 2007). A number of recessive potyviral resistance genes from pepper, tomato, lettuce, barley and pea have been determined to be the genes encoding eIF-4E or eIF-4E(iso) (Table 5.1). In some cases the mutant eIF-4E proteins have been shown to bind poorly or not at all to the VPg proteins of the viruses to which they confer resistance (Kang *et al.*, 2005; Yeam *et al.*, 2007; Charron *et al.*, 2008). Additionally, melon necrotic spot virus (MNSV), belonging to the *Carmovirus* genus, is also restricted by recessive alleles of eIF-4E in melon (Nieto *et al.*, 2006). Like the genomic RNA of the potyviruses, the carmovirus genome is not capped; however, the absence of a VPg protein encoded within the MNSV genome suggests a different interaction with eIF-4E from the potyviruses (Nieto *et al.*, 2006). Indeed, the determinant of MNSV avirulence appears to reside within the 3'-untranslated region of the viral genome (Diaz *et al.*, 2004), suggesting a profound difference in the molecular interaction with

**Table 5.1.** Recessive resistance genes.

| *R* gene | Plant | Protein encoded | Pathogen | Pathogen type | Reference(s) |
|---|---|---|---|---|---|
| *mlo* | Barley | 7-transmembrane protein | *Blumeria graminis* | Fungus | Buschges *et al.* (1997) |
| *mol* | Lettuce | eIF-4E | Lettuce mosaic virus | Potyvirus | Nicaise *et al.* (2003) |
| *nsv* | Melon | eIF-4E | Melon necrotic spot virus | Carmovirus | Nieto *et al.* (2006) |
| *pot-1* | Tomato | eIF-4E | Potato virus Y | Potyvirus | Ruffel *et al.* (2005) |
| *pvr1* | Pepper | eIF-4E | Tobacco etch virus | Potyvirus | Kang *et al.* (2005) |
| *pvr2* | Pepper | eIF-4E | Potato virus Y | Potyvirus | Ruffel *et al.* (2002) |
| *pvr6* | Pepper | eIF-4E(iso) | Pepper veinal mottle virus | Potyvirus | Ruffel *et al.* (2006) |
| *rym4, rym5* | Barley | eIF-4E | Barley yellow mosaic virus | Bymovirus | Stein *et al.* (2005) |
| *sbm1* | Pea | eIF-4E | Bean yellow mosaic virus/pea seedborne mosaic virus | Potyvirus | Gao *et al.* (2004), Bruun-Rasmussen *et al.* (2007) |
| *xa5* | Rice | Transcription factor IIAγ | *Xanthomonas oryzae* | Bacteria | Iyer and McCouch (2004) |
| *xa13* | Rice | Homology with nodulin MtN3 | *X. oryzae* | Bacteria | Chu *et al.* (2006) |

eIF-4E that determines resistance. In addition, induced mutations in the *Arabidopsis* genes encoding the translation factors eIF-4E and eIF-4G restrict infection by another *Carmovirus*, turnip crinkle virus (TCV), as well as cucumber mosaic cucumovirus (CMV) (Yoshii *et al.*, 2004). It is not clear to what extent differences in translation factors might play a role in restricting viruses on non-host plants. However, variations in eIF-4E are a common source of resistance to potyviruses and it has been proposed that a coevolutionary 'arms race' with potyviruses may have shaped the evolution of plant eIF4E proteins (Charron *et al.*, 2008).

Of the 30 different known resistance specificities in rice against the bacterial blight-causing pathogen *Xanthomonas oryzae* pv. *oryzae*, nine genes have been determined to be recessive in nature; two of which, the rice *xa5* and *xa13* genes, have recently been identified (Iyer and McCouch, 2004; Chu *et al.*, 2006). The *xa5* gene encodes the gamma subunit of transcription factor IIA (TFIIAγ) (Iyer and McCouch, 2004). The *xa13* gene encodes a protein required for bacterial growth that shows similarity to the nodulin MtN3 family protein that is induced in the legume *Medicago truncatula* by *Rhizobium* during nodulation (Chu *et al.*, 2006). While *xa5* differs from the dominant allele by encoding two amino acid substitutions, *xa13* differences are confined to the promoter sequence and render the promoter unresponsive to *X. oryzae*-induced upregulation (Iyer and McCouch, 2004; Chu *et al.*, 2006; Yang *et al.*, 2006). Like the eIF-4E variants, these genes are also likely to be required for the bacteria to be able to undergo a compatible interaction with the host. Thus, it is perhaps more precise to view many recessive *R* genes as conferring passive resistance via a loss of susceptibility rather than through the induction of active defence responses.

The *Mlo* genes of barley (*HvMlo*) and tomato (*SlMlo1*) encode integral transmembrane domain proteins and loss of function in these genes result in recessive resistance to fungal pathogens causing powdery mildew (Buschges *et al.*, 1997; Bai *et al.*, 2008). The resistance conferred by *mlo* alleles is specific to powdery mildews, suggesting that it is required for compatibility. At the same time, however, some *mlo* mutants that confer resistance by inducing necrosis upon infection also show spontaneous necrotic lesions and premature senescence in the absence of infection. Thus it is possible that the lack of Mlo protein results in a low level induction of defence responses such that the signalling threshold to induce cell death is reduced. Thus, when combined with PTI mechanisms, *mlo*-induced responses are sufficient to confer resistance. Consistent with this possibility, *mlo* mutations in barley result in increased susceptibility to the necrotrophic fungus *Bipolaris sorokiniana* and the hemibiotrophic fungus *Magnoporthe grisea* (Jarosch *et al.*, 1999; Kumar *et al.*, 2001). Susceptibility to necrotrophs and resistance to biotrophs caused by the same gene is also seen for dominant resistance genes (see below) and is due to the fact that a common defence response, induction of cell death, is deleterious to biotrophs but beneficial to necrotrophs.

## Dominant *R* genes

In a series of publications starting in the 1940s, Flor documented the highly specific genetic interaction between different varieties of flax (*Linum usitatissimum*) and races of its fungal pathogen (*Melampsora linii*) (Flor, 1942, 1946, 1947). The initial concept of gene-for-gene resistance described dominant disease *resistance* (*R*) genes in the host plant that conditioned a resistance response to the pathogen only when the pathogen possessed a counterpart gene that rendered it avirulent: that is, unable to cause an infection. These *avirulence* (*Avr*) genes were also found to be dominant whereas virulence in the presence of the *R* gene was recessive. These studies led to the formation of the modern gene-for-gene resistance model, wherein resistance or susceptibility is determined by the genotypes of both the plant and the pathogen. In this genetic paradigm, a plant with a given *R* gene is an incompatible host for a particular pathogen, only if that pathogen encodes a matching *Avr* gene. If either host or pathogen lacks the appropriate *R* or *Avr* gene (or encodes a non-matching allele thereof), infection ensues.

Despite the diversity of parasitism strategies used by different pathogens, one defence mechanism mediated by dominant *R* genes has been observed to function against pathogens/parasites representing the entire taxonomical range. The presence of matching *R* gene and *Avr* gene products is usually sufficient to induce a programmed cell death response known as the hypersensitive response (HR) (Heath, 2000). Although it is not an unconditional outcome in gene-for-gene resistance, the HR is a typical characteristic of dominant *R* gene-mediated responses. Dominant *R* genes typically confer resistance to biotrophic pathogens and the death of cells at the site of attempted infection is thought to deprive the pathogen of the living tissue it requires and prevents it from moving to new sites of infection. Exceptions have been observed where a pathogen infection is halted in the absence of HR. It is not clear in these cases, known as extreme resistance (ER), how pathogen spread is limited (Bendahmane *et al.*, 1999). However, it is probable that similar initial mechanisms are commonly induced and that the difference between ER and HR may represent a spectrum of responses with cell death representing a 'last line of defence' required if earlier responses are not sufficient to quickly eliminate the pathogen and the source of defence induction.

Over 80 dominant *R* genes with known resistance specificities have been cloned from a number of different plants (Tables 5.2, 5.3 and 5.4). These *R* genes confer resistance to the gamut of plant pathogens including viruses, bacteria, oomycetes, fungi, nematodes and insects. While some *R* genes appear to be unique, the vast majority belong to a small set of protein classes, indicating that resistance to different pathogens is achieved through similar mechanisms.

**Table 5.2.** Receptor-like and non-canonical R proteins.

| Class | R protein | Plant | Unique domain(s) | Pathogen/parasite[a] | Effector (Avr) | Reference(s) |
|---|---|---|---|---|---|---|
| Extracellular LRR | Cf-2 | Tomato | | *Cladosporium fulvum* (F) | Avr2 | Dixon *et al.* (1996), Luderer *et al.* (2002) |
| | Cf-4 | Tomato | | *C. fulvum* (F) | Avr4 | Joosten *et al.* (1994), Thomas *et al.* (1997) |
| | Cf-5 | Tomato | | *C. fulvum* (F) | | Dixon *et al.* (1998) |
| | Cf-9 | Tomato | | *C. fulvum* (F) | Avr9 | Vankan *et al.* (1991), Hammond-Kosack *et al.* (1994) |
| | $HS1^{pro\text{-}1}$ | Sugarbeet | | *Heterodera schachtii* Schmidt (N) | | Cai *et al.* (1997) |
| | LeEix2 | Tomato | C-term. endocytosis signals | *Trichoderma viride* (F) | EIX (ethylene-inducing xylanase) | Ron and Avni (2004) |
| | Ve1 | Tomato | C-term. endocytosis signals | *Verticillium albo-atrum* (F) | | Kawchuk *et al.* (2001) |
| | Ve2 | Tomato | C-term. endocytosis signals | *V. albo-atrum* (F) | | Kawchuk *et al.* (2001) |
| | Vfa1 | Crabapple | | *Venturia inaequalis* (F) | | Belfanti *et al.* (2004), Malnoy *et al.* (2008) |
| | Vfa2 | Crabapple | | *V. inaequalis* (F) | | Belfanti *et al.* (2004), Malnoy *et al.* (2008) |
| LRR-RLK | Xa21 | Rice | | *Xanthomonas oryzae* (B) | | Song *et al.* (1995) |
| | Xa3/Xa26 | Rice | | *X. oryzae* (B) | | Xiang *et al.* (2006) |
| Non-LRR RLK | RFO1 | *Arabidopsis* | | *Fusarium oxysporum* (F) | | Diener and Ausubel (2005) |
| | Rpg1 | Barley | | *Puccinia graminis* (F) | | Rostoks *et al.* (2002) |
| | Pi-d2 | Rice | *B*-lectin ectodomain | *Magnoporthe grisea* (F) | | Chen *et al.* (2006) |
| TIR-NB/TIR-X | RLM3 | *Arabidopsis* | RCC1, HMG and WAP homology (alternative splice product) | *Leptosphaeria maculans* (F) | | Staal *et al.* (2008) |
| Unique | Bs3 | Pepper | Flavin mono-oxygenase | *Xanthomonas campestris* (B) | AvrBs3 | Romer *et al.* (2007) |
| | Hm1 | Maize | HC-toxin reductase | *Cochliobolus carbonum* (F) | HC-toxin | Johal and Briggs (1992), Meeley *et al.* (1992) |
| | RPW8 | *Arabidopsis* | TM and CC | *Erysiphe cruciferarum* (F) | | Xiao *et al.* (2001) |
| | RTM1 | *Arabidopsis* | Jacalin-like repeats | Tobacco etch virus (V) | | Chisholm *et al.* (2000) |
| | RTM2 | *Arabidopsis* | Small heat shock protein-like | Tobacco etch virus (V) | | Whitham *et al.* (2000) |
| | Tm-1 | Tomato | Novel structure | Tomato mosaic virus (V) | RNA-dependent RNA polymerase | Meshi *et al.* (1988), Ishibashi *et al.* (2007) |
| | Xa27 | Rice | Novel structure | *X. oryzae* (B) | AvrXa27 | Gurlebeck *et al.* (2005) |

[a] Letters in brackets indicate: (B), bacterial pathogen; (F), fungal pathogen; (N), nematode; (V), viral pathogen.

**Table 5.3.** R proteins encoding NB-LRR proteins of the non-TIR class.

| R protein | Plant | Pathogen/parasite[a] | Effector (Avr) | Reference(s) |
|---|---|---|---|---|
| Ctv | *Poncirus trifoliata* | Citrus tristeza virus (V) | | Yang *et al.* (2003), Rai (2006) |
| Dm3 | Lettuce | *Bremia lactucae* (F) | | Meyers *et al.* (1998) |
| Dm14 | Lettuce | *B. lactucae* (F) | | Wroblewski *et al.* (2007) |
| Dm16 | Lettuce | *B. lactucae* (F) | | Wroblewski *et al.* (2007) |
| Dm18 | Lettuce | *B. lactucae* (F) | | Wroblewski *et al.* (2007) |
| Fom-2 | Melon | *Fusarium oxysporum* (F) | | Joobeur *et al.* (2004) |
| Gpa2 | Potato | *Globodera pallida* (N) | | van der Vossen *et al.* (2000) |
| Hero | Potato | *G. pallida* (N), *Globodera rostochiensis* (N) | | Ernst *et al.* (2002) |
| HRT | *Arabidopsis* | Turnip crinkle virus (V) | Coat protein | Cooley *et al.* (2000) |
| I-2 | Tomato | *F. oxysporum* (F) | | Ori *et al.* (1997), Simons *et al.* (1998) |
| Lov1 | *Arabidopsis* | *Cochliobolus victoriae* (F) | Victorin | Sweat *et al.* (2008) |
| Lr1 | *Triticum aestivum* | *Puccinia triticina* (F) | | Cloutier *et al.* (2007) |
| Lr10 | *T. aestivum* | *P. triticina* (F) | | Feuillet *et al.* (2003) |
| Lr21 | *T. aestivum* | *P. triticina* (F) | | Huang *et al.* (2003) |
| Mi | Tomato | *Meloidogyne incognita* (N), *Macrosiphum euphorbiae* (I), *Bemisia tabaci* (I) | | Milligan *et al.* (1998), Vos *et al.* (1998), Nombela *et al.* (2003) |
| Mla1 (Mla6, Mla7, Mla10, Mla12, Mla13) | Barley | *Blumeria graminis* (F) | $Avr_{A10}$ | Halterman *et al.* (2001), Zhou *et al.* (2001), Shen *et al.* (2003), Halterman and Wise (2004), Ridout *et al.* (2006) |
| Pib | Rice | *Magnaporthe grisea* (F) | | Wang *et al.* (1999) |
| Pi2[b] | Rice | *M. grisea* (F) | | Zhou *et al.* (2006) |
| Pi9 | Rice | *M. grisea* (F) | | Qu *et al.* (2006) |
| Pi36 | Rice | *M. grisea* (F) | | Liu *et al.* (2007) |
| Pi37 | Rice | *M. grisea* (F) | | Lin *et al.* (2007) |
| Pi-ta | Rice | *M. grisea* (F) | AVR-Pita | Bryan *et al.* (2000), Orbach *et al.* (2000) |
| Piz-t[b] | Rice | *M. grisea* (F) | | Zhou *et al.* (2006) |
| Pm3 | *T. aestivum* | *B. graminis* (F) | | Srichumpa *et al.* (2005) |
| Prf | Tomato | *Psuedomonas syringae* (B) | AvrPto, AvrPtoB | Ronald *et al.* (1992), Salmeron *et al.* (1996), Kim *et al.* (2002) |
| R1 | Potato | *Phytophthora infestans* (O) | | Ballvora *et al.* (2002) |

| | | | | |
|---|---|---|---|---|
| R3a | Potato | *P. infestans* (O) | Avr3a | Armstrong *et al.* (2005), Huang *et al.* (2005) |
| RB | *Solanum bulbocastanum* | *P. infestans* (O) | | Song *et al.* (2003) |
| RCY1[c] | *Arabidopsis* | Cucumber mosaic virus (V) | Coat protein | Takahashi *et al.* (2002) |
| Rp1 | Maize | *Puccinia sorghi* (F) | | Collins *et al.* (1999) |
| Rp3 | Maize | *P. sorghi* (F) | | Webb *et al.* (2002) |
| Rpg1-b | Soybean | *P. syringae* (B) | AvrB | Ashfield *et al.* (2004) |
| RPM1 | *Arabidopsis* | *P. syringae* (B) | AvrRPM1, AvrB | Tamaki *et al.* (1988), Debener *et al.* (1991), Grant *et al.* (1995) |
| RPP8[c] | *Arabidopsis* | *Hyaloperonospora parasitica* (O) | ATR13 | McDowell *et al.* (1998) |
| RPP13 | *Arabidopsis* | *H. parasitica* (O) | ATR13 | Bittner-Eddy *et al.* (2000), Allen *et al.* (2004) |
| Rps1-k | Soybean | *Phytophthora sojae* (O) | | Gao *et al.* (2005), Gao and Bhattacharyya (2008) |
| RPS2 | *Arabidopsis* | *P. syringae* (B) | AvrRpt2 | Whalen *et al.* (1991), Bent *et al.* (1994), Mindrinos *et al.* (1994) |
| RPS5 | *Arabidopsis* | *P. syringae* (B) | AvrPphB | Jenner *et al.* (1991), Warren *et al.* (1998) |
| Rx | Potato | Potato virus X (V) | Coat protein | Bendahmane *et al.* (1995, 1999) |
| Rx2 | Potato | Potato virus X (V) | Coat protein | Querci *et al.* (1995), Bendahmane *et al.* (2000) |
| Rxo1 | Maize | *X. oryzae* (B), *Burkholderia andropogonis* (B) | AvrRxo1 | Zhao *et al.* (2004, 2005) |
| Sw-5 | Tomato | Tomato spotted wilt virus (V) | | Brommonschenkel *et al.* (2000) |
| Tm-2, Tm-2[2] | Tomato | Tomato mosaic virus (V) | Movement protein | Weber *et al.* (1993), Lanfermeijer *et al.* (2003, 2005) |
| Xa1 | Rice | *X. oryzae* (B) | | Yoshimura *et al.* (1998) |

[a] Letters in brackets indicate: (B), bacterial pathogen; (F), fungal pathogen; (I), insect; (N), nematode; (O), oomycete; (V), viral pathogen.
[b] Alleles of the same locus in rice.
[c] Alleles of the same gene from *Arabidopsis*.

**Table 5.4.** R proteins encoding NB-LRR proteins of the TIR class.

| R protein | Plant | Pathogen/parasite[a] | Effector (Avr) | Reference(s) |
|---|---|---|---|---|
| Bs4 | Tomato | *X. campestris* (B) | AvrBs4 | Bonas *et al.* (1993), Ballvora *et al.* (2001), Schornack *et al.* (2004) |
| Gro1 | Potato | *G. rostonchiensis* (N) | | Paal *et al.* (2004) |
| L | Flax | *Melampsora lini* (F) | AvrL567 | Lawrence *et al.* (1995), Dodds *et al.* (2004) |
| M | Flax | *M. lini* (F) | AvrM | Anderson *et al.* (1997), Catanzariti *et al.* (2006) |
| N | *Nicotiana glutinosa* | Tobacco mosaic virus (V) | Helicase (P50) | Whitham *et al.* (1994), Erickson *et al.* (1999) |
| N | Flax | *M. lini* (F) | | Dodds *et al.* (2001a) |
| P | Flax | *M. lini* (F) | AvrP123, AvrP4 | Dodds *et al.* (2001b), Catanzariti *et al.* (2006) |
| RAC1 | *Arabidopsis* | *Albugo candida* (O) | | Borhan *et al.* (2004) |
| RLM1 | *Arabidopsis* | *Leptosphaeria maculans* (F) | | Staal *et al.* (2006) |
| RLM2 | *Arabidopsis* | *L. maculans* (F) | | Staal *et al.* (2006) |
| RPP1 | *Arabidopsis* | *H. parasitica* (O) | ATR1 | Botella *et al.* (1998), Rehmany *et al.* (2005) |
| RPP2A/RPP2B[b] | *Arabidopsis* | *H. parasitica* (O) | | Sinapidou *et al.* (2004) |
| RPP4 | *Arabidopsis* | *H. parasitica* (O) | | van der Biezen *et al.* (2002) |
| RPP5 | *Arabidopsis* | *H. parasitica* (O) | | Parker *et al.* (1997) |
| RPS4 | *Arabidopsis* | *P. syringae* (B) | AvrRps4 | Hinsch and Staskawicz (1996), Gassmann *et al.* (1999) |
| RRS1-R | *Arabidopsis* | *Ralstonia solanacearum* (B) | PopP2 | Deslandes *et al.* (2002, 2003) |
| Tao1 | *Arabidopsis* | *P. syringae* (B) | AvrB | Eitas *et al.* (2008) |
| Y-1 | Potato | Potato virus Y (V) | | Vidal *et al.* (2002) |

[a] Letters in brackets indicate: (B), bacterial pathogen; (F), fungal pathogen; (N), nematode; (O), oomycete; (V), viral pathogen.
[b] Both proteins are required for resistance.

### *R* genes encoding extracellular domains

The tomato *Cf-2*, *Cf-4*, *Cf-5* and *Cf-9* genes provide resistance to the fungus *Cladosporium fulvum* and encode large extracellular leucine-rich repeat (LRR) domain proteins that are membrane-anchored through a transmembrane (TM) domain, with a short carboxy-terminal cytoplasmic peptide (Fig. 5.1) (Hammond-Kosack *et al.*, 1994; Dixon *et al.*, 1996, 1998). Given that LRRs often serve as ligand-binding domains, these proteins are referred to as receptor-like proteins (RLPs). The LRR domain is composed of repeats with a consensus of LxxLxxLxxLxLxxNxLxGxIPxx (Jones and Jones, 1997). The LRR domain of different alleles of the various members of the *Cf* gene family, is highly variable both in number of repeats, which ranges from 25 to 38 LRRs, as well as in the primary amino acid sequence, suggesting a very high degree of natural variation in the LRRs of these proteins (Dixon *et al.*, 1998; Parniske *et al.*, 1999; Caicedo and Schaal, 2004; Caicedo, 2008). In agreement, domain-swapping experiments with different *Cf* gene homologues have demonstrated that the LRR domain is the determinant of recognition specificity (Wulff *et al.*, 2001). The extracellular nature of the RLPs provides some indication of the kinds of *Avr* gene products that they might recognize. Indeed, the Avr determinants for Cf-2, Cf-4 and Cf-9 have been shown to be extracellular proteins secreted into the plant apoplast (de Wit and Spikman, 1982).

The rice Xa21 and Xa3/Xa26 proteins have a structure very similar to the RLPs (Fig. 5.1), with the addition of a cytoplasmic kinase domain and are known as receptor-like kinases (RLKs, Fig. 5.1) (Gomez-Gomez and Boller, 2000; Xiang *et al.*, 2006). In a similar arrangement, the rice *Pi-d2* gene encodes an RLK protein with an extracellular β-lectin domain instead of LRRs (Chen *et al.*, 2006). Curiously, the PAMP receptors FLS2 and EFR1, which mediate PTI in response to bacterial flagellin and Ef-Tu respectively, are also RLK proteins (Gomez-Gomez and Boller, 2000; Zipfel *et al.*, 2006). Genome analyses have identified dozens of RLP and RLK-encoding genes with homology to known *R* genes (Table 5.2) and the genes encoding RLKs are among the most polymorphic among *Arabidopsis* accessions (Fritz-Laylin *et al.*, 2005; Clark *et al.*, 2007). However, since relatively few *R* genes encode RLPs or RLKs, many of these may act either as PAMP receptors or perform other functions. The modes of signalling by R proteins with extracellular domains are unknown, however, RLKs could signal through their kinase domains. It will be interesting to see if this signalling will rely on heterodimerization with other RLP or RLK proteins in a manner similar to the observed flagellin-induced dimerization of *Arabidopsis* FLS2 with the protein BAK1 (Chinchilla *et al.*, 2007).

## 5.3 The NB-LRR Protein Class

Plant NB-LRR proteins share structural and mechanistic similarities with members of the NOD-like receptor (NLR) family that function in innate immune

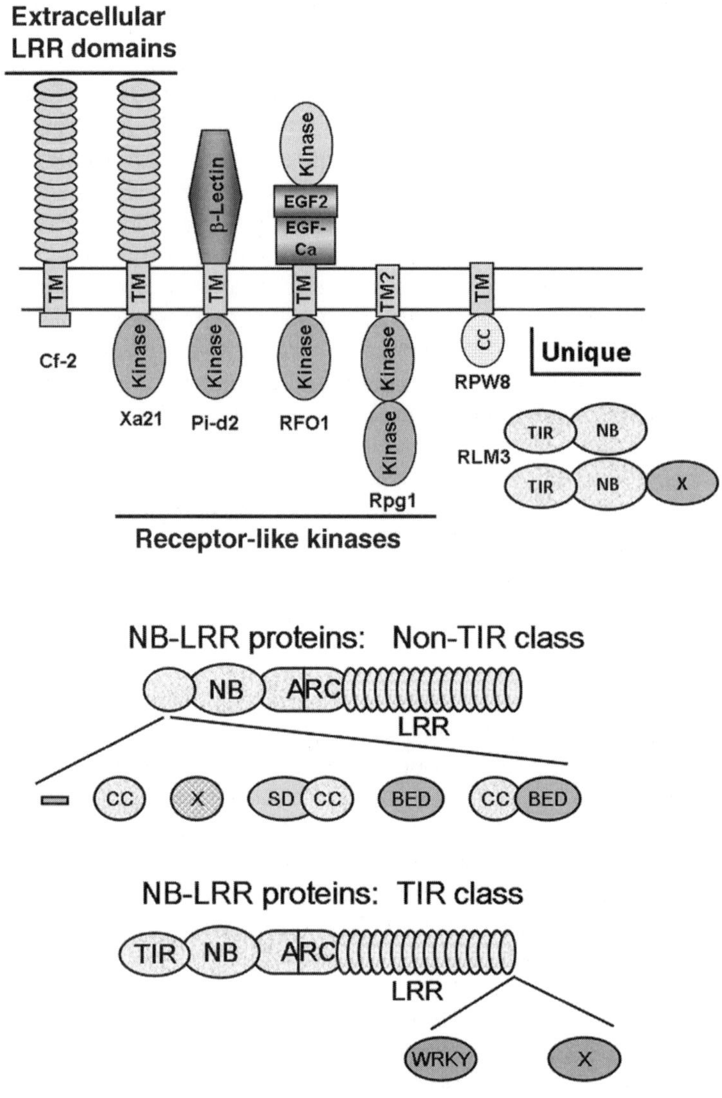


**Fig. 5.1.** Multi-domain structures of the proteins encoded by major classes of dominant *R* genes. Representative receptor-like proteins (RLPs) and receptor-like kinases (RLKs) are shown, which associate with the plasma membrane through a transmembrane (TM) domain and have both intra- and extracellular moieties as described in the text. Note the two protein products predicted to be encoded by the different splice variants of *RLM3*. The two main classes of NB-LRR proteins are depicted schematically showing the NB, ARC and LRR domains. Proteins in the non-TIR class have a high degree of variability at the amino terminus compared to the TIR class, with various amino-terminal domains encoding CC, BED, SD domains, or no assigned structure (X). Alternatively, some proteins of this class have no domain N-terminal to the NB domain (represented by a small rectangle). A subset of proteins within the TIR-NB-LRR family have been found with additional C-terminal domains homologous to WRKY transcription factors or may have a large domain of unknown structure (X) with possible nuclear localization signals (NLS) such as in the RPS4 protein.

responses in animals (Rairdan and Moffett, 2007). Plant NB-LRR proteins are so named because they possess central nucleotide-binding (NB) and carboxy-terminal LRR domains (Fig. 5.1). The LRR domains of the NB-LRR class, however are evolutionarily distinct from the extracellular LRR domains found in receptor-like R proteins, with a consensus of LxxLxxLxxLxLxx(N/C/T)x(x) LxxIPxx (Jones and Jones, 1997; Kajava, 1998). The NB-LRR proteins can be classified into separate clades that have been defined in part by the domain encoded at their amino termini. The primary subdivisions of NB-LRR proteins are based on whether or not the amino-terminal domain shares homology with the cytoplasmic domain of the animal Toll and interleukin-1 receptors (TIR domain; TIR-NB-LRR class) (Whitham *et al.*, 1994). TIR-NB-LRR proteins appear to all belong to a single ancient clade, whereas those NB-LRR lacking a TIR domain appear to show greater diversity in the domains present at their amino termini and can be grouped into at least four different clades (Meyers *et al.*, 1999). Since many of the first non-TIR NB-LRR proteins identified were predicted to encode a coiled-coil (CC) domain at their amino terminus, this class is historically referred to as the CC-NB-LRRs regardless of whether or not they actually conform to CC prediction programmes. The difference between the TIR and CC (non-TIR) NB-LRR proteins however is best defined by consensus motifs present in the NB and ARC domains that show distinct variations between the two classes of NB-LRR proteins (Meyers *et al.*, 1999; Cannon *et al.*, 2002).

Over 60 *R* genes of defined specificity encoding NB-LRR proteins have been cloned from a variety of plant species (Tables 5.3 and 5.4). However, bioinformatic analyses of sequenced plant genomes and expressed sequence tags have revealed the presence of vast numbers of genes encoding NB-LRR proteins. In the *Arabidopsis thaliana* ecotype Columbia-0, 149 genes encode NB-LRR proteins, 94 of which have an amino-terminal TIR domain (Meyers *et al.*, 2003). The number of genes encoding NB-LRR proteins is even larger in plants with larger genomes, with 333 non-redundant genes identified in the incomplete draft sequence of *Medicago truncatula*, and 399 in black cottonwood, *Populus trichocarpa* (Tuskan *et al.*, 2006; Ameline-Torregrosa *et al.*, 2008; Kohler *et al.*, 2008). From the genome of a Pinot Noir variety of grape (*Vitis vinifera*), 233 genes encoding NB-LRR proteins have been identified (Velasco *et al.*, 2007), whereas the genome of a Cabernet Sauvignon variety appears to encode a considerably larger number of NB-LRR proteins (Moroldo *et al.*, 2008). In rice, the number of genes encoding NB-LRR proteins numbers is in excess of 400 genes (Monosi *et al.*, 2004; Zhou *et al.*, 2004), none of which are predicted to encode TIR-NB-LRR proteins (Bai *et al.*, 2002). Like *Arabidopsis*, the grape *R* genes encode both CC-NB-LRR and TIR-NB-LRR proteins (Velasco *et al.*, 2007); however, there has been differential amplification of CC versus TIR classes. On the other hand, the *Populus* genome shows expansion of a class of NB-LRR proteins with an amino-terminal zinc-finger DNA-binding homology domain, known as a BED finger domain, a structure that is absent from the *Arabidopsis* genome but conserved in the rice Xa1 and two Xa1-like proteins (Bai *et al.*, 2002; Kohler *et al.*, 2008). In contrast to the large repertoires of NB-LRR genes identified in

*Arabidopsis*, rice, poplar, grape and *Medicago*, the papaya genome has a mere 58 NB-LRR genes with a predominance of CC-NB-LRRs (Ming *et al.*, 2008).

In addition to genes with structures resembling NB-LRR proteins, a number of genes in several genomes have been identified encoding proteins with alternate domain configurations (Fig. 5.1). One example with additional domains and known resistance function is the protein RRS1-R from *Arabidopsis* that recognizes the PopP2 protein from the bacterium *Ralstonia solanacearum*, and has the structure TIR-NB-LRR-NLS-WRKY. The latter domain encodes a nuclear localization signal (NLS) and resembles a family of transcription factors sharing an amino acid signature motif (WRKY), some of which have been implicated in disease resistance signalling (Deslandes *et al.*, 2002). The alternatively configured proteins are not abundant or conserved between plant genomes and it is unclear whether all are functional genes or whether some may represent pseudogene remnants of recombination events. Plant genomes also encode proteins without LRR domains, consisting of CC-NB, TIR and TIR-NB configurations (Bai *et al.*, 2002; Meyers *et al.*, 2002, 2003). These proteins are present even in rice, which lack TIR-NB-LRR proteins (Bai *et al.*, 2002), suggesting that they may have an evolutionarily conserved function such as acting as signalling adapters analogous to the mammalian TIR-containing immune adapter proteins MyD88 and Mal (Jebanathirajah *et al.*, 2002; Meyers *et al.*, 2002). At the same time, resistance functions have been described for two *Arabidopsis* loci with unusual domain arrangements, and these are discussed below in the section '5.7 Atypical Dominant *R* Genes'.

An interesting observation can be made by comparing the entire NB-LRR gene complements of the sequenced plant genomes, as well as the growing sequence collections for *R* gene candidates amplified by PCR from many other plant species. In addition to the diversity of *R* gene structures within a given plant genome, there is considerable diversity in how different gene families have expanded and evolved in independent plant lineages. The most dramatic of these is the expansion or loss of the TIR-NB-LRR class. In *Arabidopsis*, these are the most numerous, while this NB-LRR class has been lost in the monocot genome. Detection of TIR-NB-LRR genes in pine species, however, demonstrates the antiquity of this class of genes, which must have existed in an ancestral plant before the divergence of gymnosperms and angiosperms (Liu and Ekramoddoullah, 2003). Furthermore, attempts to amplify genes of the TIR class from sugarbeet have failed so far (Tian *et al.*, 2004), suggesting that this gene family was lost independently in two distant plant lineages. Thus, genes encoding NB-LRR genes may expand and diversify differentially upon speciation, explaining why the repertoires of genes encoding NB-LRRs from unrelated species are so different.

Exploration of the genomic distribution of *R* gene candidates and isolation of resistance loci has shown that the NB-LRR proteins often exist as clusters. These clusters are thought to be generated by ancient duplication events and divergence accompanied by selection for new recognition specificities (reviewed in Michelmore and Meyers, 1998). Additional divergence and variation in gene copy numbers at a given locus are likely to have arisen by unequal crossing-over events. The shared origin and tight linkage of paralogous *R* genes in a locus allows for coordinated regulation of their transcriptional activity, a characteristic shown recently for the *Arabidopsis* genes found within the *RPP5* locus (Yi and Richards, 2007).

While *R* genes that are closely related (either by common descent in closely related species, or by duplication within a locus) may be highly similar, the pathogens recognized by the different paralogues may be very different (Grube *et al.*, 2000). This is exemplified well by the *Rx* and *Gpa2* genes from potato that are located in the same *R* gene cluster, but confer resistance against a virus and nematode, respectively (Bakker *et al.*, 2003). In addition to the diversity afforded by different genes within a locus, there are also examples of considerable allelic diversity of a single *R* gene. In some cases, this allelic diversity matches similar variation of the pathogen *Avr* gene, a relationship illustrated by the highly polymorphic *Mla* locus of barley and the corresponding *Avr* genes from different isolates of the powdery mildew-causing *Blumeria graminis* f. sp. *hordei* (Halterman and Wise, 2004). The allelic diversity of an *R* gene has the potential to specify recognition of different pathogens as well, a circumstance documented for three *R* genes (*HRT*, *RCY1* and *RPP8*) that are in fact alleles of the same locus from different *Arabidopsis* ecotypes that confer resistance to viruses from two different genera, TCV and CMV, and to the oomycete *Hyaloperonospora parasitica*, respectively (Cooley *et al.*, 2000; Takahashi *et al.*, 2002). In addition to different alleles from an *R* gene conferring distinct recognition specificities, examples have also been found where a single allele provides recognition of multiple Avr determinants. In one scenario, specificity may be directed towards distinct Avr effectors that originate from the same species of pathogen, such as the recognition by *Arabidopsis* RPM1 of two *Pseudomonas syringae* effectors, AvrRPM1 and AvrB (Grant *et al.*, 1995). Likewise, a single *R* gene can mediate recognition of effectors from different types of organisms, as seen by the resistance mediated by *Mi-1* gene against a root-knot nematode, a white fly and the potato aphid (Rossi *et al.*, 1998). Contrasting the diversity at a single locus is the convergent evolution observed for the *Arabidopsis RPM1* and the soybean *Rgp1-b* genes that both recognize the *P. syringae* effector AvrB and encode two CC-NB-LRR proteins with limited sequence similarity and likely distinct ancestral origins (Ashfield *et al.*, 2004). In addition, the *Arabidopsis Tao1* gene, which encodes a TIR-NB-LRR, responds weakly to AvrB (Eitas *et al.*, 2008).

## 5.4 NB-LRR Protein Domain Functions

As the most numerous of characterized *R* genes, it is not surprising that the proteins encoded by the NB-LRR class have been the best characterized at a genetic and molecular level. The genes encoding NB-LRR proteins show the highest degree of polymorphism of all *Arabidopsis* genes (Clark *et al.*, 2007). Furthermore, *R* genes show high levels of polymorphism manifested as a presence or absence of a given *R* gene both within and between populations

(Grant *et al.*, 1998; Shen *et al.*, 2006; Ding *et al.*, 2007a, b, c). Much of the variation seen between *R* genes is present in the LRR domain and experiments where these regions were exchanged between closely related *R* genes have shown that the LRR domain is responsible for pathogen recognition specificity (Ellis *et al.*, 2000; Shen *et al.*, 2003; Dodds *et al.*, 2006; Qu *et al.*, 2006; Rairdan and Moffett, 2006). Furthermore, the spectrum of viruses recognized by the Rx protein could be extended by directed evolution of only the LRR domain (Farnham and Baulcombe, 2006).

The signalling moiety of NB-LRRs was traditionally thought to be the amino-terminal domain based on the signalling role of animal TIR domains and the strong conservation of sequence within this domain in the plant TIR class of proteins (Whitham *et al.*, 1994). Indications suggesting that the plant TIR domains are involved in signalling come from studies showing that fragments of the flax L10 TIR-NB-LRR protein encoding the TIR plus 39 residues of the NB domain, as well as several *Arabidopsis* TIR domain plus 45 residues of their NB domains, induce an Avr-independent cell death when transiently expressed in tobacco leaves (Frost *et al.*, 2004; Swiderski *et al.*, 2009). Signalling through animal TIR domains occurs via protein–protein interactions with other TIR domain-containing adapters (homotypic interactions) (Meyers *et al.*, 2002). Although the isolated N TIR domain is able to self-associate (Mestre and Baulcombe, 2006), no TIR-containing signal adapter proteins have been identified.

Unlike the highly conserved TIR domain, the non-TIR amino termini are not well conserved, although a large number of characterized CC domains possess a conserved EDVID motif that mediates an intramolecular interaction (Rairdan *et al.*, 2008). Like the TIR domain, the amino termini of the non-TIR NB-LRR proteins with known resistance functions have also been posited to interact to act as protein–protein interaction domains. This prediction has been substantiated by the identification of a handful of proteins that interact with the amino termini of very specific CC-NB-LRR proteins (discussed below). However, as with the TIR class, none of these interacting proteins appear to be obvious signal adaptor proteins. Within the family *Solanaceae*, a number of CC-NB-LRR proteins have additional sequences at their amino termini, including a homology domain that is conserved among the proteins Prf, Mi-1, Hero and R1, coined the Solanaceous domain (SD), in addition to unique sequences that show no similarity to other R proteins (Mucyn *et al.*, 2006).

The central NB domain is the most conserved sequence among all NB-LRR classes and suggests a conserved functional activity (Tameling *et al.*, 2002). The NB and LRR domains are separated by a region known as the ARC domain since it is shared by the Apaf-1, plant R and Ced-4 proteins (van der Biezen and Jones, 1998a). Recent studies have shown that within the ARC domain are two distinct functional units that have been further delineated as the ARC1 and ARC2 domains (Albrecht and Takken, 2006; McHale *et al.*, 2006; Rairdan and Moffett, 2006). Phylogenetic analysis of the NB-ARC regions (or NBS) of NB-LRR proteins demonstrated that the TIR and non-TIR were split into two distinct clades, with no individuals clustering with members of the other class (Meyers *et al.*, 1999). Thus, sequence motifs within the NB and ARC domains can allow tentative classification of *R* gene analogues (RGAs) as likely TIR or non-TIR type proteins without knowledge of the amino-terminal sequence. Within the non-TIR clade, it is interesting that NB-LRRs from the same clade may have different amino termini in different species, as exemplified by the fusion of either amino-terminal CC or BED domains to related NB-ARC domains (Kohler *et al.*, 2008).

The NB domains from the tomato CC-NB-LRR proteins I-2 and Mi-1 have been shown to bind ATP and have functional ATPase activity *in vitro* when purified from bacteria (Tameling *et al.*, 2002). Transient over-expression of the NB domain of Rx in tobacco leaves was shown to trigger a robust Avr-independent programmed cell death that resembled HR (Rairdan *et al.*, 2008). Thus, unlike the results seen with TIR-NB-LRR proteins this observation supports a role for the NB domain as a signalling domain. Functional ATPase activity within this domain appears to be unnecessary for signalling and may serve a role in regulating the intact R protein (Rairdan *et al.*, 2008).

The conserved NB-ARC arrangement in plant NB-LRR proteins and animal proteins involved in innate immunity or apoptosis suggests functional similarities for these domains in signalling (van der Biezen and Jones, 1998a). When aligned with the mammalian Apaf-1 and *C. elegans* CED-4, there are eight conserved motifs shared with plant NB-LRR proteins, three of which (kinase 1 or P-loop, kinase 2 and kinase 3) form the nucleotide-binding pocket (van der Biezen and Jones, 1998a). Similarities among the R and Apaf-1 proteins in the NB-ARC domains has allowed three-dimensional models of the plant NB-ARC region to be predicted using the resolved crystal structure of Apaf-1, which comprises a three-layered $\alpha/\beta$ domain, a helical domain I, and a winged helix domain (Chattopadhyaya and Pal, 2008; van Ooijen *et al.*, 2008). The winged helix domain is structurally the most conserved (Chattopadhyaya and Pal, 2008). Validation of models with mutagenesis data shows that autoactivation mutations cluster on the side opposite to the ARC2/NB interface within the three-dimensional structure (van Ooijen *et al.*, 2008).

For the non-TIR class, intramolecular interactions between domains have been detected by coimmunoprecipitation and functional complementation studies using the pepper Bs2 and Rx proteins. Reconstitution of full-length protein activity was achieved by coexpression of fragments encoding CC plus NB-ARC-LRR or CC-NB-ARC and LRR (Moffett *et al.*, 2002; Leister *et al.*, 2005). Complementation of the latter two fragments has been ascribed to an interaction between the LRR and the ARC1 component of the ARC domain (Rairdan and Moffett, 2006). For Rx functional complementation by CC and NB-ARC-LRR fragments, mutational analysis suggests a role for the EDVID motif within the CC domain as the interaction interface. Moreover, since the consensus sequence EDVID is the only identifiable conserved motif within CC domains, this motif may form a common interface in the amino-terminal domain for interaction with the rest of the protein for the non-TIR class (Rairdan *et al.*, 2008). A similar complementation study of *in planta* intramolecular interactions within the TIR-NB-LRR protein N failed to demonstrate interactions between different domains, although Avr-elicited oligomerization of N was observed in addition to the homotypic interactions between TIR domains (Mestre and Baulcombe, 2006). However, an interaction between the N LRR domain and TIR-NB-ARC fragment could be demonstrated through an *in vitro* pull-down assay with bacterially-expressed recombinant proteins, and in a yeast two-hybrid assay (Ueda *et al.*, 2006). It is unclear whether the different observations made for the N, Bs2 and Rx proteins reflect fundamental differences in function between TIR and non-TIR NB-LRR classes, or whether the apparent differences may simply reflect subtleties in the natures of these proteins that make them more or less amenable to observing aspects of a common mechanism of NB-LRR activation and signalling.

NB-LRR proteins are thought to be held in an auto-inhibited conformation prior to recognition via intramolecular interactions involving the ARC and LRR domains (Moffett *et al.*, 2002; Rairdan and Moffett, 2006; Rairdan *et al.*, 2008). Expression of several NB-LRR proteins with deleted ARC and/or LRR domains or with point mutations within these regions has been shown in many cases to result in Avr-independent HR (Hwang *et al.*, 2000; Bendahmane *et al.*, 2002; Shirano *et al.*, 2002; Zhang *et al.*, 2003; Howles *et al.*, 2005; Tameling *et al.*, 2006). Amino acid residue substitutions resulting in autoactivity have so far been described within the ARC2 and LRR regions. A well conserved three amino acid motif (MHD) within the ARC2 domain appears to play a major role in negative regulation of NB-LRR activation as substitutions for H or D in the proteins Rx, I-2, Mi-1, L6 and NRC1 have resulted in Avr-independent cell death (Bendahmane *et al.*, 2002; De la Fuente van Bentem *et al.*, 2005; Howles *et al.*, 2005; Gabriels *et al.*, 2007; van Ooijen *et al.*, 2008). The importance of this motif as a key negative regulator of NB-ARC proteins is also evident from its conservation in the mammalian Apaf-1 (van der Biezen and Jones, 1998a). The MHD motif of NB-LRR and Apaf-1 proteins has been proposed to play a role analogous to the sensor II motif found in other ATPases, functioning to coordinate the bound nucleotide and intramolecular interactions (van Ooijen *et al.*, 2008). The RNBS-D motif is another of the eight conserved NB-ARC motifs within the ARC2 for which inactivating and autoactivating mutations have been identified (Bendahmane *et al.*, 2002), further supporting a role for ARC2 as the switch point for converting recognition into activation (Rairdan and Moffett, 2006).

Autoactivation appears to also result from an incompatibility between different domains of NB-LRR proteins. A natural example of this has been reported for the maize *Rp1* locus, for which unequal crossing over involving paralogues has resulted in recombinant genes, including one with an autoactive phenotype (Sun *et al.*, 2001). Experimental exchanges of NB-LRR domain segments have demonstrated Avr-independent activation of Rx, Mi-1 and L6 due to inappropriate pairings of domains from closely related paralogues (Hwang and Williamson, 2003; Howles *et al.*, 2005; Rairdan and Moffett, 2006). One study has shown that autoactivation is caused by incompatibilities between the ARC2 and LRR domains (Rairdan and Moffett, 2006), consistent with a role for the ARC2 domain as a molecular switch that relays alterations in the LRR/ARC interface to the signalling moieties of the protein.

Elicitation of the Rx protein CC-NB-ARC and LRR fragments by the cognate Avr protein, the potato virus X (PVX) coat protein (CP), has been shown to induce dissociation of the *trans*-complementing fragments; this observation suggests that an intramolecular interaction occurs in the inactive NB-LRR state, and an Avr-induced conformation change activating the protein disrupts this interaction (Moffett *et al.*, 2002; Rairdan and Moffett, 2006). In autoactive mutation variants of Rx, the interaction between CC-NB-ARC and LRR is retained, as well as the CP-dependent dissociation, suggesting that the intramolecular interaction is not required for signal initiation per se, but may rather be involved in 're-setting' the molecule to allow for additional rounds of recognition and signal initiation (Rairdan and Moffett, 2006; Rairdan *et al.*, 2008).

## 5.5 Modes of Indirect and Direct Avr Recognition by NB-LRR Proteins

The most predictable mechanism of gene-for-gene resistance would be through a direct receptor–ligand interaction between R protein and Avr effector. However, direct interactions have only been demonstrated in a few cases. Such interactions have been shown in yeast two-hybrid experiments for the flax rust resistance L5 and L6 proteins with the corresponding *M. linii* AvrL567 proteins. These studies showed that the TIR-NB-LRR proteins encoded by different alleles of the flax *L* gene interacted with the gene products of the corresponding alleles of the flax rust *AvrL567* gene, and that these interactions were specified by the LRR domain (Dodds *et al.*, 2006). Additional yeast two-hybrid interactions were demonstrated for the *Arabidopsis* RRS1-R protein and its cognate Avr, the *R. solanacearum* PopP2 protein, for the rice Pi-ta protein and its Avr from the rice blast fungus *M. grisea* (AvrPi-ta), and for the P50 subunit of the tobacco mosaic virus (TMV) replicase and the *N* gene product (Jia *et al.*, 2000; Deslandes *et al.*, 2003; Dodds *et al.*, 2006; Ueda *et al.*, 2006). The P50 interaction was observed with full-length N protein or with a construct having the TIR domain deleted and interaction between P50 and the latter construct was supported by an *in vitro* pull-down assay (Ueda *et al.*, 2006). The Pi-ta interaction was narrowed down to the C-terminal LRR domain by a deletion series tested in a yeast two-hybrid assay, and direct interaction between AvrPi-ta and full-length Pi-ta was supported by probing AvrPi-ta immobilized on membranes with Pi-ta purified from bacteria (Jia *et al.*, 2000). No direct interactions have been demonstrated either *in planta* or *in vitro* with proteins purified from plant cells, suggesting that other regions within the NB-LRR proteins or additional host factors could regulate R–Avr interactions when the proteins are expressed within the correct cellular context.

For many *R* genes, there have been (largely unpublished) unsuccessful attempts to demonstrate direct interactions between NB-LRR and Avr proteins through numerous approaches. Several findings have instead supported a model of indirect recognition, whereby recognition is mediated by a second host protein that physically interacts with the NB-LRR protein (Dangl and

Jones, 2001). One model of indirect recognition, the guard hypothesis, proposes that the NB-LRR monitors the status of a host protein that is the target of the virulence activity of the Avr effector protein: the NB-LRR protein guard detects the changes effected by the activity of the cognate Avr determinant on the target protein under surveillance (guardee) (van der Biezen and Jones, 1998b). However, of the proteins identified so far as cofactors in indirect recognition (Table 5.5), none have been shown to be key targets to promote virulence in a susceptible background.

Interestingly, recognition cofactors interact with the amino-terminal domains of their cognate NB-LRR partners (Table 5.5). One of the best studied of these proteins is RIN4 from *Arabidopsis*, which interacts with the CC domain of the CC-NB-LRR protein RPM1 (Mackey *et al.*, 2002; Axtell and Staskawicz, 2003). RIN4 also interacts with the two Avr proteins, AvrB and AvrRpm1, and this interaction appears to activate the RPM1 protein (Mackey *et al.*, 2002). RIN4 also interacts with the CC-NB-LRR protein RPS2 (Axtell and Staskawicz, 2003). The Avr determinant for RPS2 is the cysteine protease AvrRpt2, which activates RPS2 upon cleavage of RIN4 (Axtell *et al.*, 2003). Cleavage of RIN4 also weakly activates RPM1, suggesting that both proteins respond to alterations of their recognition cofactor, RIN4, by Avr proteins (Axtell *et al.*, 2003; Mackey *et al.*, 2003). Possibly similar to RPS2 activation, the PBS1 kinase, which interacts with the CC domain of the *Arabidopsis* CC-NB-LRR protein RPS5, is also cleaved by the proteolytic activity of AvrPphB, the cognate Avr of RPS5 (Shao *et al.*, 2003).

Another well-studied indirect interaction is that of two *P. syringae* effector proteins, AvrPto and AvrPtoB, with the tomato protein Prf, which makes use of the protein kinase Pto as recognition cofactor (Pedley and Martin, 2003). Pto has historically been labelled an R protein because its polymorphism in tomato cultivars results in an apparent gene-for-gene relationship with AvrPto (Martin *et al.*, 1993, 2003). Subsequent findings have shown that the tomato Prf protein is required for Pto-mediated resistance and shares homology with the typical R proteins of the non-TIR NB-LRR class, making Prf the actual R protein participating in this gene-for-gene interaction (Salmeron *et al.*, 1996). Like the above examples, Prf interacts with Pto through its amino-terminal domain (Mucyn *et al.*, 2006). Observations noting that virulence activities of AvrPtoB and AvrPto occur independently of Pto suggest that recognition of these effectors does not strictly conform to the guard hypothesis model (Shan *et al.*, 2000a; Abramovitch *et al.*, 2003).

Using a biochemical approach, the Ran GTPase-activating protein (RanGAP2) from potato and *Nicotiana benthamiana* was shown to interact with the amino-terminal CC domain of the potato protein Rx (Sacco *et al.*, 2007; Tameling and Baulcombe, 2007). This interaction was shown to be required for Rx-mediated resistance and possibly provides an example of a protein with a known function in other cellular processes that has been co-opted for pathogen recognition (Sacco *et al.*, 2007; Tameling and Baulcombe, 2007). RanGAP2 also interacts with CC domains from the related proteins Rx2 and Gpa2. Since Gpa2 recognizes a different Avr determinant, this example suggests that the CC domain, through its associated protein, provides

**Table 5.5.** NB-LRR interacting proteins.

| Protein | Biochemical/putative activity | NB-LRR | NB-LRR interacting domain | Effect of Avr on NB-LRR interactor | Reference(s) |
|---|---|---|---|---|---|
| NRIP1 | Thiosulfate sulfur transferase | N | TIR | Cellular localization altered | Caplan *et al.* (2008b) |
| Pto | Kinase | Prf | CC | | Martin *et al.* (1993), Mucyn *et al.* (2006) |
| PBS1 | Kinase | RPS5 | CC | Cleavage | Swiderski and Innes (2001), Shao *et al.* (2003), Ade *et al.* (2007) |
| RanGAP2 | Ran GTPase activation | Rx, Rx2 | CC | Unknown | Sacco *et al.* (2007) |
| | | Gpa2 | CC | Unknown | Tameling and Baulcombe (2007) |
| RIN4 | Unknown | RPM1 | CC | Phosphorylation | Mackey *et al.* (2002) |
| | | RPS2 | Not determined | Cleavage | Axtell *et al.* (2003) |
| WRKY1/2 | Transcription factor | Mla | CC | Unknown | Shen *et al.* (2007) |
| CRT1 | ATPase/chaperone | HRT (Rx, RPS2, SSI4) | NB | Unknown | Kang *et al.* (2008) |
| Hsp90 | APTase/chaperone | N | LRR | Unknown | Hubert *et al.* (2003), Bieri *et al.* (2004), Liu *et al.* (2004), De la Fuente van Bentem *et al.* (2005) |
| | | RPM1 | Not determined | | |
| | | Mla1 | LRR | | |
| | | Mla6 | LRR | | |
| | | I-2 | LRR | | |
| PP5 | Protein phosphatase | I-2 | LRR | Unknown | De la Fuente van Bentem *et al.* (2005) |
| RIN13 | Unknown | RPM1 | NB-ARC | Unknown | Al-Daoude *et al.* (2005) |
| Sgt1 | Chaperone | Bs2 | LRR | Unknown | Bieri *et al.* (2004) |
| | | Mla1 | LRR | Unknown | Leister *et al.* (2005) |

an initial access point for Avr recognition, with the specific LRR determining which interacting Avr proteins will activate a given NB-LRR protein (Sacco *et al.*, 2007; Rairdan *et al.*, 2008).

Further evidence for an initial association of the Avr with the NB-LRR amino terminus through an interacting host protein cofactor may come from studies on the N protein, a TIR-NB-LRR from tobacco. The N protein was recently shown to interact through its TIR domain with a chloroplast protein (NRIP1) with sulfur transferase activity that in turn interacts with the Avr determinant, TMV P50 (Caplan *et al.*, 2008b). Taken together with the reported interaction between the NB-LRR of N and P50 in yeast (Ueda *et al.*, 2006), it is possible that *in planta* interactions with NB-LRR proteins are initially indirect, mediated by an amino-terminal binding protein that acts as a scaffold for R and Avr proteins to form a ternary complex that allows a direct R–Avr interaction through the LRR domains. Such a scenario may explain why Avr/R protein interactions can be detected when brought together in a heterologous system.

Lastly, CC domain-interacting proteins were described that associate with barley Mla1 and Mla6 and are members of the WRKY family of transcription factors (Shen *et al.*, 2007). This is a striking observation since the RRS1-R protein has a WRKY domain fused C-terminal to its LRR domain (Deslandes *et al.*, 2002), suggesting similar connections between transcriptional regulation by WRKY transcription factors and NB-LRR proteins. The association of Mla10 with HvWRKY2 appears to be dependent on coexpression $Avr_{A10}$, although it remains to be shown whether Avr recognition is direct or indirect in this case (Shen *et al.*, 2007).

Outside of the NB-LRR classes, reliance on a third protein by the RLK protein Cf-2 for Avr detection has also been observed (Kruger *et al.*, 2002). The *C. fulvum* protein Avr2 interacts directly with the tomato Rcr3 protease and inhibits its activity. This physical interaction has been shown to be required for induction of Cf-2-mediated HR, although an interaction between Cf-2 and Rcr3 has not been demonstrated (Kruger *et al.*, 2002; Rooney *et al.*, 2005). These observations suggest that indirect models of recognition developed for a few model pathosystems might be applicable to different classes of R proteins.

Additional NB-LRR-interacting proteins have been identified that are likely to serve different functions other than as acting as recognition cofactors. Some of these additional interacting proteins can be conceptually grouped as putative chaperones of NB-LRR proteins (Table 5.5). The protein Hsp90 has been well defined as a substrate-specific chaperone in other eukaryotic systems where it has been studied. Observations that plant Hsp90 and the proteins Rar1 and Sgt1 appear to be differentially required for the accumulation of a number of NB-LRR proteins, including Rx, RPM1, Mla1 and Mla6, and demonstrated interactions between these three proteins implicate them as R protein chaperones that may be part of ternary or higher order folding complexes (Tornero *et al.*, 2002; Hubert *et al.*, 2003; Lu *et al.*, 2003; Azevedo *et al.*, 2006). The CRT1 protein identified from *Arabidopsis* interacts with the NB domain of a number of NB-LRR proteins from both the TIR and the non-TIR class, and has an ATPase domain that most closely resembles that of Hsp90, suggesting a possible chaperone role (Kang *et al.*, 2008). The protein phosphatase 5 (PP5), a co-chaperone of Hsp90, has been shown to interact with the tomato I-2, although an essential role for PP5 in I-2 function has not been established genetically (De la Fuente van Bentem *et al.*, 2005). So far, the data accumulated for the above group of NB-LRR-interacting proteins suggests that they are likely to be differentially required chaperones whose collective roles are to facilitate the proper folding of the R proteins with which they interact, rather than to function themselves in signalling. None the less, the importance of these proteins for enabling Avr recognition and/or signalling must be acknowledged as they represent the majority of known essential factors required for the general function of *R* genes.

Indirect models of interactions between R proteins and Avr determinants provide an alternative evolutionary mode for adaptation of novel specificities from what would be expected in directly interacting protein systems. By detecting perturbations in an associated protein, such as in the guard model, or by detecting perturbations in intramolecular interactions that are stabilized by the amino-terminal interacting cofactor, it would be possible for R proteins to evolve to recognize activities of pathogen effectors that are not sequence specific. This relationship between R and Avr proteins, rather than recognition of specific 'antigen' epitopes, could afford a tolerance for more sequence diversity in Avr proteins, reducing the repertoire of pathogen 'receptors' required in plants compared to the extreme immune receptor diversification required in animal adaptive immunity. Moreover, the existence of two stages of recognition could allow expansion of specificity repertoires. An initiator interaction between an Avr and a host protein cofactor anchored at the NB-LRR protein amino terminus, followed by a recognition interaction between the LRR and Avr provides two interfaces with the pathogen elicitor protein for diversification.

## Signal initiation by NB-LRR proteins

Models of NB-LRR activation have emerged from detailed mutagenesis studies wherein mutants of NB-LRR protein are transiently expressed, usually in the model plant *N. benthamiana*, and sometimes as fragments. Expression of several NB-LRR proteins lacking their LRR domains results in an Avr-independent HR, suggesting that in the absence of pathogen elicitation, this domain plays an inhibitory role for NB-LRR signalling (Bendahmane *et al.*, 2002). However, the observations that deletion of the LRR does not always result in Avr-independent activation and that the LRR is required for point mutation induced auto-activity supports a positive role for the LRR in activation (Bendahmane *et al.*, 2002; Moffett *et al.*, 2002; Hwang and Williamson, 2003). The structural similarity of NB-LRR proteins to the well-characterized animal protein Apaf-1 and mutagenesis studies of the NB and ARC domains have led to the concept of the NB-ARC module functioning as a 'molecular switch' for activation (Takken *et al.*, 2006). In this model, recognition of the pathogen Avr induces a conformational change in the ARC, which causes an alteration of the nucleotide-binding state of the NB domain, resulting in a further conformation change that initiates signalling (Takken *et al.*, 2006). This model is supported by two NB mutations that alter *in vitro* ATPase activity of the NB domain, but result in I-2 autoactivation *in planta* (Tameling *et al.*, 2006). Further studies are needed however to determine at what stage ATP is hydrolysed in the context of full-length proteins *in planta*.

A 'perfect-fit' model of NB-LRR activation has been put forward that was developed through studies of the Rx protein, and builds on the molecular switch model (Rairdan and Moffett, 2006; Rairdan *et al.*, 2008). In this model, in the absence of Avr recognition, NB-LRR proteins are held in an inactive hair-trigger state through intramolecular interactions. A key observation from experiments with Rx was the elicitation of Avr-independent programmed cell death by a protein fragment encompassing the NB domain alone in the absence of the amino-terminal CC domain (Rairdan *et al.*, 2008), which was previously thought to be the signalling moiety of the NB-LRR proteins (Takken *et al.*, 2006). In this model, the ARC1 domain plays a role in recruiting the LRR to interact with the protein amino terminus. The NB-LRR protein remains in a constrained inactive state through the perfect fit of its ARC and LRR interactions. Elicitor recognition results in perturbation of the LRR that alters the interface between the LRR and ARC2; thus LRR perception of the Avr is sensed through the ARC2 domain, which acts as the switch that allows progression to an active state, with conformation changes resulting in release of the inhibitory intramolecular interactions and subsequent signalling through the NB domain (Rairdan and Moffett, 2006; Rairdan *et al.*, 2008).

Unlike the signalling pathways that have been defined for transmission of signals initiated at the plasma membrane to the nucleus through mitogen-activated protein kinase (MAPK) signalling cascades (see Song *et al.*, Chapter 2, this volume), the signalling pathways that lead to extreme resistance and HR remain to be defined. A number of proteins that act downstream of Avr recognition have been identified and defence activation is known to involve transcriptional reprogramming through a number of transcriptions factors (see Boyle *et al.*, Chapter 4, and Parent *et al.*, Chapter 6, this volume). However, since no direct link from an R protein to a signalling pathway is known as of yet, detailed discussion of these downstream players is beyond the scope of this chapter.

## 5.6 Localization in Function: Recognizing Avrs Where They Attack and Transmitting the Signal to the Nucleus

Recent studies have been developing a model in which cellular localization of NB-LRR plays a key role in function. Analysis of the sequences of candidate R proteins with protein localization prediction programmes suggests that various subcellular sites and organelles are destinations for NB-LRR proteins, including the cytoplasm, plasma membrane, chloroplast and nucleus (Caplan *et al.*,

2008a). Due to the poor reliability of these predictions, the true sites of R protein localization need to be determined experimentally. The proteins RIN4 and Pto have been shown to be post-translationally modified by acylation or myristoylation, respectively, directing these proteins to the plasma membrane (Kim *et al.*, 2005; de Vries *et al.*, 2006). These modifications provide a means of concentrating the associated NB-LRR proteins RPM1 and Prf at the plasma membrane, the ultimate destination of their cognate Avr determinants, AvrRPM1, AvrB and AvrPto, which are myristoylated within the host cell (Dixon *et al.*, 2000; Nimchuk *et al.*, 2000; Shan *et al.*, 2000b).

A clear example of a nuclear localized NB-LRR protein is the *Arabidopsis* RRS1-R protein, which interacts with its Avr in yeast (Deslandes *et al.*, 2003). It is possible that RRS1-R interacts with PopP2 in a different manner from the other NB-LRR proteins as its C-terminal WRKY domain extension is unique and the interacting domain is unknown. Also, RRS1-S encoded by the susceptible allele also interacts with PopP2. Both RRS1-R and PopP2 have identifiable NLS, however, the interaction appears to be required for accumulation of RRS1-R in the nucleus instead of in the cytoplasm (Deslandes *et al.*, 2003). The *Arabidopsis* protein RPS4 has a canonical bipartite NLS motif within a domain found C-terminal to the LRR, and localization to the nucleus was demonstrated to be necessary for the HR induced by RPS4 upon over-expression in tobacco leaves (Wirthmueller *et al.*, 2007). In addition, a small fraction of the N and Mla10 proteins localize to the nucleus and the fusion of a nuclear export sequence (NES) to these proteins renders them inactive (Shen *et al.*, 2007; Caplan *et al.*, 2008b). In the case of Mla10, nuclear localization may be required in order to interact with the WRKY transcription factor that is its binding partner (Shen *et al.*, 2007). However, whether this localization is required for recognition of its Avr, or for interacting with downstream signalling components remains to be elucidated. A general role for nucleocytoplasmic trafficking comes from studies of a constitutive gain-of-function mutant in SNC1, a TIR-NB-LRR protein with unknown resistance specificity. Suppressor mutants have been identified with lesions in the genes encoding members of the nucleoporin and importin-alpha families, although it is still unclear if these proteins are directly involved in NB-LRR function (Palma *et al.*, 2005; Zhang and Li, 2005; Shen and Schulze-Lefert, 2007).

## 5.7 Atypical Dominant *R* Genes

The majority of characterized *R* genes belong to defined protein classes. This in turn allows for the identification of *R* gene homologues in the absence of a defined gene-for-gene relationship. However, a number of genes have been identified that, despite showing polymorphism, do not belong to the 'typical' classes of *R* genes. This is best exemplified by the maize *Hm1* gene, the first resistance gene to be cloned, which confers resistance to the fungal ascomycete *Cochliobolus carbonum* race 1 (CCR1) that causes lethal leaf blight and ear mould disease. The *Hm1*-mediated resistance in maize is dictated by the presence of the host-specific (host-selective) toxin from CCR1, HC-toxin. The protein encoded by *Hm1* has demonstrated HC-toxin reductase (HCTR) activity, detectable only in maize cultivars that are resistant to CCR1 (Johal and Briggs, 1992; Meeley *et al.*, 1992). HC-toxin is required for pathogenicity of CCR1; thus the biochemical activity of HCTR detoxifies the HC-toxin, rendering CCR1 non-pathogenic (Johal and Briggs, 1992; Meeley *et al.*, 1992). The *Hm1* gene may be a case of a gene that normally mediates 'non-host' resistance as *Hm1* homologues appear to have evolved early and exclusively in grasses. These *Hm1* homologues mediate resistance to CCR1 in other grasses and thus the recessive non-functional version found in certain maize cultivars may be an exception rather than the rule (Sindhu *et al.*, 2008).

The pepper *Bs3* gene confers resistance to *Xanthomonas campestris* pv. *vesicatoria* strains expressing the AvrBs3 protein. The *Bs3* gene encodes a protein with homology to flavin mono-oxygenases (FMOs) (Romer *et al.*, 2007). The AvrBs3, delivered to the host cell by the type III secretion system, acts as a transcriptional activator and induces transcription of *Bs3*, whereas the *bs3* allele possesses a polymorphism in its promoter such that it is not bound to or activated by AvrBs3. Upon expression, the encoded FMO induces cell death, limiting pathogen spread (Marois *et al.*, 2002; Szurek *et al.*, 2002; Gurlebeck *et al.*, 2005).

Like *Bs3*, the rice *Xa27* gene is induced in the presence of its cognate Avr, AvrXa27 of *Xanthomonas*, which belongs to the same class of transcriptional activator as AvrBs3. Also similar to Bs3, the polymorphism responsible for susceptibility is present in the *Xa27* promoter (Gurlebeck *et al.*, 2005). The protein encoded by *Xa27* is an intronless gene encoding a protein of 113 amino acids with no sequence or structural similarity to known proteins, and identifiable homologues are found only rice (Gurlebeck *et al.*, 2005). Thus its function remains unknown, although like Bs3, it may simply induce cell death to limit pathogen spread.

Additional atypical proteins identified from *Arabidopsis* include RPW8 and RLM3 (Fig. 5.1). RPW8 is a complex of two genes (*RPW8.1* and *RPW8.2*) that confer broad-spectrum resistance to powdery mildew (Xiao *et al.*, 2001). The proteins encoded at this locus are small, largely basic proteins with a predicted TM region and signal peptide and coiled-coil region (Xiao *et al.*, 2001). The atypical *R* gene *RLM3* from *Arabidopsis* confers resistance to the fungus *Leptosphaeria maculans*, and encodes proteins with the structure TIR-NB-ARC or TIR-X that are expressed from alternatively spliced transcripts, providing the first example of a truncated NB-LRR protein of the TIR class that can be assigned a function (Staal *et al.*, 2008). Although RLM3 shares some domain structures with the dominant NB-LRR *R* genes, it is likely that it functions downstream of a primary R protein(s) since it is also required for resistance to *Botrytis cinerea* and *Alternaria* species.

The intracellular moieties of the proteins encoded by the *RFO1* and *Rpg1* genes share the RLK domain with the membrane-spanning R proteins, but lack the extracellular LRR domains (Fig. 5.1) (Rostoks *et al.*, 2002; Diener and Ausubel, 2005). Instead, RFO1 has a putative extracellular domain belonging to the wall-associated kinases (WAK), and had been previously annotated as Wall-Associated Kinase-Like 22 (WAKL22) (Diener and Ausubel, 2005). Rpg1 appears to lack any extracellular domain, but does have a predicted TM and two tandem kinase domains (Brueggeman *et al.*, 2002). The involvement of RFO1 signalling in *Arabidopsis* resistance to *Verticillium longisporum* (Johansson *et al.*, 2006) and the unique structure of Rpg1 apparently comprising only signalling and not perception domains suggests that, like RLM3, these proteins may function downstream of another R protein that mediates recognition and activates signalling; this could also be in a manner akin to the functions of the NB-LRR proteins NRC1 and NRG1 (see next section). Alternatively, RFO1 and Rpg1b could be R protein cofactors that happen to be polymorphic, analogous to the initial identification of the tomato Pto kinase as a candidate R protein.

## 5.8 The Buddy System: NB-LRR Proteins Working Together

A few cases of two NB-LRR proteins participating in signalling have been documented, suggesting that some of the candidate *R* genes in plant genomes do not encode unique recognition specificity. Instead they encode a protein that collaborates with an R-Avr pair. Two such potential partner NB-LRR proteins have been identified by virus-induced gene silencing (VIGS) screens in *N. benthamiana*. NRC1, a typical CC-NB-LRR protein, was shown to be required for HR induced through several different R proteins by either Avr proteins or auto-active mutations (Gabriels *et al.*, 2007). Intriguingly, the protein pathways involving NRC1 were induced through the extracellular LRR proteins Cf-4, Cf-9 and LeEix2, and through three intracellular NB-LRR proteins from the non-TIR class, Rx, Mi-1 and Prf (Pto) (Gabriels *et al.*, 2007). Also a CC-NB-LRR protein unrelated to NRC1, NRG1, was shown to participate in resistance mediated by the TIR-class protein N against TMV (Peart *et al.*, 2005). In contrast to NRG1, NRC1 silencing only attenuated cell death induced through the tested R proteins and did not compromise pathogen resistance to *P. syringae*, PVX or TMV (Gabriels *et al.*, 2007). NRG1 was shown to contribute towards the N-mediated and Avr-dependent resistance response to TMV, and furthermore, induced defences targeting TMV when over-expressed in the absence of N induction (Peart *et al.*, 2005).

From *Arabidopsis*, examples of Avr recognition and/or signalling involving multiple R proteins have also been described. The RPP2A and RPP2B are encoded by neighbouring genes within a cluster specifying resistance to the *H. parasitica* (At) isolate Cala2 (Sinapidou *et al.*, 2004). Both genes are required for resistance conferred by the RPP2 locus. The RPP2B protein encodes a typical TIR-class NB-LRR protein. RPP2A, however, has an unusual domain arrangement that includes two TIR-NB domains, a DUF640 domain that is unique to this R protein, and insertion of a short CC motif between the ARC and LRR, which only comprises seven repeats (TIR-NB-DUF640-TIR-NB-ARC-CC-LRR) (Sinapidou *et al.*, 2004). The *RLM3* locus involved in resistance against the fungus *L. maculans* also encodes multiple proteins whose collaboration is required for full resistance (Staal *et al.*, 2008). Like NRC1, RLM3 appears to be involved in resistance to multiple pathogens, although in the cases examined for RLM3, all the pathogens targeted were necrotrophs. Other cases of alternatively spliced transcripts reported for the TIR class include the flax *M* and *L6* genes, the tomato *Bs4*, the *Arabidopsis RPS4* gene and the *N. glutinosa N* gene. While some of the alternative transcripts generated from *R* genes may have no role in defence, as suggested for *Bs4* and *L6* (Ayliffe *et al.*, 1999; Schornack *et al.*, 2004), *RPS4* has been experimentally shown to express transcript variants when defence responses are induced. Moreover, expression of an *RPS4* gene with either its second or third intron deleted compromised its function, demonstrating a biological role for the alternative splice products (Zhang and Gassmann, 2007). In tobacco, two splice variants appear to be required for function of the *N* gene, with the second transcript accumulating in an Avr-dependent manner (Whitham *et al.*, 1994; Dinesh-Kumar *et al.*, 2000; Takabatake *et al.*, 2006). These examples might reflect a role for differentially spliced mRNAs in either regulation of expression or in more widespread cooperation of full-length NB-LRR proteins and products of splice variants in recognition and/or signalling complexes.

## 5.9 Pathogens Fight Back: Co-option and Suppression of R Proteins

The interplay between pathogen *Avr* and host *R* genes is often portrayed as a biological arms race. Continued immune pressure on pathogens by R proteins selects for mutations in *Avr* alleles that allow escape from recognition, while *R* genes duplicated within the genome diverge to evolve new recognition specificities (de Wit, 2007). Escalation of the arms race is apparent from evolution of virulence activities that suppress components of plant defence systems activated by other Avr proteins, and from counter-evolution of resistance specificities that recognize the effectors that interfere with resistance pathways.

On the pathogen side, this evolved virulence is exemplified by the *P. syringae* effectors. From *P. syringae* pv. *tomato*, the effector AvrPtoB suppresses types of programmed cell death that include HR in *N. benthamiana* (Abramovitch *et al.*, 2003), while the virulence activity of AvrPphC from *P. syringae* pv. *phaseolicola* masks the avirulence activity of the protein AvrPphF on bean (Tsiamis *et al.*, 2000). The protein AvrPtoB is modular, in which the amino terminus functions in suppression of host basal defence responses and elicits resistance through Pto/Prf, while the carboxyl terminus functions as an E3 ubiquitin ligase and inhibits cell death (Abramovitch *et al.*, 2006; de Torres *et al.*, 2006; He *et al.*, 2006; Xiao *et al.*, 2007). Deletion of the carboxyl terminus was shown to allow truncated AvrPtoB to elicit Pto-independent cell death in tomato, referred to as Rsb (Resistance suppressed by AvrPtoB C terminus), which was subsequently shown to be encoded by the *Fen* gene, a paralogue of *Pto*, which is targeted for degradation by the AvrPtoB E3 ubiquitin ligase activity (Abramovitch *et al.*, 2006; Rosebrock *et al.*, 2007). These observations suggest that tomato may have evolved Prf-dependent resistance through Fen to an ancestral AvrPtoB capable of promoting virulence by suppressing basal defences. Subsequently, AvrPtoB acquired a new function at its carboxyl terminus to suppress recognition by targeting Fen for degradation. Finally, recognition of AvrPtoB was re-established through the evolution of Pto, which detects AvrPtoB without being a target for its E3 ubiquitin ligase activity (Rosebrock *et al.*, 2007).

Recent reports have brought to light an alternative strategy for pathogens with necrotrophic lifestyles in this evolutionary battle. The host-selective toxin victorin from the necrotrophic fungus *Cochliobolus victoriae* that causes Victoria blight on oats causes symptoms that include a type of programmed cell death. This sensitivity to victorin, conferred by a single dominant gene designated *Vb*, has long been known to be genetically linked with a resistance locus *Pc-2* directed at the crown rust-causing fungus *Puccinia coronata* (Litzenberger, 1949). Victorin sensitivity has subsequently been studied in *Arabidopsis*. These studies have identified *Arabidopsis* ecotypes with victorin sensitivity conferred by a single dominant gene (Lorang *et al.*, 2004). Victorin sensitivity was shown to be dependent on a CC-NB-LRR protein called LOV1 (locus orchestrating victorin) (Lorang *et al.*, 2007; Sweat *et al.*, 2008). The evolutionary distance between *Arabidopsis* and oats strongly implies independent evolution of victorin sensitivity and this example suggests that most plants have the intrinsic ability to respond to a wide variety of pathogen-derived molecules. In this case, what is typically considered a dominant resistance gene is in fact a dominant susceptibility gene (or a recessive resistance gene) in what is sometimes referred to as an inverse gene-for-gene relationship. Single gene-mediated susceptibility to host-selective toxins may be widespread (Wolpert *et al.*, 2002) and at least one other such gene, the *Pc* gene of sorghum, maps to a cluster of genes encoding NB-LRR proteins (Nagy *et al.*, 2007). Furthermore, silencing of *Sgt1* and *Eds1*, both of which are required for NB-LRR function, decreases the susceptibility of *N. benthamiana* to the necrotroph *B. cinerea* (El Oirdi and Bouarab, 2007). Thus '*R* genes' can be both beneficial and detrimental to the plant and balancing selection between these two selection pressures may explain the presence/absence of polymorphisms seen in some *R* genes.

## 5.10 Autoimmune Responses in Hybrid Incompatibility: Consequences of Resistance Protein Diversity

Hybrid necrosis is a phenomenon that has been observed in offspring of plants with different genetic backgrounds, whether from the same or different species, owing to a genetic incompatibility in the hybrid progeny (Bomblies *et al.*, 2007). Hybrid necrosis presents a variety of symptoms including the characteristic phenotypes induced in gene-for-gene resistance. Recent observations of inappropriate R protein activation in interspecific crosses have pointed towards the involvement of incompatible interactions between the product of a polymorphic gene and the R protein with which it associates. This was first exemplified by the Cf-2 interaction with the *Ne/Rcr3* gene in hybrid offspring of *Lycopersicon esculentum* (now *Solanum lycopersicum*) and *Lycopersicon pimpinellifolium* (now *Solanum pimpinellifolium*) (Kruger *et al.*, 2002; Wulff *et al.*, 2004). These two species carry alleles that encode seven amino acid differences in the necrosis (*Ne*) gene, which was found to be the same gene that encodes Rcr3, the Cf-2 cofactor involved in indirect recognition of Avr2 (Kruger *et al.*, 2002). The combination of Cf-2 and Rcr3 proteins from different plant species that have had a period of divergent evolution appears to result in an incompatibility that causes autoactivation of Cf-2, resulting in Avr-independent cell death. The concept of hybrid incompatibility occurring by autoimmunity has been most recently demonstrated in the *Arabidopsis* model system, where a low proportion of intraspecific crosses yield progeny demonstrating hybrid necrosis. In two cases, the incompatibility was shown to involve a member of the TIR class of NB-LRR proteins (Bomblies *et al.*, 2007; Alcazar *et al.*, 2009). The deleterious effects of hybrid necrosis suggest that R proteins could contribute to the gene-flow barriers between and, to a lesser extent, within species (Bomblies and Weigel, 2007).

## 5.11 Future Prospects

Intensive research efforts have led to an explosion of characterized *R* genes and the effector proteins encoded by viral, bacterial, fungal and oomycete pathogens to which they confer recognition (Catanzariti *et al.*, 2006; Kamoun, 2006; Stavrinides *et al.*, 2008). At the same time, however, defining how Avr proteins trigger R protein-mediated resistance has proved to be challenging, as typical R proteins are difficult biochemical subjects. Moreover, genetic efforts to identify important components of R protein signalling have yielded few players, suggesting that the proteins involved in resistance might be important for viability by performing essential cellular functions. Further insight into how plants are able to resist pathogens may be gleaned from the studies that define how effectors function at a molecular level to enhance pathogen virulence by inhibiting resistance mechanisms. By identifying targets of virulence function, it may be possible to discover proteins involved in the initial interactions that lead to gene-for-gene resistance. Although some of these initial Avr targets are known, little is known about how pathogen perception is transduced into a signal that produces a response. The varied structures of the different *R* gene classes suggests that there may be numerous different initiation pathways (input) that seemingly converge upon a similar output pathway that leads to programmed cell death. The existence of distinct classes of pathogen Avr receptors are likely to exclude the possibility of finding a 'Rosetta stone' for deciphering R protein activation; however, it is possible that there will be a few key proteins for each class of *R* gene that will lead to a downstream player that will be a nexus for resistance signalling.

## References

Abramovitch, R.B., Kim, Y.J., Chen, S., Dickman, M.B. and Martin, G.B. (2003) *Pseudomonas* type III effector AvrPtoB induces plant disease susceptibility by inhibition of host programmed cell death. *EMBO Journal* 22, 60–69.

Abramovitch, R.B., Janjusevic, R., Stebbins, C.E. and Martin, G.B. (2006) Type III effector AvrPtoB requires intrinsic E3 ubiquitin ligase activity to suppress plant cell death and immunity. *Proceedings of the National Academy of Sciences, USA* 103, 2851–2856.

Ade, J., DeYoung, B.J., Golstein, C. and Innes, R.W. (2007) Indirect activation of a plant nucleotide binding site-leucine-rich repeat protein by a bacterial protease. *Proceedings of the National Academy of Sciences, USA* 104, 2531–2536.

Albrecht, M. and Takken, F.L. (2006) Update on the domain architectures of NLRs and R proteins. *Biochemical and Biophysical Research Communications* 339, 459–462.

Alcazar, R., Garcia, A.V., Parker, J.E. and Reymond, M. (2009) Incremental steps toward incompatibility revealed by *Arabidopsis* epistatic interactions modulating salicylic acid pathway activation. *Proceedings of the National Academy of Sciences, USA* 106, 334–339.

Al-Daoude, A., de Torres Zabala, M., Ko, J.H. and Grant, M. (2005) RIN13 is a positive regulator of the plant disease resistance protein RPM1. *The Plant Cell* 17, 1016–1028.

Allen, R.L., Bittner-Eddy, P.D., Grenville-Briggs, L.J., Meitz, J.C., Rehmany, A.P., Rose, L.E. and Beynon, J.L. (2004) Host–parasite coevolutionary conflict between *Arabidopsis* and downy mildew. *Science* 306, 1957–1960.

Ameline-Torregrosa, C., Wang, B.B., O'Bleness, M.S., Deshpande, S., Zhu, H., Roe, B., Young, N.D. and Cannon, S.B. (2008) Identification and characterization of nucleotide-binding site-leucine-rich repeat genes in the model plant *Medicago truncatula*. *Plant Physiology* 146, 5–21.

Anderson, P.A., Lawrence, G.J., Morrish, B.C., Ayliffe, M.A., Finnegan, E.J. and Ellis, J.G. (1997) Inactivation of the flax rust resistance gene M associated with loss of a repeated unit within the leucine-rich repeat coding region. *The Plant Cell* 9, 641–651.

Armstrong, M.R., Whisson, S.C., Pritchard, L., Bos, J.I., Venter, E., Avrova, A.O., Rehmany, A.P., Bohme, U., Brooks, K., Cherevach, I., Hamlin, N., White, B., Fraser, A., Lord, A., Quail, M.A., Churcher, C., Hall, N., Berriman, M., Huang, S., Kamoun, S., Beynon, J.L. and Birch, P.R. (2005) An ancestral oomycete locus contains late blight avirulence gene Avr3a, encoding a protein that is recognized in the host cytoplasm. *Proceedings of the National Academy of Sciences, USA* 102, 7766–7771.

Ashfield, T., Ong, L.E., Nobuta, K., Schneider, C.M. and Innes, R.W. (2004) Convergent evolution of disease resistance gene specificity in two flowering plant families. *The Plant Cell* 16, 309–318.

Axtell, M.J. and Staskawicz, B.J. (2003) Initiation of RPS2-specified disease resistance in *Arabidopsis* is coupled to the AvrRpt2-directed elimination of RIN4. *Cell* 112, 369–377.

Axtell, M.J., Chisholm, S.T., Dahlbeck, D. and Staskawicz, B.J. (2003) Genetic and molecular evidence that the *Pseudomonas syringae* type III effector protein AvrRpt2 is a cysteine protease. *Molecular Microbiology* 49, 1537–1546.

Ayliffe, M.A., Frost, D.V., Finnegan, E.J., Lawrence, G.J., Anderson, P.A. and Ellis, J.G. (1999) Analysis of alternative transcripts of the flax L6 rust resistance gene. *The Plant Journal* 17, 287–292.

Azevedo, C., Betsuyaku, S., Peart, J., Takahashi, A., Noel, L., Sadanandom, A., Casais, C., Parker, J. and Shirasu, K. (2006) Role of SGT1 in resistance protein accumulation in plant immunity. *EMBO Journal* 25, 2007–2016.

Bai, J., Pennill, L.A., Ning, J., Lee, S.W., Ramalingam, J., Webb, C.A., Zhao, B., Sun, Q., Nelson, J.C., Leach, J.E. and Hulbert, S.H. (2002) Diversity in nucleotide binding site-leucine-rich repeat genes in cereals. *Genome Research* 12, 1871–1884.

Bai, Y., Pavan, S., Zheng, Z., Zappel, N.F., Reinstadler, A., Lotti, C., De Giovanni, C., Ricciardi, L., Lindhout, P., Visser, R., Theres, K. and Panstruga, R. (2008) Naturally occurring broad-spectrum powdery mildew resistance in a Central American tomato accession is caused by loss of mlo function. *Molecular Plant–Microbe Interactions* 21, 30–39.

Bakker, E., Butterbach, P., Rouppe van der Voort, J., van der Vossen, E., van Vliet, J., Bakker, J. and Goverse, A. (2003) Genetic and physical mapping of homologues of the virus resistance gene Rx1 and the cyst nematode resistance gene Gpa2 in potato. *Theoreical and Applied Genetics* 106, 1524–1531.

Ballvora, A., Pierre, M., van den Ackerveken, G., Schornack, S., Rossier, O., Ganal, M., Lahaye, T. and Bonas, U. (2001) Genetic mapping and functional analysis of the tomato Bs4 locus governing recognition of the *Xanthomonas campestris* pv. *vesicatoria* AvrBs4 protein. *Molecular Plant–Microbe Interactions* 14, 629–638.

Ballvora, A., Ercolano, M.R., Weiss, J., Meksem, K., Bormann, C.A., Oberhagemann, P., Salamini, F. and Gebhardt, C. (2002) The R1 gene for potato resistance to late blight (*Phytophthora infestans*) belongs to the leucine zipper/NBS/LRR class of plant resistance genes. *The Plant Journal* 30, 361–371.

Beauchemin, C., Boutet, N. and Laliberte, J.F. (2007) Visualization of the interaction between the precursors of VPg, the viral protein linked to the genome of turnip mosaic virus, and the translation eukaryotic initiation factor iso 4E *in planta*. *Journal of Virology* 81, 775–782.

Belfanti, E., Silfverberg-Dilworth, E., Tartarini, S., Patocchi, A., Barbieri, M., Zhu, J., Vinatzer, B.A., Gianfranceschi, L., Gessler, C. and Sansavini, S. (2004) The HcrVf2 gene from a wild apple confers scab resistance to a transgenic cultivated variety. *Proceedings of the National Academy of Sciences, USA* 101, 886–890.

Bendahmane, A., Kohm, B.A., Dedi, C. and Baulcombe, D.C. (1995) The coat protein of potato virus X is a strain-specific elicitor of Rx1-mediated virus resistance in potato. *The Plant Journal* 8, 933–941.

Bendahmane, A., Kanyuka, K. and Baulcombe, D.C. (1999) The Rx gene from potato controls separate virus resistance and cell death responses. *The Plant Cell* 11, 781–792.

Bendahmane, A., Querci, M., Kanyuka, K. and Baulcombe, D.C. (2000) *Agrobacterium* transient expression system as a tool for the isolation of disease resistance genes: application to the Rx2 locus in potato. *The Plant Journal* 21, 73–81.

Bendahmane, A., Farnham, G., Moffett, P. and Baulcombe, D.C. (2002) Constitutive gain-of-function mutants in a nucleotide binding site-leucine rich repeat protein encoded at the Rx locus of potato. *The Plant Journal* 32, 195–204.

Bent, A.F., Kunkel, B.N., Dahlbeck, D., Brown, K.L., Schmidt, R., Giraudat, J., Leung, J. and Staskawicz, B.J. (1994) RPS2 of *Arabidopsis thaliana*: a leucine-rich repeat class of plant disease resistance genes. *Science* 265, 1856–1860.

Bieri, S., Mauch, S., Shen, Q.H., Peart, J., Devoto, A., Casais, C., Ceron, F., Schulze, S., Steinbiss, H.H., Shirasu, K. and Schulze-Lefert, P. (2004) RAR1 positively controls steady state levels of barley MLA resistance proteins and enables sufficient MLA6 accumulation for effective resistance. *The Plant Cell* 16, 3480–3495.

Biffen, R.H. (1905) Mendel's laws of inheritance and wheat breeding. *Journal of Agricultural Science* 1, 4–48.

Bittner-Eddy, P.D., Crute, I.R., Holub, E.B. and Beynon, J.L. (2000) RPP13 is a simple locus in

*Arabidopsis thaliana* for alleles that specify downy mildew resistance to different avirulence determinants in *Peronospora parasitica*. *The Plant Journal* 21, 177–188.

Bomblies, K. and Weigel, D. (2007) Hybrid necrosis: autoimmunity as a potential gene-flow barrier in plant species. *Nature Reviews Genetics* 8, 382–393.

Bomblies, K., Lempe, J., Epple, P., Warthmann, N., Lanz, C., Dangl, J.L. and Weigel, D. (2007) Autoimmune response as a mechanism for a Dobzhansky-Muller-type incompatibility syndrome in plants. *Plos Biology* 5, e236.

Bonas, U., Conrads-Strauch, J. and Balbo, I. (1993) Resistance in tomato to *Xanthomonas campestris* pv *vesicatoria* is determined by alleles of the pepper-specific avirulence gene avrBs3. *Molecular Genetics and Genomics* 238, 261–269.

Borhan, M.H., Holub, E.B., Beynon, J.L., Rozwadowski, K. and Rimmer, S.R. (2004) The *Arabidopsis* TIR-NB-LRR gene RAC1 confers resistance to *Albugo candida* (white rust) and is dependent on EDS1 but not PAD4. *Molecular Plant–Microbe Interactions* 17, 711–719.

Botella, M.A., Parker, J.E., Frost, L.N., Bittner-Eddy, P.D., Beynon, J.L., Daniels, M.J., Holub, E.B. and Jones, J.D. (1998) Three genes of the *Arabidopsis* RPP1 complex resistance locus recognize distinct *Peronospora parasitica* avirulence determinants. *The Plant Cell* 10, 1847–1860.

Brommonschenkel, S.H., Frary, A. and Tanksley, S.D. (2000) The broad-spectrum tospovirus resistance gene Sw-5 of tomato is a homolog of the root-knot nematode resistance gene Mi. *Molecular Plant–Microbe Interactions* 13, 1130–1138.

Brueggeman, R., Rostoks, N., Kudrna, D., Kilian, A., Han, F., Chen, J., Druka, A., Steffenson, B. and Kleinhofs, A. (2002) The barley stem rust-resistance gene Rpg1 is a novel disease-resistance gene with homology to receptor kinases. *Proceedings of the National Academy of Sciences, USA* 99, 9328–9333.

Bruun-Rasmussen, M., Moller, I.S., Tulinius, G., Hansen, J.K., Lund, O.S. and Johansen, I.E. (2007) The same allele of translation initiation factor 4E mediates resistance against two potyvirus spp. in *Pisum sativum*. *Molecular Plant–Microbe Interactions* 20, 1075–1082.

Bryan, G.T., Wu, K.S., Farrall, L., Jia, Y., Hershey, H.P., McAdams, S.A., Faulk, K.N., Donaldson, G.K., Tarchini, R. and Valent, B. (2000) tA single amino acid difference distinguishes resistant and susceptible alleles of the rice blast resistance gene Pi-ta. *The Plant Cell* 12, 2033–2046.

Buschges, R., Hollricher, K., Panstruga, R., Simons, G., Wolter, M., Frijters, A., van Daelen, R., van der Lee, T., Diergaarde, P., Groenendijk, J., Topsch, S., Vos, P., Salamini, F. and Schulze-Lefert, P. (1997) The barley Mlo gene: a novel control element of plant pathogen resistance. *Cell* 88, 695–705.

Cai, D., Kleine, M., Kifle, S., Harloff, H.J., Sandal, N.N., Marcker, K.A., Klein-Lankhorst, R.M., Salentijn, E.M., Lange, W., Stiekema, W.J., Wyss, U., Grundler, F.M. and Jung, C. (1997) Positional cloning of a gene for nematode resistance in sugar beet. *Science* 275, 832–834.

Caicedo, A.L. (2008) Geographic diversity cline of *R* gene homologs in wild populations of *Solanum pimpinellifolium* (Solanaceae). *American Journal of Botany* 95, 393–398.

Caicedo, A.L. and Schaal, B.A. (2004) Heterogeneous evolutionary processes affect *R* gene diversity in natural populations of *Solanum pimpinellifolium*. *Proceedings of the National Academy of Sciences, USA* 101, 17444–17449.

Cannon, S.B., Zhu, H., Baumgarten, A.M., Spangler, R., May, G., Cook, D.R. and Young, N.D. (2002) Diversity, distribution, and ancient taxonomic relationships within the TIR and non-TIR NBS-LRR resistance gene subfamilies. *Journal of Molecular Evolution* 54, 548–562.

Caplan, J., Padmanabhan, M. and Dinesh-Kumar, S.P. (2008a) Plant NB-LRR immune receptors: from recognition to transcriptional reprogramming. *Cell Host & Microbe* 3, 126–135.

Caplan, J.L., Mamillapalli, P., Burch-Smith, T.M., Czymmek, K. and Dinesh-Kumar, S.P. (2008b) Chloroplastic protein NRIP1 mediates innate immune receptor recognition of a viral effector. *Cell* 132, 449–462.

Catanzariti, A.M., Dodds, P.N., Lawrence, G.J., Ayliffe, M.A. and Ellis, J.G. (2006) Haustorially expressed secreted proteins from flax rust are highly enriched for avirulence elicitors. *The Plant Cell* 18, 243–256.

Charron, C., Nicolai, M., Gallois, J.L., Robaglia, C., Moury, B., Palloix, A. and Caranta, C. (2008) Natural variation and functional analyses provide evidence for co-evolution between plant eIF4E and potyviral VPg. *The Plant Journal* 54, 56–68.

Chattopadhyaya, R. and Pal, A. (2008) Three-dimensional models of NB-ARC domains of disease resistance proteins in tomato, *Arabidopsis*, and flax. *Journal of Biomolecular Structure and Dynamics* 25, 357–371.

Chen, X., Shang, J., Chen, D., Lei, C., Zou, Y., Zhai, W., Liu, G., Xu, J., Ling, Z., Cao, G., Ma, B., Wang, Y., Zhao, X., Li, S. and Zhu, L. (2006) A B-lectin receptor kinase gene conferring rice blast resistance. *The Plant Journal* 46, 794–804.

Chinchilla, D., Zipfel, C., Robatzek, S., Kemmerling, B., Nurnberger, T., Jones, J.D., Felix, G. and Boller, T. (2007) A flagellin-induced complex of the receptor FLS2 and BAK1 initiates plant defence. *Nature* 448, 497–500.

Chisholm, S.T., Mahajan, S.K., Whitham, S.A., Yamamoto, M.L. and Carrington, J.C. (2000) Cloning of the *Arabidopsis RTM1* gene, which controls restriction of long-distance movement of tobacco etch virus. *Proceedings of the National Academy of Sciences, USA* 97, 489–494.

Chisholm, S.T., Coaker, G., Day, B. and Staskawicz, B.J. (2006) Host–microbe interactions: shaping the evolution of the plant immune response. *Cell* 124, 803–814.

Chu, Z., Yuan, M., Yao, J., Ge, X., Yuan, B., Xu, C., Li, X., Fu, B., Li, Z., Bennetzen, J.L., Zhang, Q. and Wang, S. (2006) Promoter mutations of an essential gene for pollen development result in disease resistance in rice. *Genes & Development* 20, 1250–1255.

Clark, R.M., Schweikert, G., Toomajian, C., Ossowski, S., Zeller, G., Shinn, P., Warthmann, N., Hu, T.T., Fu, G., Hinds, D.A., Chen, H., Frazer, K.A., Huson, D.H., Scholkopf, B., Nordborg, M., Ratsch, G., Ecker, J.R. and Weigel, D. (2007) Common sequence polymorphisms shaping genetic diversity in *Arabidopsis thaliana*. *Science* 317, 338–342.

Cloutier, S., McCallum, B.D., Loutre, C., Banks, T.W., Wicker, T., Feuillet, C., Keller, B. and Jordan, M.C. (2007) Leaf rust resistance gene *Lr1*, isolated from bread wheat (*Triticum aestivum* L.) is a member of the large psr567 gene family. *Plant Molecular Biology* 65, 93–106.

Collins, N., Drake, J., Ayliffe, M., Sun, Q., Ellis, J., Hulbert, S. and Pryor, T. (1999) Molecular characterization of the maize Rp1-D rust resistance haplotype and its mutants. *The Plant Cell* 11, 1365–1376.

Cooley, M.B., Pathirana, S., Wu, H.J., Kachroo, P. and Klessig, D.F. (2000) Members of the *Arabidopsis* HRT/RPP8 family of resistance genes confer resistance to both viral and oomycete pathogens. *The Plant Cell* 12, 663–676.

Dangl, J.L. and Jones, J.D. (2001) Plant pathogens and integrated defence responses to infection. *Nature* 411, 826–833.

De la Fuente van Bentem, S., Vossen, J.H., de Vries, K.J., van Wees, S., Tameling, W.I., Dekker, H.L., de Koster, C.G., Haring, M.A., Takken, F.L. and Cornelissen, B.J. (2005) Heat shock protein 90 and its co-chaperone protein phosphatase 5 interact with distinct regions of the tomato I-2 disease resistance protein. *The Plant Journal* 43, 284–298.

de Torres, M., Mansfield, J.W., Grabov, N., Brown, I.R., Ammouneh, H., Tsiamis, G., Forsyth, A., Robatzek, S., Grant, M. and Boch, J. (2006) *Pseudomonas syringae* effector AvrPtoB suppresses basal defence in *Arabidopsis*. *The Plant Journal* 47, 368–382.

de Vries, J.S., Andriotis, V.M., Wu, A.J. and Rathjen, J.P. (2006) Tomato Pto encodes a functional N-myristoylation motif that is required for signal transduction in *Nicotiana benthamiana*. *The Plant Journal* 45, 31–45.

de Wit, P.J. (2007) How plants recognize pathogens and defend themselves. *Cell and Molecular Life Sciences* 64, 2726–2732.

de Wit, P.J.G.M. and Spikman, G. (1982) Evidence for the occurrence of race and cultivar-specific elicitors of necrosis in inter-cellular fluids of compatible interactions of *Cladosporium fulvum* and tomato. *Physiological Plant Pathology* 21, 1–8.

Debener, T., Lehnackers, H., Arnold, M. and Dangl, J.L. (1991) Identification and molecular mapping of a single *Arabidopsis thaliana* locus determining resistance to a phytopathogenic *Pseudomonas syringae* isolate. *The Plant Journal* 1, 289–302.

Deslandes, L., Olivier, J., Theulieres, F., Hirsch, J., Feng, D.X., Bittner-Eddy, P., Beynon, J. and Marco, Y. (2002) Resistance to *Ralstonia solanacearum* in *Arabidopsis thaliana* is conferred by the recessive *RRS1-R* gene, a member of a novel family of resistance genes. *Proceedings of the National Academy of Sciences, USA* 99, 2404–2409.

Deslandes, L., Olivier, J., Peeters, N., Feng, D.X., Khounlotham, M., Boucher, C., Somssich, I., Genin, S. and Marco, Y. (2003) Physical interaction between RRS1-R, a protein conferring resistance to bacterial wilt, and PopP2, a type III effector targeted to the plant nucleus. *Proceedings of the National Academy of Sciences, USA* 100, 8024–8029.

Diaz, J.A., Nieto, C., Moriones, E., Truniger, V. and Aranda, M.A. (2004) Molecular characterization of a melon necrotic spot virus strain that overcomes the resistance in melon and nonhost plants. *Molecular Plant–Microbe Interactions* 17, 668–675.

Diener, A.C. and Ausubel, F.M. (2005) RESISTANCE TO FUSARIUM OXYSPORUM 1, a dominant *Arabidopsis* disease-resistance gene, is not race specific. *Genetics* 171, 305–321.

Dinesh-Kumar, S.P., Tham, W.H. and Baker, B.J. (2000) Structure-function analysis of the tobacco mosaic virus resistance gene *N*. *Proceedings of the National Academy of Sciences, USA* 97, 14789–14794.

Ding, J., Araki, H., Wang, Q., Zhang, P., Yang, S., Chen, J.Q. and Tian, D. (2007a) Highly asymmetric rice genomes. *BMC Genomics* 8, 154.

Ding, J., Cheng, H., Jin, X., Araki, H., Yang, Y. and Tian, D. (2007b) Contrasting patterns of evolution between allelic groups at a single locus in *Arabidopsis*. *Genetica* 129, 235–242.

Ding, J., Zhang, W., Jing, Z., Chen, J.Q. and Tian, D. (2007c) Unique pattern of *R*-gene variation within populations in *Arabidopsis*. *Molecular Genetics and Genomics* 277, 619–629.

Dixon, M.S., Jones, D.A., Keddie, J.S., Thomas, C.M., Harrison, K. and Jones, J.D. (1996) The tomato Cf-2 disease resistance locus comprises two functional genes encoding leucine-rich repeat proteins. *Cell* 84, 451–459.

Dixon, M.S., Hatzixanthis, K., Jones, D.A., Harrison, K. and Jones, J.D. (1998) The tomato Cf-5 disease resistance gene and six homologs show pronounced allelic variation in leucine-rich repeat copy number. *The Plant Cell* 10, 1915–1925.

Dixon, M.S., Golstein, C., Thomas, C.M., van Der Biezen, E.A. and Jones, J.D. (2000) Genetic complexity of pathogen perception by plants: the example of *Rcr3*, a tomato gene required specifically by Cf-2. *Proceedings of the National Academy of Sciences, USA* 97, 8807–8814.

Dodds, P.N., Lawrence, G.J. and Ellis, J.G. (2001a) Contrasting modes of evolution acting on the complex N locus for rust resistance in flax. *The Plant Journal* 27, 439–453.

Dodds, P.N., Lawrence, G.J. and Ellis, J.G. (2001b) Six amino acid changes confined to the leucine-rich repeat beta-strand/beta-turn motif determine the difference between the P and P2 rust resistance specificities in flax. *The Plant Cell* 13, 163–178.

Dodds, P.N., Lawrence, G.J., Catanzariti, A.M., Ayliffe, M.A. and Ellis, J.G. (2004) The *Melampsora lini* AvrL567 avirulence genes are expressed in haustoria and their products are recognized inside plant cells. *The Plant Cell* 16, 755–768.

Dodds, P.N., Lawrence, G.J., Catanzariti, A.M., Teh, T., Wang, C.I., Ayliffe, M.A., Kobe, B. and Ellis, J.G. (2006) Direct protein interaction underlies gene-for-gene specificity and coevolution of the flax resistance genes and flax rust avirulence genes. *Proceedings of the National Academy of Sciences, USA* 103, 8888–8893.

Eitas, T.K., Nimchuk, Z.L. and Dangl, J.L. (2008) *Arabidopsis* TAO1 is a TIR-NB-LRR protein that contributes to disease resistance induced by the *Pseudomonas syringae* effector AvrB. *Proceedings of the National Academy of Sciences, USA* 105, 6475–6480.

El Oirdi, M. and Bouarab, K. (2007) Plant signalling components EDS1 and SGT1 enhance disease caused by the necrotrophic pathogen *Botrytis cinerea*. *New Phytologist* 175, 131–139.

Ellis, J., Dodds, P. and Pryor, T. (2000) Structure, function and evolution of plant disease resistance genes. *Current Opinion in Plant Biology* 3, 278–284.

Erickson, F.L., Holzberg, S., Calderon-Urrea, A., Handley, V., Axtell, M., Corr, C. and Baker, B. (1999) The helicase domain of the TMV replicase proteins induces the N-mediated defence response in tobacco. *The Plant Journal* 18, 67–75.

Ernst, K., Kumar, A., Kriseleit, D., Kloos, D.U., Phillips, M.S. and Ganal, M.W. (2002) The broad-spectrum potato cyst nematode resistance gene (*Hero*) from tomato is the only member of a large gene family of NBS-LRR genes with an unusual amino acid repeat in the LRR region. *The Plant Journal* 31, 127–136.

Farnham, G. and Baulcombe, D.C. (2006) Artificial evolution extends the spectrum of viruses that are targeted by a disease-resistance gene from potato. *Proceedings of the National Academy of Sciences, USA* 103, 18828–18833.

Feuillet, C., Travella, S., Stein, N., Albar, L., Nublat, A. and Keller, B. (2003) Map-based isolation of the leaf rust disease resistance gene *Lr10* from the hexaploid wheat (*Triticum aestivum* L.) genome. *Proceedings of the National Academy of Sciences, USA* 100, 15253–15258.

Flor, H.H. (1942) Inheritance of pathogenicity in *Melampsora lini*. *Phytopathology* 32, 653–669.

Flor, H.H. (1946) Genetics of pathogenicity in *Melampsora lini*. . *Journal of Agricultural Research* 73, 335–357.

Flor, H.H. (1947) Inheritance of reaction to rust in flax. *Journal of Agricultural Research* 74, 241–262.

Fritz-Laylin, L.K., Krishnamurthy, N., Tor, M., Sjolander, K.V. and Jones, J.D. (2005) Phylogenomic analysis of the receptor-like proteins of rice and *Arabidopsis*. *Plant Physiology* 138, 611–623.

Frost, D., Way, H., Howles, P., Luck, J., Manners, J., Hardham, A., Finnegan, J. and Ellis, J. (2004) Tobacco transgenic for the flax rust resistance gene *L* expresses allele-specific activation of defense responses. *Molecular Plant–Microbe Interactions* 17, 224–232.

Gabriels, S.H., Vossen, J.H., Ekengren, S.K., van Ooijen, G., Abd-El-Haliem, A.M., van den Berg, G.C., Rainey, D.Y., Martin, G.B., Takken, F.L., de Wit, P.J. and Joosten, M.H. (2007) An NB-LRR protein required for HR signalling mediated by both extra- and intracellular resistance proteins. *The Plant Journal* 50, 14–28.

Gao, H. and Bhattacharyya, M.K. (2008) The soybean-*Phytophthora* resistance locus Rps1-k encompasses coiled coil-nucleotide binding-leucine rich repeat-like genes and repetitive sequences. *BMC Plant Biology* 8, 29.

Gao, H., Narayanan, N.N., Ellison, L. and Bhattacharyya, M.K. (2005) Two classes of highly similar coiled coil-nucleotide binding-leucine rich repeat genes isolated from the Rps1-k locus encode *Phytophthora* resistance in soybean. *Molecular Plant–Microbe Interactions* 18, 1035–1045.

Gao, Z., Johansen, E., Eyers, S., Thomas, C.L., Noel Ellis, T.H. and Maule, A.J. (2004) The potyvirus recessive resistance gene, *sbm1*, identifies a novel role for translation initiation factor eIF4E in cell-to-cell trafficking. *The Plant Journal* 40, 376–385.

Gassmann, W., Hinsch, M.E. and Staskawicz, B.J. (1999) The *Arabidopsis* RPS4 bacterial-resistance gene is a member of the TIR-NBS-LRR family of disease-resistance genes. *The Plant Journal* 20, 265–277.

Gomez-Gomez, L. and Boller, T. (2000) FLS2: an LRR receptor-like kinase involved in the perception of the bacterial elicitor flagellin in *Arabidopsis*. *Mol Cell* 5, 1003–1011.

Grant, M.R., Godiard, L., Straube, E., Ashfield, T., Lewald, J., Sattler, A., Innes, R.W. and Dangl, J.L. (1995) Structure of the *Arabidopsis RPM1* gene enabling dual specificity disease resistance. *Science* 269, 843–846.

Grant, M.R., McDowell, J.M., Sharpe, A.G., de Torres Zabala, M., Lydiate, D.J. and Dangl, J.L. (1998) Independent deletions of a pathogen-resistance gene in *Brassica* and *Arabidopsis*. *Proceedings of the National Academy of Sciences, USA* 95, 15843–15848.

Grube, R.C., Radwanski, E.R. and Jahn, M. (2000) Comparative genetics of disease resistance within the Solanaceae. *Genetics* 155, 873–887.

Gurlebeck, D., Szurek, B. and Bonas, U. (2005) Dimerization of the bacterial effector protein AvrBs3 in the plant cell cytoplasm prior to nuclear import. *The Plant Journal* 42, 175–187.

Halterman, D.A. and Wise, R.P. (2004) A single-amino acid substitution in the sixth leucine-rich repeat of barley MLA6 and MLA13 alleviates dependence on RAR1 for disease resistance signaling. *The Plant Journal* 38, 215–226.

Halterman, D., Zhou, F., Wei, F., Wise, R.P. and Schulze-Lefert, P. (2001) The MLA6 coiled-coil, NBS-LRR protein confers AvrMla6-dependent resistance specificity to *Blumeria graminis* f. sp. *hordei* in barley and wheat. *The Plant Journal* 25, 335–348.

Hammond-Kosack, K.E., Jones, D.A. and Jones, J. (1994) Identification of two genes required in tomato for full Cf-9-dependent resistance to *Cladosporium fulvum*. *The Plant Cell* 6, 361–374.

He, P., Shan, L., Lin, N.C., Martin, G.B., Kemmerling, B., Nurnberger, T. and Sheen, J. (2006) Specific bacterial suppressors of MAMP signaling upstream of MAPKKK in *Arabidopsis* innate immunity. *Cell* 125, 563–575.

Heath, M.C. (2000) Hypersensitive response-related death. *Plant Molecular Biology* 44, 321–334.

Hinsch, M. and Staskawicz, B. (1996) Identification of a new *Arabidopsis* disease resistance locus, RPs4, and cloning of the corresponding avirulence gene, *avrRps4*, from *Pseudomonas syringae* pv. *pisi*. *Molecular Plant–Microbe Interactions* 9, 55–61.

Howles, P., Lawrence, G., Finnegan, J., McFadden, H., Ayliffe, M., Dodds, P. and Ellis, J. (2005) Autoactive alleles of the flax L6 rust resistance gene induce non-race-specific rust resistance associated with the hypersensitive response. *Molecular Plant–Microbe Interactions* 18, 570–582.

Huang, L., Brooks, S.A., Li, W., Fellers, J.P., Trick, H.N. and Gill, B.S. (2003) Map-based cloning of leaf rust resistance gene *Lr21* from the large and polyploid genome of bread wheat. *Genetics* 164, 655–664.

Huang, S., van der Vossen, E.A., Kuang, H., Vleeshouwers, V.G., Zhang, N., Borm, T.J., van Eck, H.J., Baker, B., Jacobsen, E. and Visser, R.G. (2005) Comparative genomics enabled the isolation of the *R3*a late blight resistance gene in potato. *The Plant Journal* 42, 251–261.

Hubert, D.A., Tornero, P., Belkhadir, Y., Krishna, P., Takahashi, A., Shirasu, K. and Dangl, J.L. (2003) Cytosolic Hsp90 associates with and modulates the *Arabidopsis* RPM1 disease resistance protein. *EMBO Journal* 22, 5679–5689.

Hwang, C.F. and Williamson, V.M. (2003) Leucine-rich repeat-mediated intramolecular interactions in nematode recognition and cell death signaling by the tomato resistance protein Mi. *The Plant Journal* 34, 585–593.

Hwang, C.F., Bhakta, A.V., Truesdell, G.M., Pudlo, W.M. and Williamson, V.M. (2000) Evidence for a role of the N terminus and leucine-rich repeat region of the Mi gene product in regulation of localized cell death. *The Plant Cell* 12, 1319–1329.

Ishibashi, K., Masuda, K., Naito, S., Meshi, T. and Ishikawa, M. (2007) An inhibitor of viral RNA replication is encoded by a plant resistance gene. *Proceedings of the National Academy of Sciences, USA* 104, 13833–13838.

Iyer, A.S. and McCouch, S.R. (2004) The rice bacterial blight resistance gene *xa5* encodes a novel form of disease resistance. *Molecular Plant–Microbe Interactions* 17, 1348–1354.

Jarosch, B., Kogel, K.H. and Schaffrath, U. (1999) The ambivalence of the barley Mlo locus: mutations conferring resistance against powdery mildew (*Blumeria graminis* f. sp. *hordei*) enhance susceptibility to the rice blast fungus *Magnaporthe grisea. Molecular Plant–Microbe Interactions* 12, 508–514.

Jebanathirajah, J.A., Peri, S. and Pandey, A. (2002) Toll and interleukin-1 receptor (TIR) domain-containing proteins in plants: a genomic perspective. *Trends in Plant Science* 7, 388–391.

Jenner, C., Hitchin, E., Mansfield, J., Walters, K., Betteridge, P., Teverson, D. and Taylor, J. (1991) Gene-for-gene interactions between *Pseudomonas syringae* pv. *phaseolicola* and *Phaseolus. Molecular Plant–Microbe Interactions* 4, 553–562.

Jia, Y., McAdams, S.A., Bryan, G.T., Hershey, H.P. and Valent, B. (2000) Direct interaction of resistance gene and avirulence gene products confers rice blast resistance. *EMBO Journal* 19, 4004–4014.

Johal, G.S. and Briggs, S.P. (1992) Reductase activity encoded by the HM1 disease resistance gene in maize. *Science* 258, 985–987.

Johansson, A., Staal, J. and Dixelius, C. (2006) Early responses in the *Arabidopsis-Verticillium longisporum* pathosystem are dependent on NDR1, JA- and ET-associated signals via cytosolic NPR1 and RFO1. *Molecular Plant–Microbe Interactions* 19, 958–969.

Jones, D.A. and Jones, J.D.G. (1997) The role of leucine-rich repeat proteins in plant defences. *Advances in Botanical Research* 24, 89–167.

Joobeur, T., King, J.J., Nolin, S.J., Thomas, C.E. and Dean, R.A. (2004) The Fusarium wilt resistance locus Fom-2 of melon contains a single resistance gene with complex features. *The Plant Journal* 39, 283–297.

Joosten, M.H., Cozijnsen, T.J. and De Wit, P.J. (1994) Host resistance to a fungal tomato pathogen lost by a single base-pair change in an avirulence gene. *Nature* 367, 384–386.

Kajava, A.V. (1998) Structural diversity of leucine-rich repeat proteins. *Journal of Molecular Biology* 277, 519–527.

Kamoun, S. (2006) A catalogue of the effector secretome of plant pathogenic oomycetes. *Annual Review of Phytopathology* 44, 41–60.

Kang, B.C., Yeam, I., Frantz, J.D., Murphy, J.F. and Jahn, M.M. (2005) The pvr1 locus in *Capsicum* encodes a translation initiation factor eIF4E that interacts with tobacco etch virus VPg. *The Plant Journal* 42, 392–405.

Kang, H.G., Kuhl, J.C., Kachroo, P. and Klessig, D.F. (2008) CRT1, an *Arabidopsis* ATPase that interacts with diverse resistance proteins and modulates disease resistance to turnip crinkle virus. *Cell Host & Microbe* 3, 48–57.

Kawchuk, L.M., Hachey, J., Lynch, D.R., Kulcsar, F., van Rooijen, G., Waterer, D.R., Robertson, A., Kokko, E., Byers, R., Howard, R.J., Fischer, R. and Prufer, D. (2001) Tomato Ve disease resistance genes encode cell surface-like receptors. *Proceedings of the National Academy of Sciences, USA* 98, 6511–6515.

Khan, M.A., Miyoshi, H., Ray, S., Natsuaki, T., Suehiro, N. and Goss, D.J. (2006) Interaction of genome-linked protein (VPg) of turnip mosaic virus with wheat germ translation initiation factors eIFiso4E and eIFiso4F. *Journal of Biological Chemistry* 281, 28002–28010.

Kim, H.S., Desveaux, D., Singer, A.U., Patel, P., Sondek, J. and Dangl, J.L. (2005) The *Pseudomonas syringae* effector AvrRpt2 cleaves its C-terminally acylated target, RIN4, from *Arabidopsis* membranes to block RPM1 activation. *Proceedings of the National Academy of Sciences, USA* 102, 6496–6501.

Kim, Y.J., Lin, N.C. and Martin, G.B. (2002) Two distinct *Pseudomonas* effector proteins interact with the Pto kinase and activate plant immunity. *Cell* 109, 589–598.

Kneller, E.L., Rakotondrafara, A.M. and Miller, W.A. (2006) Cap-independent translation of plant viral RNAs. *Virus Research* 119, 63–75.

Kohler, A., Rinaldi, C., Duplessis, S., Baucher, M., Geelen, D., Duchaussoy, F., Meyers, B.C., Boerjan, W. and Martin, F. (2008) Genome-wide identification of NBS resistance genes in *Populus trichocarpa. Plant Molecular Biology* 66, 619–636.

Kruger, J., Thomas, C.M., Golstein, C., Dixon, M.S., Smoker, M., Tang, S., Mulder, L. and Jones, J.D. (2002) A tomato cysteine protease required for Cf-2-dependent disease resistance and suppression of autonecrosis. *Science* 296, 744–747.

Kumar, J., Huckelhoven, R., Beckhove, U., Nagarajan, S. and Kogel, K.H. (2001) A compromised Mlo pathway affects the response of barley to the necrotrophic fungus *Bipolaris sorokiniana* (teleomorph: *Cochliobolus sativus*) and its toxins. *Phytopathology* 91, 127–133.

Lanfermeijer, F.C., Dijkhuis, J., Sturre, M.J., de Haan, P. and Hille, J. (2003) Cloning and characterization of the durable tomato mosaic virus resistance gene *Tm-2(2)* from *Lycopersicon esculentum. Plant Molecular Biology* 52, 1037–1049.

Lanfermeijer, F.C., Warmink, J. and Hille, J. (2005) The products of the broken *Tm-2* and the durable *Tm-2(2)* resistance genes from tomato differ in four amino acids. *Journal of Experimental Botany* 56, 2925–2933.

Lawrence, G.J., Finnegan, E.J., Ayliffe, M.A. and Ellis, J.G. (1995) The *L6* gene for flax rust resistance is related to the *Arabidopsis* bacterial resistance gene *RPS2* and the tobacco viral resistance gene *N. The Plant Cell* 7, 1195–1206.

Leister, R.T., Dahlbeck, D., Day, B., Li, Y., Chesnokova, O. and Staskawicz, B.J. (2005) Molecular genetic evidence for the role of SGT1 in the intramolecular complementation of Bs2 protein activity in *Nicotiana benthamiana. The Plant Cell* 17, 1268–1278.

Leonard, S., Plante, D., Wittmann, S., Daigneault, N., Fortin, M.G. and Laliberte, J.F. (2000) Complex formation between potyvirus VPg and translation eukaryotic initiation factor 4E correlates with virus infectivity. *Journal of Virology* 74, 7730–7737.

Leonard, S., Viel, C., Beauchemin, C., Daigneault, N., Fortin, M.G. and Laliberte, J.F. (2004) Interaction of VPg-Pro of turnip mosaic virus with the translation initiation factor 4E and the poly(A)-binding protein *in planta. Journal of General Virology* 85, 1055–1063.

Lin, F., Chen, S., Que, Z., Wang, L., Liu, X. and Pan, Q. (2007) The blast resistance gene *Pi37* encodes a nucleotide binding site leucine-rich repeat protein and is a member of a resistance gene cluster on rice chromosome 1. *Genetics* 177, 1871–1880.

Litzenberger, S.C. (1949) Nature of susceptibility to *Helminthosporium victoriae* and resistance to *Puccinia coronata* in Victoria oats. *Phytopathology* 39, 300–318.

Liu, J.J. and Ekramoddoullah, A.K. (2003) Isolation, genetic variation and expression of TIR-NBS-LRR resistance gene analogs from western white pine (*Pinus monticola* Dougl. ex. D. Don.). *Molecular Genetics and Genomics* 270, 432–441.

Liu, X., Lin, F., Wang, L. and Pan, Q. (2007) The *in silico* map-based cloning of Pi36, a rice coiled-coil nucleotide-binding site leucine-rich repeat gene that confers race-specific resistance to the blast fungus. *Genetics* 176, 2541–2549.

Liu, Y., Burch-Smith, T., Schiff, M., Feng, S. and Dinesh-Kumar, S.P. (2004) Molecular chaperone Hsp90 associates with resistance protein N and its signaling proteins SGT1 and Rar1 to modulate an innate immune response in plants. *Journal of Biological Chemistry* 279, 2101–2108.

Lorang, J.M., Carkaci-Salli, N. and Wolpert, T.J. (2004) Identification and characterization of victorin sensitivity in *Arabidopsis thaliana. Molecular Plant–Microbe Interactions* 17, 577–582.

Lorang, J.M., Sweat, T.A. and Wolpert, T.J. (2007) Plant disease susceptibility conferred by a 'resistance' gene. *Proceedings of the National Academy of Sciences, USA* 104, 14861–14866.
Lu, R., Malcuit, I., Moffett, P., Ruiz, M.T., Peart, J., Wu, A.J., Rathjen, J.P., Bendahmane, A., Day, L. and Baulcombe, D.C. (2003) High throughput virus-induced gene silencing implicates heat shock protein 90 in plant disease resistance. *EMBO Journal* 22, 5690–5699.
Luderer, R., Takken, F.L., de Wit, P.J. and Joosten, M.H. (2002) *Cladosporium fulvum* overcomes Cf-2-mediated resistance by producing truncated AVR2 elicitor proteins. *Molecular Microbiology* 45, 875–884.
Mackey, D., Holt, B.F., Wiig, A. and Dangl, J.L. (2002) RIN4 interacts with *Pseudomonas syringae* type III effector molecules and is required for RPM1-mediated resistance in *Arabidopsis*. *Cell* 108, 743–754.
Mackey, D., Belkhadir, Y., Alonso, J.M., Ecker, J.R. and Dangl, J.L. (2003) *Arabidopsis* RIN4 is a target of the type III virulence effector AvrRpt2 and modulates RPS2-mediated resistance. *Cell* 112, 379–389.
Malnoy, M., Xu, M., Borejsza-Wysocka, E., Korban, S.S. and Aldwinckle, H.S. (2008) Two receptor-like genes, Vfa1 and Vfa2, confer resistance to the fungal pathogen *Venturia inaequalis* inciting apple scab disease. *Molecular Plant–Microbe Interactions* 21, 448–458.
Marois, E., Van den Ackerveken, G. and Bonas, U. (2002) The xanthomonas type III effector protein AvrBs3 modulates plant gene expression and induces cell hypertrophy in the susceptible host. *Molecular Plant–Microbe Interactions* 15, 637–646.
Martin, G.B., Brommonschenkel, S.H., Chunwongse, J., Frary, A., Ganal, M.W., Spivey, R., Wu, T., Earle, E.D. and Tanksley, S.D. (1993) Map-based cloning of a protein kinase gene conferring disease resistance in tomato. *Science* 262, 1432–1436.
Martin, G.B., Bogdanove, A.J. and Sessa, G. (2003) Understanding the functions of plant disease resistance proteins. *Annual Review of Plant Biology* 54, 23–61.
McDowell, J.M., Dhandaydham, M., Long, T.A., Aarts, M.G., Goff, S., Holub, E.B. and Dangl, J.L. (1998) Intragenic recombination and diversifying selection contribute to the evolution of downy mildew resistance at the RPP8 locus of *Arabidopsis*. *The Plant Cell* 10, 1861–1874.
McHale, L., Tan, X., Koehl, P. and Michelmore, R.W. (2006) Plant NBS-LRR proteins: adaptable guards. *Genome Biology* 7, 212.
Meeley, R.B., Johal, G.S., Briggs, S.P. and Walton, J.D. (1992) A biochemical phenotype for a disease resistance gene of maize. *The Plant Cell* 4, 71–77.
Meshi, T., Motoyoshi, F., Adachi, A., Watanabe, Y., Takamatsu, N. and Okada, Y. (1988) Two concomitant base substitutions in the putative replicase genes of tobacco mosaic virus confer the ability to overcome the effects of a tomato resistance gene, Tm-1. *EMBO Journal* 7, 1575–1581.
Mestre, P. and Baulcombe, D.C. (2006) Elicitor-mediated oligomerization of the tobacco N disease resistance protein. *The Plant Cell* 18, 491–501.
Meyers, B.C., Chin, D.B., Shen, K.A., Sivaramakrishnan, S., Lavelle, D.O., Zhang, Z. and Michelmore, R.W. (1998) The major resistance gene cluster in lettuce is highly duplicated and spans several megabases. *The Plant Cell* 10, 1817–1832.
Meyers, B.C., Dickerman, A.W., Michelmore, R.W., Sivaramakrishnan, S., Sobral, B.W. and Young, N.D. (1999) Plant disease resistance genes encode members of an ancient and diverse protein family within the nucleotide-binding superfamily. *The Plant Journal* 20, 317–332.
Meyers, B.C., Morgante, M. and Michelmore, R.W. (2002) TIR-X and TIR-NBS proteins: two new families related to disease resistance TIR-NBS-LRR proteins encoded in *Arabidopsis* and other plant genomes. *The Plant Journal* 32, 77–92.

Meyers, B.C., Kozik, A., Griego, A., Kuang, H. and Michelmore, R.W. (2003) Genome-wide analysis of NBS-LRR-encoding genes in *Arabidopsis*. *The Plant Cell* 15, 809–834.

Michelmore, R.W. and Meyers, B.C. (1998) Clusters of resistance genes in plants evolve by divergent selection and a birth-and-death process. *Genome Research* 8, 1113–1130.

Milligan, S.B., Bodeau, J., Yaghoobi, J., Kaloshian, I., Zabel, P. and Williamson, V.M. (1998) The root knot nematode resistance gene *Mi* from tomato is a member of the leucine zipper, nucleotide binding, leucine-rich repeat family of plant genes. *The Plant Cell* 10, 1307–1319.

Mindrinos, M., Katagiri, F., Yu, G.L. and Ausubel, F.M. (1994) The *A. thaliana* disease resistance gene *RPS2* encodes a protein containing a nucleotide-binding site and leucine-rich repeats. *Cell* 78, 1089–1099.

Ming, R., Hou, S., Feng, Y., Yu, Q., Dionne-Laporte, A., Saw, J.H., Senin, P., Wang, W., Ly, B.V., Lewis, K.L., Salzberg, S.L., Feng, L., Jones, M.R., Skelton, R.L., Murray, J.E., Chen, C., Qian, W., Shen, J., Du, P., Eustice, M., Tong, E., Tang, H., Lyons, E., Paull, R.E., Michael, T.P., Wall, K., Rice, D.W., Albert, H., Wang, M.L., Zhu, Y.J., Schatz, M., Nagarajan, N., Acob, R.A., Guan, P., Blas, A., Wai, C.M., Ackerman, C.M., Ren, Y., Liu, C., Wang, J., Na, J.K., Shakirov, E.V., Haas, B., Thimmapuram, J., Nelson, D., Wang, X., Bowers, J.E., Gschwend, A.R., Delcher, A.L., Singh, R., Suzuki, J.Y., Tripathi, S., Neupane, K., Wei, H., Irikura, B., Paidi, M., Jiang, N., Zhang, W., Presting, G., Windsor, A., Navajas-Perez, R., Torres, M.J., Feltus, F.A., Porter, B., Li, Y., Burroughs, A.M., Luo, M.C., Liu, L., Christopher, D.A., Mount, S.M., Moore, P.H., Sugimura, T., Jiang, J., Schuler, M.A., Friedman, V., Mitchell-Olds, T., Shippen, D.E., dePamphilis, C.W., Palmer, J.D., Freeling, M., Paterson, A.H., Gonsalves, D., Wang, L. and Alam, M. (2008) The draft genome of the transgenic tropical fruit tree papaya (*Carica papaya* Linnaeus). *Nature* 452, 991–996.

Moffett, P., Farnham, G., Peart, J. and Baulcombe, D.C. (2002) Interaction between domains of a plant NBS-LRR protein in disease resistance-related cell death. *EMBO Journal* 21, 4511–4519.

Monosi, B., Wisser, R.J., Pennill, L. and Hulbert, S.H. (2004) Full-genome analysis of resistance gene homologues in rice. *Theoretical and Applied Genetics* 109, 1434–1447.

Moroldo, M., Paillard, S., Marconi, R., Fabrice, L., Canaguier, A., Cruaud, C., De Berardinis, V., Guichard, C., Brunaud, V., Le Clainche, I., Scalabrin, S., Testolin, R., Di Gaspero, G., Morgante, M. and Adam-Blondon, A.F. (2008) A physical map of the heterozygous grapevine 'Cabernet Sauvignon' allows mapping candidate genes for disease resistance. *BMC Plant Biology* 8, 66.

Mucyn, T.S., Clemente, A., Andriotis, V.M., Balmuth, A.L., Oldroyd, G.E., Staskawicz, B.J. and Rathjen, J.P. (2006) The tomato NBARC-LRR protein Prf interacts with Pto kinase *in vivo* to regulate specific plant immunity. *The Plant Cell* 18, 2792–2806.

Nagy, E.D., Lee, T.C., Ramakrishna, W., Xu, Z., Klein, P.E., SanMiguel, P., Cheng, C.P., Li, J., Devos, K.M., Schertz, K., Dunkle, L. and Bennetzen, J.L. (2007) Fine mapping of the Pc locus of *Sorghum bicolor*, a gene controlling the reaction to a fungal pathogen and its host-selective toxin. *Theoretical and Applied Genetics* 114, 961–970.

Nicaise, V., German-Retana, S., Sanjuan, R., Dubrana, M.P., Mazier, M., Maisonneuve, B., Candresse, T., Caranta, C. and LeGall, O. (2003) The eukaryotic translation initiation factor 4E controls lettuce susceptibility to the potyvirus lettuce mosaic virus. *Plant Physiology* 132, 1272–1282.

Nieto, C., Morales, M., Orjeda, G., Clepet, C., Monfort, A., Sturbois, B., Puigdomenech, P., Pitrat, M., Caboche, M., Dogimont, C., Garcia-Mas, J., Aranda, M.A. and Bendahmane, A. (2006) An eIF4E allele confers resistance to an uncapped and non-polyadenylated RNA virus in melon. *The Plant Journal* 48, 452–462.

Nimchuk, Z., Marois, E., Kjemtrup, S., Leister, R.T., Katagiri, F. and Dangl, J.L. (2000)

Eukaryotic fatty acylation drives plasma membrane targeting and enhances function of several type III effector proteins from *Pseudomonas syringae*. *Cell* 101, 353–363.

Nombela, G., Williamson, V.M. and Muniz, M. (2003) The root-knot nematode resistance gene Mi-1.2 of tomato is responsible for resistance against the whitefly *Bemisia tabaci*. *Molecular Plant–Microbe Interactions* 16, 645–649.

Nurnberger, T., Brunner, F., Kemmerling, B. and Piater, L. (2004) Innate immunity in plants and animals: striking similarities and obvious differences. *Immunol Review* 198, 249–266.

Orbach, M.J., Farrall, L., Sweigard, J.A., Chumley, F.G. and Valent, B. (2000) A telomeric avirulence gene determines efficacy for the rice blast resistance gene *Pi-ta*. *The Plant Cell* 12, 2019–2032.

Ori, N., Eshed, Y., Paran, I., Presting, G., Aviv, D., Tanksley, S., Zamir, D. and Fluhr, R. (1997) The I2C family from the wilt disease resistance locus I2 belongs to the nucleotide binding, leucine-rich repeat superfamily of plant resistance genes. *The Plant Cell* 9, 521–532.

Paal, J., Henselewski, H., Muth, J., Meksem, K., Menendez, C.M., Salamini, F., Ballvora, A. and Gebhardt, C. (2004) Molecular cloning of the potato *Gro1-4* gene conferring resistance to pathotype Ro1 of the root cyst nematode *Globodera rostochiensis*, based on a candidate gene approach. *The Plant Journal* 38, 285–297.

Palma, K., Zhang, Y. and Li, X. (2005) An importin alpha homolog, MOS6, plays an important role in plant innate immunity. *Current Biology* 15, 1129–1135.

Parker, J.E., Coleman, M.J., Szabo, V., Frost, L.N., Schmidt, R., van der Biezen, E.A., Moores, T., Dean, C., Daniels, M.J. and Jones, J.D. (1997) The *Arabidopsis* downy mildew resistance gene *RPP5* shares similarity to the toll and interleukin-1 receptors with N and L6. *The Plant Cell* 9, 879–894.

Parniske, M., Wulff, B.B., Bonnema, G., Thomas, C.M., Jones, D.A. and Jones, J.D. (1999) Homologues of the Cf-9 disease resistance gene (*Hcr9s*) are present at multiple loci on the short arm of tomato chromosome 1. *Molecular Plant–Microbe Interactions* 12, 93–102.

Peart, J.R., Mestre, P., Lu, R., Malcuit, I. and Baulcombe, D.C. (2005) NRG1, a CC-NB-LRR protein, together with N, a TIR-NB-LRR protein, mediates resistance against tobacco mosaic virus. *Current Biology* 15, 968–973.

Pedley, K.F. and Martin, G.B. (2003) Molecular basis of Pto-mediated resistance to bacterial speck disease in tomato. *Annual Review of Phytopathology* 41, 215–243.

Qu, S., Liu, G., Zhou, B., Bellizzi, M., Zeng, L., Dai, L., Han, B. and Wang, G.L. (2006) The broad-spectrum blast resistance gene *Pi9* encodes a nucleotide-binding site-leucine-rich repeat protein and is a member of a multigene family in rice. *Genetics* 172, 1901–1914.

Querci, M., Baulcombe, D.C., Goldbach, R.W. and Salazar, L.F. (1995) Analysis of the resistance-breaking determinants of potato-virus-X (PVX) strain Hb on different potato genotypes expressing extreme resistance to PVX. *Phytopathology* 85, 1003–1010.

Rai, M. (2006) Refinement of the citrus tristeza virus resistance gene (*Ctv*) positional map in *Poncirus trifoliata* and generation of transgenic grapefruit (*Citrus paradisi*) plant lines with candidate resistance genes in this region. *Plant Molecular Biology* 61, 399–414.

Rairdan, G.J. and Moffett, P. (2006) Distinct domains in the ARC region of the potato resistance protein Rx mediate LRR binding and inhibition of activation. *The Plant Cell* 18, 2082–2093.

Rairdan, G. and Moffett, P. (2007) Brothers in arms? Common and contrasting themes in pathogen perception by plant NB-LRR and animal NACHT-LRR proteins. *Microbes and Infection* 9, 677–686.

Rairdan, G.J., Collier, S.M., Sacco, M.A., Baldwin, T.T., Boettrich, T. and Moffett, P. (2008) The coiled-coil and nucleotide binding domains of the potato Rx disease resistance protein function in pathogen recognition and signaling. *The Plant Cell* 20, 739–751.

Rehmany, A.P., Gordon, A., Rose, L.E., Allen, R.L., Armstrong, M.R., Whisson, S.C., Kamoun,

S., Tyler, B.M., Birch, P.R. and Beynon, J.L. (2005) Differential recognition of highly divergent downy mildew avirulence gene alleles by RPP1 resistance genes from two *Arabidopsis* lines. *The Plant Cell* 17, 1839–1850.

Ridout, C.J., Skamnioti, P., Porritt, O., Sacristan, S., Jones, J.D. and Brown, J.K. (2006) Multiple avirulence paralogues in cereal powdery mildew fungi may contribute to parasite fitness and defeat of plant resistance. *The Plant Cell* 18, 2402–2414.

Romer, P., Hahn, S., Jordan, T., Strauss, T., Bonas, U. and Lahaye, T. (2007) Plant pathogen recognition mediated by promoter activation of the pepper *Bs3* resistance gene. *Science* 318, 645–648.

Ron, M. and Avni, A. (2004) The receptor for the fungal elicitor ethylene-inducing xylanase is a member of a resistance-like gene family in tomato. *The Plant Cell* 16, 1604–1615.

Ronald, P.C., Salmeron, J.M., Carland, F.M. and Staskawicz, B.J. (1992) The cloned avirulence gene avrPto induces disease resistance in tomato cultivars containing the Pto resistance gene. *Journal of Bacteriology* 174, 1604–1611.

Rooney, H.C., Van't Klooster, J.W., van der Hoorn, R.A., Joosten, M.H., Jones, J.D. and de Wit, P.J. (2005) *Cladosporium* Avr2 inhibits tomato Rcr3 protease required for Cf-2-dependent disease resistance. *Science* 308, 1783–1786.

Rosebrock, T.R., Zeng, L., Brady, J.J., Abramovitch, R.B., Xiao, F. and Martin, G.B. (2007) A bacterial E3 ubiquitin ligase targets a host protein kinase to disrupt plant immunity. *Nature* 448, 370–374.

Rossi, M., Goggin, F.L., Milligan, S.B., Kaloshian, I., Ullman, D.E. and Williamson, V.M. (1998) The nematode resistance gene *Mi* of tomato confers resistance against the potato aphid. *Proceedings of the National Academy of Sciences, USA* 95, 9750–9754.

Rostoks, N., Zale, M., Soule, J., Brueggeman, R., Druka, A., Kudrna, D., Steffenson, B. and Kleinhofs, A. (2002) A barley gene family homologous to the maize rust resistance gene *Rp1-D*. *Theoretical and Applied Genetics* 104, 1298–1306.

Roudet-Tavert, G., Michon, T., Walter, J., Delaunay, T., Redondo, E. and Le Gall, O. (2007) Central domain of a potyvirus VPg is involved in the interaction with the host translation initiation factor eIF4E and the viral protein HcPro. *Journal of General Virology* 88, 1029–1033.

Ruffel, S., Dussault, M.H., Palloix, A., Moury, B., Bendahmane, A., Robaglia, C. and Caranta, C. (2002) A natural recessive resistance gene against potato virus Y in pepper corresponds to the eukaryotic initiation factor 4E (eIF4E). *The Plant Journal* 32, 1067–1075.

Ruffel, S., Gallois, J.L., Lesage, M.L. and Caranta, C. (2005) The recessive potyvirus resistance gene *pot-1* is the tomato orthologue of the pepper *pvr2-eIF4E* gene. *Molecular Genetics and Genomics* 274, 346–353.

Ruffel, S., Gallois, J.L., Moury, B., Robaglia, C., Palloix, A. and Caranta, C. (2006) Simultaneous mutations in translation initiation factors eIF4E and eIF(iso)4E are required to prevent pepper veinal mottle virus infection of pepper. *Journal of General Virology* 87, 2089–2098.

Sacco, M.A., Mansoor, S. and Moffett, P. (2007) A RanGAP protein physically interacts with the NB-LRR protein Rx, and is required for Rx-mediated viral resistance. *The Plant Journal* 52, 82–93.

Salmeron, J.M., Oldroyd, G.E., Rommens, C.M., Scofield, S.R., Kim, H.S., Lavelle, D.T., Dahlbeck, D. and Staskawicz, B.J. (1996) Tomato Prf is a member of the leucine-rich repeat class of plant disease resistance genes and lies embedded within the Pto kinase gene cluster. *Cell* 86, 123–133.

Schornack, S., Ballvora, A., Gurlebeck, D., Peart, J., Baulcombe, D., Ganal, M., Baker, B., Bonas, U. and Lahaye, T. (2004) The tomato resistance protein Bs4 is a predicted non-nuclear TIR-NB-LRR protein that mediates defense responses to severely truncated derivatives of AvrBs4 and overexpressed AvrBs3. *The Plant Journal* 37, 46–60.

Shan, L., He, P., Zhou, J.M. and Tang, X. (2000a) A cluster of mutations disrupt the avirulence but not the virulence function of AvrPto. *Molecular Plant–Microbe Interactions* 13, 592–598.

Shan, L., Thara, V.K., Martin, G.B., Zhou, J.M. and Tang, X. (2000b) The *Pseudomonas* AvrPto protein is differentially recognized by tomato and tobacco and is localized to the plant plasma membrane. *The Plant Cell* 12, 2323–2338.

Shao, F., Golstein, C., Ade, J., Stoutemyer, M., Dixon, J.E. and Innes, R.W. (2003) Cleavage of *Arabidopsis* PBS1 by a bacterial type III effector. *Science* 301, 1230–1233.

Shen, J., Araki, H., Chen, L., Chen, J.Q. and Tian, D. (2006) Unique evolutionary mechanism in *R*-genes under the presence/absence polymorphism in *Arabidopsis thaliana*. *Genetics* 172, 1243–1250.

Shen, Q.H. and Schulze-Lefert, P. (2007) Rumble in the nuclear jungle: compartmentalization, trafficking, and nuclear action of plant immune receptors. *EMBO Journal* 26, 4293–4301.

Shen, Q.H., Zhou, F., Bieri, S., Haizel, T., Shirasu, K. and Schulze-Lefert, P. (2003) Recognition specificity and RAR1/SGT1 dependence in barley Mla disease resistance genes to the powdery mildew fungus. *The Plant Cell* 15, 732–744.

Shen, Q.H., Saijo, Y., Mauch, S., Biskup, C., Bieri, S., Keller, B., Seki, H., Ulker, B., Somssich, I.E. and Schulze-Lefert, P. (2007) Nuclear activity of MLA immune receptors links isolate-specific and basal disease-resistance responses. *Science* 315, 1098–1103.

Shirano, Y., Kachroo, P., Shah, J. and Klessig, D.F. (2002) A gain-of-function mutation in an *Arabidopsis* Toll Interleukin1 receptor-nucleotide binding site-leucine-rich repeat type *R* gene triggers defense responses and results in enhanced disease resistance. *The Plant Cell* 14, 3149–3162.

Simons, G., Groenendijk, J., Wijbrandi, J., Reijans, M., Groenen, J., Diergaarde, P., Van der Lee, T., Bleeker, M., Onstenk, J., de Both, M., Haring, M., Mes, J., Cornelissen, B., Zabeau, M. and Vos, P. (1998) Dissection of the fusarium I2 gene cluster in tomato reveals six homologs and one active gene copy. *The Plant Cell* 10, 1055–1068.

Sinapidou, E., Williams, K., Nott, L., Bahkt, S., Tor, M., Crute, I., Bittner-Eddy, P. and Beynon, J. (2004) Two TIR:NB:LRR genes are required to specify resistance to *Peronospora parasitica* isolate Cala2 in *Arabidopsis*. *The Plant Journal* 38, 898–909.

Sindhu, A., Chintamanani, S., Brandt, A.S., Zanis, M., Scofield, S.R. and Johal, G.S. (2008) A guardian of grasses: specific origin and conservation of a unique disease-resistance gene in the grass lineage. *Proceedings of the National Academy of Sciences, USA* 105, 1762–1767.

Song, J., Bradeen, J.M., Naess, S.K., Raasch, J.A., Wielgus, S.M., Haberlach, G.T., Liu, J., Kuang, H., Austin-Phillips, S., Buell, C.R., Helgeson, J.P. and Jiang, J. (2003) Gene *RB* cloned from *Solanum bulbocastanum* confers broad spectrum resistance to potato late blight. *Proceedings of the National Academy of Sciences, USA* 100, 9128–9133.

Song, W.Y., Wang, G.L., Chen, L.L., Kim, H.S., Pi, L.Y., Holsten, T., Gardner, J., Wang, B., Zhai, W.X., Zhu, L.H., Fauquet, C. and Ronald, P. (1995) A receptor kinase-like protein encoded by the rice disease resistance gene, *Xa21*. *Science* 270, 1804–1806.

Srichumpa, P., Brunner, S., Keller, B. and Yahiaoui, N. (2005) Allelic series of four powdery mildew resistance genes at the Pm3 locus in hexaploid bread wheat. *Plant Physiology* 139, 885–895.

Staal, J., Kaliff, M., Bohman, S. and Dixelius, C. (2006) Transgressive segregation reveals two *Arabidopsis* TIR-NB-LRR resistance genes effective against *Leptosphaeria maculans*, causal agent of blackleg disease. *The Plant Journal* 46, 218–230.

Staal, J., Kaliff, M., Dewaele, E., Persson, M. and Dixelius, C. (2008) *RLM3*, a TIR domain-encoding gene involved in broad-range immunity of *Arabidopsis* to necrotrophic fungal pathogens. *The Plant Journal* 55, 188–200.

Stavrinides, J., McCann, H.C. and Guttman, D.S. (2008) Host–pathogen interplay and the evolution of bacterial effectors. *Cellular Microbiology* 10, 285–292.

Stein, N., Perovic, D., Kumlehn, J., Pellio, B., Stracke, S., Streng, S., Ordon, F. and Graner, A. (2005) The eukaryotic translation initiation factor 4E confers multiallelic recessive bymovirus resistance in *Hordeum vulgare* (L.). *The Plant Journal* 42, 912–922.

Sun, Q., Collins, N.C., Ayliffe, M., Smith, S.M., Drake, J., Pryor, T. and Hulbert, S.H. (2001) Recombination between paralogues at the Rp1 rust resistance locus in maize. *Genetics* 158, 423–438.

Sweat, T.A., Lorang, J.M., Bakker, E.G. and Wolpert, T.J. (2008) Characterization of natural and induced variation in the LOV1 gene, a CC-NB-LRR gene conferring victorin sensitivity and disease susceptibility in *Arabidopsis*. *Molecular Plant–Microbe Interactions* 21, 7–19.

Swiderski, M.R. and Innes, R.W. (2001) The *Arabidopsis PBS1* resistance gene encodes a member of a novel protein kinase subfamily. *The Plant Journal* 26, 101–112.

Swiderski, M.R., Birker, D. and Jones, J.D. (2009) The TIR domain of TIR-NB-LRR resistance proteins is a signaling domain involved in cell death induction. *Molecular Plant–Microbe Interactions* 22, 157–165.

Szurek, B., Rossier, O., Hause, G. and Bonas, U. (2002) Type III-dependent translocation of the *Xanthomonas* AvrBs3 protein into the plant cell. *Molecular Microbiology* 46, 13–23.

Takabatake, R., Seo, S., Mitsuhara, I., Tsuda, S. and Ohashi, Y. (2006) Accumulation of the two transcripts of the *N* gene, conferring resistance to tobacco mosaic virus, is probably important for *N* gene-dependent hypersensitive cell death. *Plant Cell Physiology* 47, 254–261.

Takahashi, H., Miller, J., Nozaki, Y., Takeda, M., Shah, J., Hase, S., Ikegami, M., Ehara, Y. and Dinesh-Kumar, S.P. (2002) *RCY1*, an *Arabidopsis thaliana* RPP8/HRT family resistance gene, conferring resistance to cucumber mosaic virus requires salicylic acid, ethylene and a novel signal transduction mechanism. *The Plant Journal* 32, 655–667.

Takken, F.L., Albrecht, M. and Tameling, W.I. (2006) Resistance proteins: molecular switches of plant defence. *Current Opinion in Plant Biology* 9, 383–390.

Tamaki, S., Dahlbeck, D., Staskawicz, B. and Keen, N.T. (1988) Characterization and expression of two avirulence genes cloned from *Pseudomonas syringae* pv. *glycinea*. *Journal of Bacteriology* 170, 4846–4854.

Tameling, W.I. and Baulcombe, D.C. (2007) Physical association of the NB-LRR resistance protein Rx with a ran GTPase-activating protein is required for extreme resistance to potato virus X. *The Plant Cell* 19, 1682–1694.

Tameling, W.I., Elzinga, S.D., Darmin, P.S., Vossen, J.H., Takken, F.L., Haring, M.A. and Cornelissen, B.J. (2002) The tomato *R* gene products I-2 and MI-1 are functional ATP binding proteins with ATPase activity. *The Plant Cell* 14, 2929–2939.

Tameling, W.I., Vossen, J.H., Albrecht, M., Lengauer, T., Berden, J.A., Haring, M.A., Cornelissen, B.J. and Takken, F.L. (2006) Mutations in the NB-ARC domain of I-2 that impair ATP hydrolysis cause autoactivation. *Plant Physiology* 140, 1233–1245.

Thivierge, K., Nicaise, V., Dufresne, P.J., Cotton, S., Laliberte, J.F., Le Gall, O. and Fortin, M.G. (2005) Plant virus RNAs. Coordinated recruitment of conserved host functions by (+) ssRNA viruses during early infection events. *Plant Physiology* 138, 1822–1827.

Thomas, C.M., Jones, D.A., Parniske, M., Harrison, K., Balint-Kurti, P.J., Hatzixanthis, K. and Jones, J.D. (1997) Characterization of the tomato *Cf-4* gene for resistance to *Cladosporium fulvum* identifies sequences that determine recognitional specificity in Cf-4 and Cf-9. *The Plant Cell* 9, 2209–2224.

Tian, Y., Fan, L., Thurau, T., Jung, C. and Cai, D. (2004) The absence of TIR-type resistance gene analogues in the sugar beet (*Beta vulgaris* L.) genome. *Journal of Molecular Evolution* 58, 40–53.

Tornero, P., Merritt, P., Sadanandom, A., Shirasu, K., Innes, R.W. and Dangl, J.L. (2002) RAR1 and NDR1 contribute quantitatively to disease resistance in *Arabidopsis*, and their relative contributions are dependent on the *R* gene assayed. *The Plant Cell* 14, 1005–1015.

Tsiamis, G., Mansfield, J.W., Hockenhull, R., Jackson, R.W., Sesma, A., Athanassopoulos, E., Bennett, M.A., Stevens, C., Vivian, A., Taylor, J.D. and Murillo, J. (2000) Cultivar-specific avirulence and virulence functions assigned to avrPphF in *Pseudomonas syringae* pv. *phaseolicola*, the cause of bean halo-blight disease. *EMBO Journal* 19, 3204–3214.

Tuskan, G.A., Difazio, S., Jansson, S., Bohlmann, J., Grigoriev, I., Hellsten, U., Putnam, N., Ralph, S., Rombauts, S., Salamov, A., Schein, J., Sterck, L., Aerts, A., Bhalerao, R.R., Bhalerao, R.P., Blaudez, D., Boerjan, W., Brun, A., Brunner, A., Busov, V., Campbell, M., Carlson, J., Chalot, M., Chapman, J., Chen, G.L., Cooper, D., Coutinho, P.M., Couturier, J., Covert, S., Cronk, Q., Cunningham, R., Davis, J., Degroeve, S., Dejardin, A., Depamphilis, C., Detter, J., Dirks, B., Dubchak, I., Duplessis, S., Ehlting, J., Ellis, B., Gendler, K., Goodstein, D., Gribskov, M., Grimwood, J., Groover, A., Gunter, L., Hamberger, B., Heinze, B., Helariutta, Y., Henrissat, B., Holligan, D., Holt, R., Huang, W., Islam-Faridi, N., Jones, S., Jones-Rhoades, M., Jorgensen, R., Joshi, C., Kangasjarvi, J., Karlsson, J., Kelleher, C., Kirkpatrick, R., Kirst, M., Kohler, A., Kalluri, U., Larimer, F., Leebens-Mack, J., Leple, J.C., Locascio, P., Lou, Y., Lucas, S., Martin, F., Montanini, B., Napoli, C., Nelson, D.R., Nelson, C., Nieminen, K., Nilsson, O., Pereda, V., Peter, G., Philippe, R., Pilate, G., Poliakov, A., Razumovskaya, J., Richardson, P., Rinaldi, C., Ritland, K., Rouze, P., Ryaboy, D., Schmutz, J., Schrader, J., Segerman, B., Shin, H., Siddiqui, A., Sterky, F., Terry, A., Tsai, C.J., Uberbacher, E., Unneberg, P., Vahala, J., Wall, K., Wessler, S., Yang, G., Yin, T., Douglas, C., Marra, M., Sandberg, G., Van de Peer, Y. and Rokhsar, D. (2006) The genome of black cottonwood, *Populus trichocarpa* (Torr. & Gray). *Science* 313, 1596–1604.

Ueda, H., Yamaguchi, Y. and Sano, H. (2006) Direct interaction between the tobacco mosaic virus helicase domain and the ATP-bound resistance protein, N factor during the hypersensitive response in tobacco plants. *Plant Molecular Biology* 61, 31–45.

van der Biezen, E.A. and Jones, J.D. (1998a) The NB-ARC domain: a novel signalling motif shared by plant resistance gene products and regulators of cell death in animals. *Current Biology* 8, R226–R227.

van der Biezen, E.A. and Jones, J.D. (1998b) Plant disease-resistance proteins and the gene-for-gene concept. *Trends in Biochemical Science* 23, 454–456.

van der Biezen, E.A., Freddie, C.T., Kahn, K., Parker, J.E. and Jones, J.D. (2002) *Arabidopsis RPP4* is a member of the RPP5 multigene family of TIR-NB-LRR genes and confers downy mildew resistance through multiple signalling components. *The Plant Journal* 29, 439–451.

van der Vossen, E.A., van der Voort, J.N., Kanyuka, K., Bendahmane, A., Sandbrink, H., Baulcombe, D.C., Bakker, J., Stiekema, W.J. and Klein-Lankhorst, R.M. (2000) Homologues of a single resistance-gene cluster in potato confer resistance to distinct pathogens: a virus and a nematode. *The Plant Journal* 23, 567–576.

van Ooijen, G., Mayr, G., Kasiem, M.M., Albrecht, M., Cornelissen, B.J. and Takken, F.L. (2008) Structure-function analysis of the NB-ARC domain of plant disease resistance proteins. *Journal of Experimental Botany* 59, 1383–1397.

Vankan, J.A.L., Vandenackerveken, G.F.J.M. and Dewit, P.J.G.M. (1991) Cloning and characterization of Cdna of avirulence gene *Avr9* of the fungal pathogen *Cladosporium fulvum*, causal agent of tomato leaf mold. *Molecular Plant–Microbe Interactions* 4, 52–59.

Velasco, R., Zharkikh, A., Troggio, M., Cartwright, D.A., Cestaro, A., Pruss, D., Pindo, M., Fitzgerald, L.M., Vezzulli, S., Reid, J., Malacarne, G., Iliev, D., Coppola, G., Wardell, B., Micheletti, D., Macalma, T., Facci, M., Mitchell, J.T., Perazzolli, M., Eldredge, G., Gatto, P., Oyzerski, R., Moretto, M., Gutin, N., Stefanini, M., Chen, Y., Segala, C., Davenport, C., Dematte, L., Mraz, A., Battilana, J., Stormo, K., Costa, F., Tao, Q., Si-Ammour, A.,

Harkins, T., Lackey, A., Perbost, C., Taillon, B., Stella, A., Solovyev, V., Fawcett, J.A., Sterck, L., Vandepoele, K., Grando, S.M., Toppo, S., Moser, C., Lanchbury, J., Bogden, R., Skolnick, M., Sgaramella, V., Bhatnagar, S.K., Fontana, P., Gutin, A., Van de Peer, Y., Salamini, F. and Viola, R. (2007) A high quality draft consensus sequence of the genome of a heterozygous grapevine variety. *PLoS ONE* 2, e1326.

Vidal, S., Cabrera, H., Andersson, R.A., Fredriksson, A. and Valkonen, J.P.T. (2002) Potato gene *Y-1* is an *N* gene homolog that confers cell death upon infection with potato virus Y. *Molecular Plant–Microbe Interactions* 15, 717–727.

Vos, P., Simons, G., Jesse, T., Wijbrandi, J., Heinen, L., Hogers, R., Frijters, A., Groenendijk, J., Diergaarde, P., Reijans, M., Fierens-Onstenk, J., de Both, M., Peleman, J., Liharska, T., Hontelez, J. and Zabeau, M. (1998) The tomato *Mi-1* gene confers resistance to both root-knot nematodes and potato aphids. *Nature Biotechnology* 16, 1365–1369.

Wang, Z.X., Yano, M., Yamanouchi, U., Iwamoto, M., Monna, L., Hayasaka, H., Katayose, Y. and Sasaki, T. (1999) The Pib gene for rice blast resistance belongs to the nucleotide binding and leucine-rich repeat class of plant disease resistance genes. *The Plant Journal* 19, 55–64.

Warren, R.F., Henk, A., Mowery, P., Holub, E. and Innes, R.W. (1998) A mutation within the leucine-rich repeat domain of the *Arabidopsis* disease resistance gene *RPS5* partially suppresses multiple bacterial and downy mildew resistance genes. *The Plant Cell* 10, 1439–1452.

Webb, C.A., Richter, T.E., Collins, N.C., Nicolas, M., Trick, H.N., Pryor, T. and Hulbert, S.H. (2002) Genetic and molecular characterization of the maize rp3 rust resistance locus. *Genetics* 162, 381–394.

Weber, H., Schultze, S. and Pfitzner, A.J. (1993) Two amino acid substitutions in the tomato mosaic virus 30-kilodalton movement protein confer the ability to overcome the *Tm-2(2)* resistance gene in the tomato. *Journal of Virology* 67, 6432–6438.

Whalen, M.C., Innes, R.W., Bent, A.F. and Staskawicz, B.J. (1991) Identification of *Pseudomonas syringae* pathogens of *Arabidopsis* and a bacterial locus determining avirulence on both *Arabidopsis* and soybean. *The Plant Cell* 3, 49–59.

Whitham, S., Dinesh-Kumar, S.P., Choi, D., Hehl, R., Corr, C. and Baker, B. (1994) The product of the tobacco mosaic virus resistance gene *N*: similarity to toll and the interleukin-1 receptor. *Cell* 78, 1101–1115.

Whitham, S.A., Anderberg, R.J., Chisholm, S.T. and Carrington, J.C. (2000) *Arabidopsis RTM2* gene is necessary for specific restriction of tobacco etch virus and encodes an unusual small heat shock-like protein. *The Plant Cell* 12, 569–582.

Wirthmueller, L., Zhang, Y., Jones, J.D. and Parker, J.E. (2007) Nuclear accumulation of the *Arabidopsis* immune receptor RPS4 is necessary for triggering EDS1-dependent defense. *Current Biology* 17, 2023–2029.

Wolpert, T.J., Dunkle, L.D. and Ciuffetti, L.M. (2002) Host-selective toxins and avirulence determinants: what's in a name? *Annual Review of Phytopathology* 40, 251–285.

Wroblewski, T., Piskurewicz, U., Tomczak, A., Ochoa, O. and Michelmore, R.W. (2007) Silencing of the major family of NBS-LRR-encoding genes in lettuce results in the loss of multiple resistance specificities. *The Plant Journal* 51, 803–818.

Wulff, B.B., Thomas, C.M., Smoker, M., Grant, M. and Jones, J.D. (2001) Domain swapping and gene shuffling identify sequences required for induction of an Avr-dependent hypersensitive response by the tomato Cf-4 and Cf-9 proteins. *The Plant Cell* 13, 255–272.

Wulff, B.B., Kruijt, M., Collins, P.L., Thomas, C.M., Ludwig, A.A., De Wit, P.J. and Jones, J.D. (2004) Gene shuffling-generated and natural variants of the tomato resistance gene *Cf-9* exhibit different auto-necrosis-inducing activities in *Nicotiana* species. *The Plant Journal* 40, 942–956.

Xiang, Y., Cao, Y., Xu, C., Li, X. and Wang, S. (2006) Xa3, conferring resistance for rice bacterial blight and encoding a receptor kinase-like protein, is the same as Xa26. *Theoretical and Applied Genetics* 113, 1347–1355.

Xiao, F., He, P., Abramovitch, R.B., Dawson, J.E., Nicholson, L.K., Sheen, J. and Martin, G.B. (2007) The N-terminal region of *Pseudomonas* type III effector AvrPtoB elicits Pto-dependent immunity and has two distinct virulence determinants. *The Plant Journal* 52, 595–614.

Xiao, S., Ellwood, S., Calis, O., Patrick, E., Li, T., Coleman, M. and Turner, J.G. (2001) Broad-spectrum mildew resistance in *Arabidopsis thaliana* mediated by RPW8. *Science* 291, 118–120.

Yang, B., Sugio, A. and White, F.F. (2006) Os8N3 is a host disease-susceptibility gene for bacterial blight of rice. *Proceedings of the National Academy of Sciences, USA* 103, 10503–10508.

Yang, Z.N., Ye, X.R., Molina, J., Roose, M.L. and Mirkov, T.E. (2003) Sequence analysis of a 282-kilobase region surrounding the citrus tristeza virus resistance gene (*Ctv*) locus in *Poncirus trifoliata* L. Raf. *Plant Physiology* 131, 482–492.

Yeam, I., Cavatorta, J.R., Ripoll, D.R., Kang, B.C. and Jahn, M.M. (2007) Functional dissection of naturally occurring amino acid substitutions in eIF4E that confers recessive potyvirus resistance in plants. *The Plant Cell* 19, 2913–2928.

Yi, H. and Richards, E.J. (2007) A cluster of disease resistance genes in *Arabidopsis* is coordinately regulated by transcriptional activation and RNA silencing. *The Plant Cell* 19, 2929–2939.

Yoshii, M., Nishikiori, M., Tomita, K., Yoshioka, N., Kozuka, R., Naito, S. and Ishikawa, M. (2004) The *Arabidopsis* cucumovirus multiplication 1 and 2 loci encode translation initiation factors 4E and 4G. *Journal of Virology* 78, 6102–6111.

Yoshimura, S., Yamanouchi, U., Katayose, Y., Toki, S., Wang, Z.X., Kono, I., Kurata, N., Yano, M., Iwata, N. and Sasaki, T. (1998) Expression of *Xa1*, a bacterial blight-resistance gene in rice, is induced by bacterial inoculation. *Proceedings of the National Academy of Sciences, USA* 95, 1663–1668.

Zhang, X.C. and Gassmann, W. (2007) Alternative splicing and mRNA levels of the disease resistance gene *RPS4* are induced during defense responses. *Plant Physiology* 145, 1577–1587.

Zhang, Y. and Li, X. (2005) A putative nucleoporin 96 is required for both basal defense and constitutive resistance responses mediated by suppressor of npr1-1,constitutive 1. *The Plant Cell* 17, 1306–1316.

Zhang, Y., Goritschnig, S., Dong, X. and Li, X. (2003) A gain-of-function mutation in a plant disease resistance gene leads to constitutive activation of downstream signal transduction pathways in suppressor of npr1-1, constitutive 1. *The Plant Cell* 15, 2636–2646.

Zhao, B., Ardales, E.Y., Raymundo, A., Bai, J., Trick, H.N., Leach, J.E. and Hulbert, S.H. (2004) The *avrRxo1* gene from the rice pathogen *Xanthomonas oryzae* pv. *oryzicola* confers a nonhost defense reaction on maize with resistance gene *Rxo1*. *Molecular Plant–Microbe Interactions* 17, 771–779.

Zhao, B., Lin, X., Poland, J., Trick, H., Leach, J. and Hulbert, S. (2005) A maize resistance gene functions against bacterial streak disease in rice. *Proceedings of the National Academy of Sciences, USA* 102, 15383–15388.

Zhou, B., Qu, S., Liu, G., Dolan, M., Sakai, H., Lu, G., Bellizzi, M. and Wang, G.L. (2006) The eight amino-acid differences within three leucine-rich repeats between Pi2 and Piz-t resistance proteins determine the resistance specificity to *Magnaporthe grisea*. *Molecular Plant–Microbe Interactions* 19, 1216–1228.

Zhou, F., Kurth, J., Wei, F., Elliott, C., Vale, G., Yahiaoui, N., Keller, B., Somerville, S., Wise, R. and Schulze-Lefert, P. (2001) Cell-autonomous expression of barley Mla1 confers race-specific resistance to the powdery mildew fungus via a Rar1-independent signaling pathway. *The Plant Cell* 13, 337–350.

Zhou, T., Wang, Y., Chen, J.Q., Araki, H., Jing, Z., Jiang, K., Shen, J. and Tian, D. (2004) Genome-wide identification of NBS genes in japonica rice reveals significant expansion of divergent non-TIR NBS-LRR genes. *Molecular Genetics and Genomics* 271, 402–415.

Zipfel, C., Kunze, G., Chinchilla, D., Caniard, A., Jones, J.D., Boller, T. and Felix, G. (2006) Perception of the bacterial PAMP EF-Tu by the receptor EFR restricts *Agrobacterium*-mediated transformation. *Cell* 125, 749–760.

# 6 Transcription Factor Families Involved in Plant Defence: from Discovery to Structure

JEAN-SÉBASTIEN PARENT,[1] LAURENT CAPPADOCIA,[1] ALEXANDRE MARÉCHAL,[1] PIERRE R. FOBERT[2] AND NORMAND BRISSON[1]

[1]*Université de Montréal, Montréal, Québec, Canada;* [2]*Plant Biotechnology Institute, Saskatoon, Saskatchewan, Canada*

## Abstract

Transcription factors play crucial roles in most if not all aspects of plant development, physiology and response to environment. Their function is particularly important in the plant defence response to infection by pathogens. The modulation of gene expression by transcription factors involved in the defence response largely determines whether a plant will survive an attack by a pathogen. An increasing number of transcription factors involved in defence gene regulation in response to pathogen detection have been identified and are the subject of this review. This chapter will focus on well-characterized transcription factor families that have been shown to be involved in plant defence. We will summarize current knowledge about the involvement of each family in defence as well as the structure/function relationships of their various members with particular emphasis on their DNA-binding properties.

## 6.1 Introduction

Because they are unable to move, plants cannot avoid regular contact with potentially dangerous microbes. Following recognition of the pathogens, plants are able to activate defence responses. Should these responses be sufficiently potent, plant survival is guaranteed and colonization of tissue is prevented. Such a plant–pathogen interaction is described as incompatible, the plant being resistant and the pathogen avirulent. On the other hand, if the response is inappropriate, pathogens are able to proliferate by feeding on plant tissue and they eventually overcome their host. In this case, the interaction is compatible and the plant susceptible while the pathogen is said to be virulent. These continuous interactions between plants and pathogens have resulted in the acquisition on both parts of elaborate attack and defence mechanisms that are constantly being solicited. Thus the evolution of these host–microbe interactions can be thought of as a green molecular arms race that has been ongoing for hundreds of millions of years (reviewed in Chisholm *et al.*, 2006).

For an infection to be successful, the microbes must first penetrate several layers of protection surrounding plant cells. While plants are able to resist most diseases, evolution has provided solutions to ensure that some pathogens are ultimately able to feed on their host and proliferate. The interior of the leaf is usually reached by either direct penetration, through wounds or through natural openings such as stomata. Once present in the apoplast, the cell wall has to be overcome to gain access to the plasma membrane and cytosol where energy resides (reviewed in Huckelhoven, 2007). When pathogens have gained a hold in the apoplast, their conserved surface structures or secreted molecules may be detected by the plant cells and the first active defence response is turned on. During this process, pathogen-associated molecular patterns (PAMPs) are recognized by pathogen-recognition receptors (PRRs) and defence responses are induced. Again, pathogens have evolved ways of downregulating primary innate immunity and plants have responded with a second, more specific resistance (*R*)-gene mediated defence response. Resistance proteins monitor avirulence determinants encoded by pathogens either directly or via their effects on cell metabolism. This detection results in the activation of a defence response that is both quicker and stronger than the one induced by primary innate immunity. In fact, *R*-gene mediated defence response often leads to a localized programmed cell death known as the hypersensitive response (HR) that is quite effective against pathogens that feed off living tissue (i.e. biotrophs). Several reviews on the detection mechanisms evolved by plants to resist pathogen infection have been published in recent years and readers should rely on these for a thorough discussion of this particular topic (Nurnberger *et al.*, 2004; Chisholm *et al.*, 2006; Zipfel, 2008) as well as the information contained in this volume.

After these initial detection events, signal transduction cascades are activated and rely on hormones such as salicylic acid (SA), jasmonic acid (JA) and ethylene (ET) as secondary messengers to enhance plant resistance against infection. The choice of the messenger seems to be related in part to the parasite's feeding strategy, either biotroph or necrotroph. Eventually, these signals reach the nucleus where transcription factors (TFs) are recruited on to genes' regulatory regions, leading to the modulation of their expression and thus to the optimization of resistance against pathogens. Pioneering studies on the transcriptional reprogramming that takes place during an infection process in the model plant *Arabidopsis thaliana* have highlighted the large number of genes that are affected during host–microbe interactions. Indeed, the expression of as much as one-quarter of all genes in *Arabidopsis* could be modified following pathogen infection (Tao *et al.*, 2003). Numerous genes that are characterized as pathogenesis-related (*PR*) were identified among these clusters but additional genes involved in all aspects of cell metabolism were also found, underlining the complexity of plant responses to infection. Additionally, these initial reports and subsequent gene profiling experiments have demonstrated considerable overlap between the sets of genes induced by different types of pathogens triggering different plant resistance pathways (reviewed in Katagiri,

2004; and Eulgem, 2005). This suggests that similar sets of transcription factors are recruited by signalling cascades, activated following different elicitation stimuli and that, concurrently, TFs specific for each of these pathways are also at work.

This review will focus mainly on well-characterized groups of TFs that have been shown to be involved in the mediation of plant defence against pathogens. For each of these families, we first provide an overview describing the discovery of these factors and of their various functions in the plant defence process. We then summarize the available proof concerning the roles of these proteins as transcriptional modulators. Finally, we will try to link those functions with the available structural data.

## 6.2 ERF Factors

The first member of the AP2/ERF family was isolated in 1994 by Jofuku *et al.* when the floral homeotic gene *APETALA2* was cloned by transfer-DNA (T-DNA) tagging (Jofuku *et al.*, 1994). In *A. thaliana*, this family of TFs was later found to comprise 147 members that can be divided into two subclasses, namely the AP2 and ERF groups (Nakano *et al.*, 2006). So far, only the Ethylene Response Factors (ERF) subgroup has been shown to be involved in plant defence responses. The first ERF genes identified were the tobacco (*Nicotiana tabaccum*) ethylene response element binding protein-1, 2, 3 and 4 (EREBP) (Ohme-Takagi and Shinshi, 1995). In this species many defence gene promoters contain the GCC box, a consensus sequence conferring ethylene responsiveness. In an attempt to isolate an inducible factor binding to the GCC box, Ohme-Takagi and Shinshi screened an expression library prepared from mRNA isolated from ethephon-treated leaves with an oligonucleotide containing the consensus sequence and recovered a total of four clones, all sharing a certain level of identity. This gave rise to the EREBP family which was later renamed ERF and fused with the AP2 gene family that shares similar DNA-binding domains (Gutterson and Reuber, 2004). All these TFs are involved in the transcriptional regulation of a variety of biological processes including hormonal signal transduction (Ohme-Takagi and Shinshi, 1995; Berrocal-Lobo *et al.*, 2002) and biotic stress response (reviewed in Gutterson and Reuber, 2004; and Fobert, 2006). A later study analysing the *in vitro* binding properties of recombinant ERFs showed that the minimal binding site was made of the GCC box itself (AGCCGCC) plus three bases on the 3' side of the box (Hao *et al.*, 1998).

A useful approach to assess the function of a gene in defence is to obtain a plant line carrying a loss-of-function mutation and test its resistance to a given pathogen. However, few such cases have been reported for the ERF family. It is expected that with 122 genes in *Arabidopsis*, there would be a high likelihood of functional redundancy (Nakano *et al.*, 2006). In fact, to our knowledge, assignment of function for such loss-of-function mutations has been reported for only *AtERF4* and *AtERF14* (McGrath *et al.*, 2005; Oñate-

Sánchez *et al.*, 2007). In one study, *AtERF4* was identified as a negative regulator of the JA-signalling pathway (McGrath *et al.*, 2005). A knockout (KO) line was isolated for that gene and was shown to express higher levels of the defence gene *PDF1.2* and consequently to be less sensitive to the necrotrophic pathogen, *Fusarium oxysporum*. In the other study, two KO lines unable to express the gene *AtERF14* were isolated from a mutant collection (Oñate-Sánchez *et al.*, 2007). These plants were shown to be compromised for the expression of JA/ET associated defence marker genes, including *PDF1.2*, and again showed higher susceptibility to *F. oxysporum*. The same marker genes were also upregulated in plants overexpressing *AtERF14* (Oñate-Sánchez *et al.*, 2007). In the absence of a functional KO line, a useful alternative is the use of RNA-interference (RNAi) to silence a given gene. This technique was used to show that *Arabidopsis* plants with reduced expression of *ORA59* are more susceptible to infection by the necrotrophic fungus *Botrytis cinerea* while overexpression of the same gene shows the opposite effect (Pre *et al.*, 2008).

Overexpressing lines remain to this date a much-used method to link TF function and defence responses. The vast majority of experimental data linking ERF genes to plant defence have been obtained by stably overexpressing an ectopic gene *in planta* (reviewed in Fobert, 2006). Taken individually, these experiments often lead to some convincing results. However, discrepancies are often observed when overexpression of different genes from the same family is compared. For example, overexpression of the tomato (*Solanum lycopersicum*) gene *Pti4* in *Arabidopsis* leads to increased resistance to the biotroph *Pseudomonas syringae* (Gu *et al.*, 2002), while overexpression of its orthologue (*AtERF1*) in *Arabidopsis* increases susceptibility to the same pathogen (Berrocal Lobo *et al.*, 2002). These contradictory results might be due to the specialized function of the ERF in the different species or to the binding to low affinity targets as a result of the overabundance of the TF. Furthermore, overexpression of an ERF gene sometimes leads to developmental defects which might influence the plant response to pathogens. Results drawn from loss-of-function mutations, such as those reported in the studies of McGrath *et al.* (2005) and of Oñate-Sánchez *et al.* (2007), bring strong confirmation to the involvement of this type of factors in defence.

TF function is often tested through monitoring of gene expression in a mutant background. This is done by measuring the levels of mRNA by way of RNA gel blots, RT-PCR (or qRT-PCR) or even microarrays. As ERF are known to be part of the ET-signalling pathway, genes involved in this pathway have often been used as defence marker genes. Most of the time, pathogenesis-related, ET/JA-inducible genes like the defensin *PDF1.2* and the thionin *Thi2.1* were induced in the overexpressing lines (reviewed in Fobert, 2006). Some ERF were even shown to be induced by SA and accordingly, some SA-inducible genes like *PR-1*, *-2*, *-3* and *-4* were more highly expressed in some tomato and tobacco ERF-modified lines (Zhang *et al.*, 2004). As expected, a majority of the plants showing enhanced expression of genes involved in defence were more resistant to pathogen invasion. With the notable exception of the *AtERF14* KO mutant, enhanced susceptibility to pathogens has rarely been reported (Oñate-Sánchez *et al.*, 2007). Overall, enhanced expression of defence-related genes coupled with the actual increased resistance to various pathogens presents convincing evidence for the involvement of the ERF TFs in plant defence. This function would entail the direct binding to GCC boxes within defence gene promoters. Indeed, chromatin immunoprecipitation (ChIP) experiments revealed that 60% of the genes differentially expressed in plants that overexpress *Pti4* were bound by this protein (Chakravarthy *et al.*, 2003). However, this study only defined the first kilobase upstream of the ATG codon as the promoter. It is expected that if the definition was broadened, the percentage of bound targets would be even higher. However, these experiments also revealed direct binding of Pti4 to some non-GCC box-containing promoters, leading to the hypothesis that either Pti4 is able to bind to a DNA motif other than the GCC box or it interacts with other transcription factors to regulate promoter activity. This second hypothesis was confirmed for the potato *PR-10a* promoter, which contains no GCC box but is nevertheless bound by Pti4 (González-Lamothe *et al.*, 2008). In this case, it was shown that Pti4 was drafted to the promoter through its interaction with silencing element binding factor (SEBF), a negative regulator of transcription that binds to the silencing element (SE) (Boyle and Brisson, 2001). Therefore Pti4 was also shown to be an essential component of a repressosome that disassociates upon wounding or treatment with an elicitor (González-Lamothe *et al.*, 2008).

Few studies have addressed how ERF activity is regulated. Both tomato Pti4 and rice (*Oryza sativa*) OsEREBP1 have been shown to be phosphorylated, resulting in increased DNA-binding activity (Gu *et al.*, 2000; Cheong *et al.*, 2003). Indeed, Pti4 is directly phosphorylated by the serine/threonine protein kinase Pto, which confers tomato resistance to *P. syringae* pv. *tomato* strains expressing the avirulence gene *avrPto* (Zhou *et al.*, 1997). The sole presence of the bacterial factor inside the plant cell appears to be enough to activate Pti4, which is then able to bind its cognate *cis*-element inside response genes through its enhanced affinity for DNA. On the other hand, it is a mitogen-activated protein kinase (MAPK) from rice, namely BWMK1, which is responsible for the phosphorylation of OsEREBP1 (Cheong *et al.*, 2003). These results indicate that post-translational modifications are important for the activation of at least some of the ERF proteins.

The members of the ERF and AP2 families were reunited because of their common structural domain, called the AP2/ERF domain. This domain mediates the sequence-specific binding to the GCC box. The structure of the AP2/ERF domain of AtERF1 was solved by nuclear magnetic resonance (NMR) spectroscopy, both in free form and in complex with target DNA (Allen *et al.*, 1998). It is the only plant protein involved in a defence response for which the structure of the protein–DNA complex has been solved experimentally. The structure of the DNA-binding domain led to a better understanding of the DNA-binding mechanism and permitted the discrimination between the conserved residues involved in the stabilization of the structure of the domain and the residues involved in the DNA-binding activity. Figure 6.1 summarizes our knowledge on the structures of different TF DNA-binding domains with emphasis on the domain topologies and DNA-binding interfaces. The structure

| Transcription factor family name | Number of family members in *Arabidopsis* | Cognate *cis*-element | Structure of DNA-binding domain[a] | Structure-based DNA-binding insights | DNA-binding domain cartoon representation[b] |
|---|---|---|---|---|---|
| ERF | 122 | AGCCGCC **(GCC box)** | Single ~60 aa AP2/ERF domain consisting of a three-stranded $\beta$-sheet packed along an $\alpha$-helix | One face of the $\beta$-sheet running parallel to the major groove of the DNA contains important DNA binding residues; two tryptophan and four arginine residues make both sequence-specific and sequence-non-specific contacts with DNA bases | 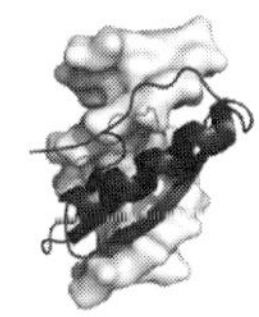 |
| WRKY | 74 | (T)GACC/T **(W box)** | Single ~60 aa WRKY domain consisting of a four-stranded or a five-stranded $\beta$-sheet and a zinc-binding pocket ($C_2H_2$ or $C_2HC$ binding motifs); a glycine residue induces a concave curvature of the WRKYGQK-containing $\beta$-strand | One edge of the $\beta$-sheet contains the residues important for DNA-binding; the $\beta$-strand, containing the invariant WRKYGQK sequence, deeply enters the major groove of the DNA | 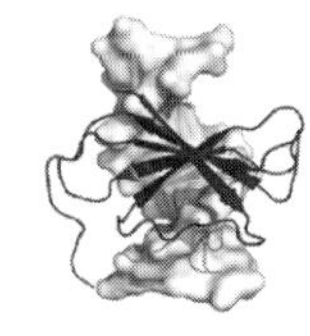 |
| R2R3-MYB | 125 | Various | Two imperfect repeats of the ~50 aa MYB domain consisting of three $\alpha$-helices; each domain is stabilized by three regularly spaced tryptophan residues | The C-terminal $\alpha$-helix of each domain inserts into the major groove of the DNA | 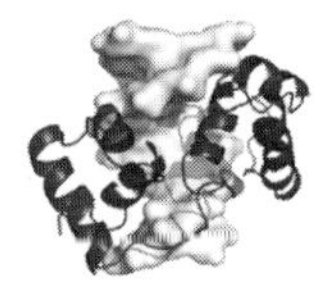 |
| TGA | 10 | TGACGT **(as-1 element)** | Single ~30 aa bZIP domain forming an $\alpha$-helix; the C-terminal part of the $\alpha$-helix contains the leucine zipper that permits dimerization | The N-terminal part of the $\alpha$-helix inserts into the major groove of the DNA | 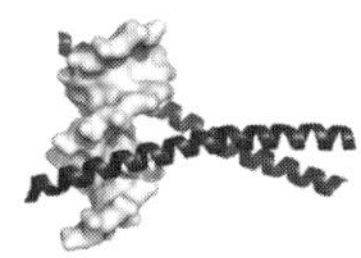 |

[a] aa, amino acids.
[b] Model preparation – ERF family: NMR structure of AtERF1 bound to DNA (Protein Data Bank (PDB) 1GCC); WRKY family: crystal structure of AtWRKY1 (PDB 1AYD) and comparative modelling with the DNA-bound crystal structure of Glial Cells Missing (GCM) (PDB 1ODH); R2R3-MYB family: crystal structure of *Mus musculus* MYB bound to DNA (PDB 1MSE); TGA family: crystal structure of cAMP response element binding (CREB) bZIP bound to DNA (PDB 1DH3).

**Fig. 6.1.** DNA-binding domains of transcription factors involved in mediating plant defence responses (modified from Fobert, 2006).

of the AP2/ERF domain consists of a three-stranded antiparallel $\beta$-sheet packed along an $\alpha$-helix (Allen *et al.*, 1998). The AP2/ERF domain binds to the GCC-box DNA as a monomer via the three-stranded $\beta$-sheet that runs parallel to the major groove of the DNA. Four arginines and two tryptophans were found to interact with the DNA nucleobases in a sequence-specific manner. Interestingly, these same amino acids also contact the phosphate-sugar moiety of the DNA. Thus, a few well-placed residues provide both affinity and specificity to the binding of the GCC box. Alignment of the *Arabidopsis* proteins has shown a high degree of conservation of those six residues (Nakano *et al.*, 2006). Thus, one would expect that most of these factors would show similar specificity towards DNA. The solution structure of the free AP2/ERF domain superimposes well with the DNA-complex domain suggesting that no change of conformation occurs upon DNA binding (Allen *et al.*, 1998). These data are all in agreement with the model of the ERF factors binding directly to cognate regulatory sequence.

## 6.3 R2R3-MYB Factors

The MYB family of TFs is found in vertebrates, insects, plants, fungi, slime moulds and viruses, from which they were first isolated (Stracke *et al.*, 2001). MYBs represent one of the largest families of TFs in plants. The first plant MYB TF was isolated in 1987 from maize (*Zea mays*) and was shown to share high similarity with the animal proto-oncogene c-MYB (Paz-Ares *et al.*, 1987). The plant MYB family has been divided into three subgroups depending on the number of times the MYB domain is repeated (Stracke *et al.*, 2001). The subgroup that includes two MYB repeats (R2 and R3), named R2R3-MYB, is the largest subgroup in plants and seems to be responsible for plant-specific processes. The first member of the subgroup that was linked to the defence response was the *Arabidopsis AtMYB30*. This gene was isolated during a screen for genes induced by *Xanthomonas campestris*, a pathogen causing a strong HR in plants (Daniel *et al.*, 1999). *Arabidopsis* cells were inoculated with *X. campestris* and a complementary DNA (cDNA) library was made from the isolated RNA. Differential screening of this library with cDNA probes from cells infected with different pathogen strains that do not cause HR led to the isolation of *AtMYB30*.

The few studies that have linked the MYB family to defence responses all identified members of the R2R3 subgroup. For example, *A. thaliana* and *N. tabaccum* plants overexpressing *AtMYB30* were shown to develop more HR-related symptoms than wild type and to be more resistant to infection by *P. syringae* (Vailleau *et al.*, 2002). Another *Arabidopsis* MYB factor, BOS1 (BOTRYTIS-SUSCEPTIBLE1), was discovered when screening for plants more susceptible to the necrotrophic pathogen *B. cinerea* (Mengiste *et al.*, 2003). The *BOS1* gene was also shown to be induced during infection and its inactivation led to increased susceptibility to yet another necrotrophic pathogen, *Alternaria brassicicola* (Mengiste *et al.*, 2003). Another MYB gene involved in HR was isolated from tobacco where virus-induced gene silencing of *MYB1* led to diminished HR and stronger infection by the tobacco mosaic virus (TMV) (Liu *et al.*, 2004). In *Arabidopsis*, HAG1/MYB28 has also been shown to be related to biotic stress (Gigolashvili *et al.*, 2007). Indeed, overexpression of *MYB28* causes accumulation of chemoprotective compounds of the glucosinolate family. These compounds are known to play a role in defence responses and accordingly, herbivore insects were less efficient in colonizing overexpressing plants (Gigolashvili *et al.*, 2007). Another study concerning genes induced during root colonization by non-pathogenic bacteria revealed that *AtMYB72* plays an important role in this process. This TF seems to be essential for mounting the Induced Systemic Resistance response that follows root colonization (Van der Ent *et al.*, 2008). As expected, a KO line of this gene was isolated and shown to be more susceptible to a subsequent infection by pathogenic *P. syringae*.

If it is generally accepted that MYB TFs are involved in plant defence responses, no study has yet highlighted the mechanism that could make this possible. It thus remains to be shown that members of the R2R3-MYB group can directly activate genes involved in defence. For example, since neither the expression of the defence gene *PDF1.2* nor *PR-1* was affected in the *bos1* mutant during infection, it is still unclear how *BOS1* contributes to resistance (Mengiste *et al.*, 2003). However, the SA-associated defence markers *ICS* (involved in the biosynthesis of SA) and *PR-1* were both shown to be upregulated in plants overexpressing *AtMYB30* (Raffaele *et al.*, 2006). In yet another study, large-scale gene expression profiling was used to identify putative targets of AtMYB30 (Raffaele *et al.*, 2008). Genes involved in the biosynthesis of very-long-chain fatty acids were not induced in the KO of *AtMYB30* after infection as compared to wild-type plants. Therefore we can conclude that some MYB proteins are at least indirectly involved in the SA-defence pathway and biosynthesis of protective compounds. However, more data are needed concerning other R2R3-MYB members in order to be able to draw a more general conclusion concerning the roles of this family.

As stated above, the members of the R2R3-MYB family possess two imperfect repeats of the MYB domain. The structure of a R2R3-MYB murine protein in complex with cognate DNA was solved by NMR spectroscopy (Ogata *et al.*, 1994). Each MYB domain consists of three α-helices in which the second and the third helices form a helix-turn-helix variant motif. The structure of each MYB domain is stabilized by three regularly spaced tryptophans that form a hydrophobic cluster, a hallmark of the MYB domain (Ogata *et al.*, 1992). Most of the residues contacting DNA are concentrated in the C-terminal α-helix of the MYB domain (Ogata *et al.*, 1994). This helix runs parallel to the major groove of the DNA (see Fig. 6.1). The R2 and R3 repeats are closely packed against each other enabling a continuous recognition of DNA. The structure of an R2R3-MYB in free form was solved by NMR spectroscopy (Ogata *et al.*, 1995). Even though the structure of the individual repeats is similar in the DNA-bound state, the relative orientation of the repeats is different (Ogata *et al.*, 1995). It thus appears that the R2 and R3 MYB interact with each other only in the presence of DNA and that the R2 and R3 MYB domains bind DNA in a cooperative manner (Ogata *et al.*, 1994).

Most of the residues contacting the DNA bases are conserved between the murine and the plant R2R3-MYB proteins, suggesting that these proteins possess similar sequence specificity. In spite of this, some R2R3-MYB plant proteins appear to possess altered sequence specificities (Martin and Paz-Ares, 1997). Interestingly, substitution of amino acids that are not directly involved in DNA binding was shown to lead to changes in the binding specificity of the protein (Solano *et al.*, 1997).

R2R3-MYB DNA-binding activity can be regulated by a reduction–oxidation mechanism (Guehmann *et al.*, 1992). A cysteine residue located in the R2 repeat seems to account for the redox regulation of the DNA-binding activity. Strikingly, this cysteine has no contact with DNA but is part of the hydrophobic core of the R2 repeat (Ogata *et al.*, 1994). Since this repeat was shown to be thermally less stable than the R3 domain (Ogata *et al.*, 1995), it has been hypothesized that the cysteine could be exposed upon R2 domain unfolding (Ogata *et al.*, 1994), consistent with the cysteine playing the role of a redox molecular sensor. A similar mechanism, implicating two cysteine residues, has been reported for some plant R2R3-MYBs (Heine *et al.*, 2004). Interestingly, the modification of many proteins during a defence response seems to be controlled by redox-mediated signalling (Jones *et al.*, 2006). The most striking example is the one of the NONEXPRESSOR OF PATHOGENESIS-RELATED GENES1 (NPR1) that requires the reduction of two cysteines to be transported to the nucleus (Mou *et al.*, 2003) and the oxidation of two others to be active (Rochon *et al.*, 2006). It is therefore easy to imagine that the defence-related R2R3-MYB could also be regulated by the same kind of redox mechanism.

## 6.4 TGA Factors

The first members of this family were identified in a screen for proteins binding a specific element (*as-1*) inside the cauliflower mosaic virus promoter. The factors TGA1a and TGA1b were thus recovered from a tobacco expression library (Katagiri *et al.*, 1989). Several years later, additional TGA factors were isolated using the yeast two-hybrid system based on their ability to interact with NPR1, a central regulator of plant defence responses (reviewed in Fobert, 2006)). These findings suggested that NPR1 could mediate at least part of its function through this family of basic region-leucine zipper (bZIP) transcription factors.

*Arabidopsis* encodes ten TGA factors that can be further divided into subgroups. Interestingly, the interaction between NPR1 and subgroup I of the TGA factors, which includes TGA1 and TGA4, is dependent on the redox status of certain cysteine residues conserved within this subgroup (Després *et al.*, 2003). These cysteines are located outside the DNA-binding domain and the DNA-binding activity of TGA1 is not directly affected by redox conditions; instead this property is conferred by interaction with NPR1 (Després *et al.*, 2003).

The most convincing evidence for the involvement of TGA factors in mediating plant defence responses comes from the analysis of a triple mutant in which all members of subgroup II (TGA2, TGA5 and TGA6) were deleted (Zhang *et al.*, 2003). This mutant was defective in systemic acquired resistance (SAR) against the biotrophic pathogens *P. syringae* and *Hyaloperonospora parasitica* and failed to accumulate *PR-1* transcripts after treatment with SA. Single or double mutants of subgroup II factors did not show altered disease resistance or *PR* gene expression, leading the authors to conclude that subgroup II TGA factors possess essential, but redundant functions for activating defence responses. Additional information on the potential mechanisms by which TGA2 and NPR1 regulate *PR* gene expression is provided in this book (see Boyle *et al.*, Chapter 4, this volume).

Although most studies to date have focused on subgroup II TGA factors, insertional KO mutants in subgroup I and III TGA factors were recently found to be compromised in resistance against *P. syringae* (Kesarwani *et al.*, 2007). Virus-induced gene silencing of tomato *TGA1* was also shown to compromise *Pto*-mediated resistance against *P. syringae* expressing *avrPto* (Ekengren *et al.*, 2003).

The structure of TGA factors has not yet been resolved. However, the crystal structures of the bZIP motif of other proteins, including GCN4, c-jun/c-fos as well as CREB, bound to DNA were solved by X-ray crystallography (Ellenberger *et al.*, 1992; Glover and Harrison, 1995; Schumacher *et al.*, 2000). In all cases, the bZIP motif forms a continuous α-helix in which the N-terminal basic region contacts the major groove of the DNA while the C-terminal leucine zipper region, which contains a repeat of hydrophobic and non-polar residues, form a coiled coil hydrophobic dimerization interface (see Fig. 6.1). The capacity to form homo and heterodimers is strongly influenced by residues flanking the hydrophobic interaction interface. These residues can promote the interaction with a specific protein partner. In the case of TGA proteins, additional factors may contribute to the formation of homo or heterodimers (Katagiri *et al.*, 1992). Many of the DNA contacting residues are conserved between the TGA factors and the CREB proteins. Consistently, the TGA factors bind to a consensus sequence TGACGT, which is related to the TGACGTCA binding site for CREB.

## 6.5 WRKY Factors

Although the very first WRKY was isolated by Ishiguro and Nakamura in 1994 (Ishiguro and Nakamura, 1994), it was only 2 years later that this family received its name, based on its conserved amino acid sequence 'WRKYGQK' (Rushton *et al.*, 1996). In this latter study, WRKY proteins were identified through a search for factors binding a particular response element of the *PR-1* gene promoter called the W box. By doing so, the group of Imre E. Somssich was looking for the factor responsible for the induction of the *PR-1* gene in elicitor-treated parsley cells (*Petroselinum crispum*). A bacteriophage cDNA library was made from total RNA extracted from parsley cells treated with a fungal oligopeptide elicitor and screened with various W-box oligonucleotides. Out of the four clones isolated, three coding sequences were recovered and named *PcWRKY1*, *2* and *3*.

As more and more of the *Arabidopsis* genome was revealed, the WRKY members were divided into three groups depending on the number and precise sequence of their DNA-binding WRKY domain (Eulgem *et al.*, 2000). With 74 genes in *Arabidopsis* and 90 in rice the WRKY family is recognized as an important family of plant TF (Ulker and Somssich, 2004). One clue that links the WRKY to the plant defence response is the induction of their expression during infection or treatment with an elicitor. Indeed, 49 WRKY genes out of 72 tested in *Arabidopsis* respond to infection by *P. syringae* or treatment with SA (Dong *et al.*, 2003). Furthermore, using an inducible version of NPR1, Wang *et al.* identified eight WRKY factors as direct targets of this defence response regulator (Wang *et al.*, 2006). However, it is only in recent years that a direct link with defence was definitely established when loss-of-function mutations of WRKY genes were shown to affect defence responses (Eulgem and Somssich, 2007). Using overexpression and antisense lines of *Arabidopsis*

*WRKY70*, it was shown that the expression of this gene is directly correlated to resistance to both *Erwinia carotovora* and *P. syringae* (Li *et al.*, 2004). Later, two insertion KO lines for this same gene were isolated and they both showed an increased susceptibility to *H. parasitica* (Knoth *et al.*, 2007). Also, an insertion KO mutant in *AtWRKY18* was reported to be defective in SAR, whereas KO plants of *AtWRKY58* were more resistant to *P. syringae* after treatment with suboptimal levels of benzothiadiazole (BTH), an analogue of SA that is used to protect plants against diseases in the fields (Wang *et al.*, 2006). This indicates that AtWRKY18 and AtWRKY58 act as positive and negative regulators, respectively, of plant defence responses. More recently, a KO of gene *AtWRKY27* was shown to have delayed symptoms when infected by the pathogen *Ralstonia solanacearum* (Mukhtar *et al.*, 2008). On the other hand, an insertion KO line for the gene *AtWRKY25* did not reveal differences in susceptibility to *P. syringae*, although disease symptoms were reduced (Zheng *et al.*, 2007).

WRKY factors have also been implicated in resistance to necrotrophic pathogens, as a KO line for *Arabidopsis WRKY33* has an increased susceptibility to *B. cinerea* as well as to *A. brassicicola* (Zheng *et al.*, 2006). In rice, *WRKY45* was shown to be induced by BTH (Shimono *et al.*, 2007). Overexpression of this gene also conferred strong resistance to the blast disease (*Magnaporthe grisea*) while silencing of the gene had the opposite effect. In barley, HvWRKY1 and HvWRKY2 proteins were shown to interact directly with the mildew A (MLA) R protein (Shen *et al.*, 2007). When these two genes are silenced using a viral vector, infection by the virulent *Blumeria graminis* is significantly reduced. They were therefore identified as repressor of the basal defence response in barley.

Studies of multiple loss-of-function lines suggest that a complex interaction network exists between the different WRKY proteins. For instance, *WRKY70* function was shown to be partially redundant with the function of *WRKY53* (Wang *et al.*, 2006). As expected, the double mutant *wrky53 wrky70* showed enhanced susceptibility to *P. syringae* as compared to the single KO plants. In another study, it was shown that the *Arabidopsis* proteins WRKY18, WRKY40 and WRKY60 physically interacted with each other in yeast (Xu *et al.*, 2006). Gel retardation was then used to show that these proteins can form hetero-complexes *in vitro* with changes in their binding affinity and/or specificity. Mutant lines for these genes were obtained and revealed that, while only the single KO *wrky18* showed increased resistance to *P. syringae*, the multiple KOs *wrky18/40*, *wrky18/60* and *wrky18/40/60* showed even more resistance. The same trend was observed in regard to infection by *B. cinerea*, illustrating the possible negative effect of these WRKY proteins on resistance as well as the partial redundancy that exists inside this family (Xu *et al.*, 2006). These results concerning the resistance of the *wrky18* line contrast with those mentioned earlier by Wang *et al.* (2006). However, this last group tested the SAR, a specific component of plant defence, while Xu *et al.* tested the basal defence of the plant (Xu *et al.*, 2006). Taken together, these two studies indicate that WRKY18 is a negative regulator of basal resistance and a positive regulator of acquired resistance which is truly interesting. In yet another study, it was shown that WRKY11 and WRKY17 also have partially redundant functions as negative regulators of defence (Journot-Catalino *et al.*, 2006). It was shown that mutation of *WRKY11* alone resulted in an increased resistance to *P. syringae*. Mutation of *WRKY17* alone did not show altered resistance, but the *wrky11/wrky17* double mutant line showed even more resistance than the *wrky11* line. These results constitute convincing evidence regarding a role for WRKYs as both negative and positive regulators of resistance. It will be interesting in the future to learn about the complex interactions controlling the balance between these two roles

The WRKY genes are usually activated during the response to pathogens so they can modulate the transcriptome of the plant (Eulgem *et al.*, 2000). Overexpression of an activator WRKY will lead to constitutive expression of SA-inducible genes and will usually be accompanied by a decrease in expression of JA/ET-inducible genes (reviewed in Fobert, 2006). As indicated above, WRKY factors are known to bind a specific sequence known as the W box (Rushton *et al.*, 1996). This was shown by a DNA–ligand binding screen as well as cotransfection assays in parsley cells. The specificity of this binding was further tested by random binding site selection (Du and Chen, 2000) and the consensus binding site TGAC-C/T was established. The direct interaction of WRKY proteins with DNA *in vivo* was shown by ChIP assays in parsley cells (Turck *et al.*, 2004). After elicitation, PcWRKY1 was shown to bind preferentially to fragments containing W boxes inside the promoters of *PcWRKY1* and *PcPR1-1*. These results were later confirmed by another study showing that the *AtWRKY33* promoter region is occupied by WRKY proteins before treatment with an elicitor and that it is occupied even more afterwards (Lippok *et al.*, 2007). While these studies answered an important question, they also raised some more as they showed that some unidentified WRKY proteins were constitutively bound to the W box inside the genes even before elicitation.

We can therefore imagine that the WRKY binding activity is regulated by the formation of different homo and heterocomplexes of WRKY proteins. The study by Xu *et al.* clearly indicated that association of different WRKYs resulted in different binding strengths and specificities *in vitro* (Xu *et al.*, 2006). The activity of a defence-related gene would then be the result of the ratio of different WRKY proteins present on the promoter. More precise ChIP studies could eventually shed some light on this matter. It is also possible that the WRKY activity is regulated by post-translational modifications. Indeed, two-dimensional Western blotting revealed that a single WRKY protein can be present in different forms and that some of these forms become more abundant after elicitation (Turck *et al.*, 2004). Another study showed that WRKY22 and WRKY29 were downstream components of a MAPK signalling cascade in *Arabidopsis* (Asai *et al.*, 2002). This shows that, just like the ERF factors, WRKYs could also be activated by specific MAPK cascades.

The WRKY domain is defined as the DNA-binding domain of the WRKY proteins (reviewed in Eulgem *et al.*, 2000). It is composed of approximately 60 amino acids that are the most conserved residues in the WRKY proteins. Importantly, the WRKY domain is sufficient for mediating sequence-specific DNA binding. A putative zinc-binding motif $C_2H_2$ or $C_2HC$ was originally found in the sequence coding for the WRKY domain. Complete loss of DNA-binding activity upon treatment of these proteins with the divalent metal chelator 1,10-*o*-phenanthroline provided evidence that zinc binding was important for proper domain folding and/or DNA binding of the WRKY domain (Rushton *et al.*, 1995; de Pater *et al.*, 1996).

Recently, the three-dimensional structures of the DNA-binding domain of AtWRKY4 and AtWRKY1 were obtained by NMR spectroscopy (Yamasaki *et al.*, 2005) and by X-ray crystallography (Duan *et al.*, 2007), respectively. The structures can nearly be superimposed, suggesting that the WRKY proteins share a common DNA-binding mechanism. The structures consist of a four-stranded (AtWRKY4) or a five-stranded (AtWRKY1) antiparallel β-sheet with a zinc-binding pocket. The WRKY domain is structurally related to the Glial Cells Missing (GCM) family of transcription factors for which a structure bound to DNA exists (Cohen *et al.*, 2003). NMR-titration experiments and DNA docking enabled elaboration of a WRKY/DNA model (Yamasaki *et al.*, 2005) (see Fig. 6.1). This model is in good agreement with a model obtained by comparative modelling using the structure of GCM/DNA as a canvas. According to these models, the β-sheet of the WRKY domain lies perpendicular to the DNA axis so that the β-strand containing the invariant WRKYGQK motif enters deeply into the major groove of the DNA making contacts with a 6-bp region. However, the structural determinants of the sequence-specific binding of the WRKY domain are still unknown.

## 6.6 Other Factors

The potato *PR-10a* gene, which is induced upon wounding, elicitor treatment or infection with the oomycete *Phytophthora infestans* contains at least two regulatory regions. One comprises a positive regulatory element, which was shown to be bound by the factor PBF-2, and a negative regulatory element (silencer), which the protein SEBF binds (Després *et al.*, 1995). The Whirly and SEBF proteins were isolated using a similar technique. The two *PR-10a* promoter regulatory elements were immobilized on magnetic beads and incubated with potato tuber extracts (Desveaux *et al.*, 2000; Boyle and Brisson, 2001). PBF-2 (now called StWhy1; *Solanum tuberosum* Whirly1) was isolated using the positive regulatory element (Desveaux *et al.*, 2000), while SEBF was isolated using the negative regulatory element (Boyle and Brisson, 2001). Both proteins were later found to belong to small families of plant proteins as compared to the large ERF, MYB and WRKY families. Interestingly, both *StWhy1* and *SEBF* genes were found to encode a plastid transit peptide at the N terminus (Boyle and Brisson, 2001; Desveaux *et al.*, 2004). Every plant species, where sufficient DNA sequence information is available, contains at least two Whirly members, one directed to plastids and one directed to mitochondria (Desveaux *et al.*, 2005), which suggests a specific role for these proteins inside organelles.

Both proteins show an uncharacteristic preference for single-stranded DNA (ssDNA) (Desveaux *et al.*, 2000; Boyle and Brisson, 2001). Furthermore, both StWhy1 and SEBF have shown sequence specificity when tested by electrophoretic mobility shift assays. The DNA-binding activity of StWhy1 is induced in potato tubers in response to wounding or an elicitor (Després *et al.*, 1995). This binding activity correlates with the induced expression of the *PR-10a* gene. Furthermore, ChIP experiments indicated that the protein is present on the promoter of the gene only when tubers are wounded or treated with an elicitor (Desveaux *et al.*, 2004; González-Lamothe *et al.*, 2008). Overexpression of StWhy1 in potato protoplasts or in yeast confirmed that the protein can activate transcription (Desveaux *et al.*, 2000). In *Arabidopsis*, two TILLING (targeted induced local lesion in genome) lines containing different point mutations in the *AtWhy1* gene were shown to be more susceptible to infection by *H. parasitica*. Another TILLING line was recently isolated that contains a point mutation in the same gene but leads to increased resistance to the same pathogen (Desveaux, D., Wilton, M., Parent, J.-S. and Brisson, N., unpublished work). Altogether these data support a role for Whirlies in defence responses.

The crystal structure of StWhy1 was solved by X-ray crystallography (Desveaux *et al.*, 2002). Like all the members of its family, StWhy1 contains a Whirly domain, which contains approximately 200 amino acids and consists of two four-stranded antiparallel β-sheets packed perpendicularly against each other and three α-helices. The Whirlies adopt a tetrameric fold in solution. The tetramerization is mediated by the α-helices whereas the β-sheets constitute the putative ssDNA-binding platform. The conserved Whirly domain is necessary and sufficient for ssDNA binding. The structure of StWhy1 is similar to the structure of the mitochondrial guide RNA-binding proteins 1 and 2 (MRP1/2). The structure of MRP1/2 has been solved in complex with guide RNA by X-ray crystallography (Schumacher *et al.*, 2006). However, since both Whirly and MRP1/2 proteins do not possess significant sequence similarity and the residues involved in the RNA binding by MRP1/2 are not conserved in the Whirlies, it seems unlikely that these proteins share a common binding mechanism. Preliminary crystallographic analysis of a StWhy2–ssDNA complex suggests that although the residues contacting ssDNA are located on the β sheets, the nucleic acid binding mechanism is different for the Whirlies and the MRPs (Cappadocia, L., Sygusch, J., Brisson, N., unpublished results).

SEBF binds ssDNA in a sequence-specific manner (Boyle and Brisson, 2001). The consensus-binding site of SEBF (called the SE element) was found to be C/TTGTCNC. Members of the SEBF family possess two consensus-sequence RNA binding domains (cs-RBDI and II; also called RNA-recognition motifs (RRM)) arranged in tandem and separated by a glycine-rich linker. SEBF binds to the SE through its cs-RBDII (González-Lamothe *et al.*, 2008). ChIP studies indicate that SEBF binds its element in the promoter of PR-10a in unstimulated cells only. SEBF is released from the promoter upon wounding or treatment with an elicitor, while the same treatment leads to the binding of the activator StWhy1 to a nearby element. Remarkably, the binding of SEBF to the promoter requires the presence of the ERF factor Pti4, which interacts with SEBF through its ERF DNA-binding domain to form the core of a repressosome (Gonzalez-Lamothe *et al.*, 2008).

The RRM is a highly plastic domain (reviewed in Maris *et al.*, 2005) capable of interacting with DNA, RNA and even proteins with a broad range of affinities and specificities. Much of our understanding of the DNA/RNA binding by the RRM domain comes from the numerous structures of RRM-containing proteins in complex with DNA or RNA that have been determined by X-ray crystallography and NMR spectroscopy. A prototypical RRM domain contains approximately 90 amino acids and consists of a four-stranded antiparallel β-sheet packed along two α-helices. The β-sheet constitutes the main DNA/RNA-binding surface while loops and/or N- and C-terminal regions contribute additional DNA/RNA-binding residues. Two RNP (ribonucleoprotein) motifs, termed RNP1 ([ILF]-[FY]-[ILV]-X-N-L) and RNP2 ([RK]-G-[FY]-[GA]-[FY]-[ILV]-x-[FY]) constitute the hallmark of the RRM domain and contain basic and aromatic residues involved in DNA/RNA binding. These motifs are located on the central strands of the β-sheet where they mediate the non-sequence-specific recognition of a pair of nucleotides. Additional non-conserved residues are responsible for the sequence-specific binding of those nucleotides. Each RRM domain can accommodate between two and eight nucleotides. Some proteins, as is the case for SEBF, possess two or multiple copies of the RRM domain arranged in tandem. In most cases, such an arrangement will permit the binding of two adjacent stretches of the same DNA/RNA molecule (Auweter *et al.*, 2006) providing an extended DNA/RNA-binding interface. In other cases, the relative orientation of the RRM domains will favour the looping of the RNA/DNA and the binding to distant sites (Maris *et al.*, 2005). The linker, in addition to its RRM domain positioning role, can also contribute amino acids that are involved in DNA/RNA binding. The plasticity of the RRM domain prevents us from building an accurate model for the binding of SEBF to ssDNA. The structural basis for sequence-specific recognition consequently awaits the structure of a SEBF–ssDNA complex.

Other TFs have been found to be involved in plant defence. However, they do not seem to be part of well-characterized groups as do the other proteins we have described above. For example, the *Arabidopsis* gene *LSD1* was found to be involved in HR-related cell death (Aviv *et al.*, 2002). Mutant plants of this gene are more susceptible to *P. syringae*. In a subsequent study, a paralogue gene, *LSD-One-Like1* (*LOL1*) was shown to function as the opposite of *LSD1* and to have the opposite effect on resistance (Epple *et al.*, 2003). These genes however could not be grouped into a coherent family and there are no structural data available for analysis.

## 6.7 Concluding Remarks

Considerable progress has been made during the last few years in our understanding of how transcription is regulated through the action of transcription factors in eukaryotes. In plants, thanks to the rapid advances in genome sequencing and the powerful genetic tools now available, we have witnessed the identification of many families of TFs that play a role in defence responses and have started unravelling the function of a few of these factors. However, when compared to other fields, progress in the study of plant TFs has been slow. A major reason for this is certainly the amazing complexity of plant TF gene families, which often requires that double or even triple KO be produced to obtain a testable phenotype. Progress has been even slower in the biochemical characterization of TFs, with only a few laboratories in the world actively involved in this area. Here again plants add an additional layer of complexity since transient expression studies, an essential tool in the study of transcription, can only be done using protoplasts or other complex approaches such as gene bombardment or *Agrobacterium* infection. There is no doubt also that the lack of a robust *in vitro* transcription system in plants has impaired progress in this field. However, one must be hopeful that the rapid development of new technologies that is taking place in biochemistry and genomic research will help not only to identify new TFs but also to understand how these factors contribute to the important remodelling of the transcriptome that takes place during defence responses.

## Acknowledgements

The authors wish to thank the Natural Sciences and Engineering Research Council of Canada and the Fonds de la Recherche sur la Nature et les Technologies du Québec for financial support.

## References

Allen, M.D., Yamasaki, K., Ohme-Takagi, M., Tateno, M. and Suzuki, M. (1998) A novel mode of DNA recognition by a beta-sheet revealed by the solution structure of the GCC-box binding domain in complex with DNA. *EMBO Journal* 17, 5484–5496.

Asai, T., Tena, G., Plotnikova, J., Willmann, M.R., Chiu, W.L., Gomez-Gomez, L., Boller, T., Ausubel, F.M. and Sheen, J. (2002) MAP kinase signalling cascade in *Arabidopsis* innate immunity. *Nature* 415, 977–983.

Auweter, S.D., Oberstrass, F.C. and Allain, F.H. (2006) Sequence-specific binding of single-stranded RNA: is there a code for recognition? *Nucleic Acids Research* 34, 4943–4959.

Aviv, D.H., Rusterucci, C., Holt, B.F., 3rd, Dietrich, R.A., Parker, J.E. and Dangl, J.L. (2002) Runaway cell death, but not basal disease resistance, in lsd1 is SA- and NIM1/NPR1-dependent. *The Plant Journal* 29, 381–391.

Berrocal-Lobo, M., Molina, A. and Solano, R. (2002) Constitutive expression of ETHYLENE-RESPONSE-FACTOR1 in *Arabidopsis* confers resistance to several necrotrophic fungi. *The Plant Journal* 29, 23–32.

Boyle, B. and Brisson, N. (2001) Repression of the defense gene *PR-10a* by the single-stranded DNA binding protein SEBF. *The Plant Cell* 13, 2525–2537.

Chakravarthy, S., Tuori, R.P., D'Ascenzo, M.D., Fobert, P.R., Després, C. and Martin, G.B. (2003) The tomato transcription factor Pti4 regulates defense-related gene expression via GCC box and non-GCC box *cis* elements. *The Plant Cell* 15, 3033–3050.

Cheong, Y.H., Moon, B.C., Kim, J.K., Kim, C.Y., Kim, M.C., Kim, I.H., Park, C.Y., Kim, J.C., Park, B.O., Koo, S.C., Yoon, H.W., Chung, W.S., Lim, C.O., Lee, S.Y. and Cho, M.J. (2003)

BWMK1, a rice mitogen-activated protein kinase, locates in the nucleus and mediates pathogenesis-related gene expression by activation of a transcription factor. *Plant Physiology* 132, 1961–1972.

Chisholm, S.T., Coaker, G., Day, B. and Staskawicz, B.J. (2006) Host–microbe interactions: shaping the evolution of the plant immune response. *Cell* 124, 803–814.

Cohen, S.X., Moulin, M., Hashemolhosseini, S., Kilian, K., Wegner, M. and Muller, C.W. (2003) Structure of the GCM domain-DNA complex: a DNA-binding domain with a novel fold and mode of target site recognition. *EMBO Journal* 22, 1835–1845.

Daniel, X., Lacomme, C., Morel, J.B. and Roby, D. (1999) A novel myb oncogene homologue in *Arabidopsis thaliana* related to hypersensitive cell death. *The Plant Journal* 20, 57–66.

de Pater, S., Greco, V., Pham, K., Memelink, J. and Kijne, J. (1996) Characterization of a zinc-dependent transcriptional activator from *Arabidopsis*. *Nucleic Acids Research* 24, 4624–4631.

Després, C., Subramaniam, R., Matton, D.P. and Brisson, N. (1995) The activation of the potato *PR-10a* gene requires the phosphorylation of the nuclear factor PBF-1. *The Plant Cell* 7, 589–598.

Després, C., Chubak, C., Rochon, A., Clark, R., Bethune, T., Desveaux, D. and Fobert, P.R. (2003) The *Arabidopsis* NPR1 disease resistance protein is a novel cofactor that confers redox regulation of DNA binding activity to the basic domain/leucine zipper transcription factor TGA1. *The Plant Cell* 15, 2181–2191.

Desveaux, D., Després, C., Joyeux, A., Subramaniam, R. and Brisson, N. (2000) PBF-2 is a novel single-stranded DNA binding factor implicated in *PR-10a* gene activation in potato. *The Plant Cell* 12, 1477–1489.

Desveaux, D., Allard, J., Brisson, N. and Sygusch, J. (2002) A new family of plant transcription factors displays a novel ssDNA-binding surface. *Nature Structural Biology* 9, 512–517.

Desveaux, D., Subramaniam, R., Després, C., Mess, J.N., Levesque, C., Fobert, P.R., Dangl, J.L. and Brisson, N. (2004) A 'Whirly' transcription factor is required for salicylic acid-dependent disease resistance in *Arabidopsis*. *Developmental Cell* 6, 229–240.

Desveaux, D., Maréchal, A. and Brisson, N. (2005) Whirly transcription factors: defense gene regulation and beyond. *Trends in Plant Science* 10, 95–102.

Dong, J., Chen, C. and Chen, Z. (2003) Expression profiles of the *Arabidopsis* WRKY gene superfamily during plant defense response. *Plant Molecular Biology* 51, 21–37.

Du, L. and Chen, Z. (2000) Identification of genes encoding receptor-like protein kinases as possible targets of pathogen- and salicylic acid-induced WRKY DNA-binding proteins in *Arabidopsis*. *The Plant Journal* 24, 837–847.

Duan, M.R., Nan, J., Liang, Y.H., Mao, P., Lu, L., Li, L., Wei, C., Lai, L., Li, Y. and Su, X.D. (2007) DNA binding mechanism revealed by high resolution crystal structure of *Arabidopsis thaliana* WRKY1 protein. *Nucleic Acids Research* 35, 1145–1154.

Ekengren, S.K., Liu, Y., Schiff, M., Dinesh-Kumar, S.P. and Martin, G.B. (2003) Two MAPK cascades, NPR1, and TGA transcription factors play a role in Pto-mediated disease resistance in tomato. *The Plant Journal* 36, 905–917.

Ellenberger, T.E., Brandl, C.J., Struhl, K. and Harrison, S.C. (1992) The GCN4 basic region leucine zipper binds DNA as a dimer of uninterrupted alpha helices: crystal structure of the protein–DNA complex. *Cell* 71, 1223–1237.

Epple, P., Mack, A.A., Morris, V.R. and Dangl, J.L. (2003) Antagonistic control of oxidative stress-induced cell death in *Arabidopsis* by two related, plant-specific zinc finger proteins. *Proceedings of the National Academy of Sciences, USA* 100, 6831–6836.

Eulgem, T. (2005) Regulation of the *Arabidopsis* defense transcriptome. *Trends in Plant Science* 10, 71–78.

Eulgem, T. and Somssich, I.E. (2007) Networks of WRKY transcription factors in defense signaling. *Current Opinion in Plant Biology* 10, 366–371.

Eulgem, T., Rushton, P.J., Robatzek, S. and Somssich, I.E. (2000) The WRKY superfamily of plant transcription factors. *Trends in Plant Science* 5, 199–206.

Fobert, P.R. (2006) Transcription factors regulating plant defense responses. *Model Plants and Crop Improvement* 8, 47.

Gigolashvili, T., Yatusevich, R., Berger, B., Müller, C. and Flügge, U.I. (2007) The R2R3-MYB transcription factor HAG1/MYB28 is a regulator of methionine-derived glucosinolate biosynthesis in *Arabidopsis thaliana*. *The Plant Journal* 51, 247–261.

Glover, J.N. and Harrison, S.C. (1995) Crystal structure of the heterodimeric bZIP transcription factor c-Fos-c-Jun bound to DNA. *Nature* 373, 257–261.

González-Lamothe, R., Boyle, P., Dulude, A., Roy, V., Lezin-Doumbou, C., Kaur, G.S., Bouarab, K., Després, C. and Brisson, N. (2008) The transcriptional activator Pti4 is required for the recruitment of a repressosome nucleated by repressor SEBF at the potato *PR-10a* gene. *The Plant Cell* 20, 3136–3147.

Gu, Y.Q., Yang, C., Thara, V.K., Zhou, J. and Martin, G.B. (2000) Pti4 is induced by ethylene and salicylic acid, and its product is phosphorylated by the Pto kinase. *The Plant Cell* 12, 771–786.

Gu, Y.Q., Wildermuth, M.C., Chakravarthy, S., Loh, Y.T., Yang, C., He, X., Han, Y. and Martin, G.B. (2002) Tomato transcription factors pti4, pti5, and pti6 activate defense responses when expressed in *Arabidopsis*. *The Plant Cell* 14, 817–831.

Guehmann, S., Vorbrueggen, G., Kalkbrenner, F. and Moelling, K. (1992) Reduction of a conserved Cys is essential for Myb DNA-binding. *Nucleic Acids Research* 20, 2279–2286.

Gutterson, N. and Reuber, T.L. (2004) Regulation of disease resistance pathways by AP2/ERF transcription factors. *Current Opinion in Plant Biology* 7, 465–471.

Hao, D., Ohme-Takagi, M. and Sarai, A. (1998) Unique mode of GCC box recognition by the DNA-binding domain of ethylene-responsive element-binding factor (ERF domain) in plants. *Journal of Biological Chemistry* 273, 26857–26861.

Heine, G.F., Hernandez, J.M. and Grotewold, E. (2004) Two cysteines in plant R2R3 MYB domains participate in REDOX-dependent DNA binding. *Journal of Biological Chemistry* 279, 37878–37885.

Huckelhoven, R. (2007) Cell wall-associated mechanisms of disease resistance and susceptibility. *Annual Review of Phytopathology* 45, 101–127.

Ishiguro, S. and Nakamura, K. (1994) Characterization of a cDNA encoding a novel DNA-binding protein, SPF1, that recognizes SP8 sequences in the 5' upstream regions of genes coding for sporamin and beta-amylase from sweet potato. *Molecular and General Genetics* 244, 563–571.

Jofuku, K.D., den Boer, B.G., Van Montagu, M. and Okamuro, J.K. (1994) Control of *Arabidopsis* flower and seed development by the homeotic gene *APETALA2*. *The Plant Cell* 6, 1211–1225.

Jones, A.M., Thomas, V., Bennett, M.H., Mansfield, J. and Grant, M. (2006) Modifications to the *Arabidopsis* defense proteome occur prior to significant transcriptional change in response to inoculation with *Pseudomonas syringae*. *Plant Physiology* 142, 1603–1620.

Journot-Catalino, N., Somssich, I.E., Roby, D. and Kroj, T. (2006) The transcription factors WRKY11 and WRKY17 act as negative regulators of basal resistance in *Arabidopsis thaliana*. *The Plant Cell* 18, 3289–3302.

Katagiri, F. (2004) A global view of defense gene expression regulation – a highly interconnected signaling network. *Current Opinion in Plant Biology* 7, 506–511.

Katagiri, F., Lam, E. and Chua, N.H. (1989) Two tobacco DNA-binding proteins with homology to the nuclear factor CREB. *Nature* 340, 727–730.

Katagiri, F., Seipel, K. and Chua, N.H. (1992) Identification of a novel dimer stabilization region in a plant bZIP transcription activator. *Molecular and Cellular Biology* 12, 4809–4816.

Kesarwani, M., Yoo, J. and Dong, X. (2007) Genetic interactions of TGA transcription factors in the regulation of pathogenesis-related genes and disease resistance in *Arabidopsis*. *Plant Physiology* 144, 336–346.

Knoth, C., Ringler, J., Dangl, J.L. and Eulgem, T. (2007) *Arabidopsis* WRKY70 is required for full RPP4-mediated disease resistance and basal defense against *Hyaloperonospora parasitica*. *Molecular Plant–Microbe Interactions* 20, 120–128.

Li, J., Brader, G. and Palva, E.T. (2004) The WRKY70 transcription factor: a node of convergence for jasmonate-mediated and salicylate-mediated signals in plant defense. *The Plant Cell* 16, 319–331.

Lippok, B., Birkenbihl, R.P., Rivory, G., Brummer, J., Schmelzer, E., Logemann, E. and Somssich, I.E. (2007) Expression of AtWRKY33 encoding a pathogen- or PAMP-responsive WRKY transcription factor is regulated by a composite DNA motif containing W box elements. *Molecular Plant–Microbe Interactions* 20, 420–429.

Liu, Y., Schiff, M. and Dinesh-Kumar, S.P. (2004) Involvement of MEK1 MAPKK, NTF6 MAPK, WRKY/MYB transcription factors, COI1 and CTR1 in N-mediated resistance to tobacco mosaic virus. *The Plant Journal* 38, 800–809.

Maris, C., Dominguez, C. and Allain, F.H. (2005) The RNA recognition motif, a plastic RNA-binding platform to regulate post-transcriptional gene expression. *FEBS Journal* 272, 2118–2131.

Martin, C. and Paz-Ares, J. (1997) MYB transcription factors in plants. *Trends in Genetics* 13, 67–73.

McGrath, K.C., Dombrecht, B., Manners, J.M., Schenk, P.M., Edgar, C.I., Maclean, D.J., Scheible, W.R., Udvardi, M.K. and Kazan, K. (2005) Repressor- and activator-type ethylene response factors functioning in jasmonate signaling and disease resistance identified via a genome-wide screen of *Arabidopsis* transcription factor gene expression. *Plant Physiology* 139, 949–959.

Mengiste, T., Chen, X., Salmeron, J. and Dietrich, R. (2003) The BOTRYTIS SUSCEPTIBLE1 gene encodes an R2R3MYB transcription factor protein that is required for biotic and abiotic stress responses in *Arabidopsis*. *The Plant Cell* 15, 2551–2565.

Mou, Z., Fan, W. and Dong, X. (2003) Inducers of plant systemic acquired resistance regulate NPR1 function through redox changes. *Cell* 113, 935–944.

Mukhtar, M.S., Deslandes, L., Auriac, M.C., Marco, Y. and Somssich, I.E. (2008) The *Arabidopsis* transcription factor WRKY27 influences wilt disease symptom development caused by *Ralstonia solanacearum*. *The Plant Journal* 56(6), 935–947.

Nakano, T., Suzuki, K., Fujimura, T. and Shinshi, H. (2006) Genome-wide analysis of the ERF gene family in *Arabidopsis* and rice. *Plant Physiology* 140, 411–432.

Nurnberger, T., Brunner, F., Kemmerling, B. and Piater, L. (2004) Innate immunity in plants and animals: striking similarities and obvious differences. *Immunological Reviews* 198, 249–266.

Ogata, K., Hojo, H., Aimoto, S., Nakai, T., Nakamura, H., Sarai, A., Ishii, S. and Nishimura, Y. (1992) Solution structure of a DNA-binding unit of Myb: a helix-turn-helix-related motif with conserved tryptophans forming a hydrophobic core. *Proceedings of the National Academy of Sciences, USA* 89, 6428–6432.

Ogata, K., Morikawa, S., Nakamura, H., Sekikawa, A., Inoue, T., Kanai, H., Sarai, A., Ishii, S. and Nishimura, Y. (1994) Solution structure of a specific DNA complex of the Myb DNA-binding domain with cooperative recognition helices. *Cell* 79, 639–648.

Ogata, K., Morikawa, S., Nakamura, H., Hojo, H., Yoshimura, S., Zhang, R., Aimoto, S., Ametani, Y., Hirata, Z., Sarai, A. and Ishii, S. and Nishimura, Y. (1995) Comparison of the free and DNA-complexed forms of the DNA-binding domain from c-Myb. *Nature Structural Biology* 2, 309–320.

Ohme-Takagi, M. and Shinshi, H. (1995) Ethylene-inducible DNA binding proteins that interact with an ethylene-responsive element. *The Plant Cell* 7, 173–182.

Oñate-Sánchez, L., Anderson, J.P., Young, J. and Singh, K.B. (2007) AtERF14, a member of the ERF family of transcription factors, plays a nonredundant role in plant defense. *Plant Physiology* 143, 400–409.

Paz-Ares, J., Ghosal, D., Wienand, U., Peterson, P.A. and Saedler, H. (1987) The regulatory *c1* locus of *Zea mays* encodes a protein with homology to myb proto-oncogene products and with structural similarities to transcriptional activators. *EMBO Journal* 6, 3553–3558.

Pre, M., Atallah, M., Champion, A., De Vos, M., Pieterse, C.M. and Memelink, J. (2008) The AP2/ERF-domain transcription factor ORA59 integrates jasmonic acid and ethylene signals in plant defense. *Plant Physiology* 147(3), 1347–1357.

Raffaele, S., Rivas, S. and Roby, D. (2006) An essential role for salicylic acid in AtMYB30-mediated control of the hypersensitive cell death program in *Arabidopsis*. *FEBS Letters* 580, 3498–3504.

Raffaele, S., Vailleau, F., Leger, A., Joubes, J., Miersch, O., Huard, C., Blee, E., Mongrand, S., Domergue, F. and Roby, D. (2008) A MYB transcription factor regulates very-long-chain fatty acid biosynthesis for activation of the hypersensitive cell death response in *Arabidopsis*. *The Plant Cell* 20(3), 752–767.

Rochon, A., Boyle, P., Wignes, T., Fobert, P.R. and Després, C. (2006) The coactivator function of *Arabidopsis* NPR1 requires the core of its BTB/POZ domain and the oxidation of C-terminal cysteines. *The Plant Cell* 18, 3670–3685.

Rushton, P.J., Macdonald, H., Huttly, A.K., Lazarus, C.M. and Hooley, R. (1995) Members of a new family of DNA-binding proteins bind to a conserved *cis*-element in the promoters of alpha-Amy2 genes. *Plant Molecular Biology* 29, 691–702.

Rushton, P.J., Torres, J.T., Parniske, M., Wernert, P., Hahlbrock, K. and Somssich, I.E. (1996) Interaction of elicitor-induced DNA-binding proteins with elicitor response elements in the promoters of parsley *PR1* genes. *EMBO Journal* 15, 5690–5700.

Schumacher, M.A., Goodman, R.H. and Brennan, R.G. (2000) The structure of a CREB bZIP. somatostatin CRE complex reveals the basis for selective dimerization and divalent cation-enhanced DNA binding. *Journal of Biological Chemistry* 275, 35242–35247.

Schumacher, M.A., Karamooz, E., Zikova, A., Trantirek, L. and Lukes, J. (2006) Crystal structures of *T. brucei* MRP1/MRP2 guide-RNA binding complex reveal RNA matchmaking mechanism. *Cell* 126, 701–711.

Shen, Q.H., Saijo, Y., Mauch, S., Biskup, C., Bieri, S., Keller, B., Seki, H., Ulker, B., Somssich, I.E. and Schulze-Lefert, P. (2007) Nuclear activity of MLA immune receptors links isolate-specific and basal disease-resistance responses. *Science* 315, 1098–1103.

Shimono, M., Sugano, S., Nakayama, A., Jiang, C.J., Ono, K., Tokl, S. and Takatsuji, H. (2007) Rice WRKY45 plays a crucial role in benzothiadiazole-inducible blast resistance. *The Plant Cell* 19, 2064–2076.

Solano, R., Fuertes, A., Sanchez-Pulido, L., Valencia, A. and Paz-Ares, J. (1997) A single residue substitution causes a switch from the dual DNA binding specificity of plant transcription factor MYB.Ph3 to the animal c-MYB specificity. *Journal of Biological Chemistry* 272, 2889–2895.

Stracke, R., Werber, M. and Weisshaar, B. (2001) The R2R3-MYB gene family in *Arabidopsis thaliana*. *Current Opinion in Plant Biology* 4, 447–456.

Tao, Y., Xie, Z., Chen, W., Glazebrook, J., Chang, H.S., Han, B., Zhu, T., Zou, G. and Katagiri, F. (2003) Quantitative nature of *Arabidopsis* responses during compatible and incompatible interactions with the bacterial pathogen *Pseudomonas syringae*. *The Plant Cell* 15, 317–330.

Turck, F., Zhou, A. and Somssich, I.E. (2004) Stimulus-dependent, promoter-specific binding of transcription factor WRKY1 to its native promoter and the defense-related gene *PcPR1-1* in parsley. *The Plant Cell* 16, 2573–2585.

Ulker, B. and Somssich, I.E. (2004) WRKY transcription factors: from DNA binding towards biological function. *Current Opinion in Plant Biology* 7, 491–498.

Vailleau, F., Daniel, X., Tronchet, M., Montillet, J.L., Triantaphylides, C. and Roby, D. (2002) A R2R3-MYB gene, AtMYB30, acts as a positive regulator of the hypersensitive cell death program in plants in response to pathogen attack. *Proceedings of the National Academy of Sciences, USA* 99, 10179–10184.

Van der Ent, S., Verhagen, B.W., Van Doorn, R., Bakker, D., Verlaan, M.G., Pel, M.J., Joosten, R.G., Proveniers, M.C., Van Loon, L.C., Ton, J. and Pieterse, C.M. (2008) MYB72 is required in early signaling steps of *Rhizobacteria*-induced systemic resistance in *Arabidopsis. Plant Physiology* 146, 1293–1304.

Wang, D., Amornsiripanitch, N. and Dong, X. (2006) A genomic approach to identify regulatory nodes in the transcriptional network of systemic acquired resistance in plants. *PLoS Pathogens* 2, e123.

Xu, X., Chen, C., Fan, B. and Chen, Z. (2006) Physical and functional interactions between pathogen-induced *Arabidopsis* WRKY18, WRKY40, and WRKY60 transcription factors. *The Plant Cell* 18, 1310–1326.

Yamasaki, K., Kigawa, T., Inoue, M., Tateno, M., Yamasaki, T., Yabuki, T., Aoki, M., Seki, E., Matsuda, T., Tomo, Y., Hayami, N., Terada, T., Shirouzu, M., Tanaka, A., Seki, M., Shinozaki, K. and Yokoyama, S. (2005) Solution structure of an *Arabidopsis* WRKY DNA binding domain. *The Plant Cell* 17, 944–956.

Zhang, H., Zhang, D., Chen, J., Yang, Y., Huang, Z., Huang, D., Wang, X.C. and Huang, R. (2004) Tomato stress-responsive factor TSRF1 interacts with ethylene responsive element GCC box and regulates pathogen resistance to *Ralstonia solanacearum. Plant Molecular Biology* 55, 825–834.

Zhang, Y., Tessaro, M.J., Lassner, M. and Li, X. (2003) Knockout analysis of *Arabidopsis* transcription factors TGA2, TGA5, and TGA6 reveals their redundant and essential roles in systemic acquired resistance. *The Plant Cell* 15, 2647–2653.

Zheng, Z., Qamar, S.A., Chen, Z. and Mengiste, T. (2006) *Arabidopsis* WRKY33 transcription factor is required for resistance to necrotrophic fungal pathogens. *The Plant Journal* 48, 592–605.

Zheng, Z., Mosher, S.L., Fan, B., Klessig, D.F. and Chen, Z. (2007) Functional analysis of *Arabidopsis* WRKY25 transcription factor in plant defense against *Pseudomonas syringae. BMC Plant Biology* 7, 2.

Zhou, J., Tang, X. and Martin, G.B. (1997) The Pto kinase conferring resistance to tomato bacterial speck disease interacts with proteins that bind a *cis*-element of pathogenesis-related genes. *EMBO Journal* 16, 3207–3218.

Zipfel, C. (2008) Pattern-recognition receptors in plant innate immunity. *Current Opinion in Immunology* 20, 10–16.

# 7 Cross Talk Between Induced Plant Immune Systems

Rocío González-Lamothe, Mohamed El Oirdi, Taha Abd El Rahman, Raphaël Sansregret, Hamed Bathily and Kamal Bouarab

*Université de Sherbrooke, Québec, Canada*

## Abstract

Plants possess different mechanisms to defend themselves against the continuous exposure to pathogenic attacks. The most elaborate defence responses are those that involve the activation of several specific antimicrobial reactions once the pathogen is detected. Thus, detection of a pathogen's component through plant receptors will unleash a defence response that will ultimately stop the pathogen spreading. Plants generally react to necrotrophic pathogens through the activation of jasmonic acid (JA)-dependent defence pathways, whereas defence responses to biotrophic pathogens are salicylic acid (SA)-dependent. Another plant immune pathway has been shown against viruses, in which the plant activates a virus-RNA degrading mechanism (RNA silencing) that is independent of the presence of receptors. To better understand the plant immune system, it is important to know how each of the previous defence mechanisms interacts with the other ones. Thus, the SA-dependent pathway regulates JA-dependent defence pathways and vice versa. Also, recent studies showed that detection of a pathogen through a plant receptor can induce RNA silencing of certain plant genes. This chapter will discuss the cross talk among some of the known plant defence pathways.

## 7.1 Introduction

How plants defend themselves from pathogenic attacks has been a subject of research for years. Plant hosts can avoid pathogenic invasions by lacking the nutrients needed by the pathogen to survive and by the presence of physical and chemical pre-existing barriers. However, plants also possess an active immune system by which they detect potential damaging invaders and induce specific defence mechanisms to stop the spreading of the intruder (Hammond-

 

Kosack and Jones, 1996). This active plant-defence response can be achieved by different pathways, specifically activated by a more or less restricted group of pathogens, giving a response tailored to the nature of the attacking organism (Jones and Dangl, 2006). Thus, an immune response is activated against biotrophic pathogens after detection of pathogen elicitors by the host receptors. As a consequence, a signalling pathway is triggered that will result into the activation of some effector proteins. Plants can recognize pathogen general elicitors through transmembrane pathogen recognition receptors (PRRs). These general elicitors are also called pathogen-associated molecular patterns (PAMPS) because they are found in a broad range of microorganisms and they are often recognized by all members of a host genus (Gordon, 2002). Plants can also detect specific elicitors or their activity through the nucleotide binding-leucine-rich repeat (NB-LRR) receptors encoded by resistance (*R*) genes. This defines what is called 'gene-for-gene' resistance (Bent and Mackey, 2007). The specific elicitors or avirulence (*avr*) genes are often found in only one strain of the microorganism, while the *R* gene is present only in certain varieties of the host. As a result of the recognition of an *avr* gene by an R protein, the plant will induce a cell death, the so-called hypersensitive response (HR), to stop the spreading of the pathogen. The signalling and effector molecules can be the same for general elicitors- and avirulence protein-activated pathways, but in general, the response produced after recognition of avirulence proteins is faster and stronger than the one induced by general elicitors, resulting in a more efficient resistance. Resistance induced by both general and specific elicitors is triggered against biotrophic pathogens and involve mainly salicylic acid (SA) as a signalling molecule.

Necrotrophic pathogens induce a different defence response that is less understood than the one triggered against biotrophs. To date, no receptors have been shown to be required for resistance against necrotrophic pathogens, but the nature of the signalling and some effector molecules have been identified. Thus, resistance against necrotrophs involves jasmonic acid (JA), and in some cases ethylene (ET) as another signalling molecule. These phytohormones are also involved in response to wounding and insect feeding. *R* gene-mediated HR is not produced by the host in response to necrotrophic pathogens, but, on the contrary, some necrotrophes such as *Botrytis cinerea* induce an HR-like response in the host to facilitate its colonization (Govrin and Levine, 2000; El Oirdi and Bouarab, 2007). Although it is now generally accepted that plant defence is mediated by SA against biotrophic pathogens and JA/ET against necrotrophic ones, it should be kept in mind that the reality is more complex, as indicated by exceptions to this rule.

A different defence mechanism not involving the detection of an elicitor by specific plant receptors is triggered in plants against certain viruses through RNA silencing. RNA-silencing mechanisms are mediated by small RNAs (sRNAs) resulting from the cleavage of double-stranded RNAs (dsRNAs) (Baulcombe, 2004). In this defence response, plant RNAseIII-like protein Dicer can recognize viral dsRNA and activate the silencing mechanism. As a result, virus-derived small interfering RNAs (viRNAs), which target viral RNA for degradation, are produced (Ding and Voinnet, 2007).

Although the first step to study a plant defence pathway requires analysis of the receptor, signalling and effector components, specific plant resistance pathways should not be considered as isolated mechanisms, but as components of the immune system network, where they extensively interact and regulate each other. Therefore, a complete understanding of the plant immune system requires knowledge of the interactions among the different pathways.

In this chapter, we will review the SA- and JA-dependent defence mechanisms both as independent pathways and as interacting responses that regulate each other. For a description of the antiviral silencing mechanism see Wadsworth and Dunoyer, Chapter 1, this volume. However, we will analyse the cross talk between receptor-mediated and silencing-mediated defence responses.

## 7.2 SA and JA: their Independent Effects and Cross Talk

### Role of JA in plant resistance against necrotrophic pathogens

JA is a cyclopentanone derivative which acts as a growth inhibitor, senescence-promoting substance. JA is synthesized from alpha-linolenic acid, a C18 polyunsaturated fatty acid present in the plant plasma membrane, by enzymes similar to lipase. Key enzymes involved in the JA synthesis pathway include lipoxygenase, allene oxide synthase and allene oxide cyclase (Agrawal *et al.*, 2004). JA synthesis is induced by some elicitors such as systemin, and in response to wounding and attack by insects and necrotrophic pathogens (Ryan, 2000). Plants can also accumulate methyl jasmonate (MeJA) in response to elicitors or infection. MeJA and JA modulate the expression of several defence genes including protein defensin (*PDF1*) in *Arabidopsis thaliana* or proteinase inhibitor I (*PI-I*) and proteinase inhibitor II (*PI-II*) in tomato (Farmer and Ryan, 1990; Karban *et al.*, 2000; Baldwin *et al.*, 2002; Pieterse and Van Loon, 2004; Pozo *et al.*, 2005). JA plays a key role in defence against necrotrophic pathogens. Mutants affected in the JA synthesis, or its signalling pathway, are more susceptible to necrotrophic pathogens compared to wild-type plants (Vijayan *et al.*, 1998; Browse and Howe, 2008). For example, the JA-insensitive mutant *coronatine insensitive gene 1* (*coi1*) shows enhanced susceptibility to the necrotrophic fungi *Alternaria brassicicola* and *B. cinerea* (Thomma *et al.*, 1998).

*Coi1* is necessary for the activation of JA-dependent defence genes (Kemal and Manners, 2007). On the other hand, overexpression of a JA carboxyl methyl transferase increased endogenous levels of MeJA leading to higher resistance against *B. cinerea* (Seo *et al.*, 2001; Xiao-Yi *et al.*, 2007).

### Role of SA in plant resistance against biotrophic pathogens

SA is a phenolic compound synthesized through the shikimic acid pathway. In *Arabidopsis*, it was shown that it can be synthesized through two pathways in both of which chorismate can be converted into SA, involving phenylalanine or isochorismate. In the first pathway, phenylalanine ammonia lyase (PAL) catalyses the first metabolic step, in which phenylalanine is converted to *trans*-cinnamic acid. The latter is subsequently converted into benzoic acid. A benzoic-acid-2-hydroxylase (BA2H) catalyses the final step, where benzoic acid is converted into SA (Shah, 2003). In the second pathway, SA is produced from chorismate through two steps that involve isochorismate synthase and isochorismate pyruvate lyase (Ogawa *et al.*, 2005).

SA plays a key role in the signal transduction pathway leading to resistance against biotrophic pathogens. SA regulates the expression of several defence genes including *PR1* (Shah, 2003). The role of SA in resistance against pathogens has been confirmed by using transgenic tobacco plants that express the *nahG* gene, encoding for salicylate hydroxylase, a SA-metabolizing enzyme from *Pseudomonas putida*. Those transgenic plants showed little or no accumulation of SA after infection with several types of pathogens including bacteria, viruses and fungi and they displayed higher levels of infection in comparison to wild-type plants (Chen *et al.*, 1995; Iris *et al.*, 1996). Plants can also accumulate methyl salicylate (MeSA) and conjugated SA in response to elicitors or infections. Finally, SA plays an important role in the induction of systemic acquired resistance (SAR) (Durrant and Dong, 2004).

## The cross talk between SA and JA

Plants often respond to attacks by insect herbivores and necrotrophic pathogens with induction of jasmonate-dependent resistance traits, but respond to attack by biotrophic pathogens with induction of salicylate-dependent resistance traits (Traw *et al.*, 2003). Equally, it has been suggested that the relative concentrations of SA and JA are important in determining the expression levels of different defence-related genes (Luis *et al.*, 2006).

Cross talk between SA- and JA-dependent defence signalling is the cornerstone in the plant pathogen interaction. The term cross talk is used in many cases to explain how two or more signalling pathways communicate (Taylor *et al.*, 2004). In some cases, different defence signal transduction pathways cooperate and enhance resistance against a pathogen attack (Spoel *et al.*, 2003).

Several studies have shown that SA antagonizes JA and vice versa (Xu *et al.*, 1994; Maleck and Dietrich, 1999; Jennifer *et al.*, 2002; Kunkel and Brooks, 2002; Spoel *et al.*, 2003; Traw *et al.*, 2003; Andrea *et al.*, 2004; Pieterse and Van Loon, 2004; Pozo *et al.*, 2005; Richard, 2005; Peng *et al.*, 2007; Vidhyasekaran, 2008). This antagonism provides the plant with an elaborate regulatory potential that leads to the activation of the most suitable defence against the invader. The antagonism between SA and JA was initially discovered using *Arabidopsis* mutants. *Arabidopsis* plants unable to accumulate SA produced higher levels of JA and showed enhanced expression of the JA-responsive genes in response to infection by *Pseudomonas syringae* pv. *tomato* strain DC3000. On the other hand, mutants unable to undergo JA synthesis accumulate higher levels of SA and display higher resistance to biotrophic pathogens (Enrique *et al*., 2003; Spoel *et al*., 2003; Pieterse and Van Loon, 2004).

## Factors controlling cross talk between SA and JA signalling pathways

### *NPR1*

NPR1 (non-expresser of *PR1* genes) is a cotranscription factor that regulates the plant defence responses downstream of the SA signalling pathway. Work with an *npr1* mutant revealed that NPR1 is a central regulator of plant defence responses including SAR, induced systemic resistance (ISR) and SA/JA cross talk (Durrant and Dong, 2004).

NPR1 contains an ankyrin-repeat domain, which is known to mediate protein–protein interactions, as demonstrated by mutations in this domain (Shah, 2003). Members of the TGA-element binding protein (TGA) family of basic-leucine-zipper (bZIP) DNA-binding proteins interact physically with NPR1 in yeast two-hybrid assays (Shah, 2003). The complex NPR1-TGA seems to be important in the regulation of the expression of several genes including *PR1* (Durrant and Dong, 2004).

Nuclear localization of NPR1, which is essential for SA-mediated defence gene expression, is not required for the suppression of JA signalling, indicating that cross talk between SA and JA is modulated through a novel function of NPR1 in the cytosol (Spoel *et al*., 2003). The cytosolic function of SA-activated NPR1 in modulating cross talk between SA- and JA-pathways is also redox-regulated. However, how SA induces changes in the cellular redox status, and which redox mediators are involved, is largely unknown (Pieterse and Van Loon, 2004). The way NPR1 coordinates these different responses and how the signalling network works downstream of NPR1 in each case, needs more investigation before it is fully understood (Durrant and Dong, 2004).

### *Mitogen-activated protein kinases (MAPKs)*

Signalling mechanisms and cellular responses acting downstream of the recognition of largely unrelated elicitors are believed to be similar, and are known to include medium alkalinization, release of $Ca^{2+}$, generation of signalling phospholipids and activation of mitogen-activated protein kinases (MAPKs). The MAPK family consists of three types of protein kinases, MAPK, MAPK kinase (MAPKK) and MAPKK kinase (MAPKKK) (Takahashi *et al*., 2007) and comprise a family of ubiquitous proline-directed, protein-serine/threonine kinases. Mitogens are the extracellular stimuli responsible for activation of MAPKs which in turn participate in signal transduction pathways that control intracellular events, including acute responses to hormones and major developmental changes in organisms such as gene expression, mitosis and differentiation (Person *et al*., 2001; Miki *et al*., 2006).

Recent studies showed that some MAPKs are involved in the control of the cross talk between SA and JA. *Arabidopsis* MPK4 leads to the activation of the JA pathway while suppressing the SA pathway (Petersen *et al.*, 2000). In addition, *mpk4* knockout plants exhibit constitutive activation of SA-dependent defences, but fail to induce JA defence marker genes in response to JA (Brodersen *et al.*, 2006).

### *WRKY transcription factors*

WRKY proteins are a superfamily of transcription factors, whose name comes from the conserved amino acid sequence WRKYGQK at the N-terminal end, together with a novel zinc-finger-like motif of the WRKY domain in the 60 amino acid region that is highly conserved among most of the family members (Thomas *et al.*, 2000). WRKY transcription factors are important regulators of SA-dependent defence responses (Wang *et al.*, 2006) and some of the WRKY members are involved in the cross talk between SA and JA (Koornneef and Pieterse, 2008), such as WRKY70 (Li *et al.*, 2006) which acts as a valve between SA and JA signalling events during plant defence (Li *et al.*, 2004). WRKY70 controls the cross talk between defence pathways, acts downstream of NPR1 in an SA-dependent signal pathway and acts as a repressor of JA-inducible genes (Li *et al.*, 2004).

In addition to WRKY70, other WRKYs including WRKY62, WRKY11 and WRKY17 play many roles in regulating the cross talk between SA and JA pathways (Koornneef and Pieterse, 2008). However, the clear mechanism behind the control of the cross talk by WRKY proteins is still not well understood.

### *Glutaredoxin*

Glutaredoxin is another factor in the regulation of signalling pathways cross talk and it is involved in the redox-dependent regulation of protein activities (Koornneef and Pieterse, 2008). Glutaredoxin belongs to a superfamily of small redox proteins (Hoog *et al.*, 1983) and at least 31 glutaredoxin genes are present in *A. thaliana* (Rouhier *et al.*, 2004). Glutaredoxins are small enzymes (about 100 amino acid residues) which are similar to thioredoxins and possess a typical glutathione-reducible CxxC or CxxS active site. They use glutathione as a cofactor (Rouhier *et al.*, 2004). The function of glutaredoxin is the reduction of ribonucleotides through electron transfer from NADPH via its disulfide (-SH) groups to deoxyribonucleotides, a process required for DNA synthesis (Hoog *et al.*, 1983; Holmgren, 1989).

As described below, TGA interacts with NPR1 in the plant nucleus in order to induce the expression of an *SA-dependent pathogenesis-related gene 1* (Shah, 2003; Spoel *et al.*, 2003). Glutaredoxin 480 induced by SA interacts with a TGA transcription factor, which is already bound to the *PDF1* promoter region and suppresses its expression (Ndamukong *et al.*, 2007). Therefore glutaredoxin 480 constitutes another protein involved in the antagonistic cross talk between SA and JA (Ndamukong *et al.*, 2007).

## 7.3 RNA silencing as a Response After Pathogen Elicitor Recognition

RNA-silencing mechanisms are mediated by cleavage of dsRNA into sRNAs, the molecule that confers the specificity of the silencing reaction. There are several classes of sRNAs based on their origin and function (Baulcombe, 2004). In order to clarify the discussion below, we will briefly describe the two best-characterized classes of sRNA, microRNA (miRNA) and small interfering RNA (siRNA). miRNAs originate from the imperfect intramolecular matches found in the secondary structure of primary miRNA transcripts (pri-miRNA). pri-miRNAs are processed into precursor miRNAs and then converted into miRNA (Ding and Voinnet, 2007). They are encoded in the genomes of multicellular eukaryotes and unicellular plants and in plants and animals they are grouped in families based on sequence similarity (Chapman and Carrington, 2007). siRNAs originate from perfectly matched dsRNA. The origin of dsRNA can be multiple: transcription of loci containing inverted or direct repeat sequences, or as discussed below, transcription from opposite promoters. siRNA production can be amplified through dsRNA synthesis by cellular RNA-dependent RNA polymerases, resulting in secondary siRNA accumulation (Chapman and Carrington, 2007).

The origin of the silencing-initiator dsRNA can be endogenous or exogenous. The former regulates different plant processes as developmental programmes, response to external stimuli or hormone signalling. The latter includes the production of virus-induced siRNAs (viRNAs) as a defence mechanism against the attack of some viruses. Until recently the activation of RNA silencing as a defence response seemed to be specific to viral pathogens, since they (but not fungi or bacteria) need to produce dsRNA to survive inside the host cell. Nevertheless, since RNA silencing can be triggered by endogenous dsRNA in response to external stimuli it was not surprising to find that plant pathogens other than viruses can also activate a host defence mechanism that involves the RNA-silencing pathway (Navarro *et al.*, 2006; Katiyar-Agarwal *et al.*, 2006, 2007; Pandey and Baldwin, 2007). We will describe here the publications that report the involvement of RNA silencing in the defence response triggered by different pathogens.

Navarro and collaborators published the first work showing the activation of the RNA-silencing pathway after detection of a pathogen elicitor (Navarro *et al.*, 2006). These authors demonstrated how *Arabidopsis* treated with a peptide (flg22) derived from the general elicitor flagellin, induces a miRNA which turns off the expression of several auxin-receptor mRNAs, rendering the plant more resistant to bacterial infections. Three well-supported lines of evidence demonstrate the hypothesis of the authors:

1. Three auxin-receptor F-box proteins (TIR1, AFB2 and AFB3) are targets of regulation by miRNA after flg22 treatment. The targets of miRNA were identified by its higher accumulation in silencing-suppressor overexpressing plants, in comparison with wild-type plants, after flg22 treatment. Two of the F-box mRNAs were previously identified as targets of the miR393 (TIR1,

from Transport Inhibitor Response 1 and AFBX, from Auxin signalling F-Box X; Jones-Rhoades and Bartel, 2004; Sunkar and Zhu, 2004) while the authors indentified a sequence that perfectly matched with this miRNA in the third F-box mRNA (AFBY). Both mRNA and protein levels of the F-box TIR1 were reduced after flg22 treatment, while miR393 accumulated under the same conditions. Promoter fusion analysis of the precursor of miR393 with eGFP also indicated an induction of the expression of eGFP after flg22 treatments.

2. Flg22 treatment leads to downregulation of the auxin-signalling pathway. Auxin signalling is a relatively simple and well-known plant pathway. The F-box proteins TIR1, AFB1, AFB2 and AFB3 are auxin receptors and they are part of the ubiquitination complex $SCF^{TIR1/AFB1/AFB2/AFB3}$. After reception of the hormone, this complex will induce the degradation of the Aux/IAA proteins, which are transcription factors that negatively regulate auxin signalling (Gray *et al.*, 2001). Based on this model, Navarro and collaborators (2006) showed the impact of flg22 in the auxin-signalling pathway, by demonstrating an increase in the stability of IAA/Aux proteins and a decrease in the expression of the auxin-signalling genes after the general elicitor treatment.

3. Auxin-signalling activation enhances disease resistance. Finally, repression of the auxin-signalling pathway by flagellin elicitation suggests a role of this pathway in conferring disease susceptibility. This has been probed in different ways. First *Pseudomonas syringae* experiments were performed in *tir-1* mutant plants that overexpress AFB1. This F box is an auxin receptor partially resistant to miR393. Therefore, after flg22 treatment, the auxin-signalling pathway cannot be downregulated and the plants are more susceptible to infection in comparison with the wild type. In a second experiment, *Arabidopsis* plants were engineered to constitutively express At-mi393a, therefore resulting in lower TIR levels. Infection experiments showed higher resistance in the transgenic plants compared to the wild type, which confirmed the role of TIR and auxin signalling in disease susceptibility.

After this work, activation of RNA silencing after recognition of an *avr* gene by an *R* gene has also been shown (Katiyar-Agarwal *et al.*, 2006, 2007). Elicitation of *Arabidopsis* plants carrying the *R* gene *RPS2* with *P. syringae* carrying the *avr* gene *avrRpt2* induced a recently discovered class of siRNA called natural antisense transcript-siRNA or nat-siRNA. nat-siRNAs are originated from the overlapping region of a pair of natural antisense transcripts or NATs (Borsani *et al.*, 2005). Thus, certain conditions can specifically induce the expression of one NAT pair transcript that will trigger, after pairing with the already expressed antisense transcript, the nat-siRNA formation. The so-called nat-siRNAATGB2 comes from the overlapping between the mRNA of a GTP-binding gene (*ATGB2*) with the mRNA of a pentatricopeptide repeat protein-like (*PPRL*) gene and it is strongly activated after infection of *Arabidopsis* RPS2 plants with *P. syringae* pv. *tomato* (*Pst*) *avrRpt2*. Since the sequence of this nat-siRNA complements the 3' untranslated region (UTR) of the antisense gene *PPRL*, a possible downregulation of *PPRL* by natsiRNAATGB2 was suggested. In fact, it was shown in this work that both the induction of the nat-siRNAATGB2 and the downregulation of the PPRL transcript depend on the expression of the NAT component *ATGB2*. More interestingly, several components of the RNA-silencing pathway, as well as the RPS2-mediated resistance pathway are shown to be essential for both the production of nat-siRNAATGB2 and the downregulation of *PPRL* mRNA, indicating that the two defence-related pathways are involved in producing the same response. Also, mutation of two different components of the RNA-silencing pathway increases bacterial susceptibility, providing direct evidence of the role of RNA silencing in this specific defence response. Finally, the question why *PPRL* mRNA needs to be downregulated after avrRpt2 recognition is answered by showing that PPRL-overexpressing lines are more susceptible to *Pst avrRpt2*, which means that this gene mediates disease susceptibility.

The two previous studies showed the role of two different classes of sRNA (miRNAs and nat-siRNA) in regulating responses against different pathogens. This suggests that the control of plant defence responses by sRNA is a general mechanism that may probably involve all the known classes of siRNA. Moreover, a new class of siRNA, the so-called long siRNA (lsiRNA), has been found to be induced after a pathogen attack (Katiyar-Agarwal *et al.*, 2007). lsiRNAs are 30- to 40-nucleotide siRNA generated from protein-coding genes that can be induced by specific developmental conditions. They can be originated from NAT, as is the case of the AtlsiRNA-1, the lsiRNA from *A. thaliana*, which was characterized by Katiyar-Agarwal *et al.* (2007). Three main questions were addressed by these authors towards understanding the role of AtlsiRNA-1: (i) how it is generated; (ii) how it mediates silencing of their targets; and (iii) how it regulates the defence response.

AtlsiRNA-1 is derived from the overlapping region between a putative leucine-rich repeat receptor-like protein kinase (RLK, also called here sRNA-generating RLK or SRRLK) and an expressed protein containing a putative RNA-binding domain, RAP (RNA-binding domain abundant in apicomplexans) domain or *AtRAP*. The sequence of AtlsiRNA-1 complements the 3'UTR of *AtRAP*, suggesting that the latter is the target of degradation. This is supported by transcription analysis showing that the expression of both SRRLK and the AtlsiRNA-1 is induced by *Pst* (*avrRpt2*) infection, and it correlates with a reduction in *AtRAP* mRNA. Moreover, the authors identified some of the proteins involved in the biogenesis of AtlsiRNA-1 by checking the *Pst* (*avrRpt2*)-dependent induction of this siRNA in the different sRNA pathway mutants, and they observed a negative correlation between the levels of AtlsiRNA-1 and *AtRAP* in those mutants. Additional confirmation of the regulation of *AtRAP* mRNA levels by AtlsiRNA-1 is obtained by generation of transgenic plants carrying a fusion of either a wild-type or mutated 3'UTR region of *AtRAP* fused to the *YFP* gene under the control of the constitutive 35S promoter. Infection of those transgenic plants with *Pst* (*avrRpt2*) leads to diminution of YFP levels fused to the wild-type 3'UTR *AtRAP*, whereas no changes in expression were observed under the same conditions when the reporter gene is fused to the mutated 3'UTR. The authors then investigated if the AtlsiRNA-

1-dependent degradation of *AtRAP* depended on the expression of the *SRRLK* gene. As expected knockout of *SRRLK* expression by T-DNA insertion led to a lack of induction of the siRNA.

The second question was answered by showing that mRNA decapping is the mechanism that leads to AtlsiRNA-1-dependent *AtRAP* degradation. This is interesting since, in plants, sRNAs mainly direct mRNA degradation through endonucleolytic cleavage, and although in animals sRNA-induced mRNA instability is well known (Valencia-Sanchez *et al.*, 2006), this is the first report of such a mechanism in plants.

Finally, the authors analysed the role of the *AtRAP* gene in resistance. Infection of *AtRAP* knockout mutants showed higher resistance to both virulent and avirulent (*avrRpt2*) *Pst*, indicating that this gene is also involved in basal resistance. On the contrary, knockout of the *SRRLK* gene did not show any differences in terms of infection. Why knocking-out *SRRLK* does not affect the defence response even when AtlsiRNA-1 is not induced was not elucidated. If this sRNA is responsible for the degradation of *AtRAP*, what leads to higher resistance to *Pst*? It should be expected that *SRRLK* mutant lines displayed a lower resistance to *Pst* (*avrRpst2*) infection.

The mechanisms just described differ from the classical virus-induced RNA silencing response in that they are initiated by the recognition of a pathogen elicitor (other than dsRNA) by a plant receptor, and that the silencing mechanism is triggered by an endogenous dsRNA. There are also important differences among the above-described elicitor-activated silencing mechanisms as to the nature of the sRNA (miRNA, nat-siRNA and AtlsiRNA) that is produced, the origin of the sRNA (expression of miRNA precursor or NAT transcription for nat-siRNA and AtlsiRNA) and the silencing mechanisms directed by the sRNA (endogenous cleavage or mRNA instability by decapping). This variety in the defence-related silencing mechanisms suggests that the regulation of the plant immune system by RNA silencing is a complex process that we are only starting to understand.

Although the cross talk between pathogen-response and silencing mechanisms has been deeply analysed only for bacterial pathogens, there is also some evidence that indicates that resistance to herbivore insects also involves the production of dsRNA and hence the RNA-silencing pathway (Pandey and Baldwin, 2007). Virus-induced gene silencing of three *Nicotiana attenuata* RNA-directed RNA polymerases (RdR), followed by a herbivore susceptibility screen, led to the identification of RdR1 as an essential protein in the defence response of this host plant to the attack by *Manduca sexta*. RdR proteins are the enzymes responsible for the production of some forms of dsRNA, the molecule common to all the RNA silencing mechanisms (Pickford and Cogoni, 2003). RdR are also involved in the spread of the silencing target within a single RNA strand. This process starts in plants with the recruitment of RdR by a long-single stranded RNA targeted with two primary siRNAs or miRNAs. The RdR will produce dsRNA followed by the cleavage by Dicer with the consistent generation of secondary siRNAs (Baulcombe, 2007). Pandey and Baldwin confirmed the role of *RdR1* in defence against herbivore attacks by stable silencing of this gene in *N. attenuata* (Pandey and Baldwin, 2007).

To identify putative defence mechanisms that are regulated by RdR, the authors performed microarray analyses. Their results showed that several alkaloid biosynthesis genes are downregulated in the *RdR1*-silencing plants after infection, suggesting that the mutant plants possess an altered nicotinic biosynthetic pathway, which was already shown to be an essential defence response (Steppuhn *et al.*, 2004). The authors then proposed that herbivore attacks activate *RdR1*, which will perform the amplification of siRNAs that will target repressors of nicotine biosynthesis.

A deeper analysis of the defence mechanisms regulated by sRNA in *N. attenuata*, after infection with *M. sexta*, has been performed (Pandey *et al.*, 2008). In this work, the sRNA transcriptome was compared between wild-type and *RdR1* mutants both before and after a herbivore attack. After identification of some miRNAs that are differentially regulated in the previous situations, the authors searched for putative targets of regulation by this miRNA. Since, as mentioned above, silencing of *RdR1* leads to a diminution of nicotine biosynthesis (Pandey and Baldwin, 2007) and the control of this defence response by phytohormones is well established (see below), the authors looked for targets of regulation by sRNA in the genes involved in phytohormone signalling, specifically JA and ET. The attack by *M. sexta* elicits a JA burst in *N. attenuata* that is essential for the defence response to be produced, since silencing of the JA-signalling cascade genes compromises host resistance to this herbivore (Halitschke *et al.*, 2003; Paschold *et al.*, 2007). The defence response, specifically the JA-dependent production of nicotine, is also negatively regulated by an ET burst triggered by *M. sexta* attack (Winz and Baldwin, 2001). Sequence analysis revealed that several of the sRNAs might potentially target the hormone-signalling pathway, and real time PCR analysis confirmed that six JA-related and two ET-related signalling and/or biosynthesis genes are differentially expressed in wild-type and *RdR1* mutants after herbivore attack. One interesting result revealed from this work is the possibility that sRNA can also activate gene expression. Thus, some of the genes identified as putative targets of the sRNAs show higher transcriptional levels in *RdR1*-silencing plants. This activity of sRNAs as positive regulators of transcription has already been shown in humans where targeting of a promoter with dsRNA increases its transcriptional activity (Li *et al.*, 2006).

By similarity with the work of flagellin and auxin signalling (Navarro *et al.*, 2006) Pandey and collaborators proposed that the F-box *coi1*, involved in JA-perception (Li *et al.*, 2004), might be the target of regulation by sRNA. But analysis of the transcription levels of the F-box *coi1* in wild-type and *RdR1* plants showed no differences, indicating that if this F box is regulated by sRNA it is done post-transcriptionally.

The transcriptional differences found in the hormone-related genes correlate with lower production of JA in *RdR1*-silenced plants after a herbivore attack, in comparison with the wild type, while the production of ET is higher in the mutant. Moreover, the exogenous application of JA restores the resistance of *N. attenuata* to *M. sexta*, suggesting that the lower level of JA in the *RdR1*-silenced plants is responsible for the higher sensitivity of the mutant to herbivore attacks. This work provided direct evidence of the role of sRNA in controlling the defence response to herbivore attacks. Another possibility not tested in this work is that RdR1 is involved in the production of sRNA in the host plant to target genes that are essential for the herbivore's development. Thus, after feeding, the insect will incorporate the sRNA, the expression of the target genes (possibly essential for the infection to be successful) will be altered and the plant will be resistant. The capacity to modulate the gene expression in a herbivore by incorporating sRNA from the host plant by feeding has been already shown in cotton (Mao *et al.*, 2007). The herbivore *Helicoverpa armigera* needs the enzyme cytochrome p450 (CYP6AE14) to attack cotton plants. This enzyme will metabolize gossypol to levels that can be tolerated by the herbivore. Infection of genetically engineered cotton plants producing dsRNA against CYP6AE14 delays the larval growth (Mao *et al.*, 2007). Nevertheless, if this mechanism exists in nature, it is still to be shown.

Recently, Navarro *et al.* (2008) identified *P. syringae* effectors that suppress transcriptional activation of some PAMP-responsive miRNAs or miRNA biogenesis, stability or activity. These results provide evidence that, like viruses, bacteria have evolved to suppress RNA silencing to cause disease. These data are novel and may help discover several targets of bacteria effectors on the RNA silencing pathway.

## 7.4 Concluding Remarks

Considerable progress has been made during the last few years in our understanding of how JA antagonizes SA and vice versa. However, few studies have elucidated the cross talk between RNA silencing and the immunity induced by elicitors. RNA silencing is involved in the resistance against viruses and viral suppressors of RNA silencing have been discovered many years ago. Remarkably, RNA silencing has been recently discovered to be important in the resistance against bacteria and nematodes. On the other hand, bacterial suppressors of RNA silencing have recently been identified too (Mosher and Baulcombe, 2008). However, there is no evidence for the involvement of RNA silencing in resistance against fungi. So the rapid development of new technologies that is taking place in biochemistry and genomic research will help not only to identify new targets involved in the cross talk between these two pathways, but also the possible involvement of RNA silencing in plant resistance against fungi and other pathogens.

## Acknowledgements

We apologize to our colleagues whose work could not be cited in this review because of space limitations. The authors wish to thank the Natural Sciences and Engineering Research Council of Canada, the Fonds de la Recherche sur la Nature et les Technologies du Québec and the University of Sherbrooke for financial support. Taha Abd El Rahman is supported by a grant from the

Egyptian government. Hamed Bathily is supported by a Canadian International Development Agency fellowship (Bourse de la Francophonie).

## References

Agrawal, G.K., Tamogami, S., Han, O., Iwahashi, H. and Rakwal, R. (2004) Rice octadecanoid pathway. *Biochemical and Biophysical Research Communications* 317, 1–15.

Andrea, A.L., Romeis, T. and Jones, J.D.G. (2004) CDPK-mediated signalling pathways: specificity and cross-talk. *Journal of Experimental Botany* 55(395), 181–188.

Baldwin, I.T., Kessler, A. and Halitschke, R. (2002) Volatile signalling in plant–plant–herbivore interactions: What is real? *Current Opinion in Plant Biology* 5(4), 351–354.

Baulcombe, D.C. (2004) RNA silencing in plants. *Nature* 431, 359–363.

Baulcombe, D.C. (2007) Molecular biology: amplified silencing. *Science* 315, 199–200.

Bent, A.F. and Mackey, D. (2007) Elicitors, effectors, and *R* genes: the new paradigm and a lifetime supply of questions. *Annual Review of Phytopathology* 45, 399–436.

Borsani, O., Zhu, J., Verslues, P., Sunkar, R. and Zhu, J. (2005) Endogenous siRNAs derived from a pair of natural *cis*-antisense transcripts regulate salt tolerance in *Arabidopsis*. *Cell* 123, 7, 1279–1291.

Brodersen. P., Petersen, M., Nielsen, H.B., Zhu, S., Newman. M.A., Shokat, K.M., Rietz, S., Parker, J. and Mundy, J. (2006) *Arabidopsis* MAP kinase 4 regulates salicylic acid- and jasmonic acid/ethylene-dependent responses via EDS1 and PAD4. *The Plant Journal* 47, 532–546.

Browse, J. and Howe, A.G. (2008) New weapons and a rapid response against insect attack. *Plant Physiology* 146, 832–838.

Chapman, E.J. and Carrington, J.C. (2007) Specialization and evolution of endogenous small RNA pathways. *Nature Reviews Genetics* 8(11), 884–896.

Chen, Z., Malamy, J., Henning, J., Conrath, U., Sánchez-Casas, P., Silva, H., Ricigliano, J. and Klessig, D.F. (1995) Induction; modification and transduction of the salicylic acid signal in plant defense responses. *Proceedings of the National Academy of Sciences, USA* 92, 4134–4137.

Ding, S.W. and Voinnet, O. (2007) Antiviral immunity directed by small RNAs. *Cell* 130, 413–426.

Durrant, W.E. and Dong, X. (2004) Systemic acquired resistance. *Annual Review of Phytopathology* 42, 185–209.

El Oirdi, M. and Bouarab, K. (2007) Plant signalling components EDS1 and SGT1 enhance disease caused by the necrotrophic pathogen *Botrytis cinerea*. *New Phytologist* 175(1), 131–139.

Enrique, R., Solano, R. and Sanchez-Serrano, J.J. (2003) Interactions between signalling compounds involved in plant defence. *Journal of Plant Growth Regulation* 22, 82–98.

Farmer, E.E. and Ryan, C.A. (1990) Interplant communication: airborne methyl jasmonate induces synthesis of proteinase inhibitors in plant leaves. *Proceedings of the National Academy of Sciences, USA* 87, 7713–7716.

Gordon, S. (2002) Pattern recognition receptors: doubling up for the innate immune response. *Cell* 111, 927–930.

Govrin, E.M. and Levine, A. (2000) The hypersensitive response facilitates plant infection by the necrotrophic pathogen *Botrytis cinerea*. *Current Biology* 10, 751–757.

Gray, W.M., Kepinski, S., Rouse, D., Leyser, O. and Estelle M. (2001) Auxin regulates SCF(TIR1)-dependent degradation of AUX/IAA proteins. *Nature* 414(6861), 271–276.

Halitschke, R., Gase, K., Hui, D., Schmidt, D. and Baldwin, I.T. (2003) Molecular interactions between the specialist herbivore *Manduca sexta* (Lepidoptera, Sphingidae) and its natural host *Nicotiana attenuata*. VI. Microarray analysis reveals that most herbivore-specific transcriptional changes are mediated by fatty acid. *Plant Physiology* 131, 1894–1902.

Hammond-Kosack, K.E., and Jones, J.D.G. (1996) Resistance gene-dependent plant defense responses. *Plant Cell* 8, 1773–1791.

Holmgren, A. (1989) T hioredoxin and glutaredoxin systems. *Journal of Biological Chemistry* 264(24), 13963–13966.

Hoog, J.O., Jornvall, H., Holmgen, A., Carlquist, M. and Persson, M. (1983) The primary structure of *Escherichia coli* glutaredoxin: distant homology with thioredoxins in a superfamily of small proteins with a redox-active cystine disulfide/cysteine dithiol. *European Journal of Biochemistry* 136, 223–232.

Iris, A.M.A., Eggermont, K., Terras, F.R.G., Thomma, B.P.H.J., De Samblanx, G.W., Buchala, A., Métraux, J.P., Manneqa, J.M. and Broekaert, W.F. (1996) Pathogen-induced systemic activation of a plant defensin gene in *Arabidopsis* follows a salicylic acid-independent pathway. *The Plant Cell* 8, 2309–2323.

Jennifer, S., Ana, L. and Richard, M. (2002) Antagonism between jasmonate and salicylate mediated induced plant resistance: effect of concentration and timing of elicitation on defence-related proteins; herbivore and pathogen performance in tomato. *Journal of Chemical Ecology* 28(6), 1131–1159.

Jones, D.G.J. and Dangl, J.L. (2006) The plant immune system. *Nature* 444, 324–329.

Jones-Rhoades, M.W. and Bartel, D.B. (2004) Computational identification of plant microRNAs and their targets; including a stress-induced miRNA. *Molecular Cell* 14(6), 787–799.

Karban, R., Baldwin, I.T., Baxter, K.J., Laue, G. and Felton, G.W. (2000) Communication between plants: induced resistance in wild tobacco plants following clipping of neighboring sagebrush. *Oecologia* 125, 66–71.

Katiyar-Agarwal, S., Morgan, R., Dahlbeck, D., Borsani, O., Villegas, A.J., Zhu, J.K., Staskawicz, B.J. and Jin, H. (2006) A pathogen-inducible endogenous siRNA in plant immunity. *Proceedings of the National Academy of Sciences, USA* 103, 18002–18007.

Katiyar-Agarwal, S., Gao, S., Vivian-Smith, A. and Jin, H. (2007) A novel class of bacteria-induced small RNAs in *Arabidopsis. Genes & Development* 21, 3123–3134.

Kemal, K. and Manners, J.M. (2007) Jasmonate signalling: toward an integrated view. *Plant Physiology* 146, 1459–1468.

Koornneef, A. and Pieterse, M.J. (2008) Cross talk in defence signalling. *Plant Physiology* 146, 839–844.

Kunkel, B.N. and Brooks, D.M. (2002) Cross talk between signalling pathways in pathogen defence. *Current Opinion in Plant Biology* 5, 325–331.

Li, J., Brader, G. and Palva, E.T. (2004) The WRKY70 transcription factor: a node of convergence for jasmonate-mediated and salicylate-mediated signals in plant defence. *The Plant Cell*, 16, 319–331.

Li, J., Brader, G., Kariola, T. and Palva, E.T. (2006) WRKY70 modulates the selection of signalling pathways in plant defence. *The Plant Journal* 46, 477–491.

Luis, A.J.M., Kenton, P., Atzorn, R., Miersch, O. and Wasternack, C. (2006) The outcomes of concentration-specific interactions between salicylate and jasmonate signalling include synergy; antagonism; and oxidative stress leading to cell death. *Plant Physiology* 140, 249–262.

Maleck, K. and Dietrich, R.A. (1999) Defence on multiple fronts: how do plants cope with diverse enemies? *Trends in Plant Science* 4(6), 215–219.

Mao, Y.B., Cai, W.J., Wang, J.W., Hong, G.J., Tao, X.Y., Wang, L.J., Huang, Y.P. and Chen, X.Y. (2007) Silencing a cotton bollworm P450 monooxygenase gene by plant-mediated RNAi impairs larval tolerance of gossip. *Nature Biotechnology* 25, 1307–1313.

Miki, F., Fujita, Y., Noutoshi, Y., Takahashi, F., Narusaka, Y., Kazuko, Y.S. and Kazuo, S. (2006) Crosstalk between abiotic and biotic stress responses: a current view from the points of convergence in the stress signalling networks. *Current Opinion in Plant Biology* 9, 436–442.

Mosher, R.A. and Baulcombe, D.C. (2008) Bacterial pathogens encode suppressors of RNA-mediated silencing. *Genome Biology* 9(10), 237.

Navarro, L., Dunoyer, P., Jay, F., Arnold, B., Dharmasini, N., Estelle, M., Vionnet, O. and Jones, J.D. (2006) A plant miRNA contributes to antibacterial resistance by repressing auxin signalling. *Science* 312, 436–439.

Navarro, L., Jay, F., Nomura, K., He, S.Y. and Voinnet, O. (2008) Suppression of the microRNA pathway by bacterial effector proteins. *Science* 321(5891), 964–967.

Ndamukong, I., Abdallat, A.A., Thurow, C., Fode, B., Zander, M., Weigel, R. and Gatz, C. (2007) SA-inducible *Arabidopsis* glutaredoxin interacts with TGA factors and suppresses JA-responsive PDF1.2 transcription. *The Plant Journal* 50, 128–139.

Ogawa, D., Nakajima, N., Sano, T., Tamaoki, M., Aono, M., Kubo, A., Kanna, M., Ioki, M., Kamada, H. and Saji, H. (2005) Salicylic acid accumulation under O3 exposure is regulated by ethylene in tobacco plants. *Plant and Cell Physiology* 46(7), 1062–1072.

Pandey, S.P. and Baldwin, I.T. (2007) RNA-directed RNA polymerase 1 (RdR1) mediates the resistance of *Nicotiana attenuata* to herbivore attack in nature. *The Plant Journal* 50, 40–53.

Pandey, S.P., Shahi, P., Gase, K. and Baldwin, I.T. (2008) Herbivory-induced changes in the small-RNA transcriptome and phytohormone signaling in *Nicotiana attenuata*. *Proceedings of the National Academy of Sciences, USA* 105, 4559–4564.

Paschold, A., Halitschke, R. and Baldwin, I.T. (2007) Co(i)-ordinating defenses: NaCOI1 mediates herbivore-induced resistance in *Nicotiana attenuata* and reveals the role of herbivore movement in avoiding defenses. *The Plant Journal* 51, 79–91.

Peng, M., Duan, M., Wei, C. and Li, Y. (2007) WRKY62 transcription factor acts downstream of cytosolic NPR1 and negatively regulates jasmonate-responsive gene expression. *Plant Cell Physiology* 48, 833–842.

Person, G., Robinson, F., Tara, B.G., Bing, E.X., Karandikar, M.K.B. and Melanie, H.C. (2001) Mitogen-activated protein (MAP) kinase. *Endocrine Reviews* 22(2), 153–183.

Petersen, M., Brodersen, P., Naested, H., Andreasson, E., Lindhart, U., Johansen, B., Nielsen, H.B., Lacy, M., Austin, M.J, Parker, J.E., Sharma, S.B., Klessig, D.F., Martienssen, R., Mattsson, O., Jensen, A.B. and Mundy, J. (2000) *Arabidopsis* MAP kinase 4 negatively regulates systemic acquired resistance. *Cell* 103, 1111–1120.

Pickford, A.S. and Cogoni, C. (2003) RNA-mediated gene silencing. *Cellular and Molecular Life Sciences* 60(5), 871–882.

Pieterse, M.J.C. and Van Loon, L.C. (2004) NPR1: the spider in the web of induced resistance signalling pathways. *Current Opinion in Plant Biology* 7, 456–464.

Pozo, J., Marıa, L.C., Van Loon, L.C. and Pieterse, C.M.J. (2005) Jasmonates-signals in plant–microbe interactions. *Plant Growth Regulation* 23, 211–222.

Richard, M.B. (2005) Signal crosstalk and induced resistant signal: straddling the line between cost and benefit. *Annual Review of Phytopathology* 43, 545–580.

Rouhier, N., Gelhaye, E. and Jacquot, J.P. (2004) Plant glutaredoxins: still mysterious reducing systems. *Cellular and Molecular Life Sciences* 61, 1266–1277.

Ryan, C.A. (2000) The systemin signalling pathway: differential activation of plant defensive genes. *Biochimica et Biophysica Acta* 1477(1–2), 112–121.

Seo, H.S., Song, J.T., Cheong, J.J., Lee, Y.H., Lee, Y.W., Hwang, I., Lee, J.S. and Chol, Y.D. (2001) Jasmonic acid carboxyl methyltransferase: a key enzyme for jasmonate-regulated plant responses. *Proceedings of the National Academy of Sciences, USA* 98, 4788–4793.

Shah, J. (2003) The salicylic acid loop in plant defence. *Current Opinion in Plant Biology* 6, 365–371.

Spoel, H.S., Koornneef, A., Susanne, M.C., Jerôme, P.K., Van Pelt, J.A., Martin, J.M., Antony, J.B., Jean-Pierre, M., Rebecca, B., Kemal, K., Van Loon, L.C., Dong, X. and Corné, M.J. (2003) NPR1 modulates cross-talk between salicylate- and jasmonate-dependent defence pathways through a novel function in the cytosol. *The Plant Cell* 15, 760–770.

Steppuhn, A., Gase, K., Krock B., Halitschke, R. and Baldwin, I.T. (2004) Nicotine's defensive function in nature. *PLoS Biology* 2(8), 1074–1080.

Sunkar, R. and Zhu, J.K. (2004) Novel and stress-regulated microRNAs and other small RNAs from *Arabidopsis*. *The Plant Cell* 16, 2001–2019.

Takahashi, F., Yoshida, R., Ichimura, K., Mizoguchi, T., Seo, S., Yonezawa, M., Maruyama, K., Yamaguchi-Shinozaki, K. and Shinozaki, K. (2007) The mitogen-activated protein kinase cascade MKK3-MPK6 is an important part of the jasmonate signal transduction pathway in *Arabidopsis*. *The Plant Cell* 19, 805–818.

Taylor, J.E., Hatcher, P.E. and Paul, N.D. (2004) Crosstalk between plant responses to pathogens and herbivores: a view from the outside in. *Journal of Experimental Botany* 55(395), 159–168.

Thomas, E., Paul, J.R., Silke, R. and Imre, E.S. (2000) The WRKY superfamily of plant transcription factors. *Trends in Plant Science* 5(5), 199–206.

Thomma, B.P.H.J., Eggermont, K., Penninckx, I.A.M.A., Mauch-Mani, B., Vogelsang, R., Cammue, B.P.A. and Broekaert, W.F. (1998) Separate jasmonate-dependent and salicylate-dependent defence-response pathways in *Arabidopsis* are essential for resistance to distinct microbial pathogens. *Proceedings of the National Academy of Sciences, USA* 95, 15107–15111.

Traw, M.B., Kim, J., Enright, S., Cipollini, D.F. and Bergelson, J. (2003) Negative cross-talk between salicylate- and jasmonate-mediated pathways in the Wassilewskija ecotype of *Arabidopsis thaliana*. *Molecular Ecology* 12, 1125–1135.

Valencia-Sanchez, M.A., Liu, J., Hannon, G.J. and Parker, R. (2006) Control of translation and mRNA degradation by miRNAs and siRNAs. *Genes & Development* 20, 515–524.

Vidhyasekaran, P. (2008) *Fungal Pathogenesis in Plant and Crops: Molecular Biology and Host Defence Mechanisms*, 2nd edn. CRC Press, Taylor & Francis Group, New York.

Vijayan, P., Shockey, J.C., Lévesque, A., James, C.R. and John, B. (1998) A role for jasmonate in pathogen defence of *Arabidopsis*. *Proceedings of the National Academy of Sciences, USA* 95, 7209–7214.

Wang, D., Amornsiripanitch, N. and Dong, X. (2006) A genomic approach to identify regulatory nodes in the transcriptional network of systemic acquired resistance in plants. *PLoS Pathogen* 2, 1042–1050.

Winz, R. and Baldwin, I.T. (2001) Molecular interactions between the specialist herbivore *Manduca sexta* (Lepidoptera, Sphingidae) and its natural host *Nicotiana attenuata*. IV. Insect-induced ethylene reduces jasmonate-induced nicotine accumulation by regulating putrescine N-methylt. *Plant Physiology* 125, 2189–2202.

Xiao-Yi, S., Zhi-Long, W. and Daoxin, X. (2007) Jasmonate signal pathway in *Arabidopsis*. *Journal of Integrative Plant Biology* 49(1), 81–86.

Xu, Y., Chang, P.L.C., Liu, D., Narasimhan, M.L., Kashchandra, G.R., Hasegawa, P.M. and Bressan, R.A. (1994) Plant defence genes are synergistically induced by ethylene and methyl jasmonate. *The Plant Cell* 6, 1077–1085.

# 8 The Needle and the Damage Done: Type III Effectors and the Plant Immune Response

JENNIFER D. LEWIS, KARL SCHREIBER AND DARRELL DESVEAUX

*University of Toronto, Toronto, Ontario, Canada*

## Abstract

The intimate interactions between plant pathogenic bacteria and their hosts have resulted in an evolutionary arms race between host immune responses and pathogen virulence strategies. Successful pathogens are continuously under pressure to diversify their mechanisms to thwart host defences and optimize nutrient availability, while at the same time avoiding recognition by host surveillance systems. In turn, these virulence mechanisms have shaped the evolution of plant innate immunity. The needle-like structure known as the type III secretion system is used by numerous Gram-negative bacterial pathogens to inject diverse sets of effector proteins into host cells where they have been demonstrated to dampen host immune responses and promote virulence. Not surprisingly, type III effector molecules are prime 'non-self' molecules and plants have evolved to recognize their presence and deploy effective defence responses. Type III effector proteins are the direct molecular interface between pathogen and host and their study has yielded invaluable information about the evolution of host–pathogen interactions. In this chapter, we discuss recent advances in understanding the diverse virulence and avirulence functions of the type III effector proteins of plant pathogenic bacteria. We discuss how their biochemical activities on host targets can contribute to the recognition of 'modified self' by the host and the activation of plant innate immunity. Further, we describe how type III effectors can usurp host proteins for their activation and also their use of structural mimicry of eukaryotic proteins as a virulence strategy. Finally, we address the evolutionary pressures and diversification mechanisms of type III effectors and the functional consequences for adaptation of pathogens and their hosts.

 

## 8.1 Introduction

Many Gram-negative pathogenic bacteria utilize a molecular syringe known as the type III secretion system to secrete and translocate effector proteins into the cells of their hosts. In this chapter, we focus on type III effectors secreted by the plant pathogenic bacteria *Pseudomonas syringae*, *Xanthomonas* spp., *Erwinia amylovora* and *Ralstonia solanacearum*. Conserved pathogen-associated molecular patterns (PAMPs) like flagellin protein or elongation factor Ef-Tu reveal the pathogen to the plant, thus inducing basal defence responses. Bacteria have overcome this basal level of recognition by evolving effectors to suppress basal resistance. Effectors are translocated and secreted through the needle-like type III pilus from the bacteria into the plant cell, where they subvert host function for their own benefit. Basal resistance, effectors' roles in defence suppression and effector evolution have been discussed recently in several excellent reviews and will not be discussed in detail here (Espinosa and Alfano, 2004; Abramovitch *et al.*, 2006a; Da Cunha *et al.*, 2006; Grant *et al.*, 2006; He *et al.*, 2006; McCann and Guttman, 2008). In response to type III effector action, plants evolved resistance genes (*R* genes) to recognize particular effectors of the pathogen (Dangl and Jones, 2001; Martin *et al.*, 2003; Chisholm *et al.*, 2006). These recognized effectors were originally termed 'avirulence' proteins for the hypersensitive response (HR) or programmed cell death they caused in their hosts. Consequently, certain type III effectors have been demonstrated to interfere with *R* gene recognition allowing the pathogen to go undetected. This alternating response of the pathogen and host is characteristic of a classic arms race, where each tries to gain supremacy over time.

While effector proteins were long believed to interact directly with resistance proteins, this has been observed in relatively few cases (e.g. *Magnaporthe grisea* AvrPi-ta and rice Pi-ta) (Martin *et al.*, 2003). Instead, evidence is mounting that resistance proteins monitor specific host targets of type III effectors (van der Biezen and Jones, 1998b; Dangl and Jones, 2001; Martin *et al.*, 2003). The 'guard hypothesis' put forth 11 years ago asserts that effector-mediated modifications to host targets are detected by the guarding resistance protein, leading to the initiation of defence responses (van der Biezen and Jones, 1998b).

R proteins contain leucine-rich repeats (LRRs) and one or more other domains (Dangl and Jones, 2001; Martin *et al.*, 2003; Espinosa and Alfano, 2004). The largest group of *R* genes is the CC-NBS-LRR class, which also carry an N-terminal coiled-coil (CC) domain, an NBS-ARC domain (nucleotide-binding and APAF-1/R gene/CED-4 homology domain) (van der Biezen and Jones, 1998a) and C-terminal LRR region. The CC-NBS-LRR class contains several *R* genes which are known to recognize bacterial effectors, like AvrRpt2, HopAR1, AvrPto and HopAB2. The other major group, the TIR-NBS-LRR class, contains an N-terminal TIR domain, named by its homology to the Toll protein and interleukin-1 receptor. TIR class *R* genes recognize oomycete, fungal, viral and bacterial effectors, like AvrRps4 and AvrBs4 (Gassmann *et al.*, 1999; Schornack *et al.*, 2004a, b). A subclass of this group has the TIR-NBS-LRR structure with a C-terminal WRKY motif and includes the *RRS-1* resistance gene that recognizes a *R. solanacearum* effector (Deslandes *et al.*, 2002). Extracellular LRR proteins have only been found to recognize *Cladosporium fulvum* fungal effectors (Dangl and Jones, 2001; Chisholm *et al.*, 2006). Specificity in the recognition of effectors appears to be conferred by the LRR region of *R* genes, which is under diversifying selection (Dangl and Jones, 2001; McHale *et al.*, 2006).

Since this review deals with the interface between type III effectors and plant defence, we will focus on bacterial type III effectors with described avirulence functions. Despite their characterized avirulence functions, these effectors have remarkably different biochemical functions, act in different compartments of the plant cell and manipulate host metabolism in complex ways. We discuss the molecular mechanisms by which type III effectors are recognized, including the *R* genes involved if known, and also any virulence functions that have been observed when they are not recognized.

## 8.2 AvrPto

AvrPto from *P. syringae* pv. *tomato* JL1065 was first identified by the avirulence phenotype it conferred in the normally virulent strain *P. syringae* pv. *maculicola* ES4326 in resistant tomato lines (Ronald *et al.*, 1992). These resistant tomato lines were later found to carry *Pto* kinase and the *Prf* resistance gene (Salmeron *et al.*, 1996). AvrPto physically interacts with Pto kinase, which interacts with and is monitored by Prf (Martin *et al.*, 1993; Scofield *et al.*, 1996; Tang *et al.*, 1996; Mucyn *et al.*, 2006). AvrPto avirulence function requires both phosphorylation (at S149) and membrane localization by myristoylation (at G2) (Shan *et al.*, 2000; Anderson *et al.*, 2006).

*Pto* kinase is found in the *Solanaceae*, including wild and cultivated varieties of potato and tomato, as well as *Nicotiana tabacum*, *Arabidopsis* and rice (Martin *et al.*, 1993; Vleeshouwers *et al.*, 2001; Rose *et al.*, 2005). *Prf* is a CC-NBS-ARC-LRR resistance gene with homologues in many plant species (Salmeron *et al.*, 1996). *Pto* and *Prf* are part of a tightly linked gene cluster that was introgressed from the wild tomato *Solanum pimpinellifolium* into *Solanum lycopersicum* cv. Rio Grande, creating near isogenic lines for the *Pto/Prf* cluster (RG-*PtoR*) (Pedley and Martin, 2003). These lines and mutants of either *Pto* (RG-*pto11* or RG-*ptoS*) or *Prf* (RG-*prf3L*) in the same background (Salmeron *et al.*, 1994) have allowed the dissection of the roles Pto and Prf play in AvrPto and HopAB2 (see '8.3 HopAB2/AvrPtoB' below) avirulence and virulence functions.

Recognition of AvrPto is maintained in some soybean cultivars, wild tomato species and *Nicotiana clevelandii* (Ronald *et al.*, 1992; Rommens *et al.*, 1995; Riely and Martin, 2001). Natural variation within Pto in wild species of tomato can affect its ability to recognize AvrPto (Rose *et al.*, 2005). Multiple studies have examined whether expression of Pto and/or Prf in normally susceptible tomato or tobacco species can restore resistance to AvrPto. Resistance depends heavily on the level of expression (endogenous promoter versus overexpression by the 35S promoter) and the species examined. Native expression of Pto is sufficient for recognition of AvrPto in susceptible transgenic tomato lines (lacking Prf) but not in transgenic *Nicotiana benthamiana* (Balmuth and Rathjen, 2007). In transgenic *N. benthamiana*, native expression of both Pto and Prf is necessary for the recognition of *P. syringae* pv. *tabaci* carrying *avrPto* (Balmuth and Rathjen, 2007). However, overexpression of Prf (Oldroyd and Staskawicz, 1998; Balmuth and Rathjen, 2007) or Pto alone (Rommens *et al.*, 1995; Thilmony *et al.*, 1995) in transgenic *N. benthamiana* confers resistance to *P. syringae* pv. *tabaci* carrying *avrPto*.

The crystal structure of the AvrPto–Pto complex was recently solved (Xing *et al.*, 2007). Two key interfaces in each protein mediate this interaction: (i) one end of an AvrPto helical bundle with a Pto loop (particularly H49 and V51); and (ii) the AvrPto GINP motif with the Pto P+1 loop (particularly T204). Pto H49E/V51G/T204N is no longer able to interact with AvrPto (Xing *et al.*, 2007). Both of the Pto loops negatively regulate Prf-mediated defences in the absence of AvrPto in tomato plants and AvrPto is believed to activate Prf-mediated defences by interacting with the two Prf-inhibiting loops (Rathjen *et al.*, 1999; Wu *et al.*, 2004; Xing *et al.*, 2007). This is consistent with previous work showing that Pto T204 is necessary for interaction with AvrPto in the yeast two-hybrid system and also for a defence response when transiently expressed by *Agrobacterium* in *N. benthamiana* (Frederick *et al.*, 1998; Rathjen *et al.*, 1999).

Phosphorylation of Pto at T199 appears to stabilize the P+1 loop and thus facilitate the interaction with AvrPto, leading to recognition by Prf and the HR (Xing *et al.*, 2007). Consistent with this, the Pto T199A mutation reduces the affinity of the interaction with AvrPto and the strength of the HR (Sessa *et al.*, 2000; Xing *et al.*, 2007). Mutations in Pto (for example T38 (Sessa *et al.*, 2000), K69 (Rathjen *et al.*, 1999) or D164 (Wu *et al.*, 2004)) which impair autophosphorylation also disrupt AvrPto-binding and the elicitation of the HR. Constitutive gain-of-function Pto mutants that likely destabilize the P+1 loop (Xing *et al.*, 2007) confer AvrPto-independent and Prf-dependent defence responses (Rathjen *et al.*, 1999; Wu *et al.*, 2004). Conversely, mutations which presumably stabilize the Pto P+1 loop (S226D, V201D, D164N) can still elicit the HR, even though they lack kinase activity and display impaired interactions with AvrPto (Pto S226D, V201D only) (Rathjen *et al.*, 1999; Xing *et al.*, 2007). This indicates that kinase activity of Pto per se is not necessary but rather phosphorylation of Pto is required for the AvrPto interaction.

The first loop of Pto (H49 and V51) also contributes to host recognition. Pto H49E/V51G or H49D/V51D mutations elicit an HR, despite their impaired interaction with AvrPto and their insensitivity to AvrPto inhibition of Pto kinase activity (Xing *et al.*, 2007). In the absence of AvrPto, the two critical loops of Pto are proposed to maintain Prf in an inactive state (Xing *et al.*, 2007). Once AvrPto binds Pto, Pto is proposed to undergo conformational changes, which then leads to the activation of Prf (Xing *et al.*, 2007).

Defence responses induced by the AvrPto–Pto interaction require both Pto kinase activity and Pto myristoylation (Mucyn *et al.*, 2006; Balmuth and Rathjen, 2007). Mitogen-activated protein kinase (MAPK) cascades are activated in response to an AvrPto–Pto interaction in tomato, and activation is dependent on Pto and Prf (Pedley and Martin, 2004). Signalling occurs initially through MAPK kinase kinase α (MAPKKKα), which is common in both resistant and susceptible interactions, but then splits to different MAPK cascades (del Pozo *et al.*, 2004). Pto interacts with and phosphorylates another serine/threonine kinase, Pti1, in an interaction that requires Pto kinase activity (Zhou *et al.*, 1995; Sessa *et al.*, 2000). Overexpression of Pti1 in *N. tabacum* confers increased resistance to *P. syringae* pv. *tabaci* carrying *avrPto* (Zhou *et al.*, 1995). Pto also interacts with and phosphorylates Adi3, part of the AGC family of protein kinases that are involved in transducing signals from second messengers like calcium and cAMP (Devarenne *et al.*, 2006). Silencing of Adi3 in tomato results in cell death dependent on MAPKKKα (Devarenne *et al.*, 2006). Taken together, these data implicate Pto kinase activity and downstream MAPK cascades in the induction of defence responses in response to AvrPto.

Pto myristoylation contributes to AvrPto recognition in a quantitative manner and differs between tomato and tobacco (Balmuth and Rathjen, 2007). Native expression of Pto G2A (myristoylation site mutant) in susceptible transgenic tomato cv. Moneymaker allows intermediate growth of *P. syringae* pv. *tomato* DC3000 carrying *avrPto* (Balmuth and Rathjen, 2007). However, overexpression of Pto G2A in transgenic tomato lines confers the same level of resistance as overexpression of wild-type Pto (Loh *et al.*, 1998; Balmuth and Rathjen, 2007). In transgenic tobacco lines, native expression of Pto G2A and Prf no longer confers resistance (Balmuth and Rathjen, 2007). Myristoylation of Pto does not contribute to its subcellular localization (de Vries *et al.*, 2006). Instead, myristate, supplied in *trans* or as myristoylated Pto, inhibits Pto kinase activity, presumably by blocking the catalytic cleft (Andriotis and Rathjen, 2006). Myristate inhibition of kinase activity appears restricted to Pto or close homologues like Fen kinase and Pti1 kinase (Andriotis and Rathjen, 2006).

Additional Pto-interacting proteins Pti4, Pti5 and Pti6 are transcription factors with similarity to ethylene-response factors (Zhou *et al.*, 1997) that are also implicated in defence responses. Resistant RG-*PtoR* tomatoes have increased Pti4 and Pti5 expression when infiltrated with *P. syringae* pv. *tomato* T1 while Pti5 expression is further upregulated in the presence of AvrPto (Thara *et al.*, 1999). Pti4/5/6 bind the GCC motif present in many pathogenesis-related (*PR*) genes (Zhou *et al.*, 1997). Consistent with this, *PR* gene expression is upregulated during incompatible interactions (Zhou *et al.*, 1997; Jia and Martin, 1999). Overexpression of Pti5 in tomato results in more rapid induction of *PR* genes during pathogen infection and increased resistance to *P. syringae* pv. *tomato* (He *et al.*, 2001). Overexpression of Pti4/5/6 in *Arabidopsis* induces *PR1* and *PR2* gene expression and overexpression of Pti4 specifically confers increased resistance to *P. syringae* pv. *tomato* and *Erysiphe orontii*, a fungal pathogen (Gu *et al.*, 2002). Although Pti4/6 may be generally involved in pathogen defences or abiotic signalling, Pti5 appears more specific to AvrPto.

AvrPto's virulence activities can be observed in tomato lines lacking Pto or Prf where AvrPto enhances host necrosis and bacterial growth of *P. syringae* pv. *tomato* T1 (Chang *et al.*, 2000). In addition, AvrPto suppresses HR by the non-host interactions of *P. syringae* pv. *tomato* in *N. benthamiana* as well as *P. syringae* pv. *tabaci* in tomato RG-*prf3* and RG-*ptoS* lines (Kang *et al.*, 2004). The flagellin receptor FLS2 is a key virulence target of AvrPto in *Arabidopsis* and tomato (Xiang *et al.*, 2008). AvrPto interacts with *Arabidopsis* FLS2, the *Arabidopsis* Ef-Tu receptor EFR and the tomato flagellin receptor LeFLS2 *in vitro* and *in vivo*, and impairs the autophosphorylation activities of FLS2 and EFR (Xiang *et al.*, 2008). The inhibition of FLS2 autophosphorylation disrupts the recognition of AvrPto by FLS2, thus affecting downstream PAMP-induced defences like the oxidative burst (Xiang *et al.*, 2008), callose deposition (Hauck *et al.*, 2003; Xiang *et al.*, 2008), MAPK activation (He *et al.*, 2006) and the induction of flg22-induced genes (Xiang *et al.*, 2008). Together this allows AvrPto to subvert FLS2-mediated resistance. The interaction of Pto and FLS2 with AvrPto occurs at similar residues, and Pto can compete with FLS2 for AvrPto binding (Xiang *et al.*, 2008). It has been suggested that Pto may have evolved as a decoy for FLS2, to enable effector-triggered immunity against AvrPto (Zipfel and Rathjen, 2008).

AvrPto is also phosphorylated *in planta*, in a Pto- and Prf-independent manner (Anderson *et al.*, 2006). Phosphorylation of AvrPto contributes to virulence since mutation of the phosphorylation sites S147A and S149A in AvrPto compromises the ability of AvrPto to cause disease symptoms and to enhance the growth of *P. syringae* pv. *tomato* in susceptible tomato RG-*prf3* (Anderson *et al.*, 2006). AvrPto inhibits both the autophosphorylation and the transphosphorylation (of Pti1) activities of Pto (Xing *et al.*, 2007). This appears to subvert host recognition by Prf because the AvrPto–Pto interaction is not stabilized and downstream defence responses are no longer induced. Different MAPK cascades are activated in the susceptible interaction, although MAPKKKα is involved in signalling for both the resistant and the susceptible interactions (del Pozo *et al.*, 2004; Pedley and Martin, 2004).

## 8.3 HopAB2/AvrPtoB

HopAB2 (also known as AvrPtoB) from *P. syringae* pv. *tomato* DC3000 is the best-characterized member of a broadly distributed effector family found in *P. syringae*, *Erwinia* spp. and *Xanthomonas* spp. (Jackson *et al.*, 1999, 2002; Oguiza and Asensio, 2005; Lin *et al.*, 2006; Sarkar *et al.*, 2006; Lin and Martin, 2007). HopAB2 homologues form the HopAB3 and HopAB1 subfamilies. The HopAB3 family includes AvrPtoB from *P. syringae* pv. *tomato* T1, PT23 and JL1065, and the truncated homologue HopPmaL from *P. syringae* pv. *maculicola* ES4326 (Lin *et al.*, 2006). The HopAB1 family includes AvrPtoB from *P. syringae* pv. *tomato* B728A, and VirPphA from pathovars *phaseolicola*, *savastanoi* and *glycinea*. HopAB2 and HopAB3 members display Pto- and Prf-dependent defence responses (Abramovitch *et al.*, 2003; Lin *et al.*, 2006). HopAB2 is also recognized by Pto-independent Prf-dependent Rsb resistance (or Resistance suppressed by AvrPtoB C terminus) (Abramovitch *et al.*, 2003). HopAB1 members cause an HR in the soybean cv. Osumi (Jackson *et al.*, 2002).

The molecular mechanism of HopAB2 function is best characterized; thus we will mainly focus on this effector. HopAB2 is a modular protein – the N-terminal region is involved in recognition by the host (Abramovitch *et al.*, 2003) while the C-terminal region is an E3 ubiquitin ligase involved in defence suppression (Janjusevic *et al.*, 2006).

Full-length HopAB2 is recognized by Pto in the resistant tomato cultivar RG-*PtoR* and the N terminus of HopAB2 ($HopAB2_{1-307}$) is sufficient for this recognition (Abramovitch *et al.*, 2003). A smaller part of this region ($HopAB2_{121-200}$) is sufficient for interaction with tomato Pto kinase (Kim *et al.*, 2002; Xiao *et al.*, 2007). Mutation of specific residues that are involved in the interaction with Pto also impair HopAB2 avirulence activity (Xiao *et al.*, 2007). Pto kinase activity but not myristoylation are needed for recognition of HopAB2 in tomato while both activities are necessary for recognition in *N. benthamiana* (Balmuth and Rathjen, 2007). Phosphorylation of HopAB2 may be involved in host recognition (Xiao *et al.*, 2007). HopAB2 can also be recognized in *N. benthamiana* expressing Pto and Prf under their native promoters (Balmuth and Rathjen, 2007). In addition, $HopAB2_{1-387}$ displays a Pto-independent avirulence function (Rsb resistance), which is still dependent on Prf (Abramovitch *et al.*, 2003). $HopAB2_{1-387}$ targets Fen kinase, a close homologue of Pto kinase, which is also found in the *Pto* family gene cluster (Rosebrock *et al.*, 2007).

The crystal structure of the C-terminal portion of HopAB2 reveals striking homology to eukaryotic E3 ubiquitin ligases (Janjusevic *et al.*, 2006). HopAB2 displays E3 ubiquitin ligase activity *in vitro* (Janjusevic *et al.*, 2006). Mutation of key residues involved in E2 substrate binding eliminate E3 ligase activity and HR suppression (Janjusevic *et al.*, 2006). Consistent with an E3 ubiquitin ligase activity, HopAB2 interacts directly with ubiquitin, requiring key lysine residues in HopAB2 that are also needed for HR suppression (Abramovitch *et al.*, 2006b). Therefore, HopAB2 has acquired a eukaryotic E3 ligase function in order to subvert host recognition, presumably by targeting for degradation the host protein that would normally recognize it.

Rsb resistance, elicited by $HopAB2_{1-387}$, is suppressed by the C terminus of HopAB2 (Abramovitch *et al.*, 2003). Fen kinase, a close homologue of Pto, interacts with the C-terminal truncation $HopAB2_{1-387}$, and is ubiquinated by HopAB2's E3 ligase activity (Rosebrock *et al.*, 2007). HopAB2 mutants in the E3 ligase activity exhibit stabilized interactions with Fen kinase and Fen is no longer ubiquinated (Rosebrock *et al.*, 2007). Ubiquitination of Fen kinase results in its degradation, thus disrupting the interaction between HopAB2 and Fen (Rosebrock *et al.*, 2007). This results in the suppression of host recognition. Fen kinase may have first evolved to recognize the truncated 'ancestral' HopAB2 ($HopAB2_{1-387}$), followed by HopAB2 acquiring an E3 ubiquitin ligase domain to suppress Fen-mediated resistance (Rosebrock *et al.*, 2007). Pto would then have evolved to recognize HopAB2 and to avoid HopAB2-mediated degradation.

HopAB2 also demonstrates a virulence function when expressed in susceptible tomatoes which lack Pto or Prf. When expressed in a *P. syringae* pv. *tomato* DC3000 Δ*avrPto*Δ*HopAB2* mutant, HopAB2 or $HopAB2_{1-307}$ increase bacterial growth in susceptible tomato RG-*pto11* or RG-*prf3* and cause more severe disease symptoms (Lin and Martin, 2005; Xiao *et al.*, 2007). $HopAB2_{1-307}$ is sufficient to induce the *S. lycopersicum* ACC oxidases involved in ethylene biosynthesis and to increase ethylene production (Xiao *et al.*, 2007). Ethylene is needed for enhanced host necrosis in response to *P. syringae* pv. *tomato* and *Xanthomonas campestris* pv. *campestris* (Lund *et al.*, 1998) and specifically in response to AvrPto and HopAB2 (Cohn and Martin, 2005). Ethylene does not appear to affect pathogen growth but instead appears to modulate later stages of disease development (Lund *et al.*, 1998).

The HopAB1 family members $VirPphA_{Pph}$, $VirPphA_{Psv}$ and $VirPphA_{Pgy}$ have virulence functions in several snap bean cultivars (Canadian Wonder, Tendergreen and Red Mexican), as seen by water-soaked lesions on inoculated bean pods and greater bacterial growth (Jackson *et al.*, 1999, 2002).

Structural approaches and deletion mutants of HopAB2 were key to elucidating its function *in planta*. Interestingly, AvrPto and HopAB2 are monitored, respectively, by the closely related Pto and Fen kinases, both of which signal through the *Prf R* gene. However, while phosphorylation plays important roles in AvrPto recognition and virulence, HopAB2 usurps the ubiquination pathway, a completely different host enzymatic activity.

## 8.4 AvrRpt2, AvrB and AvrRpm1

AvrRpt2 from *P. syringae* pv. *tomato* JL1065 has been one of the most extensively studied avirulence genes and induces an HR in *Arabidopsis* Col-0 plants expressing the RPS2 resistance protein (Dong *et al.*, 1991; Whalen *et al.*, 1991; Bent *et al.*, 1994; Mindrinos *et al.*, 1994). AvrRpt2 has also been shown to induce an HR in certain soybean cultivars (Whalen *et al.*, 1991). AvrRpt2 is a cysteine protease with a catalytic core characteristic of the staphopains (CA clan) and is activated in *Arabidopsis* by the cyclophilin ROC1 (Axtell *et al.*, 2003; Coaker *et al.*, 2005). Cyclophilins possess peptidyl-prolyl *cis/trans* isomerase activity and nuclear magnetic resonance spectroscopy has revealed that in the presence of ROC1, AvrRpt2 undergoes a structural change from an unfolded to a folded state (Coaker *et al.*, 2006). *In vitro* binding assays have identified a GPxL motif as the consensus ROC1-binding motif in AvrRpt2 (Coaker *et al.*, 2006). AvrRpt2 is produced as an inactive 28 kDa protein in bacteria and is activated in *Arabidopsis* or by eukaryotic extracts (Mudgett and Staskawicz, 1999; Jin *et al.*, 2003). Once activated, AvrRpt2 undergoes self-processing and cleaves off its amino terminal 71 amino acids to produce a 21 kDa protease (Mudgett and Staskawicz, 1999; Jin *et al.*, 2003).

Interestingly, AvrRpt2 directly cleaves the *Arabidopsis* protein RIN4 (RPM1-interacting protein 4) at two sites that are related to its autoprocessing site, known as RCS1 and RCS2 or AvrRpt2 Cleavage Sites-1 and -2 (Chisholm *et al.*, 2005; Coaker *et al.*, 2005; Day *et al.*, 2005; Kim, H.S. *et al.*, 2005; Takemoto and Jones, 2005). This situation is reminiscent of the cleavage of the kinase PBS1 by HopAR1 (also known as AvrPphB) at a sequence similar to the HopAR1 autoprocessing site (Shao *et al.*, 2003).

RIN4 interacts with the resistance protein RPM1 which recognizes the two unrelated avirulence proteins AvrB from *P. syringae* pv. *glycinea* race 0 and AvrRpm1 from *P. syringae* pv. *maculicola* race m2 (Tamaki *et al.*, 1988; Grant *et al.*, 1995; Mackey *et al.*, 2002). RPM1 is coimmunoprecipitated with RIN4 from *Arabidopsis* extracts and in yeast two-hybrid assays RIN4 interacts strongly with the N-terminal 176 amino acids of RPM1 (Mackey *et al.*, 2002). RIN4 also interacts with the R protein RPS2 (Axtell *et al.*, 2003; Mackey *et al.*, 2003). As noted above, AvrRpt2 directly cleaves RIN4 leading to its disappearance. RPS2 is thought to initiate signalling following the perception of RIN4 disappearance rather than through direct recognition of AvrRpt2 (Axtell *et al.*, 2003; Mackey *et al.*, 2003). In support of this, a *rin4* null allele is embryo lethal, a phenotype that can be largely suppressed in *rin4rps2* plants (Mackey *et al.*, 2003; Belkhadir *et al.*, 2004). *rin4rps2 Arabidopsis* plants display increased resistance against *P. syringae* pv. *tomato* DC3000 (Belkhadir *et al.*, 2004). Interestingly this resistance is due to ectopic expression of RPM1, indicating that loss of RIN4 can activate both RPM1 and RPS2.

However, RIN4 is not the only target of AvrRpt2 since the elimination of RIN4 does not diminish the virulence function of AvrRpt2 (Belkhadir *et al.*, 2004; Lim and Kunkel, 2004). Bioinformatic searches have identified at least 11 potential AvrRpt2 targets in *Arabidopsis* based on the presence of RCS sequences (Chisholm *et al.*, 2005; Kim, H.S. *et al.*, 2005). It remains to be determined if RIN4 is a virulence target of AvrRpt2 (Belkhadir *et al.*, 2004). RIN4 has been demonstrated to be a negative regulator of basal resistance and thus appears to have a dual role in modulating *R*-gene-mediated and basal resistance (Kim, M.G. *et al.*, 2005). Proteolytically digesting a negative regulator of basal defence does not seem like an efficient virulence strategy. Presumably, if RIN4 is a virulence target of AvrRpt2 in the absence of RPS2, one or more of the three RIN4 fragments produced by cleavage at the two RCS sites serves to enhance RIN4 function and mute basal resistance. Another possibility is that RIN4 has evolved to mimic the true virulence target of AvrRpt2 and through its association with RPS2, acts as a decoy to induce *R*-gene-mediated resistance.

RIN4 is also targeted by AvrB and AvrRpm1. AvrB and AvrRpm1 are both membrane localized in the host via myristoylation where they interact with RIN4 which is anchored in the membrane by C-terminal acylation (Nimchuk *et al.*, 2000; Kim, H.S. *et al.*, 2005). However, unlike the cleavage of RIN4 by AvrRpt2, AvrRpm1 and AvrB both induce RIN4 phosphorylation *in planta* (Mackey *et al.*, 2002). RIN4 is required for RPM1-mediated resistance induced by both AvrRpm1 and AvrB (Belkhadir *et al.*, 2004). RIN4 can be coimmunoprecipitated with AvrB and RIN4 also interacts with AvrB *in vitro* and in yeast two-hybrid assays (Mackey *et al.*, 2002; Ong and Innes, 2006; Desveaux *et al.*, 2007). AvrRpm1 can be coimmunoprecipitated with RIN4 but has not been shown to interact directly with RIN4 by yeast two-hybrid or *in vitro* assays (Mackey *et al.*, 2002).

AvrRpm1 or AvrB induce RIN4 phosphorylation but do not display any significant sequence or overall structural similarity to kinases (Mackey *et al.*, 2002; Lee *et al.*, 2004). However, a cocrystal structure of AvrB with a fragment of $RIN4_{142-176}$ revealed that functionally important residues in AvrB correspond to catalytic residues in Ser/Thr kinases suggesting that AvrB could potentially be a kinase (Desveaux *et al.*, 2007). In support of this, AvrB can bind to ADP and residues making important contact with this nucleotide in the cocrystal structure are also required for RPM1 recognition (Desveaux *et al.*, 2007). Furthermore, AvrB is phosphorylated by *Arabidopsis* extracts. The cocrystal structure of AvrB and $RIN4_{142-176}$ also revealed that the AvrB binding site (BBS) in $RIN4_{142-176}$ is adjacent to the RCS-2 domain. BBS-interacting residues in AvrB are required for triggering RPM1 function providing evidence for the indirect recognition of AvrB by RPM1 via RIN4. This was also shown by random mutagenesis of AvrB, whereby three of four mutations that lost the ability to trigger RPM1 function also lost RIN4 binding activity in yeast two-hybrid assays (Ong and Innes, 2006). In soybean, AvrB is recognized by the Rpg1-b resistance protein which has evolved the ability to recognize AvrB independently of RPM1 (Ashfield *et al.*, 2004). In susceptible soybean plants, AvrB contributes to an eight-fold increase in bacterial growth whereas in susceptible *Arabidopsis* plants, AvrB mediates a yellowing response (Ashfield *et al.*, 1995; Nimchuk *et al.*, 2000).

Since *rin4* null plants still display AvrRpt2- and AvrRpm1-mediated virulence functions, RIN4 may have evolved as an effective decoy for at least three type III effectors (Belkhadir *et al.*, 2004). However, as a result AvrRpt2 can interfere with RPM1 function in *Arabidopsis* (Ritter and Dangl, 1995). This has been demonstrated to be due to the activity of AvrRpt2 on RIN4 (Kim, H.S. *et al.*, 2005). Since RIN4 is required for RPM1 function, elimination of RIN4 by AvrRpt2 interferes with AvrB and AvrRpm1 recognition by RPM1. In addition, RPM1 requires RIN4 for its accumulation and therefore could be destabilized by AvrRpt2. In addition to interfering with *R*-gene-mediated resistance, AvrRpt2 as well as AvrRpm1 can block PAMP-induced signalling and thus basal resistance responses (Kim, M.G. *et al.*, 2005). The inhibition of basal resistance responses by AvrRpt2 and AvrRpm1 may occur in part via the manipulation of RIN4 activity in *Arabidopsis* plants lacking the R proteins RPS2 or RPM1. RPS2 and RPM1 presumably sense effector-induced perturbations of RIN4 in order to induce a hypersensitive response. RIN4 thereby provides a paradigm example of a mechanistic link between basal- and *R*-gene-mediated defences and how these defences relate to type III effector virulence functions.

## 8.5 HopZ/YopJ

The HopZ/YopJ family of effector proteins is a broadly distributed and evolutionarily diverse family found in both plant and animal pathogenic bacteria (Ma *et al.*, 2006). The *P. syringae* HopZ family is comprised of three homology groups, HopZ1, HopZ2 and HopZ3. The closely related alleles *hopZ1a*, *hopZ1b* and *hopZ1c* diversified by pathoadaptation, and the two more divergent alleles, *hopZ2* and *hopZ3*, were brought into the family by horizontal gene transfer (Ma *et al.*, 2006). *hopZ1a* from *P. syringae* pv. *syringae* A2 (formerly *hopPsyH*) is predicted to be most similar to the ancestral *hopZ* allele (Ma *et al.*, 2006). Other *hopZ1a* alleles are restricted to particular strains of *P. syringae* pv. *syringae* in group two (Hwang *et al.*, 2005; Ma *et al.*, 2006; Sarkar *et al.*, 2006). *hopZ1b* or related alleles are restricted to *P. syringae* pv. *glycinea* strains in group three (Hwang *et al.*, 2005; Ma *et al.*, 2006; Sarkar *et al.*, 2006). *hopZ1c* has been found only in *P. syringae* pv. *maculicola* ES4326 and YM7930, in group five (Hwang *et al.*, 2005; Ma *et al.*, 2006; Sarkar *et al.*, 2006). In contrast, *hopZ2* and *hopZ3* are distributed through many pathovars of *P. syringae*, consistent with their acquisition by horizontal gene transfer from *Xanthomonas* and *Erwinia* spp., respectively (Hwang *et al.*, 2005; Ma *et al.*, 2006; Sarkar *et al.*, 2006).

HopZ1a has avirulence functions in diverse species including *Arabidopsis thaliana*, soybean, rice and sesame (Ma *et al.*, 2006; Lewis *et al.*, 2008). HopZ1b shows weak avirulence functions in *A. thaliana*, causing an HR in 24% of the leaves of ecotype Columbia (Lewis *et al.*, 2008). HopZ1c, which lacks a C-terminal extension present in both HopZ1a and HopZ1b, is not recognized in any species tested so far (Ma *et al.*, 2006; Lewis *et al.*, 2008). The C terminus of the HopZ1 alleles is under strong positive selection (Ma *et al.*, 2006), and may contain determinants for host recognition. The HR is elicited by *Agrobacterium*-mediated transient expression of HopZ1a, HopZ1b and HopZ2 in *N. benthamiana*, and of HopZ3 in *N. tabacum* and snap bean (Ma *et al.*, 2006; Vinatzer *et al.*, 2006; Lewis *et al.*, 2008).

HopZ1a, HopZ1b, HopZ1c and HopZ2 are myristoylated proteins and are membrane localized *in planta* (Lewis *et al.*, 2008). In contrast, HopZ3 does not have a myristoylation consensus sequence and is a soluble protein (Lewis *et al.*, 2008). HopZ3 is also the only member of the family to have a chaperone (SchZ3) (Ma *et al.*, 2006). Myristoylation of HopZ1a contributes to its avirulence function, suggesting that recognition occurs at the plasma membrane in the host (Lewis *et al.*, 2008).

The HopZ family members demonstrate weak but significant protease activity using fluorescence-based protease assays (Ma *et al.*, 2006). Activity depends on a conserved catalytic cysteine also found in YopJ (Ma *et al.*, 2006). YopJ was originally demonstrated to have cysteine protease activity (Orth *et al.*, 2000), but more recently has been shown to possess acetyltransferase activity (Mittal *et al.*, 2006; Mukherjee *et al.*, 2006). Both cysteine proteases and acetyltransferases have the same catalytic triad and go through a similar reaction intermediate (Mukherjee *et al.*, 2007). It remains to be determined if the HopZ family members also possess acetyltransferase activity, in addition to protease activity. Regardless, the catalytic cysteine is necessary for the elicitation of the HR by HopZ1a. Although the *R* gene which recognizes HopZ1a has yet to be cloned, HopZ1a is not recognized by any of the known *R* genes, *RPM1*, *RPS2*, *RPS5*, *RPS4* or *RPS6* in *Arabidopsis* (Lewis *et al.*, 2008).

HopZ2 is the only member of the family for which a virulence function has been demonstrated (Lewis *et al.*, 2008). HopZ2 confers a significant growth benefit when expressed in the virulent strain *P. syringae* pv. *tomato* DC3000 or the non-host strain *P. syringae* pv. *cilantro* 0788-9 in *A. thaliana* ecotype Columbia (Lewis *et al.*, 2008). HopZ2 virulence function requires the catalytic cysteine residue (previously shown to be necessary for cysteine protease activity; Ma *et al.*, 2006) and the myristoylated glycine residue (Lewis *et al.*, 2008).

HopZ/YopJ homologues are also found in *X. campestris* pv. *vesicatoria* and *R. solanacearum*. *Xanthomonas* AvrBsT is recognized by the pepper *Bs1* resistance gene and in the *Arabidopsis* Pi-0 ecotype (Minsavage *et al.*, 1990; Escolar *et al.*, 2002; Cunnac *et al.*, 2007). The cysteine protease catalytic triad is necessary for recognition in pepper (Orth *et al.*, 2000). Recognition presumably occurs in the nucleus because AvrBsT possesses a putative nuclear localization signal (NLS) and is predicted to be nuclear localized (Ciesiolka *et al.*, 1999). AvrBsT induces defence responses in the Pi-0 ecotype because the Pi-0 ecotype lacks the active form of the SOBER1 carboxylesterase enzyme (Cunnac *et al.*, 2007). It is unclear at this time how the SOBER1 carboxylesterase may contribute to the HR. However, homologues of SOBER1, the acyl protein thioesterase/lysophospholipases, have functions as second messengers and in immune responses in mammalian cells (Cunnac *et al.*, 2007).

*Xanthomonas* AvrRxv, another YopJ homologue, is recognized in tomato cv. Hawaii 7998 by three non-dominant resistance genes (Whalen *et al.*, 1993; Wang *et al.*, 1994; Yu *et al.*, 1995; Ciesiolka *et al.*, 1999) and in bean (Whalen *et al.*, 1988). AvrRxv elicits defence responses when expressed in normally virulent *X. campestris* pathovars in their normally susceptible hosts, including *phaseoli* on the bean cultivar Sprite, *glycines* on soybean, *vignicola* on cowpea, *alfalfae* on lucerne, *holcicola* on corn, *malvacearum* on cotton (Whalen *et al.*, 1988). The cysteine protease catalytic core is needed for host recognition (Bonshtien *et al.*, 2005). Despite possessing putative NLSs, AvrRxv localizes to the plant cell cytoplasm, where it is likely to be recognized (Bonshtien *et al.*, 2005).

*Xanthomonas* AvrXv4 is recognized by Xv4 in *Solanum pennellii* (Astua-Monge *et al.*, 2000). Recognition of AvrXv4 and induction of the HR in *N. benthamiana* requires the catalytic triad (Roden *et al.*, 2004). Recognition is likely to occur in the cytoplasm since AvrXv4 is localized there, despite possessing an NLS (Roden *et al.*, 2004). Expression of AvrXv4 *in planta* leads to a reduction in small ubiquitin-like modifier (SUMO)-modified proteins (Roden *et al.*, 2004). Sumoylation is a post-translational modification which can regulate protein function (Roden *et al.*, 2004). AvrXv4 may then have a distinct enzymatic function.

*Xanthomonas* XopJ triggers an HR when transiently expressed in *N. benthamiana* or *N. clevelandii* (Thieme *et al.*, 2007). XopJ is membrane localized and possesses a myristoylation sequence (Thieme *et al.*, 2007). XopJ may be most similar to the HopZs, with recognition occurring at the membrane.

*Ralstonia solanacearum* has two YopJ homologues, PopP1 and PopP2. *Ralstonia* PopP1 has an avirulence function on petunia (Lavie *et al.*, 2002). PopP1 does not possess an NLS nor a membrane localization sequence and is predicted to localize to the cytosol (Lavie *et al.*, 2002). PopP2 has an avirulence function in the resistant *Arabidopsis* ecotype Nd-1 while it is virulent in the susceptible ecotype Col-5 (Deslandes *et al.*, 2002). Unlike many effectors, PopP2 interacts directly with the RRS1 R protein (Deslandes *et al.*, 2003). RRS1 is a TIR-NBS-LRR *R* gene and also contains a WRKY motif characteristic of the WRKY transcription factors (Deslandes *et al.*, 2003). Resistance is partially salicylic acid-dependent and Non-race Specific Disease Resistance 1 (NDR1)-dependent (Deslandes *et al.*, 2003). When PopP2 and RRS1 are coexpressed, RRS1 colocalizes with PopP2 to the nucleus (Deslandes *et al.*, 2003). As we learn more about the dynamics of R protein localization in response to effectors or defence responses, R protein localization to the nucleus may emerge as a common trend.

Hints about potential host targets have been obtained from global expression profiling experiments in tomato infected with *X. campestris* pv. *vesicatoria* carrying *avrRxv* or an inactive *avrRxv* mutant (Bonshtien *et al.*, 2005). Groups of upregulated genes included those involved in transcription, stress responses, signalling and defence while downregulated groups included those involved in transcription, stress responses, defence and protein synthesis (Bonshtien *et al.*, 2005). Virulence targets have also not yet been identified for the *Xanthomonas* or *Ralstonia* YopJ homologues. However, some of these homologues have virulence activity. AvrBsT exhibits virulence activity in the tomato cv. Walter (Minsavage *et al.*, 1990) and AvrXv4 is virulent in *S. lycopersicum* (Astua-Monge *et al.*, 2000). The identification of host targets will help dissect the signalling pathways leading to defence or disease in these diverse host plants.

The HopZ/YopJ family of bacterial effectors provides an opportunity to examine the effector function within an evolutionarily well-characterized family. Within the *P. syringae* HopZ family, we have a unique opportunity to examine how evolutionary pressures have shaped avirulence and virulence functions. For the YopJ homologues as a whole, it will be particularly interesting to determine whether homologues from different bacterial species target common or distinct host proteins to carry out their functions.

## 8.6 AvrRps4

AvrRps4 from *P. syringae* pv. *pisi* 151 was first identified by the HR it induced in the *Arabidopsis* ecotype Po-1 (Hinsch and Staskawicz, 1996). AvrRps4 also induces an HR in multiple other *Arabidopsis* ecotypes when expressed in *P. syringae* pv. *tomato* DC3000 and in the soybean cultivar Harosoy when expressed in *P. syringae* pv. *glycinea* race 4 (Hinsch and Staskawicz, 1996). AvrRps4 is found in *pisi* pathovars as well as some *glycinea* and *phaseolicola* pathovars (Hinsch and Staskawicz, 1996).

AvrRps4 is recognized by the RPS4 protein, a TIR-NBS-LRR class R protein (Gassmann *et al.*, 1999). The RPS4-induced HR requires ENHANCED DISEASE SUSCEPTIBILITY 1 (EDS1), SUPPRESSOR OF G2 ALLELE of skp1 (SGT1) and HEAT SHOCK PROTEIN 90 (HSP90), which have been shown to be involved in either TIR-class *R* gene signalling and/or plant defences (Zhang *et al.*, 2004; Wirthmueller *et al.*, 2007). Although RPS4 contains an NLS, it is found in endomembranes and nuclei in both healthy and infected *Arabidopsis* tissue (Wirthmueller *et al.*, 2007). While recognition of AvrRps4 by RPS4 does not appear to change the localization of RPS4, nuclear localization of RPS4 is needed for AvrRps4 recognition in *Arabidopsis* (Wirthmueller *et al.*, 2007). RPS4 and EDS1 are found in the nucleus, where EDS1 appears to mediate defence gene expression (Wirthmueller *et al.*, 2007). *RPS4* gene expression is also weakly induced by EDS1 (Zhang and Gassmann, 2007).

*RPS4* is dynamically regulated by gene expression and alternative splicing, and these appear to contribute to RPS4-mediated defence responses. *RPS4* transcript abundance is increased ~80% upon inoculation with *P. syringae* pv. *tomato* DC3000 carrying *avrRps4* (Zhang and Gassmann, 2007). Constructs lacking specific introns are impaired in resistance to *P. syringae* pv. *tomato* DC3000 carrying *avrRps4* (Zhang and Gassmann, 2003). It appears that these alternative transcripts contribute to resistance at the protein level, rather than to RNA regulation (Zhang and Gassmann, 2003).

Tomato cultivars and some soybean cultivars are susceptible to respectively *P. syringae* pv. *tomato* DC3000 carrying *avrRps4* or pv. *glycinea* race 4 carrying *avrRps4* (Hinsch and Staskawicz, 1996). The RLD ecotype of *Arabidopsis* carries a naturally susceptible *rps4* allele (Hinsch and Staskawicz, 1996). Substitution of amino acid polymorphisms (N195D or Y950H) from the susceptible RLD sequence to the resistant *RPS4* sequence, into the full-length resistant *RPS4* sequence caused a loss or reduction in RPS4-mediated resistance (Zhang and Gassmann, 2003).

To date, AvrRps4 is the only *P. syringae* effector that is recognized by the TIR-class of *R* genes. *RPS4* gene expression, alternative splicing and protein localization to the nucleus all contribute to defence responses in response to AvrRps4. Alternative splicing of certain transcripts has been observed in *Arabidopsis* inoculated with *P. syringae* pv. *tomato* DC3000 carrying *hopA1* (also known as *hopPsyA*) or *avrRpt2* (Zhang and Gassmann, 2003). It will be interesting to determine if alternative splicing and R protein nuclear localization emerge as themes in defence signalling.

## 8.7 HopAR1/AvrPphB

The sequence encoding HopAR1 (also known as AvrPphB) was originally isolated from *P. syringae* pv. *phaseolicola* race 3 (Jenner *et al.*, 1991), which elicits an HR on bean (*Phaseolus vulgaris*) cultivars that possess the *R3* resistance gene (Taylor *et al.*, 1996). This effector is also recognized in *N. benthamiana*, tomato, tobacco and *Arabidopsis* (Simonich and Innes, 1995; Tampakaki *et al.*, 2002). Structural analyses place HopAR1 in the papain superfamily of cysteine proteases whose membership also includes the *Yersinia pestis* effector YopT (Zhu *et al.*, 2004). YopT and its homologues possess an invariant catalytic triad comprised of cysteine, histidine and aspartic acid (Shao *et al.*, 2002). Mutation of these residues eliminates the capability of HopAR1 to elicit the HR on resistant hosts. HopAR1 is synthesized as a 35 kDa protein which is subsequently processed to yield a 28 kDa protein (Puri *et al.*, 1997). However, unlike AvrRpt2, AvrPphB does not require a eukaryotic cofactor for activation. This autoproteolysis requires an intact catalytic triad, and serves to expose an N-terminal myristoylation site that directs HopAR1 to the plasma membrane (Nimchuk *et al.*, 2000). Mutagenesis experiments suggest that the myristoylation site may be dispensable for the elicitation of the HR, depending on the host plant studied (Tampakaki *et al.*, 2002).

The specific molecular events leading to HopAR1-induced HR have been detailed most extensively in *Arabidopsis*. Recognition of HopAR1 requires RPS5, a CC-NBS-LRR resistance protein (Simonich and Innes, 1995). The LRR domain appears to negatively regulate the activity of RPS5, because transient expression of a construct encoding only the CC and NBS domains induces an HR in *N. benthamiana*, even in the absence of HopAR1 (Ade *et al.*, 2007). The CC domain interacts with the serine/threonine kinase PBS1 (Swiderski and Innes, 2001), whose autophosphorylation is essential for this interaction (Ade *et al.*, 2007). Shao *et al.* (2003) observed that PBS1 is degraded in the presence of HopAR1, and that this cleavage occurs at a similar amino acid motif at which HopAR1 autocleavage takes place. Importantly, a protease-inactive form of HopAR1 can be coimmunoprecipitated with RPS5, but only in the presence of PBS1 (Ade *et al.*, 2007). This strongly suggests that RPS5 indirectly recognizes HopAR1 via PBS1. Overall, the current model of RPS5 function posits that in the inactive state, phosphorylated PBS1 is bound to the N-terminal CC domain of RPS5 oligomers. The LRR domain is likely to be bound to the NBS domain to block RPS5 activation. When HopAR1 is introduced into the host, PBS1 is cleaved, resulting in a conformational change within RPS5 that exposes the NBS, possibly due to a portion of PBS1 binding to the LRR domain. The NBS would then be accessible for binding ATP in order to activate the protein and stimulate downstream signalling pathways (Ade *et al.*, 2007). While it is evident that RPS5 has evolved to detect the cleavage of PBS1 by HopAR1, it is less clear what function this proteolysis serves in facilitating pathogen virulence on a host lacking RPS5. Further characterization of PBS1 activity and the identification of other potential HopAR1-binding proteins would certainly be useful in addressing this issue.

## 8.8 AvrPphF/HopF

Characterization of effector function may be aided by the tools of structural biology, as illustrated by the HopF (also known as AvrPphF) family of effector proteins. In beans, the product of the *R1* resistance gene enables recognition of HopF (Taylor *et al.*, 1996). The *R1* gene remains to be cloned and characterized. An unknown *R* gene also confers resistance to HopF-expressing *P. syringae* in tomato and tobacco (Robert-Seilaniantz *et al.*, 2006). The gene encoding HopF was first isolated from races 5 and 7 of *P. syringae* pv. *phaseolicola* (Jackson *et al.*, 1999; Tsiamis *et al.*, 2000), and has subsequently been found in *P. syringae* pv. *tomato* as well as pv. *delphinii* (Fouts *et al.*, 2002; Deng *et al.*, 2003). The *hopF* locus contains two open reading frames encoding the effector and a chaperone, ShcF (Jackson *et al.*, 1999; Tsiamis *et al.*, 2000). Crystal structures have been obtained for both of these components, and SchF displays significant structural similarity to other bacterial type III chaperones (Singer *et al.*, 2004). HopF forms a 'mushroom'-like structure with distinct 'head' and 'stalk' subdomains. While there are not any obvious structural homologues of HopF, a portion of the head subdomain is somewhat similar to the catalytic domains found in various ADP-ribosyltransferase (ADP-RT) toxins, such as diphtheria toxin. *In vitro* analyses, however, did not reveal any ADP-RT activity, NAD binding, or NAD glycohydrolase activity for HopF. Despite this, a number of functionally important residues were predicted for HopF based on its homology to diphtheria toxin and sequence conservation among HopF alleles. Mutation of HopF residues in a region homologous to the NAD-binding pocket of diphtheria toxin completely eliminated the virulence function of this effector on a susceptible bean cultivar. A similar effect was observed in tobacco upon mutation of a putative N-terminal myristoylation site in HopF (Robert-Seilaniantz *et al.*, 2006). Importantly, these mutations also abolished avirulence on a resistant plant host, indicating that the activity of HopF is essential for its recognition by a certain resistance protein. Although these findings provided valuable insight into the potential function of HopF, the host proteins targeted by this effector have yet to be identified. Tsiamis *et al.* (2000) noted that AvrB2 (also known as AvrPphC) suppresses the HR induced by HopF (Tsiamis *et al.*, 2000). The targeting of RIN4 by AvrB (Mackey *et al.*, 2002) raises the intriguing possibility that the virulence function of HopF could involve an association with RIN4 or RIN4-associated proteins. Obviously, a significant amount of work remains to be done in order to characterize both the virulence function of HopF and the mechanism through which this effector elicits the HR.

## 8.9 AvrPphE/HopX

The HopX (also known as AvrPphE) family of effectors remains a relatively mysterious group of virulence factors. Homologues of HopX exist in a number of *P. syringae* pathovars, including pv. *maculicola*, pv. *tomato*, pv.

*phaseolicola*, pv. *tabaci*, pv. *glycinea*, pv. *angulata*, pv. *delphinii* and pv. *syringae* (Mansfield *et al.*, 1994; Alfano *et al.*, 2000; Fouts *et al.*, 2002; Guttman *et al.*, 2002; Charity *et al.*, 2003; Deng *et al.*, 2003; Rohmer *et al.*, 2003). HopX elicits an HR on *Arabidopsis*, tobacco and beans, the latter of which is mediated by the *R2* resistance gene (Mansfield *et al.*, 1994; Nimchuk *et al.*, 2007). Like HopAR1, HopX proteins contain a conserved catalytic triad of cysteine, histidine and aspartic acid, characteristic of the transglutaminase superfamily (Makarova *et al.*, 1999; Nimchuk *et al.*, 2007). Despite the conservation of this triad, *in vitro* assays with HopX failed to detect activity for any of the classes of transglutaminase enzymes (Nimchuk *et al.*, 2007). Mutation of the catalytic triad residues did, however, abolish HopX-induced HR on *Arabidopsis* and on resistant bean cultivars. On the other hand, some HopX alleles do not induce an HR even though the catalytic triad is intact, suggesting that multiple domains may be involved in HopX (a)virulence (Stevens *et al.*, 1998). Indeed, a comparison of various HopX alleles also revealed a conserved domain N-terminal to the catalytic triad, designated the 'A domain'. Notably, the mutation of three residues within this domain rendered HopX incapable of eliciting an HR on either beans or *Arabidopsis*. Based on secondary structure predictions, Nimchuk *et al.* (2007) speculated that the A domain may mediate interactions with host targets or bind a nucleotide or cofactor required for HopX activity.

## 8.10 *Xanthomonas* AvrBs3

The *Xanthomonas* AvrBs3/AvrBs4 family is a large family of nuclear-targeted effectors that contain multiple tandem copies of highly similar 34 amino acid repeats (Lahaye and Bonas, 2001; Schornack *et al.*, 2006). Homologues are found in *Xanthomonas campestris* pv. *vesicatoria*, pv. *armoraciae*, pv. *malvacearum*, pv. *manihotis*, *Xanthomonas oryzae* pv. *oryzae*, pv. *oryzicola*, *Xanthomonas axonopodis* pv. *citri* and *R. solanacearum* (Schornack *et al.*, 2006). In contrast to the *P. syringae* effectors discussed thus far, AvrBs3 members are nuclear localized and carry out their functions by manipulating host gene expression, rather than affecting protein function.

The *Xanthomonas campestris* pv. *vesicatoria* effector AvrBs3 is recognized by Bs3 in pepper (Bonas *et al.*, 1989; Pierre *et al.*, 2000). AvrBs3 contains a nuclear localization sequence and an acidic transcriptional activation domain, which are necessary for HR elicitation (van den Ackerveken *et al.*, 1996; Szurek *et al.*, 2001). This suggests that recognition of AvrBs3 occurs in the plant cell nucleus, where AvrBs3 is found (Szurek *et al.*, 2002). Consistent with this, AvrBs3 interacts with importin α, which is involved in the import of proteins into the nucleus (Szurek *et al.*, 2001). Binding of AvrBs3 to the *Bs3* promoter induces its transcription (Romer *et al.*, 2007). In this example the Avr protein binds and activates the promoter of its cognate *R* gene! This would appear to be a suicidal strategy, however, a virulence target of AvrBs3, *upa20* is also transcriptionally activated by AvrBs3 via a similar promoter sequence as

*Bs3*. Therefore, in this example the promoter region of the *R* gene mimics the promoter of the virulence target.

Bs3 shows some similarity to flavin-dependent monooxygenases, including FMO1 from *Arabidopsis* (Romer *et al.*, 2007). Its enzymatic activity is necessary for the HR (Romer *et al.*, 2007). FMO1 regulates the EDS1 pathway, which is involved in defence responses initiated by TIR-class *R* genes (Bartsch *et al.*, 2006). Overexpression of *Arabidopsis* FMO1 confers enhanced resistance to *P. syringae* pv. *tomato* DC3000 and the oomycete pathogen *Hyaloperonospora parasitica* (Koch *et al.*, 2006). Systemic acquired resistance is compromised in *fmo1* mutants (Mishina and Zeier, 2006), although they still accumulate salicylic acid (Bartsch *et al.*, 2006; Koch *et al.*, 2006). It is unclear whether Bs3 also mediates some aspect of broad-spectrum resistance as does *Arabidopsis* FMO1.

In susceptible pepper plants, AvrBs3 induces the transcription of cell-wall-associated genes and the small auxin up RNA (SAUR) family of auxin-induced genes (Marois *et al.*, 2002). AvrBs3 causes hypertrophy of the mesophyll tissue in pepper, *S. lycopersicum* and *S. pennellii*, and pustule formation on the leaf surface (Marois *et al.*, 2002). Hypertrophy of the leaf tissue appears to be due to the induction of UPA20, a basic helix-loop-helix (bHLH) transcription factor, by AvrBs3 (Kay *et al.*, 2007). Consistent with its role as a transcriptional activator, UPA20 localizes to the nucleus and requires both the basic region and the dimerization domain for its function (Kay *et al.*, 2007). The *Upa20* promoter and AvrBs3 interact directly *in planta*, with specificity conferred by the AvrBs3 tandem repeats (Kay *et al.*, 2007). When AvrBs3 is present with Bs3, AvrBs3 appears to affect the transcriptional start site of *Upa20*, which may affect its function (Kay *et al.*, 2007). UPA20 also induces the expression of *Upa7*, a putative $\alpha$-expansin (Kay *et al.*, 2007). Thus, the induction of UPA20 by AvrBs3 appears to initiate a signalling cascade leading to cell expansion and tissue hypertrophy.

The closely related AvrBs3-like gene *AvrBs4* is also found in *Xanthomonas campestris* pv. *vesicotoria* (Schornack *et al.*, 2006). Despite its sequence similarity to AvrBs3, AvrBs4 is not recognized by Bs3 in pepper (Bonas *et al.*, 1993; Romer *et al.*, 2007) but is recognized by Bs4 in tomato (Bonas *et al.*, 1993; Schornack *et al.*, 2004b). When delivered by *Agrobacterium*, AvrBs4 also induces an HR in *N. tabacum*, *N. clevelandii*, *N. benthamiana* and *Solanum tuberosum* (Schornack *et al.*, 2004a, b). The NLSs of AvrBs4 are not necessary for recognition by tomato Bs4 and these proteins do not interact directly (Ballvora *et al.*, 2001; Schornack *et al.*, 2004a, b). Bs4 is a TIR-NBS-LRR protein and acts through EDS1 and SGT1 (Schornack *et al.*, 2004a, b). *Bs4* also appears to undergo alternative splicing but no role has been found for this in defence signalling (Schornack *et al.*, 2004a, b). It is not clear at this time where Bs4 is present in the plant cell and whether, like RPS4, nuclear localization is necessary for defence responses.

AvrXa27 is an AvrBs3-like effector found in *X. oryzae* pv. *oryzae* (Schornack *et al.*, 2006). AvrXa27 is recognized by the Xa27 *R* gene in rice (Gu *et al.*, 2004). Like AvrBs3, the nuclear localization sequence and activation domain are necessary for avirulence function (Gu *et al.*, 2005). *Xa27* expression is induced when plants are inoculated with strains carrying *avrXa27* (Gu *et al.*, 2005). Ectopic expression of *Xa27* confers resistance to normally virulent strains, without requiring *avrXa27* (Gu *et al.*, 2005).

The *Xanthomonas* AvrBs3 family of effectors provides a unique strategy for virulence function. Nuclear targeting and manipulation of host expression appears unique to this family of effectors and is reminiscent of *Agrobacterium tumefaciens* infection (Dafny-Yelin *et al.*, 2008). Multiple AvrBs3-like effectors exist in *Xanthomonas* spp. (Schornack *et al.*, 2006). As we learn more about the functional diversity of effectors, it will be interesting to determine if nuclear targeting of effectors emerges as a common strategy or one which is restricted to certain bacteria.

## 8.11 *Erwinia amylovora*

*Erwinia amylovora* is a Gram-negative bacterium that carries a type III secretion system but unlike *P. syringae*, it is a necrotrophic pathogen (Toth and Birch, 2005). Few effectors have been characterized in *E. amylovora* but interestingly, some of these share similarity to *P. syringae* effectors.

DspA/E is a homologue of *P. syringae* AvrE. DspA/E can elicit defence responses in soybean when carried by *P. syringae* pv. *glycinea* race 4 (Bogdanove *et al.*, 1998). The hypersensitive response is elicited by *E. amylovora* carrying *dspA/E* in tobacco or by *Agrobacterium* carrying *dspA/E* in *N. benthamiana* (Bogdanove *et al.*, 1998; Oh *et al.*, 2007). SGT1, a known player in *R* gene-mediated signalling, contributes to the DspA/E-induced HR in *N. benthamiana* (Oh and Beer, 2005; Azevedo *et al.*, 2006). DspA/E also interacts with several LRR receptor-like kinases that are found in resistant and susceptible apples (Meng *et al.*, 2006). DspA/E contributes to *E. amylovora* virulence in apples, pears, soybean and tobacco (Gaudriault *et al.*, 1997; Bogdanove *et al.*, 1998; Boureau *et al.*, 2006).

An *E. amylovora* AvrRpt2 homologue, $AvrRpt2_{EA}$, was also recently identified (Zhao *et al.*, 2006). $AvrRpt2_{EA}$ can induce a weak HR when expressed in *P. syringae* pv. *tomato* DC3000 in *Arabidopsis* (Zhao *et al.*, 2006). However, if the promoter and signal sequence of *avrRpt2*$_{EA}$ are replaced with those from the *P. syringae* effector *avrRpt2*, a strong HR is induced (Zhao *et al.*, 2006). $AvrRpt2_{EA}$ has a virulence function in pear fruits and in *Arabidopsis* (Zhao *et al.*, 2006). In an *rps2* mutant, *P. syringae* carrying $AvrRpt2_{EA}$ grows 1 log better than *P. syringae* carrying AvrRpt2 (Zhao *et al.*, 2006). While the expression and secretion of $AvrRpt2_{EA}$ may be different from those of AvrRpt2, it has similar functions *in planta*. At this time, it is still unknown whether $AvrRpt2_{EA}$ also targets an apple or pear homologue of RIN4 or the cyclophilin ROC1.

Functional characterization of *E. amylovora* type III effectors is in its infancy and is an exciting area of new research. It is intriguing that, thus far, the effectors identified are homologous to those in *P. syringae*. The obvious question is whether they target similar host proteins or because of their necrotrophic lifestyle, these effectors have distinct targets. At least in the case of AvrRpt2, it appears that the *Erwinia* homologue retains similar avirulence and virulence functions. It would also be interesting to compare the evolutionary origins of the *Erwinia* effectors and their counterparts in other bacterial species since horizontal gene transfer is a common means for introducing genetic diversity (McCann and Guttman, 2008).

## 8.12 Conclusions

Numerous models have been proposed that allude to the evolutionary pressures that may have led to the sequential acquisition of basal resistance in the host, a type III secretion system and associated effectors in the pathogen, followed by *R*-gene-mediated resistance in the host (Espinosa and Alfano, 2004; Chisholm *et al.*, 2006; Jones and Dangl, 2006). As a result of this, these three aspects of plant-bacterial pathogens appear to have been tightly intertwined and in the case of RIN4, converge on one protein. It remains to be determined if RIN4 is a true virulence target of the three type III effectors that target it, or if it is a decoy that has evolved to mimic the true virulence targets of these effectors and betray their presence to R proteins. Mimicry appears to be a decoy strategy of the *Bs3* gene which contains a promoter element used by AvrBs3 to activate the expression of its virulence target *Upa20*. In both of these cases the virulence mechanisms of type III effector proteins have influenced the resistance strategy that has evolved in the host. Although type III effectors may have multiple targets, interaction with only one target that is associated with the appropriate resistance protein appears to be sufficient to induce effective defences and thwart pathogen growth. Host resistance modulators such as RIN4, Pto and PBS1 are convergence points between pathogen virulence and host resistance. They interact with type III effectors and relay this interaction to the resistance proteins that are also associated with them. RIN4 and Pto can interact with multiple type III effectors and RIN4 has also been demonstrated to interact with multiple resistance proteins. It remains to be determined if similar proteins also modulate the other R–Avr interactions described in this review.

Mimicry of type III effector virulence targets is an effective mechanism of protection on the host side, but just as important is the ability of type III effectors to usurp eukaryotic cell machinery for their activation. Examples include eukaryotic proteins such as myristoyltransferases which localize type III effectors to the plasma membrane, kinases for activation by phosphorylation as well as a cyclophilin for activation by structural reorganization. Overall structural mimicry of eukaryotic proteins is also an important strategy employed by type III effector proteins of both plant and animal pathogens (Stebbins and Galan, 2001; Desveaux *et al.*, 2006). An excellent example is the C terminus of the type III effector HopAB2 which mimics a eukaryotic E3 ligase despite sharing little sequence similarity to eukaryotic E3 ligases. This emphasizes the importance of structural approaches to understanding the functions of type III effector proteins, especially the complexes of type III effectors with their eukaryotic targets. The novel sequences employed by pathogens to usurp eukaryotic proteins promises novel tools to probe eukaryotic protein functions that would not be obtained from studies solely based on eukaryotic systems.

## 8.13 Future Perspectives

Genetic approaches have provided invaluable insight to understanding the key players involved in mediating plant defence responses. With the identification of hundreds of type III effectors from plant pathogenic bacteria by elegant genomic screens, one important future challenge will be their functional characterization (Lindeberg *et al.*, 2005). Biochemical and structural approaches will be required to tease out the functions of these enigmatic proteins. The example of HopAB2 can be used as a cautionary tale against generalizing about overall protein function without the functional dissection of individual protein domains. In addition, identification of host targets will be of paramount importance to unravelling the mechanisms of pathogen virulence and undoubtedly plant resistance. These interactions will undoubtedly differ in various host plants raising the important question: 'Does pathogen host range correlate with specific type III effector–host target interactions?' Addressing this question will be key to understanding how type III effectors contribute to host range and determine the outcome of plant–pathogen interactions.

## References

Abramovitch, R.B., Kim, Y.J., Chen, S.R., Dickman, M.B. and Martin, G.B. (2003) *Pseudomonas* type III effector AvrPtoB induces plant disease susceptibility by inhibition of host programmed cell death. *EMBO Journal* 22, 60–69.

Abramovitch, R.B., Anderson, J.C. and Martin, G.B. (2006a) Bacterial elicitation and evasion of plant innate immunity. *Nature Reviews Molecular Cell Biology* 7, 601–611.

Abramovitch, R.B., Janjusevic, R., Stebbins, C.E. and Martin, G.B. (2006b) Type III effector AvrPtoB requires intrinsic E3 ubiquitin ligase activity to suppress plant cell death and immunity. *Proceedings of the National Academy of Sciences, USA* 103, 2851–2856.

Ade, J., DeYoung, B.J., Golstein, C. and Innes, R.W. (2007) Indirect activation of a plant nucleotide binding site-leucine-rich repeat protein by a bacterial protease. *Proceedings of the National Academy of Sciences, USA* 104, 2531–2536.

Alfano, J.R., Charkowski, A.O., Deng, W.L., Badel, J.L., Petnicki-Ocwieja, T., van Dijk, K. and Collmer, A. (2000) The *Pseudomonas syringae* Hrp pathogenicity island has a tripartite mosaic structure composed of a cluster of type III secretion genes bounded by exchangeable effector and conserved effector loci that contribute to parasitic fitness and pathogenicity in plants. *Proceedings of the National Academy of Sciences, USA* 97, 4856–4861.

Anderson, J.C., Pascuzzi, P.E., Xiao, F.M., Sessa, G. and Martin, G.B. (2006) Host-mediated phosphorylation of type III effector AvrPto promotes *Pseudomonas* virulence and avirulence in tomato. *The Plant Cell* 18, 502–514.

Andriotis, V.M.E. and Rathjen, J.P. (2006) The Pto kinase of tomato, which regulates plant immunity, is repressed by its myristoylated N terminus. *Journal of Biological Chemistry* 281, 26578–26586.

Ashfield, T., Keen, N.T., Buzzell, R.I. and Innes, R.W. (1995) Soybean resistance genes specific for different *Pseudomonas syringae* avirulence genes are allelic, or closely linked, at the *Rpg1* locus. *Genetics* 141, 1597–1604.

Ashfield, T., Ong, L.E., Nobuta, K., Schneider, C.M. and Innes, R.W. (2004) Convergent evolution of disease resistance gene specificity in two flowering plant families. *The Plant Cell* 16, 309–318.

Astua-Monge, G., Minsavage, G.V., Stall, R.E., Vallejos, C.E., Davis, M.J. and Jones, J.B. (2000) *Xv4-vrxv4*: a new gene-for-gene interaction identified between *Xanthomonas campestris* pv. *vesicatoria* race T3 and the wild tomato relative *Lycopersicon pennellii. Molecular Plant–Microbe Interactions* 13, 1346–1355.

Axtell, M.J., Chisholm, S.T., Dahlbeck, D. and Staskawicz, B.J. (2003) Genetic and molecular evidence that the *Pseudomonas syringae* type III effector protein AvrRpt2 is a cysteine protease. *Molecular Microbiology* 49, 1537–1546.

Azevedo, C., Betsuyaku, S., Peart, J., Takahashi, A., Noel, L., Sadanandom, A., Casais, C., Parker, J. and Shirasu, K. (2006) Role of SGT1 in resistance protein accumulation in plant immunity. *EMBO Journal* 25, 2007–2016.

Ballvora, A., Pierre, M., van den Ackerveken, G., Schornack, S., Rossier, O., Ganal, M., Lahaye, T. and Bonas, U. (2001) Genetic mapping and functional analysis of the tomato *Bs4* locus governing recognition of the *Xanthomonas campestris* pv. *vesicatoria* AvrBs4 protein. *Molecular Plant–Microbe Interactions* 14, 629–638.

Balmuth, A. and Rathjen, J.P. (2007) Genetic and molecular requirements for function of the Pto/Prf effector recognition complex in tomato and *Nicotiana benthamiana*. *The Plant Journal* 51, 978–990.

Bartsch, M., Gobbato, E., Bednarek, P., Debey, S., Schultze, J.L., Bautor, J. and Parker, J.E. (2006) Salicylic acid-independent ENHANCED DISEASE SUSCEPTIBILITY1 signaling in *Arabidopsis* immunity and cell death is regulated by the monooxygenase *FMO1* and the Nudix hydrolase *NUDT7*. *The Plant Cell* 18, 1038–1051.

Belkhadir, Y., Nimchuk, Z., Hubert, D.A., Mackey, D. and Dangl, J.L. (2004) *Arabidopsis* RIN4 negatively regulates disease resistance mediated by RPS2 and RPM1 downstream or independent of the NDR1 signal modulator and is not required for the virulence functions of bacterial type III effectors AvrRpt2 or AvrRpm1. *The Plant Cell* 16, 2822–2835.

Bent, A.F., Kunkel, B.N., Dahlbeck, D., Brown, K.L., Schmidt, R., Giraudat, J., Leung, J. and Staskawicz, B.J. (1994) *RPS2* of *Arabidopsis thaliana*: a leucine-rich repeat class of plant disease resistance genes. *Science* 265, 1856–1860.

Bogdanove, A.J., Kim, J.F., Wei, Z.M., Kolchinsky, P., Charkowski, A.O., Conlin, A.K., Collmer, A. and Beer, S.V. (1998) Homology and functional similarity of an *hrp*-linked pathogenicity locus, *dspEF*, of *Erwinia amylovora* and the avirulence locus *avrE* of *Pseudomonas syringae* pathovar *tomato*. *Proceedings of the National Academy of Sciences, USA* 95, 1325–1330.

Bonas, U., Stall, R.E. and Staskawicz, B. (1989) Genetic and structural characterization of the avirulence gene *AvrBs3* from *Xanthomonas campestris* pv. *vesicatoria*. *Molecular and General Genetics* 218, 127–136.

Bonas, U., Conradsstrauch, J. and Balbo, I. (1993) Resistance in tomato to *Xanthomonas campestris* pv. *vesicatoria* is determined by alleles of the pepper-specific avirulence gene *AvrBs3*. *Molecular and General Genetics* 238, 261–269.

Bonshtien, A., Lev, A., Gibly, A., Debbie, P., Avni, A. and Sessa, G. (2005) Molecular properties of the *Xanthomonas* AvrRxv effector and global transcriptional changes determined by its expression in resistant tomato plants. *Molecular Plant–Microbe Interactions* 18, 300–310.

Boureau, T., El Maarouf-Bouteau, H., Garnier, A., Brisset, M.N., Perino, C., Pucheu, I. and Barny, M.A. (2006) DspA/E, a type III effector essential for *Erwinia amylovora*

pathogenicity and growth *in planta*, induces cell death in host apple and nonhost tobacco plants. *Molecular Plant–Microbe Interactions* 19, 16–24.
Chang, J.H., Rathjen, J.P., Bernal, A.J., Staskawicz, B.J. and Michelmore, R.W. (2000) *avrPto* enhances growth and necrosis caused by *Pseudomonas syringae* pv. *tomato* in tomato lines lacking either *Pto* or *Prf*. *Molecular Plant–Microbe Interactions* 13, 568–571.
Charity, J.C., Pak, K., Delwiche, C.F. and Hutcheson, S.W. (2003) Novel exchangeable effector loci associated with the *Pseudomonas syringae hrp* pathogenicity island: evidence for integron like assembly from transposed gene cassettes. *Molecular Plant–Microbe Interactions* 16, 495–507.
Chisholm, S.T., Dahlbeck, D., Krishnamurthy, N., Day, B., Sjolander, K. and Staskawicz, B.J. (2005) Molecular characterization of proteolytic cleavage sites of the *Pseudomonas syringae* effector AvrRpt2. *Proceedings of the National Academy of Sciences, USA* 102, 2087–2092.
Chisholm, S.T., Coaker, G., Day, B. and Staskawicz, B.J. (2006) Host–microbe interactions: shaping the evolution of the plant immune response. *Cell* 124, 803–814.
Ciesiolka, L.D., Hwin, T., Gearlds, J.D., Minsavage, G.V., Saenz, R., Bravo, M., Handley, V., Conover, S.M., Zhang, H., Caporgno, J., Phengrasamy, N.B., Toms, A.O., Stall, R.E. and Whalen, M.C. (1999) Regulation of expression of avirulence gene *avrRxv* and identification of a family of host interaction factors by sequence analysis of *avrBsT*. *Molecular Plant–Microbe Interactions* 12, 35–44.
Coaker, G., Falick, A. and Staskawicz, B. (2005) Activation of a phytopathogenic bacterial effector protein by a eukaryotic cyclophilin. *Science* 308, 548–550.
Coaker, G., Zhu, G., Ding, Z.F., Van Doren, S.R. and Staskawicz, B. (2006) Eukaryotic cyclophilin as a molecular switch for effector activation. *Molecular Microbiology* 61, 1485–1496.
Cohn, J.R. and Martin, G.B. (2005) *Pseudomonas syringae* pv. *tomato* type III effectors AvrPto and AvrPtoB promote ethylene-dependent cell death in tomato. *The Plant Journal* 44, 139–154.
Cunnac, S., Wilson, A., Nuwer, J., Kirik, A., Baranage, G. and Mudgett, M.B. (2007) A conserved carboxylesterase is a SUPPRESSOR OF AVRBST-ELICITED RESISTANCE in *Arabidopsis*. *The Plant Cell* 19, 688–705.
Da Cunha, L., McFall, A.J. and Mackey, D. (2006) Innate immunity in plants: a continuum of layered defenses. *Microbes and Infection* 8, 1372–1381.
Dafny-Yelin, M., Levy, A. and Tzfira, T. (2008) The ongoing saga of *Agrobacterium*–host interactions. *Trends in Plant Science* 13, 102–105.
Dangl, J.L. and Jones, J.D.G. (2001) Plant pathogens and integrated defence responses to infection. *Nature* 411, 826–833.
Day, B., Dahlbeck, D., Huang, J., Chisholm, S.T., Li, D.H. and Staskawicz, B.J. (2005) Molecular basis for the RIN4 negative regulation of RPS2 disease resistance. *The Plant Cell* 17, 1292–1305.
de Vries, J.S., Andriotis, V.M.E., Wu, A.J. and Rathjen, J.P. (2006) Tomato *Pto* encodes a functional N-myristoylation motif that is required for signal transduction in *Nicotiana benthamiana*. *The Plant Journal* 45, 31–45.
del Pozo, O., Pedley, K.F. and Martin, G.B. (2004) MAPKKK alpha is a positive regulator of cell death associated with both plant immunity and disease. *EMBO Journal* 23, 3072–3082.
Deng, W.L., Rehm, A.H., Charkowski, A.O., Rojas, C.M. and Collmer, A. (2003) *Pseudomonas syringae* exchangeable effector loci: sequence diversity in representative pathovars and virulence function in *P. syringae* pv. *syringae* B728a. *Journal of Bacteriology* 185, 2592–2602.
Deslandes, L., Olivier, J., Theulieres, F., Hirsch, J., Feng, D.X., Bittner-Eddy, P., Beynon, J.

and Marco, Y. (2002) Resistance to *Ralstonia solanacearum* in *Arabidopsis thaliana* is conferred by the recessive *RRS1-R* gene, a member of a novel family of resistance genes. *Proceedings of the National Academy of Sciences, USA* 99, 2404–2409.

Deslandes, L., Olivier, J., Peeters, N., Feng, D.X., Khounlotham, M., Boucher, C., Somssich, L., Genin, S. and Marco, Y. (2003) Physical interaction between RRS1-R, a protein conferring resistance to bacterial wilt, and PopP2, a type III effector targeted to the plant nucleus. *Proceedings of the National Academy of Sciences, USA* 100, 8024–8029.

Desveaux, D., Singer, A.U. and Dangl, J.L. (2006) Type III effector proteins: doppelgangers of bacterial virulence. *Current Opinion in Plant Biology* 9, 376–382.

Desveaux, D., Singer, A.U., Wu, A.J., McNulty, B.C., Musselwhite, L., Nimchuk, Z., Sondek, J. and Dangl, J.L. (2007) Type III effector activation via nucleotide binding, phosphorylation, and host target interaction. *PLoS Pathogens* 3, 456–469.

Devarenne, T.P., Ekengren, S.K., Pedley, K.F. and Martin, G.B. (2006) Adi3 is a Pdk1-interacting AGC kinase that negatively regulates plant cell death. *EMBO Journal* 25, 255–265.

Dong, X.N., Mindrinos, M., Davis, K.R. and Ausubel, F.M. (1991) Induction of *Arabidopsis* defense genes by virulent and avirulent *Pseudomonas syringae* strains and by a cloned avirulence gene. *The Plant Cell* 3, 61–72.

Escolar, L., van den Ackerveken, G., Pieplow, S., Rossier, O. and Bonas, U. (2002) Type III secretion and *in planta* recognition of the *Xanthomonas* avirulence proteins AvrBs1 and AvrBsT. *Molecular Plant Pathology* 2, 287–296.

Espinosa, A. and Alfano, J.R. (2004) Disabling surveillance: bacterial type III secretion system effectors that suppress innate immunity. *Cellular Microbiology* 6, 1027–1040.

Fouts, D.E., Abramovitch, R.B., Alfano, J.R., Baldo, A.M., Buell, C.R., Cartinhour, S., Chatterjee, A.K., D'Ascenzo, M., Gwinn, M.L., Lazarowitz, S.G., Lin, N.C., Martin, G.B., Rehm, A.H., Schneider, D.J., van Dijk, K., Tang, X.Y. and Collmer, A. (2002) Genome-wide identification of *Pseudomonas syringae* pv. *tomato* DC3000 promoters controlled by the HrpL alternative sigma factor. *Proceedings of the National Academy of Sciences, USA* 99, 2275–2280.

Frederick, R.D., Thilmony, R.L., Sessa, G. and Martin, G.B. (1998) Recognition specificity for the bacterial avirulence protein AvrPto is determined by Thr-204 in the activation loop of the tomato Pto kinase. *Molecular Cell* 2, 241–245.

Gassmann, W., Hinsch, M.E. and Staskawicz, B.J. (1999) The *Arabidopsis RPS4* bacterial resistance gene is a member of the TIR-NBS-LRR family of disease resistance genes. *The Plant Journal* 20, 265–277.

Gaudriault, S., Malandrin, L., Paulin, J.P. and Barny, M.A. (1997) DspA, an essential pathogenicity factor of *Erwinia amylovora* showing homology with AvrE of *Pseudomonas syringae*, is secreted via the Hrp secretion pathway in a DspB-dependent way. *Molecular Microbiology* 26, 1057–1069.

Grant, M.R., Godiard, L., Straube, E., Ashfield, T., Lewald, J., Sattler, A., Innes, R.W. and Dangl, J.L. (1995) Structure of the *Arabidopsis RPM1* gene enabling dual-specificity disease resistance. *Science* 269, 843–846.

Grant, S.R., Fisher, E.J., Chang, J.H., Mole, B.M. and Dangl, J.L. (2006) Subterfuge and manipulation: type III effector proteins of phytopathogenic bacteria. *Annual Review of Microbiology* 60, 425–449.

Gu, K., Tian, D., Yang, F., Wu, L., Sreekala, C., Wang, D., Wang, G.L. and Yin, Z. (2004) High-resolution genetic mapping of *Xa27(t)*, a new bacterial blight resistance gene in rice, *Oryza sativa* L. *Theoretical and Applied Genetics* 108, 800–807.

Gu, K.Y., Yang, B., Tian, D.S., Wu, L.F., Wang, D.J., Sreekala, C., Yang, F., Chu, Z.Q., Wang, G.L., White, F.F. and Yin, Z.C. (2005) *R* gene expression induced by a type-III effector triggers disease resistance in rice. *Nature* 435, 1122–1125.

Gu, Y.Q., Wildermuth, M.C., Chakravarthy, S., Loh, Y.T., Yang, C.M., He, X.H., Han, Y. and Martin, G.B. (2002) Tomato transcription factors Pti4, Pti5, and Pti6 activate defense responses when expressed in *Arabidopsis*. *The Plant Cell* 14, 817–831.

Guttman, D.S., Vinatzer, B.A., Sarkar, S.F., Ranall, M.V., Kettler, G. and Greenberg, J.T. (2002) A functional screen for the type III (Hrp) secretome of the plant pathogen *Pseudomonas syringae*. *Science* 295, 1722–1726.

Hauck, P., Thilmony, R. and He, S.Y. (2003) A *Pseudomonas syringae* type III effector suppresses cell wall-based extracellular defense in susceptible *Arabidopsis* plants. *Proceedings of the National Academy of Sciences, USA* 100, 8577–8582.

He, P., Warren, R.F., Zhao, T.H., Shan, L.B., Zhu, L.H., Tang, X.Y. and Zhou, J.M. (2001) Overexpression of *Pti5* in tomato potentiates pathogen-induced defense gene expression and enhances disease resistance to *Pseudomonas syringae* pv. *tomato*. *Molecular Plant–Microbe Interactions* 14, 1453–1457.

He, P., Shan, L., Lin, N.C., Martin, G.B., Kemmerling, B., Nurnberger, T. and Sheen, J. (2006) Specific bacterial suppressors of MAMP signaling upstream of MAPKKK in *Arabidopsis* innate immunity. *Cell* 125, 563–575.

Hinsch, M. and Staskawicz, B. (1996) Identification of a new *Arabidopsis* disease resistance locus, *RPS4*, and cloning of the corresponding avirulence gene, *avrRps4*, from *Pseudomonas syringae* pv. *pisi*. *Molecular Plant–Microbe Interactions* 9, 55–61.

Hwang, M.S.H., Morgan, R.L., Sarkar, S.F., Wang, P.W. and Guttman, D.S. (2005) Phylogenetic characterization of virulence and resistance phenotypes of *Pseudomonas syringae*. *Applied and Environmental Microbiology* 71, 5182–5191.

Jackson, R.W., Athanassopoulos, E., Tsiamis, G., Mansfield, J.W., Sesma, A., Arnold, D.L., Gibbon, M.J., Murillo, J., Taylor, J.D. and Vivian, A. (1999) Identification of a pathogenicity island, which contains genes for virulence and avirulence, on a large native plasmid in the bean pathogen *Pseudomonas syringae* pathovar *phaseolicola*. *Proceedings of the National Academy of Sciences, USA* 96, 10875–10880.

Jackson, R.W., Mansfield, J.W., Ammouneh, H., Dutton, L.C., Wharton, B., Ortiz-Barredo, A., Arnold, D.L., Tsiamis, G., Sesma, A., Butcher, D., Boch, J., Kim, Y.J., Martin, G.B., Tegli, S., Murillo, J. and Vivian, A. (2002) Location and activity of members of a family of *virPphA* homologues in pathovars of *Pseudomonas syringae* and *P. savastanoi*. *Molecular Plant Pathology* 3, 205–216.

Janjusevic, R., Abramovitch, R.B., Martin, G.B. and Stebbins, C.E. (2006) A bacterial inhibitor of host programmed cell death defenses is an E3 ubiquitin ligase. *Science* 311, 222–226.

Jenner, C., Hitchin, E., Mansfield, J., Walters, K., Betteridge, P., Teverson, D. and Taylor, J. (1991) Gene-for-gene interactions between *Pseudomonas syringae* pv. *phaseolicola* and *Phaseolus*. *Molecular Plant–Microbe Interactions* 4, 553–562.

Jia, Y.L. and Martin, G.B. (1999) Rapid transcript accumulation of pathogenesis-related genes during an incompatible interaction in bacterial speck disease-resistant tomato plants. *Plant Molecular Biology* 40, 455–465.

Jin, P., Wood, M.D., Wu, Y., Xie, Z.Y. and Katagiri, F. (2003) Cleavage of the *Pseudomonas syringae* type III effector AvrRpt2 requires a host factor(s) common among eukaryotes and is important for AvrRpt2 localization in the host cell. *Plant Physiology* 133, 1072–1082.

Jones, J.D.G. and Dangl, J.L. (2006) The plant immune system. *Nature* 444, 323–329.

Kang, L., Tang, X.Y. and Mysore, K.S. (2004) *Pseudomonas* type III effector AvrPto suppresses the programmed cell death induced by two nonhost pathogens in *Nicotiana benthamiana* and tomato. *Molecular Plant–Microbe Interactions* 17, 1328–1336.

Kay, S., Hahn, S., Marois, E., Hause, G. and Bonas, U. (2007) A bacterial effector acts as a plant transcription factor and induces a cell size regulator. *Science* 318, 648–651.

Kim, H.S., Desveaux, D., Singer, A.U., Patel, P., Sondek, J. and Dangl, J.L. (2005) The

*Pseudomonas syringae* effector AvrRpt2 cleaves its C-terminally acylated target, RIN4, from *Arabidopsis* membranes to block RPM1 activation. *Proceedings of the National Academy of Sciences, USA* 102, 6496–6501.

Kim, M.G., da Cunha, L., McFall, A.J., Belkhadir, Y., DebRoy, S., Dangl, J.L. and Mackey, D. (2005) Two *Pseudomonas syringae* type III effectors inhibit RIN-regulated basal defense in *Arabidopsis. Cell* 121, 749–759.

Kim, Y.J., Lin, N.C. and Martin, G.B. (2002) Two distinct *Pseudomonas* effector proteins interact with the Pto kinase and activate plant immunity. *Cell* 109, 589–598.

Koch, M., Vorwerk, S., Masur, C., Sharifi-Sirchi, G., Olivieri, N. and Schlaich, N.L. (2006) A role for a flavin-containing mono-oxygenase in resistance against microbial pathogens in *Arabidopsis. The Plant Journal* 47, 629–639.

Lahaye, T. and Bonas, U. (2001) Molecular secrets of bacterial type III effector proteins. *Trends in Plant Science* 6, 479–485.

Lavie, M., Shillington, E., Eguiluz, C., Grimsley, N. and Boucher, C. (2002) PopP1, a new member of the YopJ/AvrRxv family of type III effector proteins, acts as a host-specificity factor and modulates aggressiveness of *Ralstonia solanacearum. Molecular Plant–Microbe Interactions* 15, 1058–1068.

Lee, C.C., Wood, M.D., Ng, K., Andersen, C.B., Liu, Y., Luginbuhl, P., Spraggon, G. and Katagiri, F. (2004) Crystal structure of the type III effector AvrB from *Pseudomonas syringae. Structure* 12, 487–494.

Lewis, J.D., Abada, W., Ma, W.B., Guttman, D.S. and Desveaux, D. (2008) The HopZ family of *Pseudomonas syringae* type III effectors require myristoylation for virulence and avirulence functions in *Arabidopsis thaliana. Journal of Bacteriology* 190, 2880–2891.

Lim, M.T.S. and Kunkel, B.N. (2004) The *Pseudomonas syringae* type III effector AvrRpt2 promotes virulence independently of RIN4, a predicted virulence target in *Arabidopsis thaliana. The Plant Journal* 40, 790–798.

Lin, N.C. and Martin, G.B. (2005) An *avrPto/avrPtoB* mutant of *Pseudomonas syringae* pv. *tomato* DC3000 does not elicit Pto-mediated resistance and is less virulent on tomato. *Molecular Plant–Microbe Interactions* 18, 43–51.

Lin, N.C. and Martin, G.B. (2007) Pto- and Prf-mediated recognition of AvrPto and AvrPtoB restricts the ability of diverse *Pseudomonas syringae* pathovars to infect tomato. *Molecular Plant–Microbe Interactions* 20, 806–815.

Lin, N.C., Abramovitch, R.B., Mm, Y.J. and Martin, G.B. (2006) Diverse AvrPtoB homologs from several *Pseudomonas syringae* pathovars elicit *Pto*-dependent resistance and have similar virulence activities. *Applied and Environmental Microbiology* 72, 702–712.

Lindeberg, M., Stavrinides, J., Chang, J.H., Alfano, J.R., Collmer, A., Dangl, J.L., Greenberg, J.T., Mansfield, J.W. and Guttman, D.S. (2005) Proposed guidelines for a unified nomenclature and phylogenetic analysis of type III Hop effector proteins in the plant pathogen *Pseudomonas syringae. Molecular Plant–Microbe Interactions* 18, 275–282.

Loh, Y.T., Zhou, J.M. and Martin, G.B. (1998) The myristylation motif of Pto is not required for disease resistance. *Molecular Plant–Microbe Interactions* 11, 572–576.

Lund, S.T., Stall, R.E. and Klee, H.J. (1998) Ethylene regulates the susceptible response to pathogen infection in tomato. *The Plant Cell* 10, 371–382.

Ma, W.B., Dong, F.F.T., Stavrinides, J. and Guttman, D.S. (2006) Type III effector diversification via both pathoadaptation and horizontal transfer in response to a coevolutionary arms race. *PloS Genetics* 2, 2131–2142.

Mackey, D., Holt, B.F., Wiig, A. and Dangl, J.L. (2002) RIN4 interacts with *Pseudomonas syringae* type III effector molecules and is required for RPM1-mediated resistance in *Arabidopsis. Cell* 108, 743–754.

Mackey, D., Belkhadir, Y., Alonso, J.M., Ecker, J.R. and Dangl, J.L. (2003) *Arabidopsis* RIN4 is a target of the type III virulence effector AvrRpt2 and modulates RPS2-mediated resistance. *Cell* 112, 379–389.

Makarova, K.S., Aravind, L. and Koonin, E.V. (1999) A superfamily of archaeal, bacterial, and eukaryotic proteins homologous to animal transglutaminases. *Protein Science* 8, 1714–1719.

Mansfield, J., Jenner, C., Hockenhull, R., Bennett, M.A. and Stewart, R. (1994) Characterization of *avrPphE*, a gene for cultivar-specific avirulence from *Pseudomonas syringae* pv. *phaseolicola* which is physically linked to *hrpY*, a new *hrp* gene identified in the halo-blight bacterium. *Molecular Plant–Microbe Interactions* 7, 726–739.

Marois, E., Van den Ackerveken, G. and Bonas, U. (2002) The *Xanthomonas* type III effector protein AvrBs3 modulates plant gene expression and induces cell hypertrophy in the susceptible host. *Molecular Plant–Microbe Interactions* 15, 637–646.

Martin, G.B., Brommonschenkel, S.H., Chunwongse, J., Frary, A., Ganal, M.W., Spivey, R., Wu, T.Y., Earle, E.D. and Tanksley, S.D. (1993) Map-based cloning of a protein kinase gene conferring disease resistance in tomato. *Science* 262, 1432–1436.

Martin, G.B., Bogdanove, A.J. and Sessa, G. (2003) Understanding the functions of plant disease resistance proteins. *Annual Review of Plant Biology* 54, 23–61.

McCann, H.C. and Guttman, D.S. (2008) Evolution of the type III secretion system and its effectors in plant–microbe interactions. *New Phytologist* 177, 33–47.

McHale, L., Tan, X.P., Koehl, P. and Michelmore, R.W. (2006) Plant NBS-LRR proteins: adaptable guards. *Genome Biology* 7, 212.

Meng, X.D., Bonasera, J.M., Kim, J.F., Nissinen, R.M. and Beer, S.V. (2006) Apple proteins that interact with DspA/E, a pathogenicity effector of *Erwinia amylovora*, the fire blight pathogen. *Molecular Plant–Microbe Interactions* 19, 53–61.

Mindrinos, M., Katagiri, F., Yu, G.L. and Ausubel, F.M. (1994) The *A. thaliana* disease resistance gene *RPS2* encodes a protein containing a nucleotide-binding site and leucine-rich repeats. *Cell* 78, 1089–1099.

Minsavage, G.V., Dahlbeck, D., Whalen, M.C., Kearney, B., Bonas, U., Staskawicz, B.J. and Stall, R.E. (1990) Gene-for-gene relationships specifying disease resistance in *Xanthomonas campestris* pv. *vesicatoria* – pepper interactions. *Molecular Plant–Microbe Interactions* 3, 41–47.

Mishina, T.E. and Zeier, J. (2006) The *Arabidopsis* flavin-dependent monooxygenase FMO1 is an essential component of biologically induced systemic acquired resistance. *Plant Physiology* 141, 1666–1675.

Mittal, R., Peak-Chew, S.Y. and McMahon, H.T. (2006) Acetylation of MEK2 and I kappa B kinase (IKK) activation loop residues by YopJ inhibits signaling. *Proceedings of the National Academy of Sciences, USA* 103, 18574–18579.

Mucyn, T.S., Clemente, A., Andriotis, V.M.E., Balmuth, A.L., Oldroyd, G.E.D., Staskawicz, B.J. and Rathjen, J.P. (2006) The tomato NBARC-LRR protein Prf interacts with Pto kinase *in vivo* to regulate specific plant immunity. *The Plant Cell* 18, 2792–2806.

Mudgett, M.B. and Staskawicz, B.J. (1999) Characterization of the *Pseudomonas syringae* pv. *tomato* AvrRpt2 protein: demonstration of secretion and processing during bacterial pathogenesis. *Molecular Microbiology* 32, 927–941.

Mukherjee, S., Keitany, G., Li, Y., Wang, Y., Ball, H.L., Goldsmith, E.J. and Orth, K. (2006) *Yersinia* YopJ acetylates and inhibits kinase activation by blocking phosphorylation. *Science* 312, 1211–1214.

Mukherjee, S., Hao, Y.H. and Orth, K. (2007) A newly discovered post-translational modification: the acetylation of serine and threonine residues. *Trends in Biochemical Sciences* 32, 210–216.

Nimchuk, Z., Marois, E., Kjemtrup, S., Leister, R.T., Katagiri, F. and Dangl, J.L. (2000) Eukaryotic fatty acylation drives plasma membrane targeting and enhances function of several type III effector proteins from *Pseudomonas syringae*. *Cell* 101, 353–363.

Nimchuk, Z.L., Fisher, E.J., Desveaux, D., Chang, J.H. and Dangl, J.L. (2007) The HopX

(AvrPphE) family of *Pseudomonas syringae* type III effectors require a catalytic triad and a novel N-terminal domain for function. *Molecular Plant–Microbe Interactions* 20, 346–357.

Oguiza, J.A. and Asensio, A.C. (2005) The VirPphA/AvrPtoB family of type III effectors in *Pseudomonas syringae*. *Research in Microbiology* 156, 298–303.

Oh, C.S. and Beer, S.V. (2005) Molecular genetics of *Erwinia amylovora* involved in the development of fire blight. *FEMS Microbiology Letters* 253, 185–192.

Oh, C.S., Martin, G.B. and Beer, S.V. (2007) DspA/E, a type III effector of *Erwinia amylovora*, is required for early rapid growth in *Nicotiana benthamiana* and causes NbSGT1-dependent cell death. *Molecular Plant Pathology* 8, 255–265.

Oldroyd, G.E.D. and Staskawicz, B.J. (1998) Genetically engineered broad-spectrum disease resistance in tomato. *Proceedings of the National Academy of Sciences, USA* 95, 10300–10305.

Ong, L.E. and Innes, R.W. (2006) *AvrB* mutants lose both virulence and avirulence activities on soybean and *Arabidopsis*. *Molecular Microbiology* 60, 951–962.

Orth, K., Xu, Z.H., Mudgett, M.B., Bao, Z.Q., Palmer, L.E., Bliska, J.B., Mangel, W.F., Staskawicz, B. and Dixon, J.E. (2000) Disruption of signaling by *Yersinia* effector YopJ, a ubiquitin-like protein protease. *Science* 290, 1594–1597.

Pedley, K.F. and Martin, G.B. (2003) Molecular basis of Pto-mediated resistance to bacterial speck disease in tomato. *Annual Review of Phytopathology* 41, 215–243.

Pedley, K.F. and Martin, G.B. (2004) Identification of MAPKs and their possible MAPK kinase activators involved in the Pto-mediated defense response of tomato. *Journal of Biological Chemistry* 279, 49229–49235.

Pierre, M., Noel, L., Lahaye, T., Ballvora, A., Veuskens, J., Ganal, M. and Bonas, U. (2000) High-resolution genetic mapping of the pepper resistance locus *Bs3* governing recognition of the *Xanthomonas campestris* pv. *vesicatora* AvrBs3 protein. *Theoretical and Applied Genetics* 101, 255–263.

Puri, N., Jenner, C., Bennett, M., Stewart, R., Mansfield, J., Lyons, N. and Taylor, J. (1997) Expression of *avrPphB*, an avirulence gene from *Pseudomonas syringae* pv. *phaseolicola*, and the delivery of signals causing the hypersensitive reaction in bean. *Molecular Plant–Microbe Interactions* 10, 247–256.

Rathjen, J.P., Chang, J.H., Staskawicz, B.J. and Michelmore, R.W. (1999) Constitutively active Pto induces a Prf-dependent hypersensitive response in the absence of AvrPto. *EMBO Journal* 18, 3232–3240.

Riely, B.K. and Martin, G.B. (2001) Ancient origin of pathogen recognition specificity conferred by the tomato disease resistance gene *Pto*. *Proceedings of the National Academy of Sciences, USA* 98, 2059–2064.

Ritter, C. and Dangl, J.L. (1995) The *AvrRpm1* gene of *Pseudomonas syringae* pv. *maculicola* is required for virulence on *Arabidopsis*. *Molecular Plant–Microbe Interactions* 8, 444–453.

Robert-Seilaniantz, A., Shan, L.B., Zhou, J.M. and Tang, X.Y. (2006) The *Pseudomonas syringae* pv. *tomato* DC3000 type III effector HopF2 has a putative myristoylation site required for its avirulence and virulence functions. *Molecular Plant–Microbe Interactions* 19, 130–138.

Roden, J., Eardley, L., Hotson, A., Cao, Y.Y. and Mudgett, M.B. (2004) Characterization of the *Xanthomonas* AvrXv4 effector, a SUMO protease translocated into plant cells. *Molecular Plant–Microbe Interactions* 17, 633–643.

Rohmer, L., Kjemtrup, S., Marchesini, P. and Dangl, J.L. (2003) Nucleotide sequence, functional characterization and evolution of pFKN, a virulence plasmid in *Pseudomonas syringae* pathovar *maculicola*. *Molecular Microbiology* 47, 1545–1562.

Romer, P., Hahn, S., Jordan, T., Strauss, T., Bonas, U. and Lahaye, T. (2007) Plant pathogen recognition mediated by promoter activation of the pepper *Bs3* resistance gene. *Science* 318, 645–648.

Rommens, C.M.T., Salmeron, J.M., Oldroyd, G.E.D. and Staskawicz, B.J. (1995) Intergeneric transfer and functional expression of the tomato disease resistance gene *Pto*. *The Plant Cell* 7, 1537–1544.

Ronald, P.C., Salmeron, J.M., Carland, F.M. and Staskawicz, B.J. (1992) The cloned avirulence gene *AvrPto* induces disease resistance in tomato cultivars containing the *Pto* resistance gene. *Journal of Bacteriology* 174, 1604–1611.

Rose, L.E., Langley, C.H., Bernal, A.J. and Michelmore, R.W. (2005) Natural variation in the *Pto* pathogen resistance gene within species of wild tomato (*Lycopersicon*). I. Functional analysis of *Pto* alleles. *Genetics* 171, 345–357.

Rosebrock, T.R., Zeng, L.R., Brady, J.J., Abramovitch, R.B., Xiao, F.M. and Martin, G.B. (2007) A bacterial E3 ubiquitin ligase targets a host protein kinase to disrupt plant immunity. *Nature* 448, 370–374.

Salmeron, J.M., Barker, S.J., Carland, F.M., Mehta, A.Y. and Staskawicz, B.J. (1994) Tomato mutants altered in bacterial disease resistance provide evidence for a new locus controlling pathogen recognition. *The Plant Cell* 6, 511–520.

Salmeron, J.M., Oldroyd, G.E.D., Rommens, C.M.T., Scofield, S.R., Kim, H.S., Lavelle, D.T., Dahlbeck, D. and Staskawicz, B.J. (1996) Tomato *Prf* is a member of the leucine-rich repeat class of plant disease resistance genes and lies embedded within the *Pto* kinase gene cluster. *Cell* 86, 123–133.

Sarkar, S.F., Gordon, J.S., Martin, G.B. and Guttman, D.S. (2006) Comparative genomics of host-specific virulence in *Pseudomonas syringae*. *Genetics* 174, 1041–1056.

Schornack, S., Ballvora, A., Gurlebeck, D., Peart, J., Baulcombe, D., Ganal, M., Baker, B., Bonas, U. and Lahaye, T. (2004a) The tomato resistance protein Bs4 is a predicted non-nuclear TIR-NB-LRR protein that mediates defense responses to severely truncated derivatives of AvrBs4 and overexpressed AvrBs3. *The Plant Journal* 37, 787–787.

Schornack, S., Ballvora, A., Gurlebeck, D., Peart, J., Ganal, M., Baker, B., Bonas, U. and Lahaye, T. (2004b) The tomato resistance protein Bs4 is a predicted non-nuclear TIR-NB-LRR protein that mediates defense responses to severely truncated derivatives of AvrBs4 and overexpressed AvrBs3. *The Plant Journal* 37, 46–60.

Schornack, S., Meyer, A., Romer, P., Jordan, T. and Lahaye, T. (2006) Gene-for-gene-mediated recognition of nuclear-targeted AvrBs3-like bacterial effector proteins. *Journal of Plant Physiology* 163, 256–272.

Scofield, S.R., Tobias, C.M., Rathjen, J.P., Chang, J.H., Lavelle, D.T., Michelmore, R.W. and Staskawicz, B.J. (1996) Molecular basis of gene-for-gene specificity in bacterial speck disease of tomato. *Science* 274, 2063–2065.

Sessa, G., D'Ascenzo, M. and Martin, G.B. (2000) Thr38 and Ser198 are Pto autophosphorylation sites required for the AvrPto-Pto-mediated hypersensitive response. *EMBO Journal* 19, 2257–2269.

Shan, L.B., Thara, V.K., Martin, G.B., Zhou, J.M. and Tang, X.Y. (2000) The *Pseudomonas* AvrPto protein is differentially recognized by tomato and tobacco and is localized to the plant plasma membrane. *The Plant Cell* 12, 2323–2337.

Shao, F., Merritt, P.M., Bao, Z., Innes, R.W. and Dixon, J.E. (2002) A *Yersinia* effector and a *Pseudomonas* avirulence protein define a family of cysteine proteases functioning in bacterial pathogenesis. *Cell* 109, 575–588.

Shao, F., Golstein, C., Ade, J., Stoutemyer, M., Dixon, J.E. and Innes, R.W. (2003) Cleavage of *Arabidopsis* PBS1 by a bacterial type III effector. *Science* 301, 1230–1233.

Simonich, M.T. and Innes, R.W. (1995) A disease resistance gene in *Arabidopsis* with specificity for the *avrPph3* gene of *Pseudomonas syringae* pv. *phaseolicola*. *Molecular Plant–Microbe Interactions* 8, 637–640.

Singer, A.U., Desveaux, D., Betts, L., Chang, J.H., Nimchuk, Z., Grant, S.R., Dangl, J.L. and Sondek, J. (2004) Crystal structures of the type III effector protein AvrPphF and its chaperone reveal residues required for plant pathogenesis. *Structure* 12, 1669–1681.

Stebbins, C.E. and Galan, J.E. (2001) Structural mimicry in bacterial virulence. *Nature* 412, 701–705.

Stevens, C., Bennett, M.A., Athanassopoulos, E., Tsiamis, G., Taylor, J.D. and Mansfield, J.W. (1998) Sequence variations in alleles of the avirulence gene *avrPphE.R2* from *Pseudomonas syringae* pv. *phaseolicola* lead to loss of recognition of the AvrPphE protein within bean cells and a gain in cultivar-specific virulence. *Molecular Microbiology* 29, 165–177.

Swiderski, M.R. and Innes, R.W. (2001) The *Arabidopsis PBS1* resistance gene encodes a member of a novel protein kinase subfamily. *The Plant Journal* 26, 101–112.

Szurek, B., Marois, E., Bones, U. and Van den Ackerveken, G. (2001) Eukaryotic features of the *Xanthomonas* type III effector AvrBs3: protein domains involved in transcriptional activation and the interaction with nuclear import receptors from pepper. *The Plant Journal* 26, 523–534.

Szurek, B., Rossier, O., Hause, G. and Bonas, U. (2002) Type III-dependent translocation of the *Xanthomonas* AvrBs3 protein into the plant cell. *Molecular Microbiology* 46, 13–23.

Takemoto, D. and Jones, D.A. (2005) Membrane release and destabilization of *Arabidopsis* RIN4 following cleavage by *Pseudomonas syringae* AvrRpt2. *Molecular Plant–Microbe Interactions* 18, 1258–1268.

Tamaki, S., Dahlbeck, D., Staskawicz, B. and Keen, N.T. (1988) Characterization and expression of two avirulence genes cloned from *Pseudomonas syringae* pv. *glycinea*. *Journal of Bacteriology* 170, 4846–4854.

Tampakaki, A.P., Bastaki, M., Mansfield, J.W. and Panopoulos, N.J. (2002) Molecular determinants required for the avirulence function of AvrPphB in bean and other plants. *Molecular Plant–Microbe Interactions* 15, 292–300.

Tang, X.Y., Frederick, R.D., Zhou, J.M., Halterman, D.A., Jia, Y.L. and Martin, G.B. (1996) Initiation of plant disease resistance by physical interaction of AvrPto and Pto kinase. *Science* 274, 2060–2063.

Taylor, J.D., Teverson, D.M. and Davis, J.H.C. (1996) Sources of resistance to *Pseudomonas syringae* pv. *phaseolicola* races in *Phaseolus vulgaris*. *Plant Pathology* 45, 479–485.

Thara, V.K., Tang, X.Y., Gu, Y.Q., Martin, G.B. and Zhou, J.M. (1999) *Pseudomonas syringae* pv. *tomato* induces the expression of tomato *EREBP*-like genes *Pti4* and *Pti5* independent of ethylene, salicylate and jasmonate. *The Plant Journal* 20, 475–483.

Thieme, F., Szczesny, R., Urban, A., Kirchner, O., Hause, G. and Bonas, U. (2007) New type III effectors from *Xanthomonas campestris* pv. *vesicatoria* trigger plant reactions dependent on a conserved N-myristoylation motif. *Molecular Plant–Microbe Interactions* 20, 1250–1261.

Thilmony, R.L., Chen, Z.T., Bressan, R.A. and Martin, G.B. (1995) Expression of the tomato *Pto* gene in tobacco enhances resistance to *Pseudomonas syringae* pv. *tabaci* expressing *AvrPto*. *The Plant Cell* 7, 1529–1536.

Toth, I.K. and Birch, P.R.J. (2005) Rotting softly and stealthily. *Current Opinion in Plant Biology* 8, 424–429.

Tsiamis, G., Mansfield, J.W., Hockenhull, R., Jackson, R.W., Sesma, A., Athanassopoulos, E., Bennett, M.A., Stevens, C., Vivian, A., Taylor, J.D. and Murillo, J. (2000) Cultivar-specific avirulence and virulence functions assigned to *avrPphF* in *Pseudomonas syringae* pv. *phaseolicola*, the cause of bean halo-blight disease. *EMBO Journal* 19, 3204–3214.

van den Ackerveken, G., Marois, E. and Bonas, U. (1996) Recognition of the bacterial avirulence protein AvrBs3 occurs inside the host plant cell. *Cell* 87, 1307–1316.

van der Biezen, E.A. and Jones, J.D.G. (1998a) The NB-ARC domain: a novel signalling motif shared by plant resistance gene products and regulators of cell death in animals. *Current Biology* 8, R226–R227.

van der Biezen, E.A. and Jones, J.D.G. (1998b) Plant disease resistance proteins and the gene-for-gene concept. *Trends in Biochemical Sciences* 23, 454–456.

Vinatzer, B.A., Teitzel, G.M., Lee, M.W., Jelenska, J., Hotton, S., Fairfax, K., Jenrette, J. and Greenberg, J.T. (2006) The type III effector repertoire of *Pseudomonas syringae* pv. *syringae* B728a and its role in survival and disease on host and non-host plants. *Molecular Microbiology* 62, 26–44.

Vleeshouwers, V., Martens, A., van Dooijeweert, W., Colon, L.T., Govers, F. and Kamoun, S. (2001) Ancient diversification of the *Pto* kinase family preceded speciation in *Solanum*. *Molecular Plant–Microbe Interactions* 14, 996–1005.

Wang, J.F., Stall, R.E. and Vallejos, C.E. (1994) Genetic analysis of a complex hypersensitive reaction to bacterial spot in tomato. *Phytopathology* 84, 126–132.

Whalen, M.C., Stall, R.E. and Staskawicz, B.J. (1988) Characterization of a gene from a tomato pathogen determining hypersensitive resistance in non-host species and genetic analysis of this resistance in bean. *Proceedings of the National Academy of Sciences, USA* 85, 6743–6747.

Whalen, M.C., Innes, R.W., Bent, A.F. and Staskawicz, B.J. (1991) Identification of *Pseudomonas syringae* pathogens of *Arabidopsis* and a bacterial locus determining avirulence on both *Arabidopsis* and soybean. *The Plant Cell* 3, 49–59.

Whalen, M.C., Wang, J.F., Carland, F.M., Heiskell, M.E., Dahlbeck, D., Minsavage, G.V., Jones, J.B., Scott, J.W., Stall, R.E. and Staskawicz, B.J. (1993) Avirulence gene *avrRxv* from *Xanthomonas campestris* pv. *vesicatoria* specifies resistance on tomato line Hawaii 7998. *Molecular Plant–Microbe Interactions* 6, 616–627.

Wirthmueller, L., Zhang, Y., Jones, J.D.G. and Parker, J.E. (2007) Nuclear accumulation of the *Arabidopsis* immune receptor RPS4 is necessary for triggering EDS1-dependent defense. *Current Biology* 17, 2023–2029.

Wu, A.J., Andriotis, V.M.E., Durrant, M.C. and Rathjen, J.P. (2004) A patch of surface-exposed residues mediates negative regulation of immune signaling by tomato Pto kinase. *The Plant Cell* 16, 2809–2821.

Xiang, T.T., Zong, N., Zou, Y., Wu, Y., Zhang, J., Xing, W.M., Li, Y., Tang, X.Y., Zhu, L.H., Chai, J.J. and Zhou, J.M. (2008) *Pseudomonas syringae* effector AvrPto blocks innate immunity by targeting receptor kinases. *Current Biology* 18, 74–80.

Xiao, F.M., He, P., Abramovitch, R.B., Dawson, J.E., Nicholson, L.K., Sheen, J. and Martin, G.B. (2007) The N-terminal region of *Pseudomonas* type III effector AvrPtoB elicits Pto-dependent immunity and has two distinct virulence determinants. *The Plant Journal* 52, 595–614.

Xing, W., Zou, Y., Liu, Q., Liu, J.N., Luo, X., Huang, Q.Q., Chen, S., Zhu, L.H., Bi, R.C., Hao, Q., Wu, J.W., Zhou, J.M. and Chai, J.J. (2007) The structural basis for activation of plant immunity by bacterial effector protein AvrPto. *Nature* 449, 243–247.

Yu, Z.H., Wang, J.F., Stall, R.E. and Vallejos, C.E. (1995) Genomic localization of tomato genes that control a hypersensitive reaction to *Xanthomonas campestris* pv. *vesicatoria* (Doidge) Dye. *Genetics* 141, 675–682.

Zhang, X.C. and Gassmann, W. (2003) *RPS4*-mediated disease resistance requires the combined presence of *RPS4* transcripts with full-length and truncated open reading frames. *The Plant Cell* 15, 2333–2342.

Zhang, X.C. and Gassmann, W. (2007) Alternative splicing and mRNA levels of the disease resistance gene *RPS4* are induced during defense responses. *Plant Physiology* 145, 1577–1587.

Zhang, Y., Dorey, S., Swiderski, M. and Jones, J.D.G. (2004) Expression of *RPS4* in tobacco induces an AvrRps4-independent HR that requires EDS1, SGT1 and HSP90. *The Plant Journal* 40, 213–224.

Zhao, Y.F., He, S.Y. and Sundin, G.W. (2006) The *Erwinia amylovora avrRpt2*$_{EA}$ gene contributes to virulence on pear and *AvrRpt2*$_{EA}$ is recognized by *Arabidopsis* RPS2 when expressed in *Pseudomonas syringae*. *Molecular Plant–Microbe Interactions* 19, 644–654.

Zhou, J.M., Loh, Y.T., Bressan, R.A. and Martin, G.B. (1995) The tomato gene *Pti1* encodes a serine/threonine kinase that is phosphorylated by Pto and is involved in the hypersensitive response. *Cell* 83, 925–935.

Zhou, J.M., Tang, X.Y. and Martin, G.B. (1997) The Pto kinase conferring resistance to tomato bacterial speck disease interacts with proteins that bind a cis-element of pathogenesis-related genes. *EMBO Journal* 16, 3207–3218.

Zhu, M., Shao, F., Innes, R.W., Dixon, J.E. and Xu, Z. (2004) The crystal structure of *Pseudomonas* avirulence protein AvrPphB: a papain-like fold with a distinct substrate-binding site. *Proceedings of the National Academy of Sciences, USA* 101, 302–307.

Zipfel, C. and Rathjen, J.P. (2008) Plant immunity: AvrPto targets the frontline. *Current Biology* 18, R218–R220.

# 9 Virulence Determinants and the Global Regulation of Virulence in *Xanthomonas campestris*

Adrián A. Vojnov,[1] J. Maxwell Dow[2] and Kamal Bouarab[3]

[1]*Instituto de Ciencia y Tecnología Dr. Cesar Milstein, CONICET;* [2]*National University of Ireland, Cork, Ireland;* [3]*Université de Sherbrooke, Québec, Canada*

## Abstract

*Xanthomonas campestris* pathovar *campestris* (*Xcc*) is the causal agent of black rot disease of cruciferous plants. The ability of *Xcc* to elicit disease depends upon the synthesis of a number of factors, including the extracellular polysaccharide xanthan, extracellular plant cell wall degrading enzymes and cyclic glucan, as well as the formation of biofilms. The synthesis of these extracellular virulence factors is subject to coordinated control by genes within the *rpf* gene cluster. Some of these *rpf* genes encode elements of a cell–cell signalling system mediated by the diffusible signal factor (DSF), which is an unsaturated fatty acid. Here we review current progress on our understanding of the roles of xanthan, cyclic glucan and biofilm development in the interaction of *Xcc* with plants, and of the mechanistic basis of regulation of these processes by DSF. New roles for xanthan and cyclic glucan in disease, through the suppression of plant immune responses, have been uncovered. Xanthan induces susceptibility to *Xcc* in *Arabidopsis thaliana* and *Nicotiana benthamiana* by suppressing basal defences such as callose deposition. Unlike xanthan, which acts only locally, the effects of cyclic glucan on plant defence suppression and callose deposition occur in a systemic fashion. Xanthan is also involved in biofilm formation by *Xcc*. A fine balance of DSF signalling is required for the formation of structured biofilms in static cultures in minimal medium and for virulence to plants. Recent observations have shown that the perception of the DSF signal requires the sensor kinase RpfC and is linked to the degradation of the intracellular second messenger bis-(3-5)-cyclic di-guanosine monophosphate (cyclic di-GMP) by the HD-GYP domain regulator RpfG. The mechanisms by which cyclic di-GMP exerts its regulatory influence on xanthan, cyclic glucan and biofilm formation remain unclear.

 

## 9.1 Introduction

*Xanthomonas campestris* pv. *campestris* (*Xcc*) is the causal agent of black rot of cruciferous crops, an economically important disease worldwide (Onsando, 1992). *Xcc* is a vascular pathogen and is normally restricted to the xylem of leaves of infected plants at early stages of the disease. Bacteria enter unwounded plants through hydathodes, structures that are found where the veins impinge on the leaf margins. Bacterial progression along the xylem is followed by the typical symptoms of vein blackening and V-shaped chlorotic and necrotic lesions extending from leaf margins along the veins. *Xcc* produces a number of factors that contribute to the ability to cause disease. *Xcc* uses a type III secretion system to deliver effector proteins into the plant cell, in order to modulate host responses and promote disease. Additional factors contributing to virulence include the synthesis of a range of extracellular enzymes capable of degrading plant cell walls and other polymers, synthesis of the extracellular polysaccharide (EPS) xanthan, synthesis of cyclic glucan and the formation of biofilms. The synthesis of these latter factors is under the coordinated control of the *rpf* gene cluster (for regulation of pathogenicity factors). The *rpf* genes act in positive regulation of the synthesis of EPS, extracellular enzymes and cyclic glucan and have complex regulatory effects on biofilm formation. Mutations in *rpf* genes lead to a reduction of virulence in host plants. Several genes within the *rpf* gene cluster encode components for the synthesis and perception of a small diffusible molecule, which has been called DSF (for diffusible signal factor) (Barber *et al.*, 1997). DSF has been characterized as the unsaturated fatty acid *cis*-11-methyl-dodecenoic acid (Wang *et al.*, 2004).

Recent work has shed more light both on the role of DSF-controlled processes in promoting bacterial disease and on the mechanisms of DSF signal transduction. Here we review these findings, specifically addressing the action of xanthan and extracellular cyclic glucan in the suppression of plant defence responses, the role of bis-(3-5)-cyclic di-guanosine monophosphate (cyclic di-GMP) as a second messenger in DSF signalling, and the fine balance of DSF synthesis that is required for both biofilm formation in minimal medium and optimal virulence to plants. In this way, we highlight the newly discovered strategies that *Xcc* uses to cause disease in the host.

## 9.2 Xanthan and Cyclic Glucan as Virulence Factors in *Xanthomonas*–Plant Interactions

### Xanthan production and its role in disease

Xanthan is a polymer composed of repeats of pentasaccharide units and comprises a 'cellulose' backbone of β-1,4-linked glucose residues, with a side chain comprising the trisaccharide mannose-β-1,4-glucuronic acid-β-1,2-mannose-α-1,3 attached to every other glucose (Jansson *et al.*, 1975). The polymer is also substituted with pyruvate and acetate moieties. A complete structure of xanthan is represented in Fig. 9.1a. The synthesis of xanthan involves the assembly of the pentasaccharide repeating unit while linked to a polyprenol through a diphosphate bridge. Subsequently, the repeating unit is polymerized and the polymer secreted outside the cell body (Ielpi *et al.*, 1993). The genes that encode for the enzymes involved in the transfer of the sugars and of the non-glycosidic substituents are located in a cluster which comprises 12 predicted open-reading frames, *gumB–gumM* (Thorne *et al.*, 1987; Katzen *et al.*, 1996, 1998; Vojnov *et al.*, 2002). Transcriptional analysis has shown that the *gum* genes are mainly expressed as an operon from a promoter upstream of the first gene, *gumB* (Katzen *et al.*, 1996).

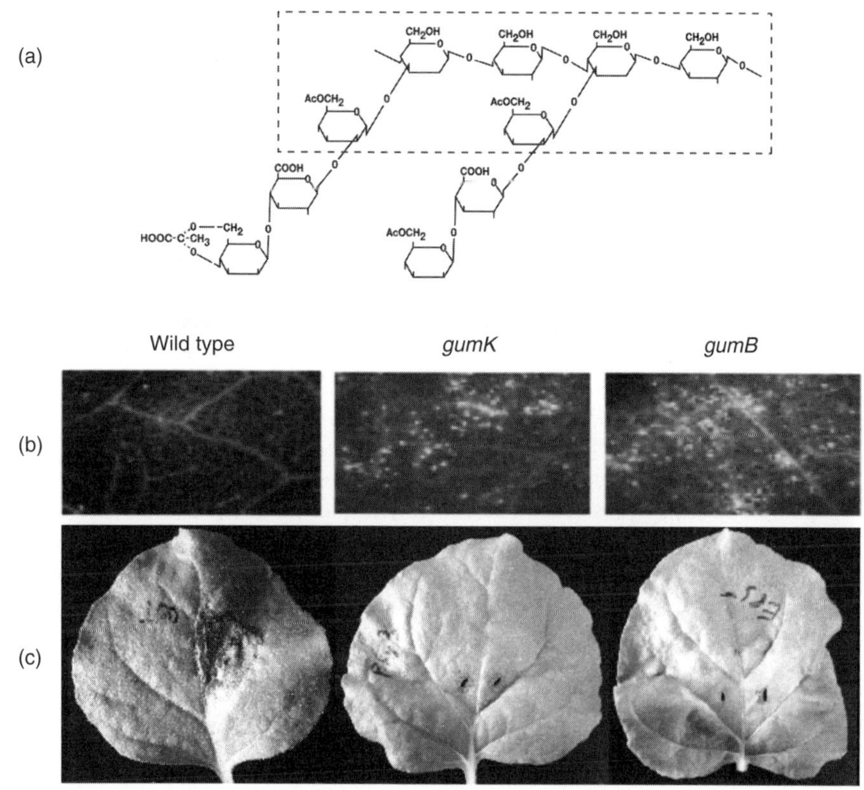


**Fig. 9.1.** The structure of xanthan and importance of the polymer in *Xcc*–plant interactions. (a) Xanthan structure. Two repeating units are represented to show the different substitutions in the mannose residues of the branches. (b, c) Mutants that synthesize truncated xanthan (*gumK*) or have no xanthan synthesis (*gumB*) have reduced virulence associated with the induction of callose deposition. (b) Callose deposition in *Nicotiana benthamiana* leaves after inoculation with wild-type *Xcc*, and derived *gumK* and *gumB* mutants. The leaves were stained for callose deposits 24 h post inoculation and observed under fluorescence microscopy where callose deposits appear as white dots. (c) Symptoms seen in *N. benthamiana* leaves after inoculation with wild-type *Xcc* strain, *gumK* or *gumB* mutants. For details see text.

The importance of xanthan has been demonstrated for disease *Xcc* on *Arabidopsis thaliana*, *Nicotiana benthamiana* and *Brassica campestris* using a mutant carrying a Tn*5* insertion in the *gumB* gene, the first gene of the gum operon. This mutant, that is unable to produce xanthan, did not incite the disease symptoms seen with the wild-type strain on these various hosts and bacterial numbers were lower than those seen with the wild type (Newman *et al*., 1994; Yun *et al*., 2006; Torres *et al*., 2007). These studies confirmed earlier conclusions of a function of xanthan in virulence in the compatible interaction of *Xcc* with plants obtained through the use of a *gumD* mutant (Thorne *et al*., 1987; Katzen *et al*., 1996, 1998; Vojnov *et al*., 2002).

The timing of xanthan production during the disease progression in host plants has been demonstrated by a reporter construct created by fusion of the region of the *gum* gene cluster that is immediately upstream of *gumB* gene with the coding sequence for β-glucuronidase of *Escherichia coli* (*gusA*). For bacteria grown in liquid culture, the expression of the *gumgusA* reporter is maximal during the stationary phase of growth, the growth period when xanthan synthesis is maximal. Furthermore expression of the *gumgusA* reporter is closely correlated with the production of xanthan in liquid medium, although a low basal level of GusA activity is found in the absence of added carbon sources when the production of xanthan was very low (Vojnov *et al*., 2001a). In bacteria inoculated into plant mesophyll tissue, *gumgusA* expression is maximal at the later phases of growth (Vojnov *et al*., 2001a), indicating that xanthan production is also maximal at this time. The level of expression of *gumgusA* is reduced in an *rpfF* mutant compared to the wild type during growth *in planta* and in liquid cultures. RpfF is an enzyme responsible for the synthesis of the DSF cell–cell signal (see below), hence these findings suggest that DSF acts to positively regulate xanthan production within the plant host as well as in culture (Vojnov *et al*., 2001a).

A similar temporal pattern of *in planta* expression has been observed for the *eps* operon of *Ralstonia solanacearum* (formerly known as *Pseudomonas solanacearum* and *Burkholderia solanacearum*) which directs the biosynthesis of EPS I, the complex exopolysaccharide of this tomato pathogen (Kang *et al*., 1999). The *eps* operon is only activated at later stages of infection. These results strongly suggest that both *Xcc* and *R. solanacearum* produce large amount of EPS only at later phases of disease. High production of EPS at later stages of disease in tissues undergoing necrosis might protect the bacteria against various stresses, such as desiccation and damage by reactive oxygen species produced as a part of the plant defence response. EPS may also contribute to the formation of biofilms by bacteria *in planta* (see below). Conversely, there are several possible reasons why bacteria would not produce large amounts of EPS during the early phases of pathogenesis. A limited EPS production may allow adherence of bacterial colonies to plant cell surfaces, thus promoting the establishment of biofilms or microcolonies, and may allow the proper functioning of type III secretion systems encoded by the *hrp* gene clusters (for hypersensitive resistance and pathogenicity), which are critical for the establishment of infection. Copious production of EPS early in pathogenesis might interfere with these processes.

More recent work examining the plant response to *Xcc* wild type and different *gum* mutants has given an insight into a further potential role for xanthan during disease: defence suppression through sequestration of apoplastic $Ca^{2+}$ (Yun *et al.*, 2006). As pointed out above, a mutant of *Xcc* with a Tn*5* transposon insertion in *gumB* is unable to produce xanthan. In contrast, a strain carrying a Tn*5* transposon insertion in *gumK*, which encodes the glucuronosyl transferase enzyme, produces a truncated xanthan with a trisaccharide (instead of a pentasaccharide) repeating unit, that is released into the growth medium (Vojnov *et al.*, 2002). This modified structure is shown within the dashed box in Fig. 9.1a. Inoculation of plants with *gumB* or *gumK* mutants leads to an enhanced deposition in the plant cell wall of callose (Fig. 9.1b, right and middle panels, respectively), a β-1,3-glucan with 1,6- modifications that is associated with increased resistance of plants against some pathogens (Stone and Clarke, 1992; Hamiduzzaman *et al.*, 2005). This effect is not seen with the wild type (Fig. 9.1b, left panel). At the macroscopic level, no symptoms are observed when the *gumB* and *gumK* mutants are inoculated to *N. benthamiana* leaves (Fig. 9.1c) (Yun *et al.*, 2006). Pre-treatment of plants with xanthan from the wild type, but not the polytrisaccharide produced by the *gumK* mutant, is able to suppress this callose deposition and allow symptom development (Yun *et al.*, 2006).

These results show a role for xanthan as a suppressor of plant defences and further suggest that the presence of the negatively charged glucuronosyl and ketal-pyruvate residues in the xanthan might be essential for this biological function during bacteria–plant interaction. It has been shown previously that local increase in $Ca^{2+}$ ions can directly activate the callose synthase enzyme and initiate callose formation (Kohle *et al.*, 1985). Indeed, $Ca^{2+}$ influx from the cell exterior to the cytosol is a prerequisite for most induced defence responses. One mechanism by which xanthan could act to suppress cell wall-based plant defences would therefore be binding of extracellular calcium ions, with consequent interference of signal transduction linked to callose synthetase activation. This ability to bind $Ca^{2+}$ will depend on its negative charge, which is conferred by the presence of glucuronosyl residues and through ketal-pyruvate substitution.

These findings present a puzzle when seen in the context of the experiments described above, describing the temporal expression of the *gum* operon during infection. If expression of the *gum* operon, and by inference xanthan production, is low at the early phases of the disease progression, is there sufficient polymer available to make a significant impact on $Ca^{2+}$ sequestration to influence callose synthesis? Measurement of the level of EPS in infected plants would provide some insight into this issue. It should also be kept in mind that *Xcc* has multiple mechanisms of defence response suppression, that include the action of cyclic glucan (see next section) and by extrapolation from work on other bacteria, certain type III-secreted effectors. Consequently, the occurrence of synergy or other forms of interplay between these different systems of suppression cannot be ruled out.

## The role of cyclic β-(1,2)-glucan in *Xcc*–plant interactions

*Xcc* synthesizes a neutral cyclic hexadecaglucoside containing 15 β-(1→2)-linkages and one α-(1→6)-linkage, which is found both in the periplasm (Talaga *et al.*, 1996) and in the culture supernatant (Amemura and Cabrera-Crespo, 1986). The gene involved in cyclic glucan biosynthesis in *Xcc* (*XC_4168, ndvB*) has been recently identified. The gene product has amino acid sequence relatedness to the C-terminal of the cyclic β-(1,2)-glucan synthetase NdvB of rhizobial species. Disruption of *ndvB* in *Xcc* abolishes the ability to synthesize cyclic β-(1,2)-glucan and the mutant is no longer able to cause disease in the model plants *N. benthamiana* and *A. thaliana* after leaf inoculation (Rigano *et al.*, 2007a). This absence of disease symptoms is associated with attenuated growth and lower final population size of the mutant, compared to the wild type (Rigano *et al.*, 2007a).

Leaves challenged with the *Xcc ndvB* mutant strain show considerably enhanced callose deposition and a faster expression of the defence-related gene *PR1*, compared to leaves inoculated with the wild-type strain. Overall, the lower bacterial numbers attained, the more rapid induction of *PR1* and alteration in the plant cell wall suggest that the host is exhibiting a resistance response to the *ndvB* strain. These findings suggest that cyclic β-(1,2)-glucan has a role in inducing host susceptibility to *Xcc* through suppression of plant defences. These conclusions are further supported by experiments which showed that pre-treatment of leaves with purified cyclic β-(1,2)-glucan suppresses *PR1* induction and callose deposition by the *ndvB* mutant and restores virulence (Rigano *et al.*, 2007a). Intriguingly, these effects are seen when cyclic glucan and bacteria are applied either to the same or to different leaves.

Cyclic β-(1,2)-glucan-induced systemic suppression is associated with the transport of the molecule throughout the plant (Rigano *et al.*, 2007a). These results suggest that the cyclic β-(1,2)-glucan generated by *Xcc* bacteria colonizing one leaf can be translocated to other leaves to induce susceptibility, thus promoting bacterial spread through the plant. The ability of the cyclic β-(1,2)-glucan of *Xcc* to act as a systemic effector suppressing host defence responses distinguishes it from the action of almost all of the suppressors so far described, which have only been reported to act locally. The exception is the *Pseudomonas syringae* phytotoxin coronatine, which induces systemic susceptibility in *A. thaliana*. It is still unknown if coronatine is itself systemically translocated, or if it exerts its effect via local activation of the jasmonic acid pathway (Cui *et al.*, 2005). The mechanism by which the cyclic β-(1,2)-glucan exerts its action, either locally or systemically, is unknown, and many questions arise from these findings. Although sequestration of $Ca^{2+}$ is a plausible mechanism for the suppression of plant defences by xanthan, does cyclic glucan act in the same way? Is there any interplay between cyclic glucan and type III-secreted effectors, some of which also act to suppress basal resistance responses? Is there a plant receptor for the cyclic glucan?

## 9.3 Regulation of Synthesis of Virulence Factors by the rpf/DSF System

As outlined above, the synthesis of EPS, extracellular enzymes and cyclic glucan in *Xcc* is controlled by the products of the *rpf* gene cluster (Tang *et al.*, 1991). This cluster comprises nine genes, *rpf A* to *I* and is located within a 21.9 kb region of the *Xcc* chromosome. The left part of this region contains six contiguous *rpf* genes with the gene order *rpf ABFCHG*. Mutations in any of these genes lead to coordinated downregulation of the synthesis of all the extracellular enzymes, cyclic glucan and EPS (Barber *et al.*, 1997; Vojnov *et al.*, 2001b). The *rpfBFGHC* genes encode components of the DSF cell–cell signalling system. RpfF and RpfB direct the production of the DSF signal molecule (Barber *et al.*, 1997), which has been characterized as the unsaturated fatty acid *cis*-11-methyl-dodecenoic acid (Fig. 9.2a) (Wang *et al.*, 2004). The synthesis of DSF is completely dependent on RpfF, which has a certain amino acid sequence similarity to enoyl-CoA hydratases, but is only partially dependent on RpfB, which is a long-chain fatty acyl CoA ligase. The *rpfB* and *rpfF* genes are cotranscribed from a promoter upstream of *rpfB*, although *rpfF* also has its own promoter (Slater *et al.*, 2000). The *rpfF* mutants can be phenotypically corrected for the production of extracellular enzymes, cyclic glucan and EPS by the exogenous addition of DSF or by growth on plates in proximity to a wild-type strain (Barber *et al.*, 1997; Vojnov *et al.*, 2001b).

Perception of the DSF signal is thought to require the two-component system comprising RpfC and RpfG, which are encoded within the *rpfGHC* operon, which is contiguous with *rpfF* but convergently transcribed (Slater *et al.*, 2000). RpfC is a complex sensor kinase with a predicted membrane-associated sensory input domain as well as histidine kinase, CheY-like receiver (REC) and C-terminal histidine phosphotransfer (HPt) domains. RpfG is a novel regulator with a REC domain and an HD-GYP domain. Although the amino acid sequence of RpfH resembles that of the sensory input domain of RpfC, no role for RpfH in DSF signalling or regulation of extracellular enzyme or xanthan synthesis is yet apparent, and *rpfH* mutants retain full virulence (Dow *et al.*, 2003; Slater *et al.*, 2000). In addition to positive regulation of virulence factors' synthesis, RpfC acts to negatively regulate DSF synthesis, a function that does not involve RpfG (Slater *et al.*, 2000).

The remaining *rpf* genes (*rpfA*, *rpfD*, *rpfE* and *rpfI*) have no apparent involvement in the DSF-dependent production of *Xcc* virulence factors and have minor regulatory roles (Barber *et al.*, 1997; Wilson *et al.*, 1998; Dow *et al.*, 2000). The function of some of these Rpf proteins has been described or can be predicted from their amino acid sequence. Accordingly, RpfA is an aconitase that may play a role in iron homeostasis (Wilson *et al.*, 1998), and RpfD has a LytTR DNA-binding domain (IPR007492) (Nikolskaya and Galperin, 2002), suggesting a role in transcriptional activation. RpfE and RpfI are conserved hypothetical proteins (Dow *et al.*, 2000).

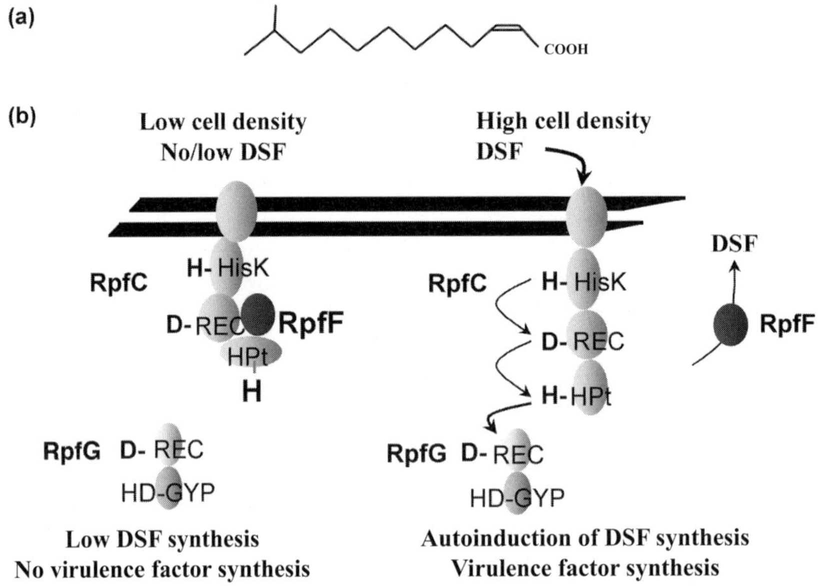


**Fig. 9.2.** Diffusible signal factor (DSF) structure, synthesis and perception in *Xcc.* (a) Structure of DSF. (b) Model for the role of Rpf proteins in DSF perception, signal transduction and autoinduction.The synthesis of the DSF signal requires RpfF whereas DSF perception and signal transduction involves the complex sensor RpfC and HD-GYP domain regulator RpfG, which is a cyclic di-GMP phosphodiesterase. At low cell density or when DSF levels are low, RpfF is bound to the REC domain of RpfC, thus sequestering it from the substrates required to synthesis DSF. At high cell density or in the presence of DSF, the binding of the signal molecule to the sensory input domain of RpfC triggers autophosphorylation of the sensor at a histidine residue followed by phosphorelay and phosphotransfer to the cognate regulator, RpfG (indicated by arrows). Phosphorylation of RpfG leads to its activation as a cyclic di-GMP phosphodiesterase, an activity associated with the HD-GYP domain. Activation of RpfG and alteration of cyclic di-GMP levels have downstream effects on the synthesis of virulence factors such as extracellular enzymes, biofilm dispersal and motility by as yet unknown mechanisms. The change in RpfC conformation (or perhaps dimerization) upon DSF binding also allows release of RpfF, and consequently an elevated synthesis of DSF in an autoinduction loop. Key to domains: REC, CheY-like two-component receiver domain; HPt, histidine phosphotransfer; HisK, histidine kinase. The residues in phosphorelay are histidine (H) and aspartic acid (D).

## Dual signalling functions of RpfC involve either phosphorelay or receiver domain–protein interactions

As outlined above, RpfC acts to positively regulate virulence factors' synthesis in response to DSF, but to negatively regulate the synthesis of DSF itself. Recent work has shown that these dual signalling functions are achieved by different mechanisms (He *et al.*, 2006a, b). Work on sensor kinases with a related domain structure (such as BvgS of *Bordetella* spp. and ArcB of *E. coli*) has involved autophosphorylation as a consequence of signal perception followed by phosphorelay via REC and HPt domains to the cognate regulator as the mechanism of signal transduction. By analogy, it was proposed that RpfC functioned in the same fashion (Slater *et al.*, 2000; He *et al.*, 2006a, b). Mutational analysis of the three conserved amino acid residues of RpfC involved in phosphorelay (H198 in the histidine kinase domain, D512 in the REC domain and H657 in the HPt domain) showed that they are essential for activation of the production of extracellular enzymes and xanthan, but not for repression of DSF biosynthesis. Domain deletion analyses revealed that the REC domain of RpfC alone was sufficient to repress DSF overproduction in an *rpfC* mutant. This may involve a physical interaction between the REC domain and RpfF, the enzyme involved in DSF biosynthesis, as suggested by coimmunoprecipitation and Western blot analyses.

These data support a model in which RpfC modulates the different functions of virulence factor synthesis and DSF synthesis by utilization of a conserved phosphorelay system and a novel domain-specific protein–protein interaction mechanism, respectively (Fig. 9.2b). In this model sequestration of RpfF by RpfC renders it inactive in DSF synthesis. Structural changes in RpfC, perhaps as a result of DSF binding and autophosphorylation, allow the release of RpfF, which is then active in DSF synthesis. In this view, perception of DSF would be autoinductive on its synthesis, but this would not involve changes in expression of the *rpfF* gene. This is consonant with the finding that transcript levels of *rpfF* are only modestly elevated over the wild type in an *rpfC* mutant, whereas DSF levels are considerably higher (Slater *et al.*, 2000). Although the model is consistent with the available data, it cannot be excluded that DSF synthesis is additionally regulated at other levels, perhaps by the supply of the substrates for RpfB/RpfF or post-transcriptional control of the expression of RpfF and RpfB proteins.

## The HD-GYP domain regulator RpfG and cyclic di-GMP degradation

Perception of the DSF signal is thought to activate the autophosphorylation of RpfC and result in phosphorelay and phosphotransfer to the REC domain of the RpfG regulatory protein. RpfG is an unusual two-component regulator in that it has an HD-GYP domain attached to the REC domain, rather than a DNA-binding domain as seen in the majority of such regulators (Slater *et al.*, 2000). The HD-GYP domain is a subset of the HD superfamily of metal-dependent phosphohydrolases (Galperin and Koonin, 1999; Galperin *et al.*, 2001). Bioinformatic studies have suggested a role for the HD-GYP domain, in the degradation of the bacterial second messenger cyclic di-GMP (Galperin and Koonin, 1999; Galperin *et al.*, 2001). Recent experimental studies have shown that the HD-GYP domain is indeed a novel cyclic di-GMP phosphodiesterase (Ryan *et al.*, 2006), thus implicating cyclic di-GMP in DSF signal transduction.

Cyclic di-GMP is an almost ubiquitous second messenger in bacteria that was first described as an allosteric activator of cellulose synthase (Ross *et al.*,

1990), but it is now recognized to regulate a range of functions, including biofilm formation, motility, developmental transitions, virulence factor synthesis and virulence in human and animal pathogens (D'Argenio and Miller, 2004; Jenal, 2004; Paul *et al.*, 2004; Romling and Amikam, 2006). Two protein domains, GGDEF (IPR000160) and EAL (IPR001633), are involved in the synthesis and degradation of cyclic di-GMP (Scarpari *et al.*, 2003; Paul *et al.*, 2004; Christen *et al.*, 2005; Ryjenkov *et al.*, 2005), respectively. Synthesis of cyclic di-GMP by the GGDEF domain occurs from GTP, whereas EAL domains are phosphodiesterases that convert cyclic di-GMP into the linear nucleotide pGpG. GGDEF and EAL domains are widely distributed in bacteria, including the ones affecting plants. The majority of proteins containing these domains have additional signalling domains, suggesting that their activities are responsive to different environmental cues (Galperin *et al.*, 2001; Romling *et al.*, 2005; Romling and Amikam, 2006). In general, high cellular levels of cyclic di-GMP promote biofilm formation and sessile growth, whereas low levels promote virulence factor synthesis and motility (Galperin *et al.*, 2001; Romling *et al.*, 2005; Romling and Amikam, 2006).

Indirect evidence for the role of RpfG in cyclic di-GMP turnover has come from experiments where GGDEF and EAL domain proteins have been ectopically expressed in *Xcc* wild type and the *rpfG* mutant. Expression of genes encoding EAL domain proteins in the *Xcc rpfG* mutant restores extracellular enzymes. In contrast, expression of genes encoding a GGDEF domain protein in wild-type *X. campestris* gives a phenocopy of the *rpfG* mutant (Ryan *et al.*, 2006). These indirect observations are consistent with a role for the HD-GYP domain in cyclic di-GMP hydrolysis. This conclusion was supported by biochemical studies that demonstrated that the isolated domain can hydrolyse cyclic di-GMP to GMP via the linear intermediate pGpG (Ryan *et al.*, 2006). Mutation of the HD residues comprising the presumed catalytic diad of the HD-GYP domain abolishes both the regulatory and the enzymatic activities against cyclic di-GMP. Further support for a role of cyclic di-GMP in DSF signal transduction has come from experiments in which the RpfC/RpfG two-component system was re-constructed in *Pseudomonas aeruginosa* and shown to confer responsiveness to exogenously-added DSF, as seen through effects on swarming motility (Ryan *et al.*, 2006). It has been proposed that phosphorylation of RpfG leads to its activation in cyclic di-GMP hydrolysis (Fouhy *et al.*, 2006) (Fig. 9.2b), although this has not been directly demonstrated.

The link of DSF signal perception to cyclic di-GMP degradation raises the related issues of whether other cyclic di-GMP signalling systems in *Xcc* regulate the same functions as RpfG and how such a system with many potential players is functionally organized. A comprehensive mutational analysis of the role of all 37 proteins with HD-GYP, GGDEF and EAL domain proteins in regulation of extracellular enzyme synthesis and motility in *Xcc* has been recently reported (Ryan *et al.*, 2007). A number of proteins, in addition to RpfG, act to regulate extracellular enzyme synthesis, although different proteins have significant roles under different growth conditions. RpfG is the only protein to have an influence under all growth conditions tested and loss of RpfG has the biggest effect on extracellular enzyme synthesis. These findings are consistent with the concept of a signalling network that includes the RpfC/RpfG system and that responds to, and integrates, information from a number of cues, including the DSF cell–cell signal. Conversely, other signalling elements in *Xcc* regulate motility but have no effect on extracellular enzyme production. This is consistent with the concept of localized action of certain elements in cyclic di-GMP signalling.

## The DSF 'regulon' and signal transduction beyond RpfG

The full extent of the DSF 'regulon' has been examined by transcriptome profiling (He *et al.*, 2006b). It is evident that DSF signalling regulates a number of functions in addition to extracellular enzymes and EPS with potential contribution to bacterial virulence; these include resistance to oxidative and other stresses, iron assimilation and motility (He *et al.*, 2006b). Several recent studies have addressed the molecular details of the DSF signal transduction beyond RpfG, processes which are currently not well understood.

The DSF/*rpf* system has been shown to activate transcription of the gene encoding the cyclic-AMP receptor-like protein Clp (He *et al.*, 2007). In *Xcc*, Clp regulates many functions including the expression of genes for extracellular enzymes and EPS synthesis, and those for the regulators Zur and FhrR. In turn, Zur regulates genes for functions such as iron uptake, the tricarboxylic acid (TCA) cycle, multi-drug resistance and detoxification, whereas FhrR regulates the expression of genes coding for flagellar synthesis and type III secretion system (He *et al.*, 2007). The evidence so far available indicates that not all of the regulatory effects of RpfG are exerted through the action of Clp. For example, Clp is apparently not involved in the regulation of biofilm dynamics in *Xcc* (He *et al.*, 2007). It is also as yet unclear how the *rpf*/DSF system exerts its influence on the expression of the *clp* gene and whether there is also an effect on the activity of the Clp protein, although this is likely to bind cyclic mononucleotides rather than cyclic di-nucleotides.

The *rpf*/DSF system certainly has an influence on the cellular level of cyclic di-GMP but the mechanism(s) by which the altered level of the nucleotide exerts its regulatory influences on *Xcc* is unknown. Work on other bacteria has involved PilZ, a cyclic di-GMP binding domain, as an adaptor in the regulatory action of cyclic di-GMP (Amikam and Galperin, 2006; Ryjenkov *et al.*, 2006). There are four PilZ domain-containing proteins in *Xcc*, whose regulatory roles have yet to be examined. The HD-GYP domain of RpfG from the related pathogen *Xanthomonas axonopodis* pv. *citri* (*Xac*) has been shown, by yeast two-hybrid analysis, to interact with a subset of GGDEF domain proteins (Andrade *et al.*, 2006). Although this may suggest an action of RpfG in modulating the activity of specific cyclic di-GMP generating systems, the biological relevance of such interactions remains to be investigated. The yeast two-hybrid analysis also revealed interactions of the HD-GYP domain of RpfG with regulatory proteins not involved in cyclic di-GMP signalling, including the $\sigma$54 sigma factor (Andrade *et al.*, 2006).

## 9.4 Biofilm Formation and Virulence in *Xanthomonas*

*Xcc* has been reported to form biofilms of different architectures under different growth conditions (Torres *et al.*, 2007). The development of each type of biofilm depends upon the synthesis of xanthan and is under the regulation of the *rpf*/DSF system, although there are marked differences in the apparent role of DSF in the two environmental conditions. In shaken rich nutrient L medium, *rpfG*, *rpfC* and *rpfF* mutants form matrix-enclosed aggregates with a reticulated structure (Fig. 9.3a), whereas the wild-type strain grows planktonically (Dow *et al.*, 2003). Addition of DSF causes dispersal of the aggregates formed by the *rpfF* mutant but not *rpfG* or *rpfC* mutants. These findings are consistent with the notion that DSF influences biofilm dispersal through an action requiring RpfG and RpfC, but has no influence on biofilm formation (Dow *et al.*, 2003). A *gumBrpfG* double mutant failed to form an aggregate and grew planktonically, indicating the essential role of xanthan in the formation of these reticulated structures.

A different scenario is evident from studies on *Xcc* biofilm formation in minimal medium in static cultures, in chambered cover slides, using confocal laser scanning microscopy (Russo *et al.*, 2006; Torres *et al.*, 2007). In the formation of a typical *Xcc* biofilm under these conditions, the bacteria contact the glass surface via the lateral cell surface and also attach to each other predominantly through lateral interactions forming microcolonies. This phase is followed with the formation by 4 days of compact aggregates of bacteria with a characteristic three-dimensional structure separated by extensive water spaces and mushroom-type biofilm structures (Russo *et al.*, 2006; Torres *et al.*, 2007) (Fig 9.3b, c). Bacteria in these structures are mostly interacting laterally (Russo *et al.*, 2006). With the *rpfF* mutant (DSF-minus), microcolonies were seen after 2 days, but these did not develop into a structured biofilm, so that after 4 days, only unstructured layers of bacteria were observed (Torres *et al.*, 2007). With the *rpfC* mutant (DSF over-producer), although the bacteria showed some aggregation at day 2, only unstructured layers of bacteria were observed at day 4 (Torres *et al.*, 2007). Overall, these results showed that DSF-mediated signalling is required for the formation of a structured biofilm in minimal medium.

The *gumB* mutant was severely affected in microcolony formation and did not form more complex structures. After 4 days, no evident biofilm architecture was observed at the base of the chamber (Torres *et al.*, 2007). Introduction of the entire *gum* cluster of genes, cloned into a cosmid vector, restores normal levels of EPS and a typical structured biofilm to the *gumB* strain 8397. These observations confirmed that xanthan synthesis in *Xcc* is crucial for the development of the structured biofilm in minimal medium (Russo *et al.*, 2006).

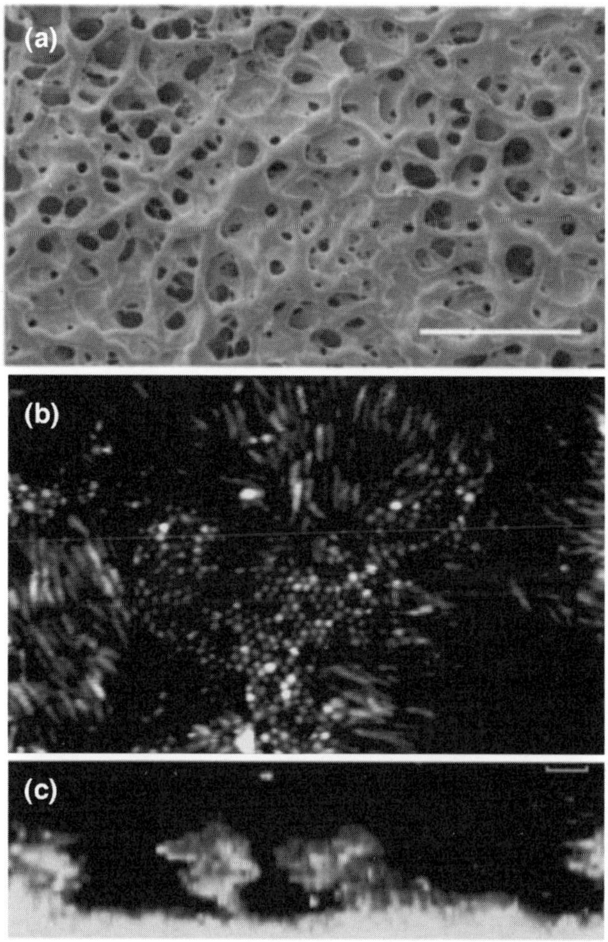


**Fig. 9.3.** Biofilm architecture and the role of *rpf* genes in biofilm formation in *Xcc* varies with growth conditions. (a) Scanning electron microscopy of the aggregates formed in L medium by *rpf* mutants showing the bacteria held in a reticular structure. Under these conditions the wild type grows planktonically with no aggregation. Scale bar is 10 μm. (b) The biofilm formed by the wild-type *Xcc* containing the GFP-expressing plasmid pRU1319 in chambered cover slides as observed by confocal laser scanning microscopy. Note the occurrence of lateral interactions between bacteria. (c) *z*-Axis projected images of wild-type *Xcc* showing the development of mushroom-shaped structures.

## DSF synthesis is fine-tuned for regulation of biofilm formation and optimal virulence

Evidence that DSF synthesis has to be fine-tuned for biofilm formation and optimal virulence has come from studies of mixed cultures of *Xcc* strains. Although mutations in *rpfF* and *gumB* genes result in the absence of a typical, structured biofilm, a mixed culture of the two strains can form a structured biofilm comprising a mixture of the two bacteria (Russo *et al.*, 2006). These results suggest that reciprocal complementation had taken place, where the lack of DSF in the *rpfF* mutant had been restored by DSF produced by the *gumB* mutant, and the xanthan produced by the *rpfF* mutant was substituting for the lack of xanthan in the *gumB* mutant. The need for regulated production of DSF for biofilm development has been indicated by mixed cultures of the wild type with different *rpf* mutants. Coinoculations of the wild-type strain with the *rpfC* (DSF over-producing) mutant result into abolition of the ability to form the wild-type structured biofilm. In contrast to the effects caused by the *rpfC* mutant, the *rpfF* mutant (DSF non-producer) does not alter the wild-type biofilm when the two strains are mixed. Taken together with the results of reciprocal complementation of *rpfF* and *gumB* mutants, these findings suggest that the amount of DSF produced has to be tightly controlled for the development of the biofilm and increased levels of DSF interfere with this process.

Phenotypic characterization of *in planta* behaviour of *rpf* mutants and *in vivo* complementation has been used to examine the role of DSF cell–cell signalling during *Xcc* pathogenesis in *N. benthamiana*. A strong correlation exists between the biofilm capacity of the strains in minimal medium and virulence (Fig. 9.4). The xanthan-deficient *gumB* mutant and the *rpfF* mutant were unable to produce a structured biofilm, in single culture, in static minimal medium (Fig. 9.4a), nor did they develop symptoms in *N. benthamiana* (Fig. 9.4b). However, in mixed cultures of the two strains, *gumB* and *rpfF* mutants developed both structured biofilm and disease symptoms. In addition, the *rpfC* mutant interfered with the growth and symptoms caused by the wild type, whereas the DSF-defective *rpfF* mutant had no effect (Torres *et al.*, 2007).

Although a close correlation was observed between the effects of DSF levels on structured biofilm formation in minimal medium and on virulence, a direct cause-and-effect relationship cannot be concluded. Although there is no evidence to suggest that addition of excess DSF negatively influences the synthesis of extracellular enzymes or xanthan, effects on the synthesis of other virulence determinants cannot be excluded. One plausible scenario for the role of natural fluctuations in DSF levels in promoting progression of *Xcc* through the xylem is as follows. At low bacterial cell density, where the production of DSF is limited, bacteria attach to the surfaces of the xylem vessels. As the microcolony forms, DSF levels rise and the bacteria start to produce virulence factors including extracellular enzymes. The latter can promote disease through interference with plant defences, provide nutrition through degradation of the xylem walls, and allow passage of bacteria between xylem elements through degraded pit membranes. In addition, the structured biofilm begins to form. Bacteria within these structures may have increased resistance to host defences. At later stages, further elevation of DSF levels promotes biofilm dispersal, so that the bacteria can be released to colonize new tissue. The presence of elevated levels of DSF at early phases prevents the formation of the structured biofilm, thus hampering bacterial survival.

Recent studies of the related bacterium *Xac* have indicated that bacteria attach to, and form, a complex, structured biofilm on glass in minimal medium

**Fig. 9.4.** Biofilm formation and virulence in *Xcc* requires xanthan synthesis and DSF signalling. (a) Biofilms formed after 4 days in static minimal medium culture by different strains of *Xcc* expressing *gfp*. The *rpfF* (DSF non-producer), *rpfC* (DSF over-producer, signal blind mutant) and *gumB* (xanthan-deficient) strains do not produce the structured biofilm of the wild type. Images were obtained by confocal scanning laser microscopy. (b) Symptom production in *N. benthamiana* by the same *Xcc* strains.

containing glucose. Similar attachment and structured biofilm formation are also seen on lemon leaves. An *Xac gumB* mutant strain does not form a structured biofilm on either abiotic or biotic surfaces and shows reduced growth and survival on leaf surfaces and reduced disease symptoms. These findings suggest an important role for the production of the xanthan and for the formation of biofilms in the epiphytic survival of *Xac* prior to development of canker disease (Rigano *et al.*, 2007b).

## 9.5 Concluding Remarks

The synthesis of virulence factors by pathogenic bacteria is tightly regulated and can occur as a response to different environmental cues. Recent years have seen an increased understanding of the molecular aspects of bacterial virulence, including the description of novel virulence determinants, the analysis of the roles of different determinants in the disease process and the dissection of the regulatory processes that link the perception of environmental cues to virulence gene expression. These studies have benefited from the determination of the full genome sequence of a number of plant pathogens, which enables comprehensive mutational analysis, development of microarrays for studies of gene expression and its regulation and development of imaging and other methods to study bacteria *in planta* as well as responses of plants to bacterial attacks.

As we have seen from the sections above, considerable progress has been made in understanding global regulation of virulence in *Xcc* mediated by the DSF cell–cell signal, in describing new roles in plant defence suppression of xanthan and cyclic glucan, and in investigating the role that biofilm formation may have in *Xcc* disease establishment and progression. It should also be borne in mind that DSF-regulated factors may have roles in other phases of the *Xcc* disease cycle, such as bacterial survival in soil on dead plant parts and epiphytic growth. The DSF system does not appear to have any significant regulatory overlap with the *hrp* (for hypersensitive reaction and pathogenesis) regulon, and it may be that the two systems operate during different phases of the disease cycle. Unrelated regulatory systems also impinge on the synthesis of virulence factors such as xanthan, which is costly in terms of metabolic energy.

The level of DSF in the immediate bacterial environment will be responsive to a number of factors, including the number of bacteria producing the signal and amount of space in which they may be confined. Bacteria confined within the xylem elements may be at a relatively high cell density, under which conditions DSF may be able to attain levels that can promote biofilm dispersal. In contrast, the levels of DSF in other environments such as on leaf surfaces may be much lower or negligible. The appreciation that DSF signal transduction is linked to alteration in the levels of the second messenger cyclic di-GMP is important since it opens the possibility that synthesis of virulence factors may be under the influence of regulatory networks of cyclic di-GMP signalling systems which respond to a range of environmental cues. By extension, this could suggest that DSF signalling may be relatively unimportant under certain environmental conditions.

Can the findings from these molecular studies be translated into new disease control measures? The demonstration that DSF synthesis is tuned for optimal virulence and biofilm formation in minimal medium suggest that such a fine balance might be readily disrupted (Russo *et al.*, 2006). This may have substantial consequences for the development of measures to control diseases caused by *Xcc* and other *Xanthomonas* spp. Similar suggestions have been made previously by Lindow and colleagues for the control of Pierce's disease of grape caused by *Xylella fastidiosa*, an organism related to *Xcc*. *Xylella fastidiosa* is a xylem-limited pathogen that uses an Rpf system and a DSF-like signal molecule to control interactions both with host plants and with its insect vector (Newman *et al.*, 1994; Chatterjee *et al.*, 2008).

Strategies for disease control through interference with DSF signalling could involve either signal quenching through enzymatic degradation, over-production of the signal, or production at inappropriate times. Such outcomes may be achieved through several methods that could include inoculation of plants with 'disarmed' xanthomonad pathogens or endophytic bacteria possessing the appropriate capabilities, or the development of transgenic plants expressing either the DSF synthase RpfF or enzymes involved in DSF degradation. Indeed, a very recent report indicates the feasibility of signal degradation by coinoculated bacteria as an approach to control *X. fastidiosa* (Newman *et al.*, 2008). Differences between the function of the Rpf/DSF signalling systems, in *Xcc* and *X. fastidiosa*, certainly occur. Nevertheless, these observations indicate potential signal interference in the wider control of diseases elicited by *Xcc* and other xanthomonads.

## Acknowledgements

Adrián Vojnov is supported by the Agencia de Promoción Científicas y tecnológica (PICT-02 No. 08-10740; PAV2003-137) and is Career Investigators of the Concejo Nacional de Investigaciones Científicas y técnicas (CONICET). Kamal Bouarab is supported by the Conseil de Recherche en Science Naturelles et Génie du Canada (CRSNG), the Fondation Canadienne pour l'Innovation (FCI) and the Université de Sherbrooke and J.M. Dow is supported by a Principal Investigator Award from the Science Foundation of Ireland.

## References

Amemura, A. and Cabrera-Crespo, J. (1986) Extracellular oligosaccharides and low-Mr polysaccharides containing (1–2)-beta-D-glucosidic linkages from strains of *Xanthomonas*, *Escherichia coli* and *Klebsiella pneumoniae*. *Journal of General Microbiology* 132, 2443–2452.

Amikam, D. and Galperin, M.Y. (2006) PilZ domain is part of the bacterial c-di-GMP binding protein. *Bioinformatics* 22, 3–6.

Andrade, M.O., Alegria, M.C., Guzzo, C.R., Docena, C., Rosa, M.C., Ramos, C.H. and Farah, C.S. (2006) The HD-GYP domain of RpfG mediates a direct linkage between the Rpf quorum-sensing pathway and a subset of diguanylate cyclase proteins in the phytopathogen *Xanthomonas axonopodis* pv. *citri*. *Molecular Microbiology* 62, 537–551.

Barber, C.E., Tang, J.L., Feng, J.X., Pan, M.Q., Wislon, T.J., Slater, H., Dow, J.M., Williams, P. and Daniels, M.J. (1997) A novel regulatory system required for pathogenicity of *Xanthomonas campestris* is mediated by a small diffusible signal molecule. *Molecular Microbiology* 24, 555–566.

Chatterjee, S., Wistrom, C. and Lindow, S.E. (2008) A cell-cell signaling sensor is required for virulence and insect transmission of *Xylella fastidiosa*. *Proceedings of the National Academy of Sciences, USA* 105, 2670–2675.

Christen, M., Christen, B., Folcher, M., Schauerte, A. and Jenal, U. (2005) Identification and characterization of a cyclic di-GMP-specific phosphodiesterase and its allosteric control by GTP. *Journal of Biological Chemistry* 280, 30829–30837.

Cui, J., Bahrami, A.K., Pringle, E.G., Hernandez-Guzman, G., Bender, C.L., Pierce, N.E. and Ausubel, F.M. (2005) *Pseudomonas syringae* manipulates systemic plant defenses against pathogens and herbivores. *Proceedings of the National Academy of Sciences, USA* 102, 1791–1796.

D'Argenio, D.A. and Miller, S.I. (2004) Cyclic di-GMP as a bacterial second messenger. *Microbiology* 150, 2497–2502.

Dow, J.M., Feng, J.X., Barber, C.E., Tang, J.L. and Daniels, M.J. (2000) Novel genes involved in the regulation of pathogenicity factor production within the *rpf* gene cluster of *Xanthomonas campestris*. *Microbiology* 146(4), 885–891.

Dow, J.M., Crossman, L., Findlay, K., He, Y.Q., Feng, J.X. and Tang, J.L. (2003) Biofilm dispersal in *Xanthomonas campestris* is controlled by cell-cell signaling and is required for full virulence to plants. *Proceedings of the National Academy of Sciences, USA* 100, 10995–11000.

Fouhy, Y., Lucey, J.F., Ryan, R.P. and Dow, J.M. (2006) Cell-cell signaling, cyclic di-GMP turnover and regulation of virulence in *Xanthomonas campestris. Research in Microbiology* 157, 899–904.

Galperin, M.Y. and Koonin, E.V. (1999) Searching for drug targets in microbial genomes. *Current Opinion in Biotechnology* 10, 571–578.

Galperin, M.Y., Nikolskaya, A.N. and Koonin, E.V. (2001) Novel domains of the prokaryotic two-component signal transduction systems. *FEMS Microbiology Letters* 203, 11–21.

Hamiduzzaman, M.M., Jakab, G., Barnavon, L., Neuhaus, J.M. and Mauch-Mani, B. (2005) beta-Aminobutyric acid-induced resistance against downy mildew in grapevine acts through the potentiation of callose formation and jasmonic acid signaling. *Molecular Plant–Microbe Interactions* 18, 819–829.

He, Y.W., Wang, C., Zhou, L., Song, H., Dow, J.M. and Zhang, L.H. (2006a) Dual signaling functions of the hybrid sensor kinase RpfC of *Xanthomonas campestris* involve either phosphorelay or receiver domain-protein interaction. *Journal of Biological Chemistry* 281, 33414–33421.

He, Y.W., Xu, M., Lin, K., Ng, Y.J., Wen, C.M., Wang, L.H., Liu, Z.D., Zhang, H.B., Dong, Y.H., Dow, J.M. and Zhang, L.H. (2006b) Genome scale analysis of diffusible signal factor regulon in *Xanthomonas campestris* pv. *campestris*: identification of novel cell-cell communication-dependent genes and functions. *Molecular Microbiology* 59, 610–622.

He, Y.W., Ng, A.Y., Xu, M., Lin, K., Wang, L.H., Dong, Y.H. and Zhang, L.H. (2007) *Xanthomonas campestris* cell-cell communication involves a putative nucleotide receptor protein Clp and a hierarchical signalling network. *Molecular Microbiology* 64, 281–292.

Ielpi, L., Couso, R.O. and Dankert, M.A. (1993) Sequential assembly and polymerization of the polyprenol-linked pentasaccharide repeating unit of the xanthan polysaccharide in *Xanthomonas campestris. Journal of Bacteriology* 175, 2490–2500.

Jansson, P.E., Kenne, L. and Lindberg, B. (1975) Structure of extracellular polysaccharide from *Xanthomonas campestris. Carbohydrate Research* 45, 275–282.

Jenal, U. (2004) Cyclic di-guanosine-monophosphate comes of age: a novel secondary messenger involved in modulating cell surface structures in bacteria? *Current Opinion in Microbiology* 7, 185–191.

Kang, Y., Saile, E., Schell, M.A. and Denny, T.P. (1999) Quantitative immunofluorescence of regulated *eps* gene expression in single cells of *Ralstonia solanacearum. Applied and Environmental Microbiology* 65, 2356–2362.

Katzen, F., Becher, A., Zorreguieta, A., Puhler, A. and Ielpi, L. (1996) Promoter analysis of the *Xanthomonas campestris* pv. *campestris* gum operon directing biosynthesis of the xanthan polysaccharide. *Journal of Bacteriology* 178, 4313–4318.

Katzen, F., Ferreiro, D.U., Oddo, C.G., Ielmini, M.V., Becker, A., Puhler, A. and Ielpi, L. (1998) *Xanthomonas campestris* pv. *campestris* gum mutants: effects on xanthan biosynthesis and plant virulence. *Journal of Bacteriology* 180, 1607–1617.

Kohle, H., Jeblick, W., Poten, F., Blaschek, W. and Kauss, H. (1985) Chitosan-elicited callose synthesis in soybean cells as a Ca-dependent process. *Plant Physiology* 77, 544–551.

Newman, K.L., Chatterjee, S., Ho, K.A. and Lindow, S.E. (2008) Virulence of plant pathogenic bacteria attenuated by degradation of fatty acid cell-to-cell signaling factors. *Molecular Plant–Microbe Interactions* 21, 326–334.

Newman, M.A., Conrads-Strauch, J., Scofield, G., Daniels, M.J. and Dow, J.M. (1994) Defense-related gene induction in *Brassica campestris* in response to defined mutants of *Xanthomonas campestris* with altered pathogenicity. *Molecular Plant–Microbe Interactions* 7, 553–563.

Nikolskaya, A.N. and Galperin, M.Y. (2002) A novel type of conserved DNA-binding domain in the transcriptional regulators of the AlgR/AgrA/LytR family. *Nucleic Acids Research* 30, 2453–2459.

Onsando, J.M. (1992) Black rot of crucifers. In: Chaube, H.S., Kumar, J., Mukhopadhyay, A.N. and Singh, U.S. (eds) *Plant Diseases of International Importance, Vol. II: Diseases of Vegetable and Oil Seed Crops.* Prentice Hall, Englewood Cliffs, New Jersey, pp. 243–252.

Paul, R., Weiser, S., Amiot, N.C., Chan, C., Schirmer, T., Giese, B. and Jenal, U. (2004) Cell cycle-dependent dynamic localization of a bacterial response regulator with a novel di-guanylate cyclase output domain. *Genes & Development* 18, 715–727.

Rigano, L.A., Payette, C., Brouillard, G., Marano, M.R., Abramowicz, L., Torres, P.S., Yun, M., Castagnaro, A.P., Oirdi, M.E., Dufour, V., Malamud, F., Dow, J.M., Bouarab, K. and Vojnov, A.A. (2007a) Bacterial cyclic {beta}-(1,2)-glucan acts in systemic suppression of plant immune responses. *The Plant Cell* 19, 2077–2089.

Rigano, L.A., Siciliano, F., Enrique, R., Sendin, L., Filippone, P., Torres, P.S., Questa, J., Dow, J.M., Castagnaro, A.P., Vojnov, A.A. and Marano, M.R. (2007b) Biofilm formation, epiphytic fitness, and canker development in *Xanthomonas axonopodis* pv. *citri. Molecular Plant–Microbe Interactions* 20, 1222–1230.

Romling, U. and Amikam, D. (2006) Cyclic di-GMP as a second messenger. *Current Opinion in Microbiology* 9, 218–228.

Romling, U., Gomelsky, M. and Galperin, M.Y. (2005) C-di-GMP: the dawning of a novel bacterial signalling system. *Molecular Microbiology* 57, 629–639.

Ross, P., Mayer, R., Weinhouse, H., Amikam, D., Huggirat, Y., Benziman, M., De Vroom, E., Fidder, A., De Paus, P., Sliedregt, L.A., van der Marel, G.A.and van Boom, J.H. (1990) The cyclic diguanylic acid regulatory system of cellulose synthesis in *Acetobacter xylinum.* Chemical synthesis and biological activity of cyclic nucleotide dimer, trimer, and phosphothioate derivatives. *Journal of Biological Chemistry* 265, 18933–18943.

Russo, D.M., Williams, A., Edwards, A., Posadas, D.M., Finnie, C., Dankert, M., Downie, J.A. and Zorreguieta, A. (2006) Proteins exported via the PrsD-PrsE type I secretion system and the acidic exopolysaccharide are involved in biofilm formation by *Rhizobium leguminosarum. Journal of Bacteriology* 188, 4474–4486.

Ryan, R.P., Fouhy, Y., Lucey, J.F., Crossman, L.C., Spiro, S., He, Y.W., Zhang, L.H., Heeb, S., Camara, M., Williams, P. and Dow, J.M. (2006) Cell-cell signaling in *Xanthomonas campestris* involves an HD-GYP domain protein that functions in cyclic di-GMP turnover. *Proceedings of the National Academy of Sciences, USA* 103, 6712–6717.

Ryan, R.P., Fouhy, Y., Lucey, J.F., Jiang, B.L., He, Y.Q., Feng, J.X., Tang, J.L. and Dow, J.M. (2007) Cyclic di-GMP signalling in the virulence and environmental adaptation of *Xanthomonas campestris. Molecular Microbiology* 63, 429–442.

Ryjenkov, D.A., Tarutina, M., Moskvin, O.V. and Gomelsky, M. (2005) Cyclic diguanylate is a ubiquitous signaling molecule in bacteria: insights into biochemistry of the GGDEF protein domain. *Journal of Bacteriology* 187, 1792–1798.

Ryjenkov, D.A., Simm, R., Romling, U. and Gomelsky, M. (2006) The PilZ domain is a receptor for the second messenger c-di-GMP: the PilZ domain protein YcgR controls motility in enterobacteria. *Journal of Biological Chemistry* 281, 30310–30314.

Scarpari, L.M., Lambais, M.R., Silva, D.S., Carraro, D.M. and Carrer, H. (2003) Expression of putative pathogenicity-related genes in *Xylella fastidiosa* grown at low and high cell density conditions *in vitro. FEMS Microbiology Letters* 222, 83–92.

Slater, H., Alvarez-Morales, A., Barber, C.E., Daniels, M.J. and Dow, J.M. (2000) A two-component system involving an HD-GYP domain protein links cell-cell signalling to pathogenicity gene expression in *Xanthomonas campestris. Molecular Microbiology* 38, 986–1003.

Stone, B.A. and Clarke, A.E. (1992) *Chemistry and Biology of (1→3)-β-D-Glucans.* La Trobe University Press, Victoria, Australia.

Talaga, P., Stahl, B., Wieruszeski, J.M., Hillenkamp, F., Tsuyumu, S., Lippens, G. and Bohin, J.P. (1996) Cell-associated glucans of *Burkholderia solanacearum* and *Xanthomonas campestris* pv. *citri*: a new family of periplasmic glucans. *Journal of Bacteriology* 178, 2263–2271.

Tang, J.L., Liu, Y.N., Barber, C.E., Dow, J.M., Wootton, J.C. and Daniels, M.J. (1991) Genetic and molecular analysis of a cluster of *rpf* genes involved in positive regulation of synthesis of extracellular enzymes and polysaccharide in *Xanthomonas campestris* pathovar *campestris. Molecular Genetics and Genomics* 226, 409–417.

Thorne, L., Tansey, L. and Pollock, T.J. (1987) Clustering of mutations blocking synthesis of xanthan gum by *Xanthomonas campestris. Journal of Bacteriology* 169, 3593–3600.

Torres, P.S., Malamud, F., Rigano, L.A., Russo, D.M., Marano, M.R., Castagnaro, A.P., Zorreguieta, A., Bouarab, K., Dow, J.M. and Vojnov, A.A. (2007) Controlled synthesis of the DSF cell-cell signal is required for biofilm formation and virulence in *Xanthomonas campestris. Environmental Microbiology* 9, 2101–2109.

Vojnov, A.A., Slater, H., Daniels, M.J. and Dow, J.M. (2001a) Expression of the gum operon directing xanthan biosynthesis in *Xanthomonas campestris* and its regulation *in planta. Molecular Plant–Microbe Interactions* 14, 768–774.

Vojnov, A.A., Slater, H., Newman, M.A., Daniels, M.J. and Dow, J.M. (2001b) Regulation of the synthesis of cyclic glucan in *Xanthomonas campestris* by a diffusible signal molecule. *Archives of Microbiology* 176, 415–420.

Vojnov, A.A., Bassi, D.E., Daniels, M.J. and Dankert, M.A. (2002) Biosynthesis of a substituted cellulose from a mutant strain of *Xanthomonas campestris. Carbohydrate Research* 337, 315–326.

Wang, L.H., He, Y., Gao, Y., Wu, J.E., Dong, Y.H., He, C., Wang, S.X., Weng, L.X., Xu, J.L., Tay, L., Fang, R.X. and Zhang, L.H. (2004) A bacterial cell-cell communication signal with cross-kingdom structural analogues. *Molecular Microbiology* 51, 903–912.

Wilson, T.J., Bertrand, N., Tang, J.L., Feng, J.X., Pan, M.Q., Barber, C.E., Dow, J.M. and Daniels, M.J. (1998) The rpfA gene of *Xanthomonas campestris* pathovar *campestris*, which is involved in the regulation of pathogenicity factor production, encodes an aconitase. *Molecular Microbiology* 28, 961–970.

Yun, M.H., Torres, P.S., El Oirdi, M., Rigano, L.A., Gonzalez-Lamothe, R., Marano, M.R., Castagnaro, A.P., Dankert, M.A., Bouarab, K. and Vojnov, A.A. (2006) Xanthan induces plant susceptibility by suppressing callose deposition. *Plant Physiology* 141, 178–187.

# 10 Suppression of Induced Plant Defence Responses by Fungal and Oomycete Pathogens

ABDELBASSET EL HADRAMI,[1] ISMAIL EL HADRAMI[2] AND FOUAD DAAYF[1]

[1]*University of Manitoba, Winnipeg, Manitoba, Canada;* [2]*University Cadi Ayyad, Marrakech, Morocco*

## Abstract

Unlike animals, plants are not mobile and do not have antibodies to mediate their resistance to pathogens. However, through their evolution, they have acquired the ability of adapting to harsh environmental conditions and a variety of defence mechanisms that allow them to stop, or at least slow down, invasion by pathogens. In parallel, plant pathogens have evolved strategies to overcome plant defence barriers. One of these strategies involves the production of plant defence suppressors. In this review, plant defence suppression by fungal pathogens is discussed in light of current knowledge about plant responses and signalling. This topic has been well documented in pathosystems involving bacteria and viruses. Therefore, this chapter focuses on interactions between plants and their fungal and fungal-like invaders.

## 10.1 Introduction

The majority of the 100,000 known fungal species are strictly saprophytic and can survive on dead organic material as a source for nutrients. Only about 10% of them, mainly filamentous ascomycetes and basidiomycetes, are able to cause disease in plants (Knogge, 1996; Agrios, 2007). The ways in which this minority evolved mechanisms to efficiently attack plants are not well understood. While attacking plants, these fungi undergo developmental and metabolic changes. Plants have acquired the ability to adapt to these pathogens through a variety of sophisticated defence mechanisms. These include pre-established physical barriers, which may stop the pathogen from accessing plant tissues, and induced responses, such as the production of molecules ranging from pathogenesis-related proteins (PR proteins), to hydroxyprolin-rich glycoproteins (HRGP) and glycin-rich glycoproteins (Akai and Fukotomi, 1980; Hahn *et al.*,

 

1989; Köller, 1991). These proteins can agglutinate and serve as a matrix for deposition of lignin and other papillae. Induced responses also include the accumulation of antifungal secondary metabolites such as phenolics, isoprenoids, saponins and alkaloids (Bennett and Wallsgrove, 1994; Osbourn, 1996a, b).

In their journey evolving defence mechanisms to fight pathogens, plants have been facing similar dynamics in their invaders, which have been developing strategies to overcome such defences. One of these strategies involves the production of defence suppressors. Such suppression by bacteria and viruses has been well documented, and therefore will not be covered here. In this chapter, we will discuss suppression of plant defence mechanisms by fungal pathogens, in light of the current knowledge in this area.

## 10.2 Infection of Plants by Fungi and Oomycetes

Fungal pathogens live on substances that are produced by their hosts. To reach these substances, plant pathogenic fungi undergo many developmental and metabolic events to ensure their establishment in/on the host tissues. These include attachment to the plant surface, germination and formation of infection structures, penetration and colonization of the host tissues, and spore production. Plant fungal pathogens specialize in infecting either the aerial or the below-ground parts of the plant. Some of them penetrate their host tissues passively through its natural openings. Others produce infection structures such as appressoria, and/or cell-wall-hydrolysing enzymes, in order to forcefully penetrate their hosts. Such differences can be seen among biotrophic, necrotrophic and hemi-biotrophic fungi.

### Infection by biotrophic fungi and oomycetes

Biotrophic fungi and oomycetes gain their way into the plant tissues using specialized infection structures called appressoria, defined as structures used by fungal pathogens to press against, and attach to, the plant surface in preparation for infection (Hawksworth *et al.*, 1995; Schulze-Lefert and Panstruga, 2003). The mechanisms associated with the appressorium formation are diverse and often species related. The action of the appressorium during the penetration can also be reinforced by the activation of several hydrolytic enzymes including endo-polygalacturonases, cellulases, glucanases and xylanases, which are involved in the digestion of the host cell wall. For example, during infection, *Phytophthora infestans*, the oomycete causing late blight on many *Solanaceae* species, produces a cocktail of cell-wall-degrading enzymes (CWDEs) including at least two types of polygalacturonases, four galactanases and two pectinesterases (Judelson and Blanco, 2005).

### Infection by necrotrophic fungi

Plant infection by necrotrophic fungi involves the secretion of copious amounts of CWDEs and toxins (Kolattukudy, 1985; Schaeffer *et al.*, 1994; Walton, 1996). As a result, certain layers of the host cells die, clearing the path for invasion by the fungus. The CWDEs secreted by fungi were first documented in *Colletrotrichum lindemuthianum* × *Phaseolus vulgaris* (Albersheim and Anderson, 1971). Hypocotyls of young seedlings were used to highlight the secretion by *C. lindemuthianum* of CWDEs, including polygalacturonases and related enzymes, that is α- and β-galactosidases, β-xylosidase and α-arabinosidase. The fragments produced upon partial degradation of the cell walls are called oligosaccharines and play an important role in plant protection against fungi (see below; Ryan, 1987). CWDEs do not completely alter the basic structure of the host cells, but lead to local perforation and the cells are kept alive. Several CWDEs produced by fungi are induced upon contact with the host plant. An external signal, generated by the host cells, seems to be required for such induction. Plants were also shown to produce specific inhibitors of CWDEs. An inhibitor of α-galactosidase, produced by *C. lindemuthianum*, was isolated from hypocotyls of *P. vulgaris* and was able to strongly inhibit (40-fold) the activity of the enzyme. This inhibitor was a glycoprotein with a high affinity for sugar residues (Albersheim and Anderson, 1971; Albersheim and Valent, 1974). Fungal pathogens *Botrytis cinerea* and *Alternaria* spp. are other necrotrophs known to cause extensive damage during the early stages of infection, by promoting cell death in the plant hosts through the secretion of phytotoxins.

### Infection by hemi-biotrophic fungi

The penetration of plant tissues by germinating fungal spores or hyphae occurs, in the simplest case, through a wound in the epidermis or cuticle or through open stomata. Some groups of fungi secrete toxins (i.e. fusicoccin) that increase the influx of potassium into the guard cells of the stomata to keep them permanently open. Hemi-biotrophic fungi usually develop as biotrophs when plant tissues are still healthy, then switch to a necrotrophic mode when their plant hosts die. This occurs often for fungi with both sexual and asexual stages.

## 10.3 Plant Defences against Fungi

### Overview

Most fungal plant pathogens grow, preferentially or exclusively, on a limited number of hosts. Several factors contribute to their specificity and host range. As soon as fungal pathogens come in contact with their hosts, they are detected and confronted with an active defence system commonly called 'basal defence' or plant 'immune' system. A successful plant defence response is then based on an effective surveillance system, enabling the early recognition of the pathogens, and the activation of further processes that will prevent them from moving forward. Successful pathogens, on the other hand, will have the ability to neutralize such plant defences. During their coevolution, plants and fungi have shaped these highly specialized plant–fungus interactions to coexist.

Besides the gene-for-gene system (Flor, 1955, 1971), the molecular basis by which a plant recognizes a fungal pathogen are still poorly understood. Plants may recognize their fungal invaders through elements present in their cell walls (e.g. chitin, glucans) or secreted (e.g. proteins) into the interplay space. Recognition can also occur through other factors such as plant cell wall fragments (e.g. oligogalacturonates) resulting from the activity of hydrolytic enzymes. Once the pathogen is recognized, a series of plant-defence-related reactions take place. These include ion fluxes across the plant plasma membrane, generation of highly reactive oxygen species (ROS, oxidative burst), phosphorylation of specific proteins, activation of enzymes involved in strengthening the cell wall, transcriptional activation of numerous defence genes, induction of phytoalexins, localized cell death at the infection site (hypersensitive response, HR), and induction of systemic acquired resistance (SAR) in distal plant organs (Baron and Zambryski, 1995; Kombrink and Somssich, 1995; Bent, 1996; Crute and Pink, 1996; Dangl *et al.*, 1996; Hammond-Kosack and Jones, 1996; Ryals, 1996). A plant defence mechanism can be effective against some, but not necessarily on other pathogens. For example, many of the plant defences that are effective against biotrophic fungi rely on programmed cell death and the activation of salicylic acid-/ethylene-dependent pathways, whereas cell death would not stop necrotrophic fungi from developing on host tissues. For the latter, plants have evolved other mechanisms mainly relying on other alternative signalling cascades such as the jasmonic acid pathway. More complexity applies in the case of hemi-biotrophs, which grow on living plant tissues until these become senescent, then switch to a necrotrophic mode where they complete the rest of their life cycle.

During the early stages of plant–fungal interactions, a number of signal molecules are released both from the host and the pathogen, thus dictating the outcome of such interactions. If the plant senses the invader's signals, it triggers defences to counter its progress, whereas the absence of such signals may lead to susceptibility. Upon sensing the fungal pathogen signals, the plant may activate defence responses, that is cell-wall reinforcement, secretion of antimicrobial proteins and/or phytoalexins (Dixon and Harrison, 1990; Bradley *et al.*, 1992; Nicholson and Hammerschmidt, 1992; Levine *et al.*, 1994; Chen *et al.*, 2000; Prell and Day, 2001; Salles *et al.*, 2002). An ultimate plant response is the HR that leads to cell death and to restriction of the pathogen from progressing further than the penetration sites. This usually involves an early generation of ROS (Jabs, 1999), predominantly ion superoxide, hydrogen peroxide, hydroxyl radicals and nitric monoxide (Lamb and Dixon, 1997; Von Tiedemann, 1997; Wojtaszek, 1997). The release of copious amounts of these ROS and the pH changes across the plasmalemma are typical characteristics of a plant tissue undergoing a HR (Doke, 1983; Wojtaszek, 1997; Dorey *et al.*, 1999). Anion superoxide is a potent toxic free radical that is able to destroy host cells, rendering them non-usable by the invading pathogen, especially biotrophs, which are dependent on living cells for their survival. This oxidative burst is considered to be a crucial part of plants' defence arsenal against fungal pathogens (Baker and Orlandi, 1995; Carver *et al.*, 1999; Baker *et al.*, 2000). On the invader side, perception of plant signals will activate a weaponry arsenal to invade the host tissues and eventually overcome plant defences (Low and Merida, 1996; Ebel and Mithöfer, 1998; Borden and Higgins, 2002).

With no mobile cells to protect them, plants launch signals from the infection site that get translocated systemically to other healthy plant parts, making them ready to better fight disease (Dangl and Jones, 2001; Ausubel, 2005; Chisholm *et al.*, 2006; Bent and Mackey, 2007). According to the gene-for-gene model (Flor, 1971) and the guard hypothesis (Van der Biezen and Jones, 1998; Dangl and Jones, 2001), plants seem to possess mechanisms by which they recognize their intruders, that is transmembrane pattern recognition receptors (PRRs) and nucleotide binding-leucine rich repeat (NB-LRR) proteins coded by most *R* genes (Dangl and Jones, 2001). The 'zigzag' model recently described by Jones and Dangl (2006) illustrates the amplitude of disease resistance or susceptibility, depending on the proportions of the pathogen-associated molecular patterns (PAMP)-triggered immunity (PTI), effector-triggered susceptibility (ETS) and effector-triggered immunity (ETI). Based on this model described with bacterial pathogens, there are four distinct phases during plant–pathogen effectors' interactions. The first phase leads to an early stoppage of the pathogen and prevention from any further colonization of the host tissues. This results from the recognition of PAMPs/ microbial-associated molecular patterns (MAMPs) by PRRs triggering a PTI. The second phase explains the success of virulent pathogens in inducing susceptibility in their hosts. This results from the deployment by the pathogen of virulence effectors that are able to interfere with the PTI and result into an ETS. In a third phase, the HR and cell death resulting from a specific recognition by one of the NB-LRRs of a given effector, either directly or indirectly triggers an ETI that gets accelerated and amplified to a PTI response. The last phase applies to some pathogens that have acquired, through their evolution, the ability to escape the vigilance of the host's ETI system or to suppress it.

## Nature of plant defences against fungi

Plant defences against pathogenic fungi can be structural, metabolic, or both. Within each type, defences can be either preformed or induced upon infection. For example, leaf waxes are structural preformed defences that plants use to form a water-repellent surface to reduce infections. In the case of fungi penetrating through stomata, plants have evolved strategies to alter this fungal activity by either modulating the circadian opening of their stomata or adapting the structure of their stomata. Once the fungal pathogens have made their way through the first structural barriers of defence, they face other structural defences, such as reinforced cell walls (i.e. by lignin, HRGP). Among cellular and cytoplasmic defences, a battery of chemicals including secondary metabolites (Hahlbrock and Scheel, 1989; Nicholson and Hammerschmidt, 1992) and PR proteins can be involved. Preformed defences that are involved in the biochemical warfare to limit the spread of fungi in the host tissues include fungitoxic exudates (i.e. protocatechuic acid from onion), phenolics/tannins (i.e. in potato and banana), and saponins (i.e. tomatine and avenacin in tomatoes and oats, respectively). The generation of a saponin-deficient mutant *sad* from the diploid oat species *Avena strigosa* (Osbourn, 2003) led to an increase in the plant susceptibility to infection, thus providing evidence that saponins are involved in the protection of oat species against fungal attacks. Preformed defences against fungi may also include lack of recognition between the host and the pathogen (i.e. non-host resistance); and lack of specific receptors on the host membrane to essential virulence factors of the fungus (i.e. HC-toxin, a cyclic tetrapeptide and a host-selective toxin from *Cochliobolus carbonum*, formerly known as *Helminthosporium carbonum* (HC)) or other substances that could sustain its growth and development (i.e. *Venturia inaequalis*) (Schulze-Lefert and Vogel, 2000; Vogel and Someville, 2000).

Induced defences include the synthesis and accumulation of fungitoxic compounds, which can be either synthesized upon infection or simply released from a non-toxic conjugated form via the action of hydrolases (Daayf *et al.*, 1997). These include phytoalexins (Greek: *phyton* – plant; *alexin* – protecting substance) (Müller and Börger, 1940) and phytoanticipins. Phytoalexins are low-molecular-weight antimicrobial compounds, actively inducible in plant tissues upon infection or elicitation, many of which appear to be involved in resistance to pathogens. Phytoalexins from the same plant families tend to be from the same chemical classes. For example, those from the *Solanaceae* and *Malvaceae* are usually sesquiterpenes, whereas those of the *Leguminosae* can be isoflavonoids or polyacetylenes. However, the same plant species can produce phytoalexins from more than one chemical class. Their mode of action includes effects on the membrane integrity of fungal cells, a blockage of the oxidative phosphorylation or damage to DNA. Despite their nature and effect, phytoalexins do not provide an absolute protection against fungal infections. Many pathogenic species have evolved mechanisms that protect them from these substances. Phytoalexin production is enhanced by inducers such as glucans (Albersheim and Anderson-Prouty, 1975), which are important components of fungal cell walls. In the presence of a slow-growing fungus, phytoalexin production can be activated by these polysaccharides, leading to an accumulation of amounts that are toxic to the fungus. However, a fast-growing fungus can spread and damage the plant tissues before enough phytoalexin is in place.

During the biochemical warfare, in which the plant attempts to defend itself from an invading fungus, a battery of proteins is also often released. These proteins target either the cell walls of the invader or its effectors. Pathogenesis-related proteins (PRs) (i.e. glucanases, chitinase and osmotin-like proteins) destroy fungal cell walls or alter their physiology (Wang *et al.*, 2004,

2005, 2006, 2008), while other enzymes help detoxify the pathogen toxins (i.e. fusaric acid) (El Hadrami *et al.*, 2005). Several enzymes such as oxidases (i.e. polyphenol oxidases and peroxidases) help generate oxidation products that are toxic to the pathogen (El Hadrami *et al.*, 1997; Daayf *et al.*, 2003; Arfaoui *et al.*, 2007).

The synthesis and accumulation of phytoalexins (Hammerschmidt, 1999) and PR proteins (Van Loon *et al.*, 2006) often occurs following a cascade of signal transduction reactions, involving factors such as hydrogen peroxide, nitric oxide, calcium, protein kinases and phosphatases, systemin, ethylene, salicylic, jasmonic and abscissic acids (Nawrath and Métraux, 1999; Romero-Puertas *et al.*, 2004; Catinot *et al.*, 2008). In spite of the established cross-talk among these pathways (Pieterse *et al.*, 2001; Kunkel and Brooks, 2002; Nandi *et al.*, 2003; Romero-Puertas *et al.*, 2004), many questions remain unanswered regarding these interactions in different crops.

## 10.4 Plant Defence Suppression by Fungi

### Fungal suppressors of plant defences

Fungi produce metabolites, including elicitors (Latin: *elegere* – to choose), which lead to recognition by their host plant. From a coevolution point of view, this represents a counterproductive strategy for fungal pathogenesis. Therefore, fungi have evolved mechanisms that could elude the recognition by the host or interfere with the plant's innate defences. Secretion of fungal suppressors of the defence responses falls under this strategy (Bushnell and Rowell, 1981).

In most documented models, the activity of elicitors has been explained by the action of a specific plant receptor that binds to the elicitor, leading to initiation of a signal transduction cascade and activation of defence responses. Given the sequence of events that occur following the recognition of the elicitor, suppressors may directly interfere with its binding, alter the signal transduction, and inhibit defence gene expression.

While elicitors from plant pathogenic fungi are able to induce active resistance through a variety of chemical and physical barriers, their suppressors have been suggested to delay or prevent such responses and/or to condition plant tissues to susceptibility in a species-specific or a race-cultivar-specific manner (Shiraishi *et al.*, 1994; Yoshioka *et al.*, 1995). Suppressors produced by fungal pathogens are then suggested as determinants of specificity. Several characterized suppressors belong to glycoproteins, glycopeptides, peptides and both anionic and non-anionic glucans (Shiraishi *et al.*, 1994; Andreu *et al.*, 1998). Their activity includes the inhibition of cell death during the HR (Doke, 1975; Storti *et al.*, 1988), of superoxide and phytoalexin accumulation (Oku *et al.*, 1977; Shiraishi *et al.*, 1978; Doke *et al.*, 1979; Doke, 1983; Ziegler and Pontzen, 1982; Kessmann and Barz, 1986; Andreu *et al.*, 1998; Ozeretskovskaya *et al.*, 2001), of the deposition of silicon-containing material

(Heath, 1981) and of infection inhibitors (Yamamoto *et al.*, 1986). Some of these suppressors are able to turn resistant/tolerant plants into hosts that become susceptible to the weakest or avirulent strains (Shiraishi *et al.*, 1978; Oku *et al.*, 1980, 1987; Kodama *et al.*, 1989). Apart from its ability to suppress plant defences and induce local susceptibility in the host plant, a suppressor is generally host specific, but with no apparent phytotoxicity to plant cells, as opposed to host-specific toxins produced by certain pathogens (Wolpert *et al.*, 2002). However, it can disturb fundamental functions of the host plasma membranes. For example, the suppressor secreted by *Mycosphaerella pinodes* is able to inhibit both the ATPase activity and the polyphosphoinositide metabolism in pea plasma membranes, causing a temporary suppression of signal transduction. This leads to a delay in the expression of genes encoding key enzymes in the biosynthetic pathway of the phytoalexin pisatin (Yoshioka *et al.*, 1990, 1992a, b; Shiraishi *et al.*, 1991a, b; Toyoda *et al.*, 1992, 1993; Kato *et al.*, 1993).

Fungal and oomycete effectors act either in the extracellular matrix or inside the host cell. In the interaction tomato × *Cladosporium fulvum*, many extracellular fungal effectors are detectable by the host receptor like-proteins (RLPs). In *Arabidopsis thaliana* × *Hyaloperonospora parasitica*, the *Atr* products seem to carry a signal peptide for secretion and probably act inside the host cell (Allen *et al.*, 2004). This was also found in *P. infestans* Avr3a proteins, which have an RxLR motif enabling them to be imported into the host cells (Bhattacharjee *et al.*, 2006) and in Avra10 from *Blumeria graminis* f. sp. *hordei* (Jones and Dangl, 2006).

Fungal pathogens produce suppressors that can lead to susceptibility in their host plant, through alteration of secondary metabolism pathways, including those leading to phytoalexins, suppression of other defence-related genes, and interference with plasma membrane ATPases and transmembrane signalling cascades (Table 10.1).

Most phytopathogenic fungi commonly infect their host using conidiospores. The initial interaction often occurs via substances secreted into the spore germination fluids. For instance, cystospores of *P. infestans* exude small anionic and non-anionic water-soluble glucans into their germination fluid, and the amounts of both types increase during incubation. These glucans were shown to suppress, in a race-cultivar-specific manner, both cell death during the HR and phytoalexin production (Doke *et al.*, 1979; Andreu *et al.*, 1998; Ozeretskovskaya *et al.*, 2001).

Suppression of host defence responses is thought to play an important role in plant–microbe interactions, especially those involving biotrophic/hemi-biotrophic pathogens, such as *P. infestans*, which require living plant tissues to establish a successful infection (Heath, 2000). To date, the nature and mode of action of plant defence suppressors, though well documented in plant–virus and plant–bacteria interactions (Bouarab *et al.*, 2002; Walton, 2002; Abramovitch *et al.*, 2003; He *et al.*, 2006), are not as well understood for interactions involving fungi or oomycetes. Examples of the production of suppressors by fungi have been shown for *M. pinodes* (Oku *et al.*, 1977; Shiraishi *et al.*, 1978; Doke *et al.*, 1979), *Ascochyta rabiei* (Daniels and

**Table 10.1.** List of known fungal suppressors.

| Pathosystem | Chemical nature of the suppressor | Origin of the suppressor | Suppressed defence responses[a] | Specificity/mode of action | References |
|---|---|---|---|---|---|
| Potato × *Phytophthora infestans* | Anionic and non-anionic glucans/ kazal-like extracellular serine proteases | Germination fluid and hyphae | Superoxide/HR/phyto-alexin/host proteases | Cultivar-race/ NADPH oxidase/ inhibits host proteases | Doke (1975), Tian *et al.* (2004) |
| Tomato × *P. infestans* | Glucans | Germination fluid | HR/phytoalexin | Cultivar-race/n.d.[b] | Storti *et al.* (1988) |
| Pea × *Mycosphaerella pinodes* | Glycopeptides | Germination fluid | Phytoalexin/PR proteins/infection | Species/ATPase | Oku *et al.* (1977), Kessmann and Barz (1986), Shiraishi *et al.* (1992) |
| Chickpea × *Ascochyta rabiei* | Glycoprotein | Culture filtrate | Phytoalexin | Cultivar-race/n.d. | Kessmann and Barz (1986) |
| Onion × *Botrytis* sp. | Peptide | Germination fluid | Infection | Species/plasma membrane | Kodama *et al.* (1989) |
| Cucumber × *Mycosphaerella melonis* | Glycopeptides | Germination fluid | Infection | Species/n.d. | Oku *et al.* (1987) |
| Soybean × *Phytophthora megasperma* f. sp. *glycinea* | Mannanglycoproteins/ invertase | Culture filtrate | Phytoalexins | Cultivar-race/n.d. | Ziegler and Pontzen (1982) |
| Chrysanthemum × *Mycosphaerella ligulicola* | Glycopeptides | Germination fluid | Infection | Species/n.d. | Oku *et al.* (1987) |
| Bean × *Uromyces phaseoli* | n.d. | Infection structures | Infection/silicon deposits | Species/n.d. | Heath *et al.* (1981) |

[a] Abbreviations used: HR, hypersensitive response; PR proteins, pathogenesis-related proteins.
[b] n.d., not determined.

Hadwiger, 1976) and *Alternaria alternata* (Hayami *et al.*, 1982; Yamamoto *et al.*, 1984). The most advanced pathosystems use elicitors of phytoalexins to illustrate the interactions with fungal suppressors. Several fungal species have evolved mechanisms to detoxify these phytoalexins (Shiraishi *et al.*, 1978; Yoshioka *et al.*, 1990). Suppressors (Doke *et al.*, 1979) can inhibit phytoalexin production and may be key pathogenicity factors for the fungi that produce them. In the majority of cases where both elicitors and suppressors have been characterized, the molecular and biochemical basis of the interactions between these two types of effectors have not been well documented. Studies of soybean responses to *Phytophthora megasperma* f. sp. *glycinea*, using $^{14}C$-labelled and unlabelled glucanase-released elicitor prepared from cell walls of the oomycete, have characterized the binding site of the elicitor, shown the correlation between the activity of the elicitor and the accumulation of glyceollin, and demonstrated that the elicitor-suppressor mycolaminaran acts at the receptor-binding site (Yoshikawa and Sugimoto, 1993).

### Induction of susceptibility

An example of suppressors that induce susceptibility is that of *ror* mutants, isolated as suppressors of *mlo*-resistance by significantly reducing resistance to *B. graminis* f. sp. *hordei* in barley (Freialdenhoven *et al.*, 1996). Resistance to powdery mildew pathogen penetration in this plant species is an important inducible defence mechanism (Thordal-Christensen *et al.*, 2000; Zeyen *et al.*, 2002). It is activated by general elicitors and leads to a local cell wall fortification by accumulation of papillae in the inner part of the plant cell walls at the fungal penetration site. The *mlo*-based resistance in barley towards *B. graminis* f. sp. *hordei* represents one of these mechanisms and gives a complete protection against the pathogen. Mutations in MLO often lead to a decrease in papillae formation. Thordal-Christensen (2003) reported on two genes, *PEN1* and *PEN2*, involved in penetration, which reduced the ability of *A. thaliana* plants to stop conidia from *B. graminis* f. sp. *hordei* by about 20%, as compared to wild-type plants. *PEN1* seemed to be involved in vesicle trafficking in the penetration resistance while *PEN2* seemed to cause constitutive cell wall changes that act as preformed barriers. *PEN1* and *PEN2* were found to be functional homologues of *ROR1* and *ROR2*, which are suppressors of *mlo*-resistance in barley. Likewise, the EDS1 (Enhanced Disease Susceptibility) was characterized as a protein that is necessary for the *R*-gene-mediated resistance to many pathogens in *A. thaliana* (Aarts *et al.*, 1998). The ubiquitin ligase-associated protein SGT1 (Suppressor of the G2 allele of *SKP1*) was also reported to mediate *R*-gene resistance in many plant species towards a variety of pathogens (Dodds and Schwechheimer, 2002; Peart *et al.*, 2002). Using virus-induced gene silencing (VIGS) of SGT1 in *Nicotiana benthamiana*, Peart *et al.* (2002) were able to show that non-host resistance against two bacterial pathogens requires the SGT1 protein. This was also the case for resistance mediated through NBS-LRR-type *R* genes and the non-LRR *R* gene *pto*.

## Interference with the elicitor-receptor activity

Oligosaccharides and glycopeptides represent some of the families of elicitors isolated from fungi and oomycetes (De Wit and Kodde, 1981; Farmer and Helgeson, 1987; Parker *et al.*, 1991). Other families include molecules such as arachidonic acid and eicosapentaenoic acid, isolated from the potato pathogen *P. infestans*, that are able to elicit phytoalexin accumulation (Bostock *et al.*, 1983). Elicitors can be also endogenous to the plant. This occurs via the activity of CWDEs from the attacking pathogen, indirectly activating plant defence responses through the release of endogenous elicitor-active fragments (Boller, 1989; Dixon and Lamb, 1990). Virulent fungal pathogens in many pathosystems seem to circumvent plant defences by secreting suppressors inhibiting the recognition of the elicitor (Bushnell and Rowell, 1981; Heath, 1981). For example, in the soybean × *P. megasperma* f. sp. *glycinea* pathosystem (Ziegler and Pontzen, 1982), accumulation of the main phytoalexin glyceollin, typically induced by a glucan elicitor derived from the cell walls of the oomycete, was reported to be suppressed by an invertase from a pathogenic race of *P. megasperma* f. sp. *glycinea*. Investigation of the suppressor activity revealed involvement of the carbohydrate moiety of the invertase since the suppressor activity was abolished by pronase and almost completely by endo-β-*N*-acetylglucosaminidase H, α-mannosidase or periodates oxidation and remained intact after heat treatment (Ziegler and Pontzen, 1982). Other fungal and oomycete species-derived suppressors include compounds from *P. infestans* that block the HR of potato and tomato (Doke *et al.*, 1979; Storti *et al.*, 1988), a protein fraction from culture filtrates of *A. rabiei* that inhibits phytoalexin accumulation in chickpea (Kessmann and Barz, 1986), and lycopeptides from germination fluids of *M. pinodes* (Yamada *et al.*, 1989) that delay the induction of phenylalanine ammonia-lyase (PAL) and phytoalexin accumulation in pea. The mode of action of most of these suppressors is not fully understood. However, it is believed that these suppressors may bind to an elicitor receptor, preventing it from binding and consequently from inducing the expression of essential defence-related genes and signalling cascades (Bushnell and Rowell, 1981; Heath, 1981). Basse and Boller (1992) provided certain evidence for the existence of a suppressor competitively inhibiting the elicitor binding. By applying elicitor-active compounds derived from a yeast extract, they were able to elicit stress responses in tomato cell suspensions (Felix *et al.*, 1991a, b; Grosskopf *et al.*, 1991). Futher characterization of the yeast extract and specifically the fraction that was active in the elicitation showed the presence of several monosaccharides such as glucose and *N*-acetylglucosamine and a high number of mannose residues and glycopeptides. Applied separately, neither the carbohydrate nor the peptide part of these molecules was able to elicit plant defences, confirming that both parts are simultaneously required for such an acivity. The oligosaccharide part, applied in the presence of the glycopeptide from which it originated, exhibited an inhibition towards its elicitor activity. This effect was reversible, dose dependent and specific, suggesting that the oligosaccharide is able to suppress the elicitor activity through competition for binding sites (Basse and Boller, 1992; Basse *et al.*, 1992). The suppressors had no effect on the response of the tomato cell suspensions to a different elicitor, derived from cell walls of *P. megasperma* f. sp. *glycinea*. This result suggests the existence of different recognition sites for different elicitors in tomato cells and that the characterized oligosaccharide suppressors act specifically on the perception of just one elicitor. The authors then put forward the hypothesis that the suppressors bind to one of the elicitor recognition sites, without producing a signal, thereby preventing induction of the stress responses by the corresponding elicitor.

### Interference with plasma membrane ATPases and signalling cascades

One of the suppression theories is based on an inhibition of the interactions between pathogen elicitors and their corresponding receptors in the host plant by blocking the binding sites (Doke *et al.*, 1979; Garas *et al.*, 1979). Several studies have also suggested that suppressors act by blocking signal(s) transduction during the elicitor-mediated activation of defence responses (Yoshioka *et al.*, 1990; Shiraishi *et al.*, 1991a, b; Toyoda *et al.*, 1992) or by affecting the formation of binding complexes in the promoter region, hence leading to the suppression of expression of specific defence-related genes at the transcription level (Wada *et al.*, 1995).

The oxidative burst and the hypersensitive cell death, in case of an incompatible interaction, are among the signalling cascades targeted for inhibition by fungal pathogens. In tomato, many fungal pathogens produce extracellular enzymes, commonly referred to as tomatinases (Roddick, 1974; Ruiz-Rubio *et al.*, 2001) that are able to detoxify the preformed antifungal steroidal glycoalkaloid α-tomatine (Fig. 10.1). This molecule is known to be the main phytoanticipin in tomato (Arneson and Durbin, 1968a, b; Roddick, 1974) and shows a uniform accumulation in tomato tissues (Arneson and Durbin, 1968a, b). When applied at high concentrations, it can inhibit a variety of microbes (Sandrock and Van Etten, 1998). *Fusarium oxysporum* f. sp. *lycopersici* is a notorious pathogen of tomato where it causes a serious vascular wilt disease. This pathogen was reported to cleave α-tomatine (Fig. 10.1) into its aglycon (tomatidine) and tetrasaccharide moieties (lycotetrose). These two by-products of the detoxification of α-tomatine have little to no antifungal activity against the pathogen, hence suggesting that the production of tomatinase is linked to the pathogenicity of the fungus (Ruiz-Rubio *et al.*, 2001). Ito *et al.* (2004) reported that both tomatidine and lycotetrose are able to inhibit the oxidative burst and the hypersensitive cell death in tomato-cell suspensions. Moreover, the authors claimed, using tomato cuttings supplemented or not with tomatidine and lycotetrose and a non-pathogenic isolate of *F. oxysporum* f. sp. *lycopersici*, that no fungal colonies were observed on the inoculated tomato cuttings in the absence of tomatidine and lycotetrose and that the pathogen developed in the xylem tissues in the presence of both products. This suggests that both molecules conditioned the cuttings for the establishment of the disease by a non-pathogenic isolate by suppressing the plant defence responses. In another study involving the tomato leaf spot pathogen *Septoria lycopersici*, which is also able to produce tomatinase and hydrolyse α-tomatine to α-2-tomatine (Fig. 10.1), it has been reported that induced defence responses of the host were suppressed upon hydrolysis (Bouarab *et al.*, 2002). Such suppression seemed to occur through a mechanism not yet determined, in which an interference with fundamental signal transduction processes renders resistant cultivars susceptible to the pathogen.

Successful pathogens are able to interfere with cascades of signalling that mediate their host defences. One of the targeted signalling pathways has been shown to be the ROS-mediated cascade. Given the ways this cascade functions, fungal suppression may occur through the secretion into the host tissues of ROS-scavenging molecules. The secretion of catalases and superoxide dismutases into plant tissues was reported in the literature (Katsuwan and Anderson, 1990; Klotz and Hutcheson, 1992; DeGroote *et al.*, 1997; San Mateo *et al.*, 1998). Other molecules such as oxalic acid and mannitol have also been reported to be secreted by pathogens to mute the ROS-signalling cascade in the host tissues (Jennings *et al.*, 1998; Cessna *et al.*, 2000). Oxalates are widely involved in fungal metabolism and it is well established that certain fungal pathogens (i.e. *Sclerotinia sclerotiorum*, *Sclerotium rolfsii*) secrete oxalic acid as part of their invasion process of plant tissues (Noyes and Hancock, 1981; Franceschi, 1989). *Sclerotinia sclerotiorum* and *S. rolfsii*, causing serious diseases in over 200 plant species, were shown to secrete substantial amounts of oxalates in infected plant tissues, suggesting a link to the pathogenicity of these two fungal species (Maxwell and Bateman, 1968; Noyes and Hancock, 1981). However, the specific role of oxalic acid in the infection process is still unclear. Keates *et al.* (1996) suggested that this molecule might have a number of functions including chelating calcium from

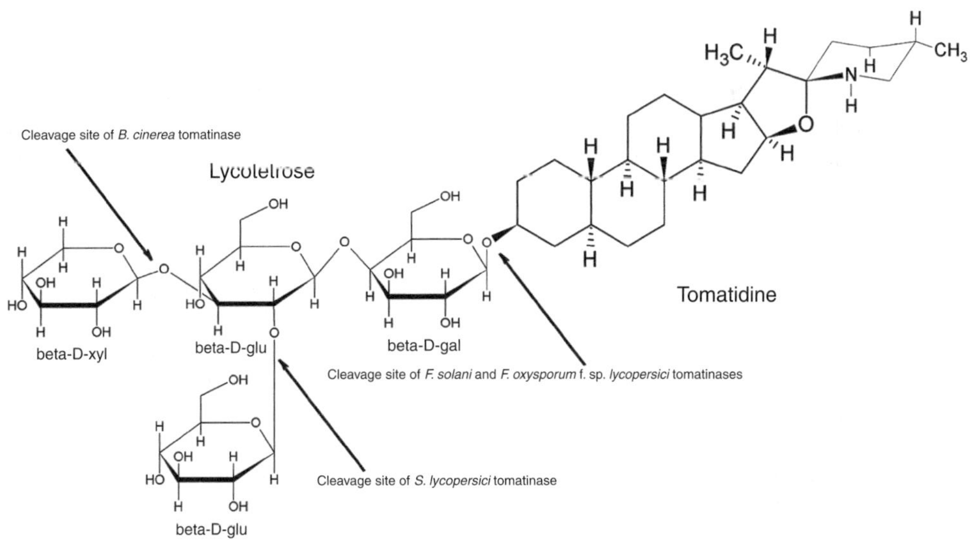


**Fig. 10.1.** Chemical structure of α-tomatine and its by-products (tomatidine linked to lycotetrose) released upon the activity of tomatinases from *Botrytis cinerea*, *Septoria lycopersici*, *Fusarium solani* and *Fusarium oxysporum* f. sp. *lycopersici*.

the cell wall thus making the pectic fraction more available to CWDEs, and providing an acidic pH required for their maximum activities. Other studies have suggested that oxalic acid produced by fungi during infection may play a key role in lignin biodegradation through its stimulation of lignin-degrading enzymes such as Mn peroxidase (Kuan and Tien, 1993). Knowing that the host plant may degrade the fungal oxalic acid through oxalate oxidase (Çalıskan and Cuming, 1998) or induce its transformation to soluble or insoluble salts, oxalic acid-producing pathogens may have evolved other mechanisms to suppress/circumvent these defence processes.

In the tomato × *F. oxysporum* f. sp. *lycopersici* interaction discussed earlier in this section, Ito *et al.* (2004) demonstrated that the fungus utilizes the main phytoanticipin of the host as a substrate by its pathogenicity factor tomatinase to produce by-products such as tomatidine and lycotetrose, which suppress host defences. This, along with results showing that mutant strains of *F. oxysporum* f. sp. *lycopersici* with low tomatinase activity exhibit low pathogenicity on tomato (Ito *et al.*, 2002), provides evidence that this pathogen had evolved a counter-defence mechanism involving the suppression of an essential defence response pre-set by the host.

The grey mould causal agent *B. cinerea* is a necrotrophic fungus that is able to infect a wide range of hosts and needs to kill plant tissues in order to feed on them (Prins *et al.*, 2000). This is due mainly to the activity of pectinolytic enzymes released by the pathogen during infection. Reports on *B. cinerea* polygalacturonase activity showed no correlation with the aggressiveness levels of several isolates, suggesting the presence of other important pathogenicity factors (Leone and Tonneijck, 1990). In this perspective, it has been shown that infections with *B. cinerea* are associated with an induction of ROS in the host tissues during the early stages of infection by the pathogen (Von Tiedemann, 1997; Unger *et al.*, 2005). Based on this observation and the discrepancies between the polygalacturonase activity and the pathogenicity of the fungus, the authors formulated a hypothesis stipulating that this pathogen forces its host plant to produce reactive oxygen intermediates, as a part of the plant's own defence responses, which in turn kill the plant tissues, enabling the establishment and spread of the pathogen. While the hypothesis was tested in other plants interacting with the same pathogen, conflicting results were reported (Govrin and Levine, 2000).

In their study, Unger *et al.* (2005) reported that a hypoaggressive isolate of *B. cinerea* was able to initiate a hypersensitive-like response on leaf tissue disks 2–3 days post-inoculation, while the aggressive isolate caused an expanding necrotic lesion that rapidly destroyed the leaf tissues. By examining the production of active oxygen intermediates and the pH of the apoplast in cell suspensions, the authors showed that aggressive isolates from the necrotrophic *B. cinerea* benefit from the suppression of plant defences to establish an infection evoking biotrophs. A biphasic oxidative burst was recorded with the non-aggressive isolate while only one phase was detected when the aggressive isolate was used for inoculation. The described biphasic phenomenon of oxidative burst and pH changes consisted of an initial superoxide burst peak with lower amplitude that was independent of the isolates' level of aggressiveness, followed by a much stronger and specific superoxide burst that was able to initiate cell death in the cell suspension. The authors showed that this second superoxide burst peak was completely suppressed when an aggressive isolate was used for inoculation. The first peak seemed to be part of the activation of pre-existing components in the plant cell wall (Baker and Orlandi, 1995). The suppressor of the second oxidative burst produced by an aggressive isolate was purified from the intercellular fluid during infection and was identified as 2-methyl succinate (2-MS) (Unger *et al.*, 2005). This suppressor seems to be an important pathogenicity factor for *B. cinerea*, since the authors reported that its secretion was proportional to the aggressiveness level of at least ten isolates. Further, adding it to the inoculation droplets enhanced lesion growth rate and significantly reduced the hypersensitive-like response to non-aggressive isolates of the pathogen. However, it is still unclear whether the purified suppressor was of fungal or plant origin. For instance, 2-MS was never isolated from pure fungal cultures and succinates are generally found in high amounts in the intercellular fluid of plants. Using succinate in different biotests, Unger *et al.* (2005) have never succeeded in showing a suppression of the superoxide burst. Therefore, the authors proposed a two-component model in which a pathogen-derived enzyme, that could be a pathogenicity factor of *B. cinerea*, will metabolize the plant-derived succinate to generate an active suppressor that could interfere with the oxidative burst-signalling pathway in the host. Yet, the action of this suppressor at the receptor level, downstream at the transduction cascade or on the reactive-oxygen-intermediate-generating oxidases has not been investigated.

In another case involving the interaction between peas and *M. pinodes*, it has been shown that orthovanadate, a suppressor produced by the fungus during infection, regulates the ATPase gene at the transcriptional level (Yoshioka *et al.*, 1992a, b). Once secreted by the pathogen, this suppressor is able to inhibit both the ATPase activity and the polyphosphoinositide metabolism in pea plasma membranes, causing a temporary suppression of signal transduction.

## Alteration of secondary metabolism pathways and phytoalexin accumulation

In the potato × *P. infestans* pathosystem, water-soluble glucans produced by *P. infestans* were reported to suppress the accumulation of sesquiterpene phytoalexins in potato tubers (Currier, 1981; Shiraishi *et al.*, 1994; Andreu *et al.*, 1998). These type of glucans can also originate from potato tissues upon activation of β-glucanases in the early stages of defence (Schröder *et al.*, 1992). Other studies have shown that *P. infestans* is capable of producing molecules other than β-glucans with an ability to suppress potato defence responses (Andreu *et al.*, 1998; Ozeretskovskaya *et al.*, 2001). Extracellular protease inhibitors such as kazal-like extracellular serine proteases have been identified in *P. infestans* and are thought to interact directly with host proteases (Tian *et al.*, 2004). Differences in the ability to produce some of these suppressors were noticed among *P. infestans* races/genotypes.

Earlier investigations of the phenylpropanoid and isoprenoid pathways in this pathosystem had suggested the suppression by *P. infestans* of potato *PAL* and *HMG* genes, controlling the early steps in each pathway, respectively (Choi *et al.*, 1992; Yoshioka *et al.*, 1996). These two genes seemed to be differentially suppressed by *P. infestans* isolates (Wang *et al.*, 2004, 2005, 2006, 2008). Highly aggressive genotypes (i.e. US8) led to a downregulation of *PAL-1* and *HMG-2* genes, resulting in a reduced accumulation of phenolic compounds and rishitin at the inoculation site. This suppression did not seem to affect PR proteins (i.e. *PR-1* and *PR-5*) (Wang *et al.*, 2008), concurring with SAR results previously described in potatoes (Cohen *et al.*, 1993). One explanation of such a specific suppression of defence-related genes *PAL* and *HMG* could be that this is due to the competitive action of suppressor(s) released by the oomycete towards the elicitors' binding activity (Doke *et al.*, 1979; Garas *et al.*, 1979). According to this model, *P. infestans* elicitors for *PR-1* and *PR-5* would be different from those for *PAL-1* and *HMG-2* and the corresponding receptors of each elicitor would be differentially affected by the action of the suppressor (Wang *et al.*, 2008). An alternative explanation would be that all these defence-related genes are activated upon the binding of the same elicitor(s) and the specific suppression of *PAL-1* and *HMG-2* is occurring only during signal transduction cascades. While this appears to be the case in other pathosystems (Yoshioka *et al.*, 1990; Shiraishi *et al.*, 1991a, b), it is not in the potato × *P. infestans* interaction, since a systemic induction of locally-suppressed *PAL-1* and *HMG-2* genes was observed (Wang *et al.*, 2008), suggesting an early translocation of SAR signal(s). The third scenario would be that suppressors are directly acting at the transcription level on the formation of binding complexes in the promoter region, hence leading to the suppression of expression of specific defence-related genes (Wada *et al.*, 1995).

In other pathosystems, studies such as the one of Yoshioka *et al.* (1992a, b) have investigated the expression patterns of several genes controlling key steps in the pisatin biosynthetic pathway (i.e. PAL, chalcone synthase (CHS)) upon application of an elicitor, in the presence or absence of orthovanadate (suppressor), both from *M. pinodes*, a notorious pathogen of pea. While a marked and rapid accumulation of a 2.8 kb PAL mRNA and 1.5 kb CHS mRNA and an enhancement of the enzymatic activities of both proteins were induced by treatment with the elicitor alone, the concomitant presence of the suppressor with the elicitor caused a delay in the synthesis/accumulation of these two defence-related genes for at least 3 h post-inoculation in pea epicotyls. This delay was followed by 6 h post-inoculation delay in the enhancement of PAL activity and a 6–9 h post-inoculation delay in the initiation of accumulation of pisatin (Yamada *et al.*, 1989; Yoshioka *et al.*, 1992a). Orthovanadate, the suppressor used, acts as a suppressor of pisatin accumulation in pea epicotyls (Yoshioka *et al.*, 1992a, b) and has also previously been reported in several cases (i.e. red bean and peanut) as an activator of plant defence mechanisms. Applied alone, orthovanadate was able to inhibit PAL and CHS mRNA accumulation in pea epicotyls. These findings suggest that orthovanadate acts by suppressing the activation of these genes at the transcriptional level, typically induced by elicitors or wounding. The recovery of the accumulation of these mRNAs in tissues treated with elicitor plus orthovanadate also shows that this inhibitor neither causes cell death nor permanently neutralizes the ability of the elicitor to induce a defence response. Furthermore, the authors (Yoshioka *et al.*, 1992a, b) have examined the effect of orthovanadate on the accumulation of mRNA encoding the P-type ATPase and showed that putative ATPase occurred at almost a constant rate in pea epicotyls, even in the presence of orthovanadate. This suggests that the regulation at the transcriptional level of the ATPase gene by orthovanadate is different from that of PAL and CHS genes. In either case, the expression of these genes has gradually been recovered as a result of the biosynthesis of new ATPase molecules and PAL and CHS enzymes from the accumulated transcripts. The mechanisms of suppression and recovery of the expression of defence-related genes may then differ among host genotypes and the interactions with different pathotypes of the pathogen. However, the authors did not provide any data to support this suggestion.

Saponins are well known as plant preformed metabolites for their protective ability of plants towards abiotic and biotic stresses (Osbourn, 2003). In the case of a fungal attack, plant saponins complex with sterols from fungal membranes causing a loss of integrity. The way by which this mechanism takes place is still poorly understood. While some fungi lack membrane sterols and may escape the effect of saponins, others have evolved ways to detoxify saponins using saponin glycosyl hydrolases. For example the avenacinase, produced by *Gaeumannomyces graminis* var. *avenae*, seems to be essential for a successful infection of saponin-producing plants (Bowyer *et al.*, 1995; Osbourn *et al.*, 1995; Sandrock *et al.*, 1995; Lairini *et al.*, 1996; Wubben *et al.*, 1996; Quidde *et al.*, 1998; Becker and Weltring, 1998).

## Detoxification of phytoalexins

One of the mechanisms by which fungal pathogens defend themselves against host plant defences is the detoxification of phytolalexins. Many pathogens were reported to catabolize the phytoalexins produced by their hosts (Van Etten *et al.*, 1982). In recent years, *Leptosphaeria maculans* and *S. sclerotiorum*, important pathogens on members of the *Brassicaceae* and the causal agents of blackleg and soft rot in canola, have been shown to be able to detoxify many phytoalexins including brassicin (Pedras and Okanga, 1999; Pedras *et al.*, 2007; Sexton *et al.*, 2009). In 1964, Uehara suggested that the pathogenicity of certain fungi might depend on their ability to detoxify the phytoalexins produced by their hosts. Van Etten *et al.* (1989), investigating this thesis on various pathosystems where genes involved in the phytoalexins detoxification have been identified and/or cloned, concluded that the genes conferring phytoalexin detoxification in fungi are always linked to the pathogenicity of the fungi harbouring them. Pisatin, believed to be the first purified and chemically identified phytoalexin (Cruickshank, 1962; Perrin and Bottomley, 1962), for a long time represented the model for studying mechanisms of phytoalexin detoxification by fungi. Cruickshank (1962) observed that this phytoalexin was less toxic to the pea pathogen *Ascochyta pisi* than to *Monilinia fructicola*, known as a non-pathogen of peas. Pathogens that can demethylate pisatin (i.e. *Fusarium solani* f. sp. *pisi*, *M. pinodes*, *Phoma pinodella*) were tolerant to both pisatin and its demethylated product hydroxymaackiain. Meanwhile, Van Etten *et al.* (1982) reported that the ability to metabolize and tolerate a phytoalexin are two independent events with no cause-to-effect relationship. For instance, *Stemphylium botryosum* was shown to be sensitive to pisatin even though it can metabolize it and the same observation applies to other fungi confronted with other phytoalexins (Van Etten *et al.*, 1980). Also, fungi do not necessarily need to catabolize their host phytoalexins and some of them have acquired other mechanisms to tolerate these molecules (Smith, 1982; Van Etten *et al.*, 1982; Denny and Van Etten, 1983; Denny *et al.*, 1987). Another set of evidence that phytoalexin degradation may be a common requirement for pathogenicity was provided by studies using the main phytoalexins of legumes, medicarpin and maakiain. *Nectria haematococca* and *A. rabiei*, pathogens of chickpea, can degrade both of these phytoalexins. These two species were reported to initiate the catabolism of medicarpin and maakiain by a variety of reactions as opposed to pisatin, where most fungi, if not all, start by a 3-*O*-demethylation reaction (Kraft *et al.*, 1987; Van Etten *et al.*, 1989). Similarly, *F. solani* f. sp. *phaseoli*, a bean pathogen, was reported to detoxify at least four major phytoalexins produced by the host *P. vulgaris* (kievitone, phaseollin, phaseollidin and phaseollinisoflavan) (Smith *et al.*, 1980; Zhang and Smith, 1983) through various mechanisms. This fungus possesses hydratases that allows it to hydrate the isopentyl side chain of isoflavanone kievitone and the pterocarpan phaseollidin (Kuhn and Smith, 1978, 1979; Smith *et al.*, 1980) and probably the phaseollinisoflavan (Zhang and Smith, 1983; Wietor-Orlandi and Smith, 1985). Phaseollin is detoxified by the fungus through hydroxylation (Kistler and Van Etten, 1981). In potato, lubimin and rishitin are the main sesquiterpenoid phytoalexins that accumulate in response to a variety of pathogens including *Fusarium sambucinum* and *P. infestans*. *Fusarium sambucinum* was reported to be able to degrade both phytoalexins (Gardner *et al.*, 1988; Desjardins and Gardener, 1989; Desjardins *et al.*, 1989). The products of degradation of rishitin by this fungal species have not been identified (Desjardins and Gardener, 1989). However, at least seven derivatives of lubimin were determined to date (Gardner *et al.*, 1988; Desjardins and Gardener, 1989; Desjardins *et al.*, 1989). *Fusarium sambucinum* was also reported to be sensitive to lubimin even though it can metabolize it relatively slowly to produce 15-dihydrolubimin and isolubimin, which are both toxic to the pathogen. Interestingly, a comparison among 26 isolates of *F. sambucinum* recovered from the field and subjected to lubimin and rishitin showed variability in the rate at which these isolates metabolize both products (Desjardins *et al.*, 1989). All the isolates that were able to rapidly degrade both phytoalexins were highly aggressive on potato and tolerant to the phytoalexins and their degradation products. Isolates that showed a slow degradation of the phytoalexins were weak pathogens. The study showed also the existence of isolates that were tolerant to either lubimin or rishitin but not to both, suggesting that the two phytoalexins were in all likelihood detoxified through different mechanisms involving different enzymes (Desjardins and Gardener, 1989;

Desjardins *et al.*, 1989). Oomycetes, on the other hand, appear to not commonly rely on phytoalexin tolerance for their pathogenicity. Some were reported to catabolize phytoalexins but little is know about the contribution of this ability to their pathogenicity (Weinstein *et al.*, 1981; Van Etten *et al.*, 1982; Sweigard *et al.*, 1986). In many studies involving oomycetes, data about phytoalexins suggest that the pathogen circumvents them. This was shown for *P. megasperma* f. sp. *glycinea*, which circumvents the phytoalexin-mediated resistance in the host by either not eliciting the phytoalexin synthesis/accumulation or repressing it (Van Etten *et al.*, 1982). However, the mechanism by which this occurs was not investigated. In the case of *P. megasperma* f. sp. *medicaginis*, the causal agent of root rots in lucerne, it has been shown that the pathogen does not elicit phytoalexin biosynthesis but if it is present escapes its toxicity through mechanisms other than catabolic degradation (Pueppke and Van Etten, 1976; Sweigard and Van Etten, 1987).

Fungal pathogens have also evolved strategies other than degradation or escape to circumvent plant secondary metabolites not categorized as phytoalexins accumulated by the plant host upon infection. These may involve spatial or temporal avoidance, minimizing the damage to plant tissues, or even direct confrontation. An example of spatio-temporal avoidance was shown in avocado, where the peel of unripe fruits is known to contain copious amounts of a preformed 1-acetoxy-2-hydroxy-4-oxo-heneicosa-12.15-diene that is able to prevent the fungal decay caused by *Colletotrichum gloeosporioides* (Prusky *et al.*, 1991; Prusky and Keen, 1993). Meanwhile, *C. gloeosporioides* spatio-temporarily avoids confrontation with this antifungal diene by inducing the germination of its spores on the surface of the fruit. Germ tubes penetrate only the outer waxy layer to produce appressoria that remain quiescent until the ripening of the fruit. Then, it takes advantage of the decrease of the diene concentration associated with the ripening process to develop on the ripe fruits. As a second strategy developed by fungal pathogens to circumvent the action of plant-preformed antifungal molecules, fungi may reduce the level of damage that they induce on the host. This allows the pathogen to avoid the biologically active preformed antifungal compounds sequestered in the host cells and/or the release of conjugated forms (Osbourn, 1996a, b). With the substantial damage they cause in their hosts, necrotrophs are likely to release the majority of both active and inactive forms of antifungal compounds. In contrast, biotrophs need their host tissues to be alive and cause only limited damage to the plants in the early stages of infection. This seems to occur by escaping the plant surveillance systems.

## Suppression of other defence-related genes

As mentioned above, evidence shows certain fungal pathogens' ability to produce suppressors that interfere with plant defences (Ziegler and Pontzen, 1982; Kessmann and Barz, 1986; Shiraishi *et al.*, 1992). Confrontational mechanisms also exist among fungi that allow them to tolerate plant antifungal compounds. These mechanisms are often dependent on their ability to secrete specific degrading enzymes and to tolerate the degradation by-products (Van Etten *et al.*, 1982, 1995; Osbourn, 1996b). Certain necrotrophic pathogens such as *B. cinerea* possess the ability to detoxify a wide range of molecules including bean phytoalexins (Deverall and Vessey, 1969), the grapevine phytoalexin resveratrol (Pezet *et al.*, 1991; Sbaghi *et al.*, 1996; Pezet, 1998), the tomato steroidal glycoalkaloid α-tomatine and other saponins (Verhoeff and Liem, 1975; Urbasch, 1986; Quidde *et al.*, 1998; Osbourn, 1999). Other fungi have evolved tolerance mechanisms that are unlikely to be based on degradation of preformed plant antifungal compounds. For example, the tolerance of *Microcyelus ulei*, a pathogen of *Hevea brasiliensis*, to HCN produced directly and indirectly by the plant as a defence response, has been ascribed to cyanide-resistant respiration (Leiberei, 1988). This pathogen was also reported to be resistant to saponins thanks to the sterol content in its membranes (Arneson and Durbin, 1968a, b; Défago and Kern, 1983).

Fungal chitin is a target of the host hydrolytic enzymes to minimize/ suppress infection (Boller, 1987). Both endogenous and exogenous chitinases were reported to be effective in degrading the infection structures of fungi and inducing resistance in many plants (Toyoda *et al.*, 1991; Chet *et al.*, 1993; Kogel *et al.*, 1994; Ikeda *et al.*, 1996; Van Loon *et al.*, 2006). Fungal pathogens, especially biotrophs, have evolved mechanisms by which they can suppress the activity of such antifungal enzymes. Using protoplasm of barley epidermal cells infected with *B. graminis* f. sp. *hordei*, Fujita *et al.* (2004) demonstrated a selective suppression of chitinase gene expression in the invaded cells. In this pathosystem, it has been shown that for the pathogen to gain access to the host plant tissues, it has to circumvent the constitutive expression of chitinases that are excreted as extracellular barriers. The activity of chitinases usually increases in the aleurone layer, providing the seeds with a defence mechanism against fungi (Swegle *et al.*, 1989; Jacobsen *et al.*, 1990). Preconditioning barley leaves with a compatible race of *B. graminis* f. sp. *hordei* can lead to subsequent infection by normally incompatible races (Colhoun, 1979; Callow, 1987). This phenomenon has been ascribed to the ability of the compatible race to suppress/modulate the host defence responses at infection sites. The successful demonstration of an efficient and selective suppression of chitinase gene expression in pathogen-invaded cells is in accordance with this hypothesis (Fujita *et al.*, 2004). Taking into account that haustoria of *B. graminis* f. sp. *hordei* are produced in the interspace between the cell wall and plasmalemma of infected cells (Manners, 1993), Fujita *et al.* (2004) hypothesized that the pathogen is producing signal molecule(s) that suppress(es) the expression of the chitinase gene in the nuclei of host cells.

In other pathosystems such as potato × late blight pathogen, it has been shown that *P. infestans* is capable of producing extracellular protease inhibitors (i.e. kazal-like extracellular serine proteases) that directly interact with, or inhibit, host proteases (Tian *et al.*, 2004). Extracellular proteases are known for their inhibitory activity of other proteases in various fungal species and many of them were reported as virulence factors for bacteria. These proteases were shown to be able to cleave many proteins *in vitro* with a high specificity (Van der Hoorn, 2008), suggesting that they can be potent suppressors of the host proteases involved in the defences (Tian *et al.*, 2004). On the other hand, proteases require a colocalization in both time and space with their substrates (Van der Hoorn, 2008) but the regulation of their activity is poorly understood. Characterization of the protease cleavage specificity and the subcellular localization are still needed in order to select candidate protein substrates on the basis of their predicted colocalization with these proteases, expression and putative cleavage sites. Approaches including the yeast double-hybrid, immobilized proteins arrays and differential proteomics (Sakamoto *et al.*, 2003) may help to identify the substrates of these proteases in the future. Many proteases contain at least an auto-inhibitory domain that is proteolytically hydrolysed to allow activation. Nevertheless, the molecular mechanism by which this activation occurs is not fully understood. Likewise, the presence of endogenous inhibitors of protease activity seems to be possible, but the identity of such inhibitors is unknown. The multi-localization of proteases in different subcellular compartments suggests also that activity is regulated by pH, $Ca^{2+}$, energy and redox status.

## Suppression at the promoter region

Transcriptional regulation of plant defence genes in response to phytopathogens is mediated by the binding of transcription factors (*trans*-acting factors) to specific sequence elements (*cis*-regulatory elements) present in their promoters. The gene that encodes PAL represents one of the major plant defence-related genes in many plant species (Cramer *et al.*, 1989; Lois *et al.*, 1989; Gowri *et al.*, 1991; Wang *et al.*, 2005, 2008). This key enzyme controls the first step leading to the biosynthesis of phenolics, many phytoalexins, lignins and suberins deemed to be the major components of cell-wall fortification mechanisms during infection, and of salicylic acid, a signal molecule for SAR. Given its ubiquitous upregulation upon fungal attack it is believed that its activation is common among plant species. The *in vivo* footprinting analysis of regulatory elements in the promoter of PAL1 of pea (*PSPAL1*) in response to non-pathogenic attack showed the existence of AC-rich sequences (Box I and Box-II) that were conserved at similar positions in the phenylpropanoid gene promoters from several plants (Imura *et al.*, 2000). After verifying that the *GUS* reporter gene was expressed under the *PSPAL1* promoter at its maximal level 24 h after the inoculation with *Phytophthora capsici*, a non-pathogenic fungus of peas, the authors were able to test constructs harbouring deletion in the two boxes identified in the *PSPAL1* promoter region (Imura *et al.*, 2000). Using the *GUS* reporter gene and tobacco plants the authors revealed the function of these AC-rich sequences (Imura *et al.*, 2000). Deletion in Box-I (dB-1) led to a reduced basal expression of *PSPAL1* and a complete loss of induced defence responses to non-pathogenic fungus *P. capcisi,* suggesting the importance of such a region of the promoter to the activation of *PSPAL1*. This Box-I may be interacting with other nuclear factors leading to the expression of the defence response upon interaction with the non-pathogen oomycete since the authors reported that they were able to observe a complex between Box-I and nuclear protein(s) using an electrophoretic mobility shift assay. Box-I was also reported to be an indispensable *cis*-regulatory element required for elicitor-mediated transcriptional activation of *PsCHS1* in the transient transfection assay in pea (Seki *et al.*, 1996, 1997). This box (Box-I) is a necessary element in the promoter region although its only apparent function seems to be ensuring the full activation of *PSPAL1*. Other conserved boxes in the promoter region may be necessary for the coordinated regulation of the gene in response to pathogen invasions.

Similarly, studies on CHS have shown transcriptional regulation of the defence-related gene (*PsCHS*) that codes for this enzyme in peas in response to fungal pathogens. This key rate-limiting enzyme in the phenylpropanoid pathway plays an important role in plant defence response against pathogens (i.e. *M. pinodes*) and controls, among other things, the synthesis of pisatin, the main phytoalexin in peas (Kato *et al.*, 1995). *PsCHS* in peas is a small multigene family in which at least eight genes have been identified (An *et al.*, 1993). Based on their elicitor-inducibility, these eight genes can be subdivided into two major groups, the elicitor-inducible (*PsCHS1*, *2*, *3*, *4*, *5* and *8*) and the non-inducible groups (*PsCHS6* and *7*) (An *et al.*, 1993). In the promoter region of *PsCHS1*, an AT-rich element (ATRE) was detected and seemed to be required for the maximal elicitor-mediated activation. However, regulatory mechanisms of this element are still not fully understood. Qian *et al.* (2007) investigated the transcription and binding factors of the ATRE using an elicitor-induced pea cDNA expression library. The authors were successful in isolating an ATRE-binding factor PsATF1 and demonstrated that the PsATF1 has an ATRE-specific binding activity. Using the yeast one-hybrid system and a β-galactosidase assay, the authors showed that PsATF1 possesses a transcription-activating activity. It acts as a complete transcription activator and its activity can be combined with other *cis*-regulatory factors such as PsGBF (*Pisum sativum* G-box binding factor) in the activation of the *PsCHS1* promoter. The ATRE-binding factor PsATF1 belongs to the bZIP transcription factors family and exists in many plant gene promoters where it binds other proteins to regulate the expression of downstream genes. In peas, the PsATF1 possesses DNA specific-binding and transcription-activating/coactivating characteristics when acting with the ATRE in the *PsCHS1* promoter.

Control of gene expression involves basal transcription factors, which position RNA polymerase at the start site of transcription. These basal transcription factors are essential to the transcription but do not increase its rate. To increase the rate of transcription, the so-called activators come into play and determine which genes are going to be transcribed and at what rate. Activators and the basal transcription factors communicate via coactivators (Tansey, 2001; Braganca *et al.*, 2002, 2003). For example in peas, the two activators PsATF1 and PsGBF (Qian *et al.*, 2007) have been shown to act on the activation of the *PsCHS1* promoter. In the case of a dual activation of the *PsCHS1* promoter (both PsATF1 and PsGBF), the expression level of the reporter gene was higher than when only PsATF1 or PsGBF was activated, suggesting that PsATF1 and PsGBF interact with the basal transcription factors through different coactivators and coactivate the expression of the *PsCHS1* gene after elicitor induction. The presence of an array of activators in the surroundings of the *PsCHS1* promoter and the possibility of coactivation by several coactivators spatio-temporally increases the level of accumulation of *PsCHS1* transcripts in response to the elicitation or fungal attack. On the other hand, suppressors that target the promoter region of essential defence-related genes may interfere with the actions of activators and coactivators or even with the basal transcription factors. These suppressor factors can also reduce or temporarily abolish the transcription of defence-related genes by competing with the binding of *trans*-acting factors to specific *cis*-regulatory elements present in the promoter regions of these genes. In this perspective Schmidt *et al.* (2004) showed that the suppression of PAL in sugarbeet by the fungal pathogen *Cercospora beticola* is mediated at the core promoter of the gene through a repression signal.

In another study using a functional analysis approach of the promoter of PAL genes in peas, Yamada *et al.* (1994) identified several *cis*-regulatory regions that were necessary for the induction by fungal elicitor or UV light, and for the suppression by the fungal suppressor isolated from the germination fluid of pycnidiospores of *M. pinodes*. The *cis*-regulatory elements were identified on the basis of a transient transformation of the chimeric genes with various parts of the promoters of *PSPAL1* and *PSPAL2* into pea protoplasts. Assaying chloramphenicol acetyltransferase (CAT) transient expression mediated by these promoters' constructs had indicated that the region of *PSPAL1* spanning from −149 to +140 was responsible for almost the entire induction of the *PSPAL1* gene in response to fungal elicitor. Deletions in the 5'-upstream region up to −149 still allowed a several-fold increase in CAT activity upon treatment with fungal elicitor. In *PSPAL2*, the promoter region located between −406 and +110 appeared to be responsible for nearly a complete induction by fungal elicitor and the extent of induction upon elicitation decreased with the extent of the 5'-upstream deletions. In these promoter regions, Boxes 1 and 2 were originally identified by *in vivo* methylation footprinting analysis of *PAL-1* (Lois *et al.*, 1989; Hauffe *et al.*, 1991) and the *CHS* gene (Schulze-Lefert *et al.*, 1989) in parsley suspension cultured cells and the *CHS* promoter in *Antirrhinum majus* (Staiger *et al.*, 1989). Both Box 1 and 2 were found to be elicitor- or UV light-responsive regulatory elements and located in similar positions in *PSPAL1* and *PSPAL2* (Yamada *et al.*, 1994). These sequences appeared also to be present at a similar location in the parsley *4CL-1* gene (Douglas *et al.*, 1991), the bean *PAL2* promoter (Cramer *et al.*, 1989), the bean *CHS15* promoter (Dron *et al.*, 1988), the soybean *CHS* promoter (Wingender *et al.*, 1989) and the *Arabidopsis PAL* promoter (Ohl, 1990). The 5'-upstream regions of both *PSPAL* genes is also known to contain a series of AT-rich sequence motifs that were reported as enhancers in the sunflower gene coding for helianthinin (Jordano *et al.*, 1989) and a gene involved in the phaseollin synthesis in French bean (Bustos *et al.*, 1989).

Using transient CAT activity in pea protoplasts that had been electroporated with chimeric gene constructs driven by promoters of *PSPAL1* and *PSPAL2* having elicitor-responsive *cis*-regulatory elements, Yamada *et al.* (1994)

showed a partial suppression of the transient CAT activity in response to the application of fungal suppressor. The *cis*-regulatory sequences involved in this negative regulation mediated by the fungal suppressor were identified. Interestingly, the CAT activity was not completely suppressed upon application of the fungal suppressor, suggesting that the site of action of the suppressor is apoplastic rather than in the plasmalemma and that the *cis*-regulatory elements required for the full suppression of CAT activity mediated by the fungal suppressor are located in other parts of the promoter region. In this perspective, Yamada *et al.* (1992) have shown that *PSPAL1* and *PSPAL2* were coordinately induced by *M. pinodes* elicitor and suppressed by suppressors derived from the pycnidiospores' germination fluid orthovanadate (Yamada *et al.*, 1992). Conserved sequence motifs such as Box 1, 2 and/or 4 seem to be involved in such coordinated induction/suppression of these two defence-related genes. However, each individual *PSPAL* gene seems to exhibit specific responses to environmental stimuli, such as fungal elicitor, fungal suppressor and UV light. Along with this observation, the enhancer sequence motif corresponding to the AGC box typically found in the promoter of other defence-related genes such as β-1,3-glucanase, chitinase and *PR-1* protein reported in many plant species (Hart *et al.*, 1993) was found to be absent in peas (Yamada *et al.*, 1994). This result suggests that these defence-related genes are activated through the activity of different *cis*-regulatory factors (Séguin *et al.*, 2004) and go along with our recent finding about the suppression of potato defence mechanisms that targeted *PAL-1 HMG-2* but not *PR-1*, *2*, *3*, *5* and *9* (Wang *et al.*, 2006, 2008).

## 10.5 Concluding Remarks and Future Prospects

During their coevolutionary journey, plants and pathogens have adapted and shaped strategies to either attack, defend and counterattack. Plants sense the presence of pathogens by interacting with their elicitors, hence leading to an activation of their innate defences. Suppressors, on the other hand, can mute these defences and manipulate the physiology of the host to give an advantage to the pathogen and optimize its infectious cycle. Evidence about suppression of plant defences by pathogens keeps growing in many pathosystems involving fungi. However, the specific mechanism by which this phenomenon occurs has been studied only in a few of these systems. Fungal suppressors can target many plant defence processes to induce susceptibility, interfere with the elicitor-receptor bindings or with the plasma membrane ATPases and transmembrane signalling cascades, alter the secondary metabolism pathways leading to phytoalexin accumulation or even detoxify these defence molecules. They can also suppress defence-related genes up- or downstream of their transcription. Investigations in this field have been revealing but many questions still remain unanswered:

**1.** How do suppressors inhibit some plant defence processes such as those involving ATPase and phosphoinositol metabolism in a species-specific manner? Further research is required to unravel the mechanisms determining such specificity.

**2.** It is largely admitted, but poorly demonstrated, that suppressors compete with elicitors' receptors. Evidence gathered in several pathosytems, in terms of affinity and transient suppression, suggests this inhibition, but in many other pathosystems, it is likely not to be the case.

**3.** How do suppressors, as determinants of specificity, evolve within a fungal species and how does such an ability of producing them affect fitness? In other words, could a fungal species produce more than one suppressor – evidence for this exists in many fungi and oomycetes – and still compete in the natural environment?

**4.** Why does suppression occur only transiently and what does this provide to the pathogen within such a narrow window of time? Does this temporary involvement of the suppression have anything to do with the fitness cost?

**5.** Knowing that disease resistance is a collective response of the plant tissues, and not only of single cells, how do plant cells, undergoing suppression, communicate with healthy cells? Does suppression occur only locally or could it get its way systemically through certain signalling pathway(s) yet to be identified?

**6.** Finally, could the action of a suppressor be predicted such that it could be utilized in molecular breeding approaches to improve plant resistance to diseases?

## References

Aarts, N., Metz, M., Holub, E., Staskawicz, B.J., Daniels, M.J. and Parker, J.E. (1998) Different requirements for EDS1 and NDR1 by disease resistance genes define at least two *R* gene-mediated signaling pathways in *Arabidopsis*. *Proceedings of the National Academy of Sciences, USA* 95, 10306–10311.

Abramovitch, R.B., Kim, Y.J., Chen, S., Dickman, M.B. and Martin, G.B. (2003) Pseudomonas type III effector AvrPtoB induces plant disease susceptibility by inhibition of host programmed cell death. *EMBO Journal* 22, 60–69.

Agrios, G.N. (2007) *Plant Pathology*, 5th edn. Academic Press, San Diego, California.

Akai, S. and Fukotomi, M. (1980) Preformed internal physical defenses. In: Horsfall J.G. and Cowling E.B. (eds) *Plant Disease: an Advanced Treatise*. Academic Press, New York, pp. 139–159.

Albersheim, P. and Anderson, A.J. (1971) Proteins from plant cell walls inhibit polygalacturonases secreted by plant pathogens. *Proceedings of the National Academy of Sciences, USA* 68, 1815–1819.

Albersheim, P. and Anderson-Prouty, A.J. (1975) Carbohydrates, proteins, cell surfaces, and the biochemistry of pathogenesis. *Annual Review of Plant Physiology* 26, 31–52.

Albersheim, P. and Valent, B. (1974) Host–pathogen interactions. VII. Plant pathogens secrete proteins which inhibit enzymes of the host capable of attacking the pathogen. *Plant Physiology* 43, 684–687.

Allen, R.L., Bittner-Eddy, P.D., Grenville-Briggs, L.J., Meitz, J.C., Rehmany, A.P., Rose, L.E. and Beynon, J.L. (2004) Host–parasite co-evolutionary conflict between *Arabidopsis* and downy mildew. *Science* 306, 1957–1960.

An, C., Ichinose, Y., Yamada, T., Tanaka, Y., Shiraishi, T. and Oku, H. (1993) Structure and organization of the genes encoding chalcone synthase in *Pisum sativum. Plant Molecular Biology* 21, 789–803.

Andreu, A., Tonón, C., Van Damme, M., Huarte, M. and Daleo, G. (1998) Effect of glucans from different races of *Phytophthora infestans* on defense reactions in potato tuber. *European Journal of Plant Pathology* 104, 777–783.

Arfaoui, A., El Hadrami, A., Mabrouk, Y., Sifi, B., Boudabous, A., El Hadrami, I., Daayf, F. and Chérif, M. (2007) Treatment of chickpea with *Rhizobium* isolates enhances the expression of phenylpropanoid defense-related genes in response to infection by *Fusarium oxysporum* f. sp. *ciceris. Plant Physiology and Biochemistry* 45(6–7), 470–479.

Arneson, P.A. and Durbin, R.D. (1968a) The sensitivity of fungi to $\alpha$-tomatine. *Phytopathology* 58, 536–537.

Arneson, P.A. and Durbin, R.D. (1968b) Studies on the mode of action of tomatine as a fungitoxic agent. *Plant Physiology* 43, 683–686.

Ausubel, F.M. (2005) Are innate immune signaling pathways in plants and animals conserved? *Nature Immunology* 6(10), 973–979.

Baker, C.J. and Orlandi, E.W. (1995) Active oxygen in plant pathogenesis. *Annual Review of Phytopathology* 33, 299–321.

Baker, C.J., Orlandi, E.W. and Deahl, K.L. (2000) Oxygen metabolism in plant/bacteria interactions: characterization of the oxygen uptake response of plant suspension cells. *Physiological and Molecular Plant Pathology* 57, 159–167.

Baron, C. and Zambryski, P.C. (1995) The plant response in pathogenesis, symbiosis, and wounding: variations on a common theme? *Annual Review of Genetics* 29, 107–129.

Basse, C.W. and Boller, T. (1992) Glycopetide elicitors of stress responses in tomato cells: *N*-linked glycans are essential for activity but act as suppressors of the same activity when released from the glycopeptides. *Plant Physiology* 98, 1239–1247.

Basse, C.W., Bock, K. and Boller, T. (1992) Elicitors and suppressors of the defense response in tomato cells. Purification and characterization of glycopeptides elicitors and glycan suppressors generated by enzymatic cleavage of yeast invertase. *Journal of Biological Chemistry* 267(15), 10258–10265.

Becker, P. and Weltring, K.M. (1998) Purification and characterization of alpha-chaconinase of *Gibberella pulicaris. FEMS Microbiology Letters* 167, 197–202.

Bennett, R.N. and Wallsgrove, R.M. (1994) Secondary metabolites in plant defence mechanisms. *New Phytologist* 127, 617–633.

Bent, A.F. (1996) Plant disease resistance genes: function meets structure. *The Plant Cell* 8, 1757–1771.

Bent, A.F. and Mackey, D. (2007) Elicitors, effectors, and *R* genes: the new paradigm and a lifetime supply of questions. *Annual Review of Phytopathology* 45, 399–436.

Bhattacharjee, S., Hiller, N.L., Liolios, K., Win, J., Kanneganti, T.D., Young, C., Kamoun, S. and Haldar, K. (2006) The malarial host-targeting signal is conserved in the Irish potato famine pathogen. *PLoS Pathogens* 2, e50.

Boller, T. (1987) Hydrolytic enzymes in plant disease resistance. In: Kosuge, T. and Nester, E.W. (eds) *Plant–Microbe Interaction.* Vol. 2. Macmillan, New York, pp. 385–413.

Boller, T. (1989) Primary signals and second messengers in the reaction of plants to pathogens. In: Boss, W.F. and Morre, D.J. (eds) *Second Messengers in Plant Growth and Development.* A.R. Liss, New York, pp. 227–255.

Borden, S. and Higgins, V.J. (2002) Hydrogen peroxide plays a critical role in the defense response of tomato to *Cladosporium fulvum. Physiological and Molecular Plant Pathology* 61, 227–236.

Bostock, R.M., Nuckles, E., Henfling, J.W.D.M. and Kúc, J.A. (1983) Effects of potato age and storage on sesquiterpenoid stress metabolite accumulation, steroid glycoalkaloid accumulation, and response to abscisic and arachidonic acids. *Phytopathology* 73, 435–438.

Bouarab, K., Melton, R., Peart, J.R., Baulcombe, D. and Osbourn, A.E. (2002) A saponin-detoxifying enzyme mediates suppression of plant defences. *Nature* 418, 889–892.

Bowyer, P., Clarke, B.R., Lunness, P., Daniels, M.J. and Osbourn, A.E. (1995) Host range of a plant pathogenic fungus determined by a saponin detoxifying enzyme. *Science* 267, 371–374.

Bradley, D.J., Kjellbom, P. and Lamb, C.J. (1992) Elicitor- and wound-induced oxidative cross-linking of a proline-rich plant cell wall protein: a novel, rapid defense response. *Cell* 70, 21–30.

Braganca, J., Swingler, T., Marques, F.I., Jones, T., Eloranta, J.J., Hurst, H.C., Shioda, T. and Bhattacharya, S. (2002) Human CREB-binding protein/p300-interacting transactivator with ED-rich tail (CITED) 4, a new member of the CITED family, functions as a co-activator for transcription factor AP-2. *Journal of Biological Chemistry* 277, 8559–8565.

Braganca, J., Eloranta, J.J., Bamforth, S.D., Ibbitt, J.C., Hurst, H.C. and Bhattacharya, S. (2003) Physical and functional interactions among AP-2 transcription factors, p300/CREBbinding protein, and CITED2. *Journal of Biological Chemistry* 278, 16021–16029.

Bushnell, W.R. and Rowell, J.B. (1981) Suppressors of defense reactions: a model for roles in specificity. *Phytopathology* 71, 1012–1014.

Bustos, M., Guiltinan, M.J., Jordano, J., Begum, D., Kalkan, F.A. and Hall, T.C. (1989) Regulation of β-glucuronidase expression in transgenic tobacco plants by an A/T-rich, *cis*-acting sequence found upstream of a French bean β-phaseolin gene. *The Plant Cell* 1, 839–853.

Çalıskan, M. and Cuming, A.C. (1998) Special specificity of $H_2O_2$-generating oxalate oxidase gene expression during wheat embryo germination. *The Plant Journal* 15(2), 165–171.

Callow, J.A. (1987) Models for host–pathogen interaction. In: Day, P.R. and Jellis, G.J. (eds) *Genetics and Plant Pathogenesis*. Blackwell, Oxford, pp. 283–295.

Carver, T.L.W., Lyngkjaer, M.F., Neyron, L. and Strudwicke, C.C. (1999) Induction of cellular accessibility and inaccessibility and suppression and potentiation of cell death in oat attacked by *Blumeria graminis* f. sp. *avenae*. *Physiological and Molecular Plant Pathology* 55, 183–196.

Catinot, J., Buchala, A., Abou-Mansour, E. and Métraux, J.P. (2008) Salicylic acid production in response to biotic and abiotic stress depends on isochorismate in *Nicotiana benthamiana*. *FEBS Letters* 582(4), 473–478.

Cessna, S.G., Sears, V.E., Dickma, M.B. and Low, P.S. (2000) Oxalic acid, a pathogenicity factor for *Sclerotinia sclerotiorum*, suppresses the oxidative burst of the host plant. *The Plant Cell* 12, 2191–2199.

Chen, C., Bélanger, R.R., Benhamou, N. and Paulitz, T. (2000) Defense enzymes induced in cucumber roots by treatment with growth-promoting rhizobacteria (PGPR) and *Pythium aphanidermatum*. *Physiological and Molecular Plant Pathology* 56, 13–23.

Chet, I., Barak, Z. and Oppenheim, A. (1993) Genetic engineering of microorganisms for improved biocontrol activity. In: Chet, I. (ed.) *Biotechnology in Plant Disease Control*. Wiley-Liss, New York, pp. 211–235.

Chisholm, S.T., Coaker, G., Day, B. and Staskawicz, B.J. (2006) Host–microbe interactions: shaping the evolution of the plant immune response. *Cell* 124, 803–814.

Choi, D., Ward, B.L. and Bostock, R.M. (1992) Differential induction and suppression of potato 3-hydroxy-3-methylglutaryl coenzyme A reductase genes in response to *Phytophthora infestans* and to its elicitor arachidonic acid. *The Plant Cell* 4, 1333–1344.

Cohen, Y., Gisi, U. and Niderman, T. (1993) Local and systemic protection against *Phytophthora infestans* induced in potato and tomato plants by jasmonic acid and jasmonic-methyl-ester. *Phytopathology* 83, 1054–1062.

Colhoun, J. (1979) Predisposition by the environment. In: Horsfall, J.G. and Cowling, E.B. (eds) *Plant Disease*. Vol. 4. Academic Press, New York, pp. 75–96.

Cramer, C.L., Edwards, K., Dron, M., Liang, X., Dildine, S.L., Bolwell, G.P., Dixon, R.A., Lamb, C.J. and Schuch, W. (1989) Phenylalanine ammonia-lyase gene organization and structure. *Plant Molecular Biology* 12, 367–383.

Cruickshank, I.A.M. (1962) Studies on phytoalexins, IV. The antimicrobial spectrum of pisatin. *Australian Journal of Biological Sciences* 15, 147–159.

Crute, I.R. and Pink, D.A.C. (1996) Genetics and utilization of pathogen resistance in plants. *The Plant Cell* 8, 1747–1755.

Currier, W.W. (1981) Molecular controls in the resistance of potato to late blight. *Trends in Biochemical Sciences* 6, 191–194.

Daayf, F., Schmitt, A. and Bélanger, R.R. (1997) Evidence of phytoalexins in cucumber leaves infected with powdery mildew following treatment with leaf extracts of *Reynoutria sacchalinensis*. *Plant Physiology* 113, 719–727.

Daayf, F., El Bellaj, M., El Hassni, M., J'Aiti, F. and El Hadrami, I. (2003) Elicitation of soluble phenolics in date palm (*Phoenix dactylifera* L.) callus by *Fusarium oxysporum* f. sp. *albedinis* culture medium. *Environmental and Experimental Botany* 49, 41–47.

Dangl, J.L. and Jones, J.D.G. (2001) Plant pathogens and integrated defence responses to infection. *Nature* 411, 826–833.

Dangl, J.L., Dietrich, R.A. and Richberg, M.H. (1996) Death don't have no mercy: cell death programs in plant–microbe interactions. *The Plant Cell* 8, 1793–1807.

Daniels, D.L. and Hadwiger, L.A. (1976) Pisatin-inducing components in filtrates of virulent and avirulent *Fusarium solani* cultures. *Physiological Plant Pathology* 8, 9–19.

De Wit, P.J.G.M. and Kodde, E. (1981) Further characterization and cultivar specificity of glycoprotein elicitors from culture filtrates and cell walls of *Cladosporium fulvum* (syn. *Fulvia fulva*). *Physiological Plant Pathology* 18, 297–314.

Défago, G. and Kern, H. (1983) Induction of *Fusarium solani* mutants insensitive to tomatine, their pathogenicity and aggressiveness to tomato fruits and pea plants. *Physiological and Molecular Plant Pathology* 22, 29–37.

DeGroote, M.A., Ochsner, U.A., Shiloh, M.U., Nathan, C., McCord, J.M., Libby, S.J., Vazquez-Torres, A., Xu, Y. and Fang, F.C. (1997) Periplasmic superoxide dismutase protects *Salmonella* from products of phagocyte NADPH-oxidase and nitric oxide synthase. *Proceedings of the National Academy of Sciences, USA* 94, 13997–14001.

Denny, T.P. and Ven Etten, H.D. (1983) Characterization of an inducible, nondegradative tolerance of *Nectria haematococca* MP VI to phytoalexins. *Journal of General Microbiology* 129, 2903–2913.

Denny, T.P., Matthews, P.S. and Van Etten, H.D. (1987) A possible mechanism of nondegradative tolerance of pisatin in *Nectria haematococca* MP VI. *Physiological and Molecular Plant Pathology* 30, 93–107.

Desjardins, A.E. and Gardner, H.W. (1989) Genetic analysis in *Gibberella pulicaris*. *Molecular Plant–Microbe Interactions* 2, 26–34.

Desjardins, A.E., Gardner, H.W. and Plattner, R.D. (1989) Detoxification of the potato phytoalexin lubimin by *Gibberella pulicaris*. *Phytochemistry* 37(4), 1001–1005.

Deverall, B.J. and Vessey, J.C. (1969) Role of a phytoalexin in controlling lesion development in leaves of *Vicia faba* after infection by *Botrytis* spp. *Annals of Applied Biology* 63, 449–458.

Dixon, R.A. and Harrison, M.J. (1990) Activation, structure, and organization of genes involved in microbial defense in plants. *Advances in Genetics* 28, 165–234.

Dixon, R.A. and Lamb, C.J. (1990) Molecular communication in interactions between plants and microbial pathogens. *Annual Review of Plant Physiology and Plant Molecular Biology* 41, 339–367.

Dodds, P.N. and Schwechheimer, C. (2002) A breakdown in defense signaling. *The Plant Cell* (Supplement), S5–S8.

Doke, N. (1975) Prevention of the hypersensitive reaction of potato cells to infection with an incompatible race of *Phytophthora infestans* by constituents of the zoospores. *Physiological Plant Pathology* 7, 1–7.

Doke, N. (1983) Involvement of superoxide anion generation in the hypersensitive response of potato tuber tissue to infection with an incompatible race of *Phytophthora infestans* and to the hyphal wall components. *Physiological Plant Pathology* 23, 345–357.

Doke, N., Garas, N. and Kúc, J. (1979) Partial characterization and aspects of the mode of action of a hypersensitivity-inhibiting factor (HIF) from *Phytophthora infestans*. *Physiological Plant Pathology* 15, 127–140.

Dorey, S., Kopp, M., Geoffroy, P., Fritig, B. and Kaufmann, S. (1999) Hydrogen peroxide from the oxidative burst is neither necessary nor sufficient for hypersensitive cell death induction, phenylalanine ammonia lyase stimulation, salicylic acid accumulation, or scopoletin consumption in cultured tobacco cells treated with elicitin. *Plant Physiology* 121, 163–171.

Douglas, C.J., Hauffe, K.D., Itesmorales, E.D., Ellard, M., Parzkowki, U., Hahlbrock, K. and Dangl, J.L. (1991) Exonic sequences are required for elicitor and light activation of a plant defense gene, but promoter sequences are sufficient for tissue specific expression. *EMBO Journal* 10, 1767–1776.

Dron, M., Clouse, S.D., Dixon, R.A., Lawton, M.A. and Lamb, C.J. (1988) Glutathione and fungal elicitor regulation of a pkmt defense gene promoter in electroporated protoplasts. *Proceedings of the National Academy of Sciences, USA* 85, 6738–6742.

Ebel, J. and Mithöfer, A. (1998) Early events in the elicitation of plant defense. *Planta* 206, 335–348.

El Hadrami, I., Ramos, T., El Bellaj, M., El Idrissi-Tourane, A. and Macheix, J.J. (1997) A sinapic derivative as an induced defense compound of date palm against *Fusarium oxysporum* f. sp. *albedinis*, the agent causing bayoud disease. *Journal of Phytopathology* 145, 329–333.

El Hadrami, A., El Idrissi-Tourane, A., El Hassni, M., Daayf, F. and El Hadrami, I. (2005) Toxin-based *in-vitro* selection and its potential application to date palm for resistance to the bayoud Fusarium wilt. *Comptes Rendus Biologies* 328(8), 732–744.

Farmer, E.E. and Helgeson, J.P. (1987) An extracellular protein from *Phytophthora parasitica* var. *nicotianae* is associated with stress metabolite accumulation in tobacco callus. *Plant Physiology* 85, 733–740.

Felix, G., Grosskopf, D.G., Regenass, M., Basse, C.W. and Boller, T. (1991a) Elicitor-induced ethylene biosynthesis in tomato cells: characterization and use as a bioassay for elicitor action. *Plant Physiology* 97, 19–25.

Felix, G., Grosskopf, D.G., Regenass, M. and Boller, T. (1991b) Rapid changes of protein phosphorylation are involved in transduction of the elicitor signal in plant cells. *Proceedings of the National Academy of Sciences, USA* 88, 8831–8834.

Flor, H.H. (1955) Host–parasite interactions in flax rust – its genetics and other implications. *Phytopathology* 45, 680–685.

Flor, H.H. (1971) Current status of the gene-for-gene concept. *Annual Review of Phytopathology* 9, 275–296.

Franceschi, V.R. (1989) Calcium oxalate formation is a rapid and reversible process in *Lemna minor* L. *Protoplasma* 148, 130–137.

Freialdenhoven, A., Peterhänsel, C., Kurth, J., Kreuzaler, F. and Schulze-Lefert, P. (1996) Identification of genes required for the function of non-race-specific *mlo* resistance to powdery mildew in barley. *The Plant Cell* 8, 5–14.

Fujita, K., Matsuda, Y., Wada, M., Hirai, Y., Mori, K., Moriura, N., Nonomura, T., Kakutani, K.

and Toyoda, H. (2004) Powdery mildew pathogens can suppress the chitinase gene expression induced in detached inner epidermis of barley coleoptile. *Plant Cell Reports* 23, 504–511.

Garas, N.A., Doke, N. and Kúc, J. (1979) Suppression of the hypersensitive reaction in potato tubers by mycelial components from *Phytophthora infestans*. *Physiological Plant Pathology* 15, 117–126.

Gardner, H.W., Desjardins, A.E., Weisleder, D. and Plattner, R.D. (1988) Biotransformation of the potato phytoalexin, lubimin, by *Gibberella pulicaris*. Identification of major products. *Biochimica et Biophysica Acta* 966, 347–356.

Govrin, E.M. and Levine, A. (2000) The hypersensitive response facilitates plant infection by the necrotrophic pathogen *Botrytis cinerea*. *Current Biology* 10, 751–757.

Gowri, G., Paiva, N.L. and Dixon, R.A. (1991) Stress responses in alfalfa (*Medicago sativa* L.) 12. Sequence analysis of phenylalanine ammonia-lyase (PAL) cDNA clones and appearance of PAL transcripts in elicitor-treated cell cultures and developing plants. *Plant Molecular Biology* 17, 415–429.

Grosskopf, D.G., Felix, G. and Boller, T. (1991) Yeast-derived glycopeptides elicitors and chitosan or digitonin differentially induce ethylene biosynthesis, phenylalanine ammonia-lyase and callose formation in suspension-cultured tomato cells. *Journal of Plant Physiology* 138, 741–746.

Hahlbrock, K. and Scheel, D. (1989) Physiology and molecular biology of phenylpropanoid metabolism. *Annual Review of Plant Physiology and Plant Molecular Biology* 40, 347–369.

Hahn, M.G., Bucheli, P., Cervone, F., Doares, S.H., O'Neill, R.A., Darvill, A. and Albersheim, P. (1989) Roles of cell wall constituents in plant–pathogen interactions. In: Kosuge, T. and Nester, E.W. (eds) *Plant–Microbe Interactions. Molecular and Genetic Perspectives*. Vol. 3. McGraw Hill Publishing Co., New York, pp. 131–181.

Hammerschmidt, R. (1999) Induced disease resistance: how do induced plants stop pathogens? *Physiological and Molecular Plant Pathology* 55, 77–84.

Hammond-Kosack, K.E. and Jones, J.D.G. (1996) Resistance gene-dependent plant defense responses. *The Plant Cell* 8, 1773–1791.

Hart, C.M., Nagy, F. and Meins, F., Jr (1993) A 61-bp enhancer element of the tobacco β-1,3-glucanase B gene interacts with a regulated nuclear protein(s). *Plant Molecular Biology* 21, 121–131.

Hauffe, K.D., Paszkowski, U., Schulze-Lefert, P., Hahlbrock, K., Dangl, J.L. and Douglas C.J. (1991) A parsley 4CL-2 promoter fragment specifies complex expression patterns in transgenic tobacco. *The Plant Cell* 3, 435–443.

Hawksworth, D.L., Kirk, P.M., Sutton, B.C. and Pegler, D.N. (1995) *Ainsworth and Bisby's Dictionary of the Fungi*, 8th edn. CAB International, Wallingford, Oxon, UK, 616 pp.

Hayami, C., Otani, H., Nishimura, S. and Kohmoto, K. (1982) Induced resistance in pear leaves by spore germination fluids of nonpathogens to *Altermaria alternata*, Japanese pear pathotype and suppression of the induction by AK-toxin. *Journal of the Faculty of Agriculture at Tottori University* 17, 9–18.

He, P., Shan, L., Lin, N.C., Martin, G.B., Kemmerling, B., Nürnberger, T. and Sheen, J. (2006) Specific bacterial suppressors of MAMP signaling upstream of MAPKKK in *Arabidopsis* innate immunity. *Cell* 125, 563–575.

Heath, M.C. (1981) The suppression of the development of silicon-containing deposits in French bean leaves by exudates of the bean rust fungus and extracts from bean rust-infected tissue. *Physiological Plant Pathology* 18, 149–155.

Heath, M.C. (2000) Advances in imaging the cell biology of plant–microbe interactions. *Annual Review of Phytopathology* 38, 443–459.

Ikeda, S., Toyoda, H., Matsuda, Y., Kurokawa, M., Tamal, T., Yoshida, K., Kami, C., Ikemoto, T., Enomoto, M., Shiraishi, K., Miyamoto, S., Hanaoka, M. and Ouchi, S. (1996) Cloning of a chitinases gene chiSH1 cloned from Gram-positive bacterium *Kurthia zopfii* and control of powdery mildew of barley. *Annals of the Phytopathological Society of Japan* 62, 11–16.

Imura, Y., Iguchi, S., Toyoda, K., Ichinose, Y., Shiraishi, T. and Yamada, T. (2000) Importance of AC-rich element on pea phenylalanine ammonia-lyase gene 1 promoter for expression induced by nonpathogenic attack. *Journal of General Plant Pathology* 66, 123–127.

Ito, S., Takahara, H., Kawaguchi, T., Tanaka, S. and Kameya-Iwaki, M. (2002) Post-transcriptional silencing of the tomatinase gene in *Fusarium oxysporum* f. sp. *lycopersici*. *Journal of Phytopathology* 150, 474–480.

Ito, S., Eto, T., Tanaka, S., Yamauchi, N., Takahara, H. and Ikeda, T. (2004) Tomatidine and lycotetraose, hydrolysis products of alpha-tomatine by *Fusarium oxysporum* tomatinase, suppress induced defense responses in tomato cells. *FEBS Letters* 571, 31–34.

Jabs, T. (1999) Reactive oxygen intermediates as mediators of programmed cell death in plants and animals. *Biochemical Pharmacology* 57, 231–245.

Jacobsen, S., Mikkelsen, J.D. and Hejgaard, J. (1990) Characterization of two antifungal endochitinases from barley grain. *Physiologia Plantarum* 79, 554–562.

Jennings, D.B., Ehrenshaft, M., Mason Pharr, D. and Williamson, J.D. (1998) Roles for mannitol and mannitol dehydrogenase in active oxygen-mediated plant defense. *Proceedings of the National Academy of Sciences, USA* 95, 15129–15133.

Jones, J.D.G. and Dangl, J.L. (2006) The plant immune system. *Nature* 444(16), 323–329.

Jordano, J., Almonguera, C. and Thomas, T.L. (1989) A sunflower helianthinin gene upstream sequence ensemble contains an enhancer and sites of nuclear protein interaction. *The Plant Cell* 1, 855–866.

Judelson, H.S. and Blanco, F. (2005) The spores of *Phytophthora*: weapons of the plant destroyer. *Nature Reviews Microbiology* 3, 47–58.

Kato, H., Wada, M., Muraya, K., Malik, K., Shiraishi, T., Ichinose, Y. and Yamada, T. (1995) Characterizations of nuclear factors for elicitor-mediated activation of the promoter of the pea phenylalanine ammonia-lyase gene 1. *Plant Physiology* 108(1), 129–139.

Kato, T., Shiraishi, T., Toyoda, K., Saitoh, K., Satoh, Y., Tahara, M., Yamada, T. and Oku, H. (1993) Inhibition of ATPase activity in pea plasma membranes by fungal suppressors from *Mycosphaerella pinodes* and their peptide moieties. *Plant Cell Physiology* 34, 439–445.

Katsuwan, J. and Anderson, A.J. (1990) Catalase and superoxide dismutase of root-colonizing saprophytic fluorescent *Pseudomonas*. *Applied and Environmental Microbiology* 56, 3576–3582.

Keates, S.A., Zhang, D., Loewus, F.A. and Franceschi, V.R. (1996) Oxalate oxidase is synthesized and secreted from bean leaf cells in response to fungal infection. *Plant Physiology* 111, 311.

Kessmann, H. and Barz, W. (1986) Elicitation and suppression of phytoalexin and isoflavone accumulation in cotyledons of *Cicer arietinum* L. as caused by wounding and by polymeric components from the fungus *Ascochyta rabiei*. *Journal of Phytopathology* 117, 321–335.

Kistler, H.C. and Van Etten, H.D. (1981) Phaseollin metabolism and tolerance in *Fusarium solani* f. sp. *phaseoli*. *Physiological Plant Pathology* 19, 257–271.

Klotz, M.G. and Hutcheson, S.W. (1992) Multiple periplasmic catalases in phytopathogenic strains of *Pseudomonas syringae*. *Applied and Environmental Microbiology* 58, 2468–2473.

Knogge, W. (1996) Fungal infection of plants. *The Plant Cell* 8, 1711–1722.

Kodama, M., Kajiwara, K., Otani, H. and Kohmoto, K. (1989) A host-recognition factor from *Botrytis* affecting scallion. In: Kohmoto, K. and Durbin, R.D. (eds) *Host-specific Toxins, Recognition and Specificity Factors in Plant Disease*. Tottori University Press, Tottori, Japan, pp. 33–44.

Kogel, K.H., Beckhove, U., Dreschers, J., Munch, S. and Romme, Y. (1994) Acquired resistance in barley. The resistance mechanism induced by 2,6-dichloroisonicotinic acid is a phenocopy of a genetically based mechanism governing race-specific powdery mildew resistance. *Plant Physiology* 106, 1269–1277.

Kolattukudy, P.E. (1985) Enzymatic penetration of the plant cuticle by fungal pathogens. *Annual Review of Phytopathology* 23, 233–250.

Köller, W. (1991) The plant cuticle: a first barrier to be overcome by fungal plant pathogens. In: Cole, G.T. and Hoch, H.C. (eds) *The Fungal Spore and Disease Initiation in Plants and Animals*. Plenum Press, New York, pp. 219–246.

Kombrink, E. and Somssich, I.E. (1995) Defense responses of plants to pathogens. In: Andrews, J.H. and Tommerup, I.C. (eds) *Advances in Botanical Research* 21. Academic Press, New York, pp. 1–34.

Kraft, B., Schwenen, L., Stockl, D. and Barz, W. (1987) Degradation of the pterocarpan phytoalexin medicarpin by *Ascochyta rabiei*. *Archives of Microbiology* 147, 201–206.

Kuan, I.C. and Tien, M. (1993) Stimulation of Mn peroxidase activity: a possible role of oxalate in lignin biodegradation. *Proceedings of the National Academy of Sciences, USA* 90, 1242–1246.

Kuhn, P.J. and Smith, D.A. (1978) Detoxification of the phytoalexin, kievitone, by *Fusarium solani* f. sp. *phaseoli*. *Annals of Applied Biology* 89, 362–366.

Kuhn, P.J. and Smith, D.A. (1979) Isolation from *Fusarium solani* f. sp. *phaseoli* of an enzymic system responsible for kievitone and phaseollidin detoxification. *Physiological Plant Pathology* 14, 179–190.

Kunkel, B.N. and Brooks, D.M. (2002) Cross talk between signaling pathways in pathogen defense. *Current Opinion in Plant Biology* 5(4), 325–331.

Lairini, K., Perez-Espenosa, A., Pineda, M. and Ruiz-Rubio, M. (1996) Purification and characterization of tomatinase from *Fusarium oxysporum* f. sp. *lycopersici*. *Applied and Environmental Microbiology* 62, 1604–1609.

Lamb, C.J. and Dixon, R.A. (1997) The oxidative burst in plant disease resistance. *Annual Review of Plant Physiology and Plant Molecular Biology* 48, 251–275.

Leiberei, R. (1988) Relationship of cyanogenic capacity of the rubber tree *Hevea brasiliensis* to susceptibility to *Microcyclus ulei*, the agent causing South American leaf blight. *Journal of Phytopathology* 122, 54–67.

Leone, G. and Tonneijck, A.E.G. (1990) A rapid procedure for screening the resistance of band cultivars (*Phaseolus vulgaris* L.) to *Botrytis cinerea* and *Sclerotinia sclerotiorum*. *Euphytica* 48(1), 87–90.

Levine, A., Tenhaken, R., Dixon, R.A. and Lamb, C.J. (1994) $H_2O_2$ from the oxidative burst orchestrates the plant hypersensitive disease resistance response. *Cell* 79 583–593.

Lois, R., Dietrich, A., Hahlbrock, K. and Schulz, W. (1989) A phenylalanine ammonia-lyase gene from parsley: structure, regulation and identification of elicitor and light responsive cis-acting elements. *EMBO Journal* 8, 1641–1648.

Low, P.S. and Merida, J.R. (1996) The oxidative burst in plant defense: function and signal transduction. *Physiologia Plantarum* 96(3), 533–542.

Manners, J.G. (1993) Fungi and bacteria. In: Manners, J.G. (ed.) *Principles of Plant Pathology*, 2nd edn. Cambridge University Press, Cambridge, pp. 16–29.

Maxwell, D.P. and Bateman, D.F. (1968) Influence of carbon source and pH on oxalate accumulation in culture filtrates of *Sclerotium rolfsii*. *Phytopathology* 58, 1351–1355.

Müller, K.O. and Börger, H. (1940) Experimentelle untersuchungen über die *Phytophthora*-resistem der kartoffel. *Arb Biologischen Reichsasnstalt Landw Forstw Berlin* 23, 189–231.

Nandi, A., Kachroo, P., Fukushige, H., Hildebrand, D., Klessig, D.F. and Shah, J. (2003) Ethylene and jasmonic acid signaling pathways affect NPR1-independent expression of defense genes without impacting resistance to *Pseudomonas syringae* and *Peronospora parasitica* in the *Arabidopsis* ssi1 mutant. *Molecular Plant–Microbe Interactions* 16, 588–599.

Nawrath, C. and Métraux, J.P. (1999) Salicylic acid induction-deficient mutants of *Arabidopsis* express PR-2 and PR-5 and accumulate high levels of camalexin after pathogen inoculation. *The Plant Cell* 11, 1393–1404.

Nicholson, R.L. and Hammerschmidt, R. (1992) Phenolic compounds and their role in disease resistance. *Annual Review of Phytopathology* 30, 369–389.

Noyes, R.D. and Hancock, J.G. (1981) Role of oxalic acid in the *Sclerotinia* wilt of sunflower. *Physiological Plant Pathology* 18, 123–132.

Ohl, S., Hedrick, S.A., Choy, J. and Lamb, C.J. (1990) Functional properties of a phenylalanine ammonia-lyase promoter from *Arabidopsis*. *The Plant Cell* 2, 837–848.

Oku, H., Shiraishi, T. and Ouchi, S. (1977) Suppression of induction of phytoalexin, pisatin, by low-molecular-weight substances from spore germination fluid of pea pathogen, *Mycosphaerella pinodes*. *Naturwissenschaften* 64, 643.

Oku, H., Shiraishi, T., Ouchi, S., Ishiura, M. and Matsueda, R. (1980) A new determinant of pathogenicity in plant disease. *Naturwissenschaften* 67, 310.

Oku, H., Shiraishi, T. and Ouchi, S. (1987) Role of specific suppressors in pathogenesis of *Mycosphaerella pinodes*. In: Nishimura, S., Asada, T. and Doke N. (eds) *Molecular Determinants of Plant Diseases*. Japan Science Society Press, Tokyo, pp. 145–156.

Osbourn, A.E. (1996a) Preformed antimicrobial compounds and plant defence against fungal attack. *The Plant Cell* 8, 1821–1831.

Osbourn, A.E. (1996b) Saponins and plant defence – a soap story. *Trends in Plant Science* 1, 4–9.

Osbourn, A.E. (1999) Antimicrobial phytoprotectants and fungal pathogens: a commentary. *Fungal Genetics and Biology* 26, 163–168.

Osbourn, A.E. (2003) Saponins in cereals. *Phytochemistry* 62, 1–4.

Osbourn, A.E., Bowyer, P., Lunnes, P., Clarke, B.R. and Daniels, M. (1995) Fungal pathogens of oat roots and tomato leaves employ closely related enzymes to detoxify different host plant saponins. *Molecular Plant–Microbe Interactions* 8, 971–987.

Ozeretskovskaya, O.L., Vasyukova, N.I., Perekhod, E.A., Chalenko, G.I., Il'inskaya, L.I. and Gerasimova, N.G. (2001) Plant resistance suppressors in the pathosystem formed by potato and the causal agent of late blight. *Applied Biochemistry and Microbiology* 37(5), 506–511.

Parker, J.E., Schulte, W., Hahlbrock, K. and Scheel, D. (1991) An extracellular glycoprotein from *Phytophthora megasperma* f. sp. *glycinea* elicits phytoalexin synthesis in cultured parsley cells and protoplasts. *Molecular Plant Microbe Interactions* 4, 19–27.

Peart, J.R., Lui, R., Sadanandom, A., Malcuit, I., Moffett, P., Brice, D.C., Schauser, L., Jaggard, D.A.W., Xiao, S., Coleman, M.J., Dow, M., Jones, J.D.G., Shirasu, K. and Baulcombe, D.C. (2002) The ubiquitin ligase-associated protein SGT1 is required for host and nonhost disease resistance in plants. *Proceedings of the National Academy of Sciences, USA* 99, 10865–10869.

Pedras, M.S. and Okanga, F.I. (1999) Strategies of cruciferous pathogenic fungi: detoxification of the phytoalexin cyclobrassinin by mimicry. *Journal of Agricultural Food Chemistry* 47(3), 1196–1202.

Pedras, M.S.C., Gadagi, R.S., Jha, M. and Sarma-Mamillapalle, V.K. (2007) Detoxification of the phytoalexin brassinin by isolates of *Leptosphaeria maculans* pathogenic on brown mustard involves an inducible hydrolase. *Phytochemistry* 68(11), 1572–1578.

Perrin, D.R. and Bottomley, W. (1962) Studies on phytoalexins. V. The structure of pisatin from *Pisum sativum* L. *Journal of the American Chemical Society* 84, 1919–1922.

Pezet, R. (1998) Purification and characterization of a 32-kDa laccase-like stilbene oxidase produced by *Botrytis cinerea*. *FEMS Microbiology Letters* 167, 203–208.

Pezet, R., Pont, V. and Hoang-Van, K. (1991) Evidence for oxidative detoxification of pterostilbene and resveratrol by a laccase-like stilbene oxidase produced by *Botrytis cinerea*. *Physiological and Molecular Plant Pathology* 39, 441–450.

Pieterse, C.M.J., Ton, J. and van Loon, L.C. (2001) Cross-talk between plant defence signaling pathways: boost or burden? *AgBiotechNet* 068 (3 June), 1–8.

Prell, H.H. and Day, P.R. (2001) *Plant–Fungal Pathogen Interaction*. Springer-Verlag, Berlin.

Prins, T.W., Tudzynski, P., von Tiedemann, A., Tudzynski, B., Ten Have, A., Hansen, M.E., Tenberge, K. and van Kan, J.A.L. (2000) Infection strategies of *Botrytis cinerea* and related necrotrophic pathogens. In: Kronstad, J.W. (ed.) *Fungal Pathology*. Kluwer Academic, Dordrecht, pp. 33–64.

Prusky, D. and Keen, N.T. (1993) Involvement of preformed antifungal compounds in the resistance of subtropical fruit decay. *Plant Disease* 77, 114–119.

Prusky, D., Plumbley, R.A. and Kobiler, I. (1991) The relationship between the antifungal diene levels and fungal inhibition during quiescent infections of *Colletotrichum gloeosporioides* in unripe avocado fruits. *Plant Pathology* 40, 45–52.

Pueppke, S.G. and Van Etten, H.D. (1976) The relation between pisatin and the development of *Aphanomyces euteiches* in diseased *Pisum sativum*. *Phytopathology* 66, 1174–1185.

Qian, W., Tan, G., Liu, H., He, S., Gao, Y. and An, C. (2007) Identification of a bHLH-type G-box binding factor and its regulation activity with G-box and Box I elements of the PsCHS1 promoter. *Plant Cell Reports* 26(1), 85–93.

Quidde, T., Osbourn, A.E. and Tudzynski, P. (1998) Detoxification of alpha-tomatine by *Botrytis cinerea*. *Physiological and Molecular Plant Pathology* 52, 151–165.

Roddick, J.G. (1974) The steroidal glycoalkaloid tomatine. *Phytochemistry* 13, 9–25.

Romero-Puertas, M.C., Perazzolli, M., Zago, E.D. and Delledonne, M. (2004) Nitric oxide signaling functions in plant–pathogen interactions. *Cellular Microbiology* 6, 795–803.

Ruiz-Rubio, M., Pérez-Espinosa, A., Lairini, K., Roldán-Arjona, T., Di Pietro, A. and Anaya, N. (2001) Metabolism of the tomato saponin α-tomatine by phytopathogenic fungi. In: Atta-ur-Rahman (ed.) *Studies in Natural Products Chemistry*. Vol. 25. Elsevier Science, Amsterdam, pp. 293–326.

Ryals, J.A., Neuenschwander, U.H., Willii, M.G., Molha, A., Steiner, H.Y. and Hunt, M.D. (1996) Systemic acquired resistance. *The Plant Cell* 8, 1809–1819.

Ryan, C.A. (1987) Oligosaccharides signalling in plants. *Annual Review of Cell Biology* 3, 295–317.

Sakamoto, W., Yaltsman, A., Adam, Z. and Takahashi, Y. (2003) Coordinated regulation and complex formation of YELLOW VARIEGATED1 and YELLOW VARIEGATED2, chloroplastic FtsH metalloproteases involved in the repair cycle of photosystem II in *Arabidopsis* thylakoid membranes. *The Plant Cell* 15, 2843–2855.

Salles, I.I., Blount, J.W., Dixon, R.A. and Schubert, K. (2002) Phytoalexin production and β-1,3-glucanase activities in *Colletotrichum trifolii* infected alfalfa (*Medicago sativa* L.). *Physiological and Molecular Plant Pathology* 61, 89–101.

San Mateo, L.R., Hobbs, M.M. and Kawula, T.H. (1998) Periplasmic copper-zinc superoxide dismutase protects *Haemophilus ducreyi* exogenous superoxide. *Molecular Microbiology* 27, 391–404.

Sandrock, R.W. and Van Etten, H.D. (1998) Fungal sensitivity to and enzymatic degradation of the phytoanticipin α-tomatine. *Phytopathology* 88, 137–143.

Sandrock, R.W., DellaPenna, D. and Van Etten, H.D. (1995) Purification and characterization of β-2-tomatinase, an enzyme involved in the degradation of α-tomatine and isolation of the gene encoding β-2-tomatinase from *Septoria lycopersici*. *Molecular Plant–Microbe Interactions* 8, 960–970.

Sbaghi, M., Jeandet, P., Bessis, R. and Leroux, P. (1996) Degradation of stilbene type phytoalexins in relation to the pathogenicity of *Botrytis cinerea* to grapevines. *Plant Pathology* 45, 139–144.

Schaeffer, H.J., Leykam, J. and Walton, J.D. (1994) Cloning and targeted gene disruption of EXG1, encoding exo-β-1,3-glucanase, in the phytopathogenic fungus *Cochliobolus carbonum*. *Applied and Environmental Microbiology* 60, 594–598.

Schmidt, K., Heberle, B., Kurrasch, J., Nehls, R. and Stahl, D.J. (2004) Suppression of phenylalanine ammonia lyase expression in sugar beet by the fungal pathogen *Cercospora beticola* is mediated at the core promoter of the gene. *Plant Molecular Biology* 55, 835–852.

Schröder, M., Hahlbrock, K. and Kombrink, E. (1992) Temporal and spatial patterns of 1,3-β-glucanase and chitinase induction in potato leaves infected by *Phytophthora infestans*. *The Plant Journal* 2, 161–172.

Schulze-Lefert, P. and Panstruga, R. (2003) Establishment of biotrophy by parasitic fungi and reprogramming of host cells for disease resistance. *Annual Review of Phytopathology* 41, 641–667.

Schulze-Lefert, P. and Vogel, J. (2000) Closing the ranks to attack by powdery mildew. *Trends in Plant Science* 5, 343–348.

Schulze-Lefert, P., Dangl, J.L., Becker-Andre, M., Hahlbrock, K. and Schulz, W. (1989) Inducible *in vivo* DNA foot prints define sequences necessary for UV-light activation of the parsley chalcone synthase gene. *EMBO Journal* 8, 651–656.

Séguin, A., Laible, G., Leyva, A., Dixon, R.A. and Lamb, C.J. (2004) Characterization of a gene encoding a DNA-binding protein that interacts *in vitro* with vascular specific *cis* elements of the phenylalanine ammonia-lyase promoter. *Plant Molecular Biology* 35, 281–291.

Seki, H., Ichinose, Y., Kato, H., Shiraishi, T. and Yamada, T. (1996) Analysis of *cis*-regulatory elements involved in the activation of a member of chalcone synthase gene family (*PsChsI*) in pea. *Plant Molecular Biology* 31, 479–491.

Seki, H., Ichinose, Y., Ito, M., Shiraishi, T. and Yamada, T. (1997) Combined effects of multiple *cis*-acting elements in elicitor-mediated activation of *PSCHS1* gene. *Plant and Cell Physiology* 38(1), 96–100.

Sexton, A.C., Minic, Z., Cozijnsen, A.J., Pedras, S.C. and Howlett, B.J. (2009) Cloning, purification and characterisation of brassinin glucosyltransferase, a phytoalexin-detoxifying enzyme from the plant pathogen *Sclerotinia sclerotiorum*. *Fungal Genetics and Biology* 46(2), 201–209.

Shiraishi, T., Oku, H., Yamashita, M. and Ouchi, S. (1978) Elicitor and suppressor of pisatin induction in spore germination fluid of pea pathogen, *Mycosphaerella pinodes*. *Annals of the Phytopathological Society of Japan* 44, 659–665.

Shiraishi, T., Araki, M., Yoshioka, H., Kobayashi, I., Yamada, T., Ichinose, Y., Kunoh, H. and Oku, H. (1991a) Inhibition of ATPase activity in plasma membranes *in situ* by a suppressor from a pea pathogen, *Mycosphaerella pinodes*. *Plant and Cell Physiology* 32, 1067–1075.

Shiraishi, T., Yamada, T., Oku, H. and Yoshioka, H. (1991b) Suppressor production as a key factor for fungal pathogenesis. In: Patil, S.S., Ouchi, S., Mills, D. and Vance, D. (eds) *Molecular Strategies of Pathogens and Host Plants*. Springer-Verlag, New York, pp. 151–162.

Shiraishi, T., Saito, K., Kim, K., Kato, K., Tahara, M., Oku, H., Yamada, T. and Ichinose, Y. (1992) Two suppressors, Supprescin A and B, secreted by a pea pathogen, *Mycosphaerella pinodes*. *Plant and Cell Physiology* 33, 663–667.

Shiraishi, T., Yamada, T., Saitoh, K., Kato, T., Toyoda, K., Yoshioka, H., Kim, H.M., Ichinose, Y., Tahara, M. and Oku, H. (1994) Suppressor: determinants of specificity produced by plant pathogens. *Plant and Cell Physiology* 35, 1107–1119.

Smith, D.A. (1982) Toxicity of phytoalexins. In: Bailey, J.A. and Mansfield, J.W. (eds) *Phytoalexins*. Blackie, London, pp. 218–252.

Smith, D.A., Kuhn, P.J., Bailey, J.A. and Burden, R.S. (1980) Detoxification of phaseollidin by *Fusarium solani* f. sp. *phaseoli*. *Phytochemistry* 19, 1673–1675.

Staiger, D., Kauren, H. and Schell, J. (1989) A CACGTG motif of the *Antirrhinum mujus* chalcone synthase promoter is recognized by an evolutionary conserved nuclear protein. *Proceedings of the National Academy of Sciences, USA* 6, 6930–6934.

Storti, E., Pelucchini, D., Tegli, S. and Scala, A. (1988) A potential defense mechanism of tomato against the late blight disease is suppressed by germinating sporangia-derived substances from *Phytophthora infestans*. *Journal of Phytopathology* 121, 275–282.

Swegle, M., Huang, J.K., Lee, G. and Muthukrishnan, S. (1989) Identification of an endochitinase cDNA clone from barley aleurone cells. *Plant Molecular Biology* 12, 403–412.

Sweigard, J. and Van Etten, H.D. (1987) Reduction in pisatin sensitivity of *Aphanomyces euteiches* by polar lipid extracts. *Phytopathology* 77, 771–775.

Sweigard, J.A., Matthews, D.E. and Van Etten, H.D. (1986) Synthesis of the phytoalexin pisatin by a methyltransferase from pea. *Plant Physiology* 80, 277–279.

Tansey, W.P. (2001) Transcriptional activation: risky business. *Genes & Development*, 15, 1045–1050.

Thordal-Christensen, H. (2003) Fresh insights into processes of nonhost resistance. *Current Opinion in Plant Biology* 6, 351–357.

Thordal-Christensen, H., Gregersen, P.L. and Collinge, D.B. (2000) The barley/*Blumeria* (syn. *Erysiphe*) *graminis* interaction. In: Slusarenko, A., Fraser, R. and van Loon, L.C. (eds) *Mechanisms of Resistance to Plant Diseases*. Kluwer Academic, Dordrecht.

Tian, M., Huitema, E., da Cunha, L., Torto-Alalibo, T. and Kamoun, S. (2004) A kazal-like extracellular serine protease inhibitor from *Phytophthora infestans* targets the tomato pathogenesis-related protease P69B. *Journal of Biological Chemistry* 279(25), 26370–26377.

Toyoda, H., Matsuda, Y., Yamaga, T., Ikeda, S., Morita, M., Tamai, T. and Ouchi, S. (1991) Suppression of the powdery mildew pathogen by chitinase microinjected into barley coleoptile epidermal cells. *Plant Cell Reports* 10, 217–220.

Toyoda, K., Shiraishi, T., Yoshioka, H., Yamada, T., Ichinose, Y. and Oku, H. (1992) Regulation of polyphosphoinositide metabolism in pea plasma membranes by elicitor and suppressor from a pea pathogen, *Mycosphaerella pinodes*. *Plant and Cell Physiology* 33, 445–452.

Toyoda, K., Shiraishi, T., Yamada, T., Ichinose, Y. and Oku, H. (1993) Rapid changes in polyphosphoinositide metabolism in pea in response to fungal signals. *Plant and Cell Physiology* 34, 729–735.

Uehara, K. (1964) Relationship between host specificity of pathogen and phytoalexin. *Annals of the Phytopathological Society of Japan* 29, 103–110.

Unger, C., Kleta, S., Jandl, G. and Von Tiedemann, A. (2005) Suppression of the defence-related oxidative burst in bean leaf tissue and bean suspension cells by the necrotrophic pathogen *Botrytis cinerea*. *Journal of Phytopathology* 153, 15–26.

Urbasch, I. (1986) Transformation von $\alpha$-tomatin durch *Botrytis cinerea*. *Planta Mediterranea* 2, 114–118.

Van der Biezen, E.A. and Jones, J.D.G. (1998) Plant disease resistance proteins and the gene-for-gene concept. *Trends in Biochemical Sciences* 23, 454–456.

Van der Hoorn, R.A.L. (2008) Plant proteases: from phenotypes to molecular mechanisms. *Annual Review of Plant Biology* 59, 191–223.

Van Etten, H.D., Matthews, P.S., Tegtmeier, K.J., Dietert, M.F. and Stein, J.I. (1980) The association of pisatin tolerance and demethylation with virulence on pea in *Nectria haematococca*. *Physiological Plant Pathology* 16, 257–268.

Van Etten, H.D., Matthews, D.E. and Smith, D.A. (1982) Metabolism of phytoalexins. In: Bailey, J.A. and Mansfield J.W. (eds) *Phytoalexins*. Blackie, London, pp. 181–217.

Van Etten, H.D., Matthews, D.E. and Matthews, P.S. (1989) Phytoalexin detoxification: importance for pathogenicity and practical implications. *Annual Review of Phytopathology* 27, 143–164.

Van Etten, H.D., Sandrock, R.W., Wasmann, C.C., Soby, S.D., McCluskey, K. and Wang, P. (1995) Detoxification of phytoanticipins and phytoalexins by phytopathogenic fungi. *Canadian Journal of Botany* 73(S1), S518–S525.

Van Loon, L.C., Rep, M. and Pieterse, C.M.J. (2006) Significance of inducible defense-related proteins in infected plants. *Annual Review of Phytopathology* 44, 135–162.

Verhoeff, K. and Liem, J.I. (1975) Toxicity of tomatine to *Botrytis cinerea*, in relation to latency. *Phytopathologische Zeitschrift* 82, 333–338.

Vogel, J. and Someville, S. (2000) Isolation and characterization of powdery mildew-resistant *Arabidopsis* mutants. *Proceedings of the National Academy of Sciences, USA* 97, 1897–1902.

Von Tiedemann, A. (1997) Evidence for a primary role of oxygen species in induction of host cell death during infection of bean leaves with *Botrytis cinerea*. *Physiological and Molecular Plant Pathology* 50, 151–166.

Wada, M., Kato, H., Malik, K., Sriprasertsak, P., Ichinose, Y., Shiraishi, T. and Yamada, T. (1995) A supprescin from a phytopathogenic fungus deactivates transcription of a plant defense gene encoding phenylalanine ammonia-lyase. *Journal of Molecular Biology* 249, 513–519.

Walton, J.D. (1996) Host-selective toxins: agents of compatibility. *The Plant Cell* 8, 1723–1733.

Walton, J.D. (2002) Coming clean with plant disease suppression. *BioScience* 52(10), 873–874.

Wang, X., El Hadrami, A., Adam, L.R. and Daayf, F. (2004) US-1 and US-8 genotypes of *Phytophthora infestans* differentially affect local, proximal and distal gene expression of phenylalanine ammonia-lyase and 3-hydroxy, 3-methylglutaryl CoA reductase in potato leaves. *Physiological and Molecular Plant Pathology* 65(3), 157–167.

Wang, X., El Hadrami, A., Adam, L.R. and Daayf, F. (2005) Genes encoding pathogenesis-related proteins PR-2, PR-3 and PR-9, are differentially regulated in potato leaves inoculated with isolates from US-1 and US-8 genotypes of *Phytophthora infestans* (Mont.) de Bary. *Physiological and Molecular Plant Pathology* 67(1), 49–56.

Wang, X., El Hadrami, A., Adam, L.R. and Daayf, F. (2006) Local and distal gene expression of *pr*-1 and *pr*-5 in potato leaves inoculated with isolates from the old (US-1) and the new (US-8) genotypes of *Phytophthora infestans* (Mont.) de Bary. *Environmental and Experimental Botany* 57(1–2), 70–79.

Wang, X., El Hadrami, A., Adam, L.R. and Daayf, F. (2008) *Phytophthora infestans* US-8 suppresses potato defense responses earlier and stronger than US-1. *Plant Pathology* 57, 1026–1037.

Weinstein, L.I., Hahn, M.G. and Albersheim, P. (1981) Host–pathogen interactions. XVIII. Isolation and biological activity of glycinol, a pterocarpan phytoalexin synthesized by soybeans. *Plant Physiology* 68, 358–363.

Wietor-Orlandi, E.A. and Smith, D.A. (1985) Metabolism of the phytoalexin, phaseollinisoflavan, by *Fusarium solani* f. sp. *phaseoli*. *Physiological Plant Pathology* 27, 197–207.

Wingender, R., Rohrig, H., Wing, D., Horicke, C. and Schell, J. (1989) Differential regulation of soybean chalcone synthase genes in plant defense, symbiosis and upon environmental stimuli. *Molecular and General Genetics* 218, 315–322.

Wojtaszek, P. (1997) Oxidative burst: an early plant response to pathogen infection. *Biochemistry Journal* 322, 681–692.

Wolpert, T.J., Dunkle, L.D. and Ciuffetti, L.M. (2002) Host-selective toxins and avirulence determinants: what's in a name? *Annual Review of Phytopathology* 40, 251–285.

Wubben, J.P., Price, K.R., Daniels, M.J. and Osbourn, A.E. (1996) Detoxification of oat leaf saponins by *Septoria avenae*. *Phytopathology* 86, 986–992.

Yamada, T., Hashimoto, H., Shiraishi, T. and Oku, H. (1989) Suppression of pisatin, phenylalanine ammonia-lyase mRNA, and chalcone synthase mRNA accumulation by a putative pathogenicity factor from the fungus *Mycosphaella pinodes*. *Molecular Plant–Microbe Interactions* 2, 256–261.

Yamada, T., Tanaka, Y., Sriprasertsak, P., Kato, H., Hashimoto, T., Kawamata, S., Ichinose, Y., Kato, H., Shiraishi, T. and Oku, H. (1992) Phenylalanine ammonia-lyase genes from *Pisum sativum*: structure, organ-specific expression and regulation by fungal elicitor and suppressor. *Plant and Cell Physiology* 33(6), 715–725.

Yamada, T., Sriprasertsak, P., Kato, H., Hashimoto, T., Shimizu, H. and Shiraishi, T. (1994) Functional analysis of the promoters of phenylalanine ammonia-lyase genes in pea. *Plant and Cell Physiology* 35, 917–926.

Yamamoto, M., Nishimura, S., Kohmoto, K. and Otani, H. (1984) Studies on host-specific AF-toxins produced by *Alternaria alternata* strawberry pathotype causing *Alternaria* black spot of strawberry. (2) Role of toxins in pathogenesis. *Annals of the Phytopathologial Society of Japan* 50, 610–619.

Yamamoto, Y., Oku, H., Shiraishi, T., Ouchi, S. and Koshizawa, K. (1986) Non-specific induction of pisatin and local resistance in pea leaves by elicitor from *Mycosphaerella pinodes*, *M. melonis* and *M. ligulicola* and effect of suppressor from *M. pinodes*. *Journal of Phytopathology* 117, 136–143.

Yoshikawa, M. and Sugimoto, K. (1993) A fungal suppressor of phytoalexin production competes for the elicitor-receptor binding. *Naturwissenschaften* 80, 374–376.

Yoshioka, H., Shiraishi, T., Yamada, T., Ichinose, Y. and Oku, H. (1990) Suppression of pisatin production and ATPase activity in pea plasma membranes by orthovanadate, verapamil and a suppressor from *Mycosphaerella pinodes*. *Plant and Cell Physiology* 31, 1139–1146.

Yoshioka, H., Shiraishi, T., Kawamata, S., Nasu, K., Yamada, T., Ichinose, Y. and Oku, H. (1992a) Orthovanadate suppresses accumulation of phenylalanine ammonia-lyase mRNA and chalcone synthase mRNA in pea epicotyls induced by elicitor from *Mycosphaerella pinodes*. *Plant and Cell Physiology* 33, 201–204.

Yoshioka, H., Shiraishi, T., Nasu, K., Yamada, T., Ichinose, Y. and Oku, H. (1992b) Suppression of the activation of chitinase and β-1,3-glucanase in pea epicotyls by orthovanadate and a suppressor from *Mycosphaerella pinodes*. *Annals of the Phytopathologial Society of Japan* 58, 405–410.

Yoshioka, H., Hayakawa, Y. and Doke, N. (1995) Suppression of phenylalanine ammonia-lyase mRNA accumulation by suppressor from *Phytophthora infestans*. *Annals of the Phytopathologial Society of Japan* 61, 7–12.

Yoshioka, H., Miyabe, M., Hayakawa, Y. and Doke, N. (1996) Expression of genes for phenylalanine ammonia-lyase and 3-hydroxy-3-methylglutaryl CoA reductase in aged potato tubers infected with *Phytophthora infestans*. *Plant and Cell Physiology* 37, 81–90.

Zeyen, R.J., Carver, T.L.W. and Lyngkjaer, M.F. (2002) The formation and role of papillae. In: Bélanger, R.R., Bushnell, W.R., Dik, A.J. and Carver, T.L.W. (eds) *The Powdery Mildews*. American Phytopathological Society (APS) Press, St Paul, Minnesota, pp. 107–125.

Zhang, Y. and Smith, D.A. (1983) Concurrent metabolism of the phytoalexins phaseollin, kievitone and phaseollinisoflavan by *Fusarium solani* f. sp. *phaseoli*. *Physiological Plant Pathology* 23, 89–100.

Ziegler, E. and Pontzen, R. (1982) Specific inhibition of glucan-elicited glyceollin accumulation in soybeans by an extracellular mannan-glycoprotein of *Phytophthora megasperma* f. sp. *glycinea*. *Physiological Plant Pathology* 20, 321–331.

# 11 Sustainable Agriculture and the Multigenomic Model: How Advances in the Genetics of Arbuscular Mycorrhizal Fungi will Change Soil Management Practices

ERIN ZIMMERMAN, MARC ST-ARNAUD AND MOHAMED HIJRI

*Université de Montréal, Montréal, Québec, Canada*

## Abstract

Arbuscular mycorrhizal (AM) fungi are important both in agriculture and in natural ecosystems due to their effects on the fitness of their plant hosts. As symbionts, AM fungi improve plant uptake of water and nutrients, and help to protect against pathogens. The study of these organisms has been obstructed in part by difficulty in identifying and quantifying them in the field, a problem springing from our poor understanding of their unusual genetic structure.

AM fungi have been shown to be multigenomic, possessing a large amount of genetic variation not only between individuals, but among nuclei within an individual. In order for simple, reliable identification and quantification techniques to be developed for large-scale use, this genetic diversity must be quantified and marker genes found for which the amount of variation is manageable. Once this has been accomplished, growers can work knowledgably with the existing strains of AM fungi in their fields, or select an appropriate commercial inoculum.

In this chapter, we will discuss the current state of knowledge in AM fungal genetics and how it can be applied to develop molecular tools which permit the management of natural AM fungal populations in agricultural fields. Assessment of AM fungal biodiversity in natural and modified ecosystems, as well as the estimation of the mycorrhizal potential of agricultural soil, are bottlenecks that greatly limit our understanding of the ecology of these cornerstone organisms. The development of efficient and inexpensive diagnostic techniques will enable us to use them to their full potential in sustainable agriculture systems.

 

## 11.1 Introduction

The arbuscular mycorrhizal (AM) fungi are a group of asexual, root-inhabiting, symbiotic fungi which make up the phylum Glomeromycota (Schüßler *et al.*, 2001). This phylum is made up of five families: Paraglomaceae, Archeosporaceae, Glomeraceae, Acaulosporaceae and Gigasporaceae, which together make up approximately 160 species (Sanders, 2002). AM fungi are among the fungi most frequently found in soil and are widely distributed geographically (Smith and Read, 1997). They form symbioses with the roots of more than 80% of all vascular plant species (Smith and Read, 1997). The AM symbiosis is believed to be over 400 million years old, and fossils show that the earliest land plants contained these endophytes (Simon *et al.*, 1993; Remy *et al.*, 1994; Redecker *et al.*, 2000). The AM symbiosis has, in fact, been proposed to be at the origin of land plants (Pirozynski and Malloch, 1975).

One of the principal host benefits of the AM symbiosis is the increased uptake of phosphorus. Phosphate ions in soil are largely unavailable to plants because they form insoluble complexes with naturally occurring metal cations. Fungal hyphae are able to extend beyond the root depletion zone, taking up bioavailable phosphate which is outside the reach of the plant (Helgason and Fitter, 2005). Plant hosts also experience improved uptake of water and nitrogen, both in the form of $NH_4^+$ released through mineralization (Subramanian and Charest, 1999; Hamel, 2004; Toussaint *et al.*, 2004; Tanaka and Yano, 2005) and as organic molecules (Hawkins *et al.*, 2000) of some other cations such as $Cu^{2+}$, $Zn^{2+}$, $K^+$ and $Fe^{2+}$ (Liu *et al.*, 2000) via this extensive network of fungal hyphae, improving both nutrition and drought tolerance. Soil structure is improved in regions with well-developed AM fungal populations because of the ability of glomalin, a fungal protein, to bind soil particles, forming macroaggregates and helping to decrease soil erosion (Rillig and Mummey, 2006). Yet another benefit conferred upon a plant host by its AM symbiont is an improved level of resistance to pathogens, in particular root-infecting fungi and nematodes (Helgason and Fitter, 2005; St-Arnaud and Vujanovic, 2007). It has been variously suggested that AM fungi may protect their hosts through competition for colonization sites, through improved nutrition, stimulated plant defence responses, or through changes in the microbial community structure of the surrounding soil (Pozo and Azcón-Aguilar, 2007; St-Arnaud and Vujanovic, 2007). Finally, AM fungi are associated with several types of plant beneficial microorganisms, including nitrogen-fixing bacteria and phosphorus-solubilizing bacteria and fungi (Barea *et al.*, 2002). In fact, in the case of nitrogen-fixing bacteria, common signalling cascades suggest that the bacteria at one point simply modified recognition pathways which originally evolved to enable AM fungal colonization (Kistner and Parniske, 2002). Research into the tripartite symbiosis – composed of a legume, an AM fungus, and bacteria of the genus *Rhizobium* – has shown that the presence of the fungus improves both the nodulation and the nitrogen-fixing capabilities of the bacteria (Barea *et al.*, 2002). As with their plant hosts, AM fungi can also provide a measure of drought resistance to the bacteria, protecting nodules from oxidative damage due to stress (Ruiz-Lozano *et al.*, 2001). With regard to phosphorus-solubilizing bacteria, one study using radio-labelled phosphorus in the form of rock phosphate showed that inoculation with AM fungi allowed the uptake of sources of phosphorus not available to the plant when only the bacteria are present. The dual inoculation significantly increased both the biomass of the plant and its accumulation of nitrogen and phosphorus (Toro *et al.*, 1997). With so many potential host benefits, it has long been recognized that managing the AM symbiosis could prove valuable in an agricultural setting.

While the use of AM fungi in agriculture has not yet been widely embraced by the intensive operations typical in Europe and North America, countries such as Cuba and India, where large amounts of chemical inputs are prohibitively expensive, have made impressive advances in this area. Following an intensive research programme in the 1990s, Cuban scientists developed an inoculum mix specific to soils in that region, as well as recommendations regarding the conditions under which it should be used. The researchers found that strains varied in their effectiveness depending on the type of soil in which they were used. Today the inoculum, called EcoMic, is recommended for a wide variety of regional crops and is used in many growing operations. Even in high-input systems, it has proven to be a very effective biofertilizer; used as a seedcoating, it produces yield increases of 10–80%/ha with a 6–10% application rate, by weight, depending on the crop species (Rivera *et al.*, 2007).

In India, where a great deal of applied mycorrhizal research has also taken place, commercial inocula are used on a large scale, with close to 2500 t produced in 2006 by four different Indian companies. Used with rice crops, researchers found that mycorrhizal inoculation could produce a modest yield increase of around 10%, but with a 25–50% reduction in the amount of fertilizer required – real savings in India's low-phosphorous soils (Sharma *et al.*, 2007).

For a prime example of how the large-scale development of a commercial symbiotic inoculum can produce economic change in North America, we need only look to the adoption of specialty legume crops, such as peas, lentils and chickpeas in the Canadian prairies, a region which is naturally poor in the *Rhizobium* bacteria required for the plants to fix atmospheric nitrogen. The development by the Saskatoon-based MicroBio RhizoGen Corporation in the late 1980s of *Rhizobium* inoculants appropriate to prairie soils made the production of these crops profitable through large savings in nitrogen fertilizers which dwarfed the cost of the inoculum itself. Over the past two decades, the area sown in Saskatchewan as specialty crops, the bulk of which are legumes, has increased from 1.2 to 16.1% of the total agricultural land in the province (Carlyle, 2004).

In order for the use of AM fungi to be widely adopted in North America, methods must be in place to quickly and cheaply analyse their status in fields so that producers can respond accordingly. We need a comprehensive picture of what taxa are present, how they are interacting with the host crop, and how they react to various soil treatments. One major roadblock in the development of these tests has been the peculiar and poorly understood genetic structure the fungi seem to possess. In the following section, we will discuss the discoveries surrounding this unusual genetic organization and some of its implications.

## 11.2 Genetic Structure and Heterokaryosis

AM fungi are primarily made up of vast, branching networks of hyphae (Fig. 11.1a). Unlike most higher fungi, these hyphae are coenocytic, that is, lacking in discrete cellular divisions. Hyphal walls form long, tube-like structures through which cytoplasm and organelles can migrate freely (Fig. 11.1b). When an AM fungus sporulates, asexual spores are formed on the terminus of a mother hypha, and large numbers of nuclei simply migrate into the spores via that connection (Fig. 11.1c).

Because Glomeralean spores are formed containing hundreds or even thousands of nuclei from the mother hypha, it is thought that there never exists a time in the AM fungal life cycle at which the organism is uninucleate (Sanders, 2002). In most life forms, the uninucleate stage acts as a genetic bottleneck, ensuring that each somatic cell which follows will possess a nucleus that is an identical mitotic product of the original. In an organism lacking both sexual recombination and a uninucleate stage, mutations which occur in individual nuclei are allowed to pass on to the next generation, potentially allowing the development of individuals containing any number of varied genomes (Sanders, 2002). While a great deal of research has been done looking into the ecology and physiology of AM fungi, only relatively recently have inquiries begun to be made into their genetic structure.

Suspicions that AM fungi possess an atypical genetic structure began with the investigation of their enzymes; different isolates of the same species were found to possess different enzymatic isoforms (Hepper *et al.*, 1988; Rosendahl and Sen, 1992). Exploration then turned to ribosomal genes. These genes are present in multiple copies occurring in tandem arrays. The sequences within an array are normally kept very similar through the process of concerted evolution, a mechanism by which repeats within a gene family exchange sequence information, thereby maintaining a high level of homogeneity and allowing the family to evolve as a unit. Concerted evolution is thought to be driven primarily by gene conversion and unequal crossing-over during meiosis. Early investigations of the internal transcribed spacer (ITS) regions flanking the 5.8S ribosomal RNA (rRNA) in *Glomus mosseae*, however, showed that individual spores contained multiple distinct ITS sequences (Sanders *et al.*, 1995). Researchers investigating rRNA genes within the AM fungal species *Scutellospora castanea* also found a high level of polymorphism (Hosny *et al.*, 1999). A later study of the same species using specific fluorescent DNA-DNA *in situ* hybridization (FISH) investigated the frequency of two divergent sequences of the ITS2 region, referred to as T2 and T4 (Kuhn *et al.*, 2001). These sequences, which had been previously demonstrated to co-occur within individual spores (Hijri *et al.*, 1999), were shown to in fact occur in different frequencies from nucleus to nucleus within a spore. The researchers used a

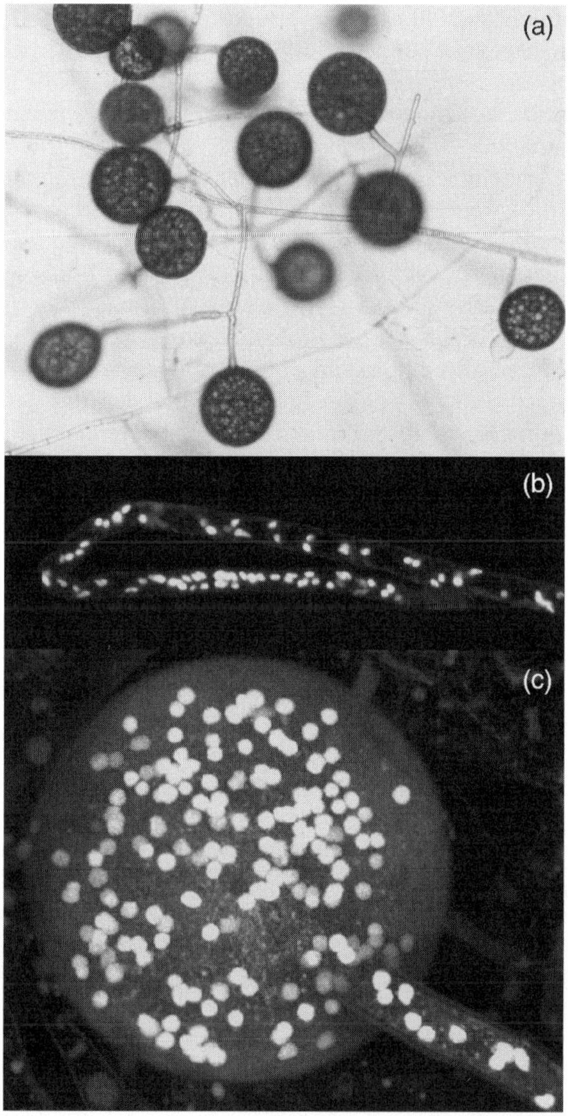


**Fig. 11.1.** (a) Spores of *Glomus intraradices* grown in *in vitro* culture with transformed carrot roots and observed under a stereomicroscope. (b) Hypha of germinating spore of *Gigaspora rosea*. (c) *Glomus diaphanum* spore showing typical multi-nucleate stage. In (b) and (c) nuclei were stained using SYTO Green fluorescent dye and observed under a confocal microscope.

phylogenetic approach to examine whether the differences between nuclei were likely to have been caused by recombination or by an accumulation of mutations in successive clonal generations. Calculated probabilities were significantly different than those expected in recombining organisms, leading the authors to believe that most of the observed variation was caused by mutation. In the same study, a binding protein-encoding gene, *BiP*, which is single-copy in other fungi, low-copy in AM fungi, and highly conserved in eukaryotes, was analysed to reveal 15 different variants in the genomic DNA of a single *Glomus intraradices* isolate. Collectively, all the research conducted on intraspecific polymorphism in AM fungi indicates a unique and complicated genetic structure.

The implications of a multigenomic arrangement are many. The discovery of this heterokaryotic structure in AM fungi has necessitated a new way of thinking about what constitutes an individual, as well as many questions about our concepts of species and populations as they apply to these organisms (Rosendahl, 2008). Although, for the sake of simplicity, a single spore and the mycelium which grows from it is conventionally referred to as an individual, a single spore can also be thought of as containing a population of nuclei, each of which *may* be capable of functioning as an individual. That is to say, if each nucleus within a fungal spore contains a full and potentially differing complement of genetic material, then the nucleus itself, rather than the spore, could be considered an individual. Sanders (2002) advances two different possibilities for the coexistence of these differing nuclei. First, it is suggested that each nucleus does, in fact, possess a full quota of required genes (Fig. 11.2a). This arrangement could then lead to competition among the genomes. Second, it is suggested that all the necessary genes coding for various functions are spread across numerous genomes, forcing their cooperation (Fig. 11.2b). Under this arrangement, the individual must be defined as the aggregate of the genomes required to form a fully functioning organism.

There is also the question of how genetic diversity arises and is maintained. Because the nuclei which comprise a new spore simply migrate from the subtending hypha into the spore, any heterogeneity in the distribution of different nuclear types across the mycelium will cause spores to arise which differ genetically from others borne on the same mycelium (Koch *et al.*, 2004). This would suggest an ongoing loss of genotypes with each passing generation. These genetic differences could also translate into functional differences, affecting the mycorrhiza.

Here, the question of genome segregation within the hyphal network becomes of interest. If a full complement of functional genes, spread across numerous nuclei, must remain in a given region of the mycelium in order for it to perform properly, segregation according to selection will be heavily restrained within a microenvironment, and a certain level of homogeneity can therefore be expected. Alternatively, differing nuclei may be unevenly distributed within the mycelium according to their fitness within a particular microenvironment (Fig. 11.3), or simply by random chance. In either scenario, a heterogeneous arrangement of nuclei would quickly lead to the loss of all but one type of genome or all but one particular group of genomes, if nuclei are cooperating. Nuclear exchange through hyphal fusion, referred to as anastomosis, may be the means by which this is prevented.

Anastomosis, the fusion of two fungal hyphae of the same or different mycelia, has been shown to occur in some species of AM fungi, particularly those belonging to the genus *Glomus*. While this joining is not believed to occur among individuals of different species or geographical origins, it has been

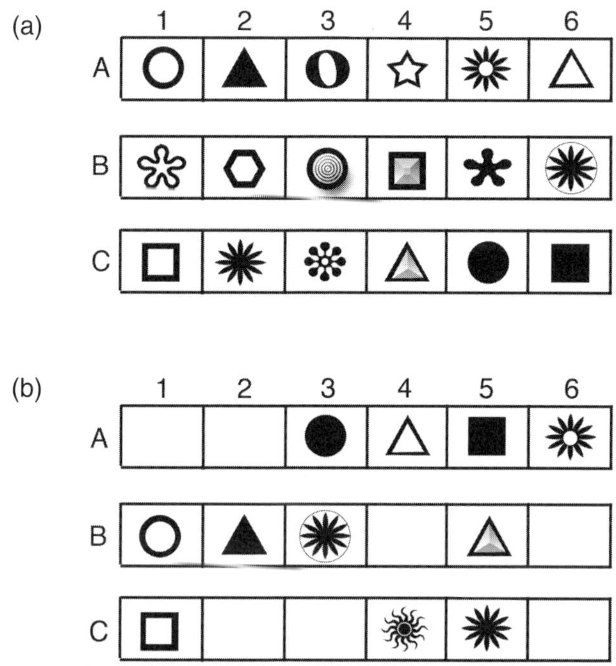


**Fig. 11.2.** Diagram showing two possibilities for the arrangement of heterokaryotic structure among nuclei of AM fungi. (a) Each nucleus possesses a full complement of required genes, with variants differing between nuclei. (b) No single nucleus possesses a full complement of functional genes, so nuclei are forced to 'cooperate'. Three different nuclei, represented by A, B, C, have six loci representing genes required for function. Shapes represent different possible variants of each of the six genes. Empty boxes represent non-functional alleles.

observed between hyphae originating from two different spores in the same isolate. One study found that in three *Glomus* species, between 34 and 90% of hyphal contacts between different germlings of the same isolate resulted in anastomoses (Giovannetti *et al.*, 1999). Interestingly, in the same experiments, no anastomoses of any sort were observed in either *Gigaspora rosea* or *S. castanea*. Anastomoses allow damaged hyphae to re-establish a protoplasmic link as well as allowing nuclear exchange between mycelia (Fig. 11.4). This type of joining may also allow the re-homogenization of nuclei which have become heterogeneous across the mycelial network, thereby maintaining some level of genetic consistency throughout (Bever and Morton, 1999). Depending on the level of compatibility necessary for a given fungal species to anastomose, this phenomenon could also lead to the formation of a single, joined mycelium covering a large area and containing all available genotypes for that taxon.

Phenomena such as differential genetic segregation and homogenization via anastomosis may seem far removed from the business of managing AM fungi in the context of an agricultural operation, but as we will see, these unusual features have real implications for the efficacy of a given inoculum and our ongoing ability to control plant–fungus interactions in the field.

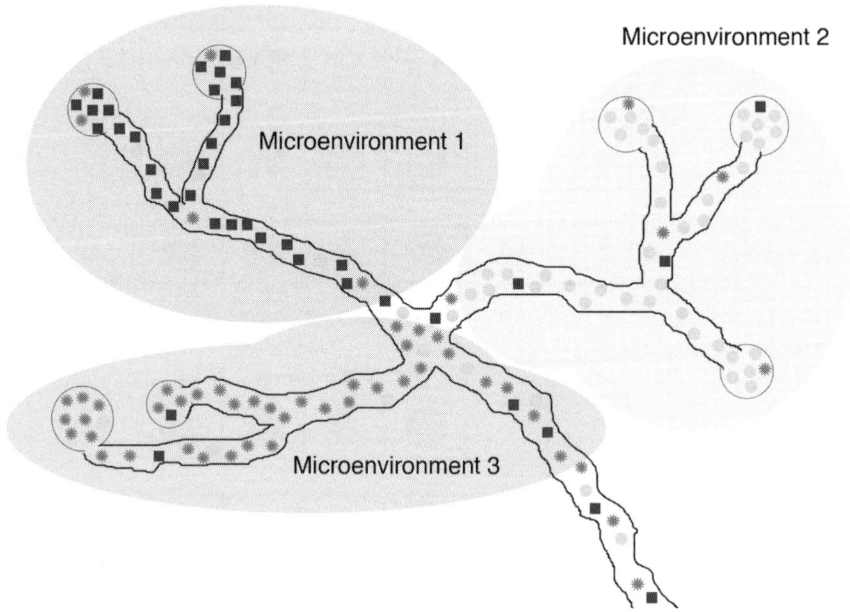


**Fig. 11.3.** Diagram showing the possibility for genetically differing spores to arise on a single mycelium. Certain genotypes, represented by different shapes, are better suited to different microenvironments and are selected for within those regions, affecting the proportions of different nuclei which enter the developing spore.

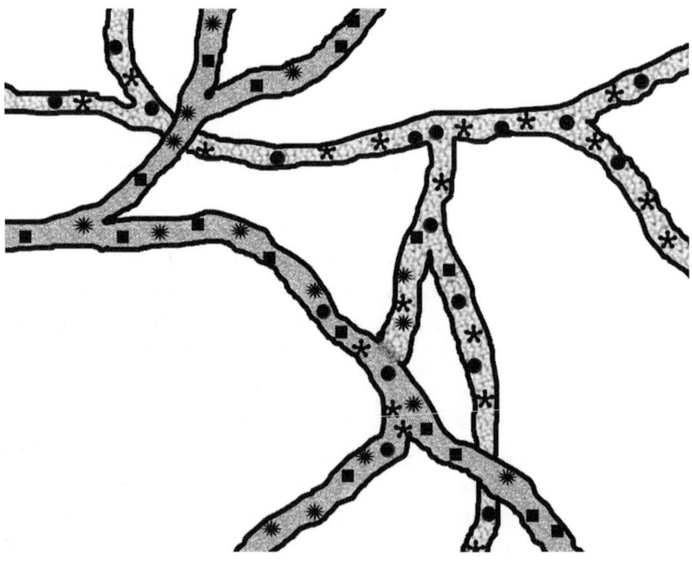

**Fig. 11.4.** Diagram showing anastomosis of two genetically differing hyphae, allowing mixing and homogenization of different nuclear types (shapes) in the vicinity of the connection.

## 11.3 Importance and Management of AM Fungi in Agriculture

While the abilities of arbuscular mycorrhizas to aid in plant fitness are not new discoveries, farming in the 21st century faces challenges which bring new importance to the effective use of AM fungi in agricultural operations. As awareness of the need for sustainable practices grows, efforts are being made to reduce our usage of chemical fertilizers and potentially harmful biocides. Furthermore, with the effects of climate change leading to increased plant stress in the form of drought, heat waves, pest problems and invasive species, the benefits of arbuscular mycorrhizas beyond that of plant nutrition also gain importance (Gavito, 2007). In the near future, the improved water uptake of mycorrhizal crops may prove more critical than any improvement in nutrition, particularly in western North America, where the most intensive grain production is carried out.

The managed use of AM fungi in agricultural operations has the potential to benefit not only the health of the crop plants, but the health of the soil itself. The long-term use of phosphorous fertilizer is associated with soil degradation and the pollution of nearby bodies of water, where runoff can cause algal blooms and disrupt plant and animal communities (Beauchemin and Simard, 1999). Soil phosphate saturation has reached problematic levels in many agricultural areas. Forecasting of crop phosphorous requirements is imprecise, and crop responses show a poor correlation with soil phosphorous test values. It is thought that this problem is in part due to functional variations in the naturally occurring AM fungi, which are not accounted for in forecasts (McKenzie and Bremer, 2003). Maintenance of a healthy and efficient AM fungal community in the field would allow a decrease in applied phosphate. This decrease would in turn benefit the AM symbiosis, as an overabundance of soil phosphorus is inhibitory to root colonization (Smith and Read, 1997). Furthermore, taking into account mycorrhizal activity, abundance and seasonality, more precise forecasts would be possible.

Current complications involved in managing mycorrhizal populations largely spring from our poor understanding of AM fungal biology. One important issue is the variability seen in the effect of a given fungus on different hosts. Studies have shown not only that AM fungal species are a factor in the derived benefit of the host plant (van der Heijden *et al*., 1998), but that different isolates of the same species can have varying effects as well. Some host/symbiont combinations seem to result in an increase in fitness for only one partner, and it can be difficult to predict which partner this will be due to the occurrence of both positive and negative feedback in natural communities (Bever, 2002). One study was conducted in which different isolates of the AM fungus *G. intraradices* were grown under identical conditions for several generations to negate environmental maternal effects (Koch *et al*., 2006). The authors then showed that a given isolate could in fact have a negative effect on one host while causing no harm in another, and that positive effects varied in intensity from isolate to isolate. A study by Bever and co-workers (1996) showed large amounts of variation in spore production by a given fungal isolate from one plant host to another. These results highlight the importance of genetic variability in choosing appropriate fungi for large-scale inoculation, as well as assessing the efficiency of AM fungal strains in a natural community in order to favour useful strains.

It can be difficult to know whether a fungal species present in a field is helping or hindering a crop, especially since even those fungi which produce neither positive nor negative growth effects in their host may still make a significant contribution in terms of phosphate uptake, as well as having an associated carbon cost (Li *et al.*, 2006). Furthermore, benefits brought about by AM fungal colonization can take different forms: one AM fungal species may provide good pathogen protection, but offer no appreciable increase in growth or nutrient uptake (Caron *et al.*, 1986; St-Arnaud *et al.*, 1994). Another may have the opposite effect. The best way to avoid these problems, it seems, is to maintain high fungal biodiversity within the field, increasing the chance of including one very effective strain (van der Heijden *et al.*, 1998). A recent study (Jansa *et al.*, 2008) found that colonization by multiple AM fungi, in that case *G. intraradices* and *Glomus claroideum*, could have synergistic effects, providing greater host benefit than any one of the component taxa would singly, so long as no one taxon dominated the others. However, maintaining high fungal diversity in a high-input agricultural field may be easier said than done. Since some species of AM fungi function much better on a particular crop, monocultures and certain rotations, especially those including non-host crops, discourage diversity. Many common agricultural practices, such as tillage and the application of phosphates and fungicides, may kill off all but the most 'robust' species, and these are not necessarily the taxa which are most beneficial to their hosts.

Furthermore, due to the possibility of a heterogeneous distribution of multiple genomes in a fungal mycelium, spores may arise containing alleles of certain protein-coding genes which are completely absent in other spores borne on the same mycelium (Koch *et al.*, 2004). These genetic differences could translate into functional differences, affecting the efficacy of a commercially developed inoculum. Maintaining and encouraging fungi with desirable attributes could prove impossible if certain genomes are being lost with each new round of sporulation and not necessarily retrieved via anastomosis. This type of genetic drift could render carefully selected commercial inocula ineffective after just a few generations. Indeed, Koch and co-workers (2004) found fivefold differences in hyphal length among isolates originally grown from single spore cultures. The authors felt that the phenotypic variation observed was great enough to cause corresponding variation in plant host growth and nutrition. A better understanding of the way in which genomes segregate in the mycelium is therefore vital to those who would develop commercial inocula or manage the AM symbiosis in agriculture.

All of these complications could be improved upon with a better understanding of AM fungal populations in the field. What is needed, as a first step, is an efficient, inexpensive means of measuring both the diversity and the overall abundance of these fungi in the field.

## 11.4 What Has Been Done in Identification and Quantification

Well-meaning recommendations made by optimistic researchers working in laboratories and test plots are often rejected by growers because they do not take into account the pressures and uncertainty inherent in agrobusiness. To be worth the risk, new farming practices must be convenient, simple, cost effective, and come with a reasonable guarantee of improvement over established methods. The management and monitoring of AM fungi in Western intensive agriculture has failed on several of these counts. While increases in drought and pest problems, as well as potential crackdowns on phosphate use may well solve any issues of cost effectiveness, it will still be up to researchers to develop methods of managing AM fungi which are convenient and straightforward without being overly expensive. Ideally, a grower could send off soil and root samples and receive a profile of what AM fungal species are present and in what amount, allowing her to add inoculum, adjust fertilization regimes, or otherwise respond accordingly. In order for this to happen, testing protocols must be developed which are standardized, require only commonly used laboratory equipment, and do not require a great amount of expertise to interpret. Molecular PCR-based methods fit this description and will, in the coming years, be key to practical mycorrhizal analysis.

### Identification

A wide variety of methods have been employed in attempting to accurately and conclusively identify different species of AM fungi. Traditionally, species were identified based on the morphological characteristics of spores. This required a great amount of expertise on the part of the researcher and often proved to be inexact due to an insufficient number of clear, informative characters. Molecular data, properly interpreted, would have the advantage of identification directly from host roots. This would provide a more accurate picture of the fungal community, since spore analysis does not necessarily predict future colonization. The first attempts at identification and phylogenetic placement based on genetic markers used variation in ribosomal DNA (Simon *et al.*, 1993; Sanders *et al.*, 1995; Lloyd-MacGilp *et al.*, 1996); however, the extremely high variation present in these sequences made this very difficult. Sequences were found which had no known species correlation, and sequences found in one species would frequently cluster with those of another species or even another genus (Clapp *et al.*, 1999; Rodriguez *et al.*, 2004). One early study found that ITS sequences amplified from within a single spore often showed more variation than those from different isolates (Lloyd-MacGilp *et al.*, 1996). Rodriguez and co-workers (2005) examined several isolates of two different AM fungal morphotypes and found them to contain extremely variable ribosomal gene sequences, as may have been expected. However, the two morphotypes were also found to contain several identical sequences. Finally, a study examining copy number in the ribosomal genes of *G. intraradices* found that the number of genes could vary between two- and fourfold among isolates of just this one species (Corradi *et al.*, 2007), making ribosomal genes inappropriate for any attempt at quantification of AM fungi via the quantification of amplified genetic material.

More recently, several protein-coding genes have been investigated for possible use as markers. This type of gene has the advantage of typically possessing lower levels of variation than ribosomal sequences, potentially making for clearer partitions among species and isolates. The lower copy numbers found for coding genes, however, can make amplification more difficult.

One group searching for such markers examined actin and elongation factor 1-alpha genes and found that amino acid sequences were highly conserved across multiple spores within an isolate (Helgason *et al.*, 2003). Pending further investigation to determine the level at which the sequences vary, the authors felt these genes had potential for use as markers in identification. Alpha- and beta-tubulin, as well as the $H^+$-ATPase gene were also evaluated for this purpose (Corradi *et al.*, 2004a, b). Beta-tubulins, in particular, may hold promise as useful markers. The authors were able to design primers which amplified only Glomeromycotan sequences from pot cultures, and found ample variation between species, but very little within, making them ideal for interspecific identification. Conversely, the $H^+$-ATPase gene was found to contain high levels of variation and did not resolve fungi as monophyletic. Two RNA polymerase II subunits, *RPB1* and *RPB2*, were also used recently in the phylogenetic reconstruction of the early evolution of the fungal kingdom, and may prove useful for identification purposes within the Glomeromycota (James *et al.*, 2006).

Where the above genes fall short is in their ability to distinguish between different isolates of a single species. Given the large variation seen in the host-benefit from one isolate to another, making this distinction would prove valuable to those attempting to manage their AM fungal communities. This knowledge would be particularly useful in quantifying relative amounts of a commercial inoculum versus a native fungus of the same species, for example the ubiquitous species, *G. intraradices*. In the case of geographically close isolates, a mere change in the relative proportions of given sequences may be all that is available to separate a more desirable strain from a less desirable one. This type of determination cannot be made until we have gained a much better understanding of the amount, and arrangement of, intra- and inter-isolate variation in AM fungi. In some geographically well-separated isolates, however, enough variation has been found in the simple sequence repeats (SSR) and the introns of nuclear and mitochondrial genes for a combination of these markers to clearly distinguish between them (Croll *et al.*, 2008).

While several genes show potential for being able to consistently distinguish between AM fungal species, in most cases, larger sequence libraries need to be built up to ensure that variation is not undersampled. Furthermore, any marker gene will need to be sequenced in a wide variety of taxa in the fungal kingdom so that contaminant sequences can be easily recognized.

In terms of the actual methodology used in retrieving this genetic information, several different approaches have been successful, although there is often a trade-off between cost and accuracy. Denaturing gradient gel electrophoresis (DGGE), wherein DNA run through a gel with a gradient of a denaturing compound can be separated based on single base-pair differences, is probably the most accurate and reliable method currently in use. This technique uses ribosomal sequence heterogeneity to its advantage, since a unique banding pattern is created as variants are separated, and as such has been able to consistently distinguish even between certain isolates of a given species, although multiple sequences are required to compensate for variant overlap (de Souza *et al.*, 2004). This trait may even circumvent problems of extreme variation when using ribosomal genes, so long as heterogeneity is consistent at different taxonomic levels. One drawback of this method, however, is that it requires costly equipment not commonly available in molecular biology laboratories, and each different type of test can require a long optimization period to establish protocols. Direct sequencing of PCR products has also been carried out successfully (Hijri *et al.*, 2006), but at prohibitively high cost for use on a large scale.

Perhaps the best compromise of cost, simplicity and effectiveness is an approach coupling PCR amplification and digestion using restriction enzymes. Terminal restriction fragment length polymorphism (T-RFLP), for example, uses amplification with fluorescently labelled primers followed by restriction digests to obtain fragments of variable lengths depending on the presence and location of cutting sites. These fragments appear as peaks on an electropherogram. Under conditions such as the analysis of field samples, where multiple species are likely to be amplified, databases of peak profiles, each corresponding to a known species, can be maintained in what is referred to as 'database T-RFLP' (Dickie and FitzJohn, 2007). Some potential difficulties with this technique are the need to have all AM fungal taxa which may be present in a sample already represented in the database, and the inability to distinguish peaks as being different taxa or simply different intra-isolate genotypes. Both of these problems could perhaps be circumvented through the extensive sampling of known species in order to build up a complete profile of frequently occurring peaks within that species. As with most other techniques, the effectiveness of T-RFLP as a means of identification depends heavily on the use of AM fungus-specific primers which can amplify a broad swath of the Glomeromycota.

## Quantification

Another challenge for researchers working with AM fungi is the development of methods which will allow the overall quantification of AM fungal abundance. The ability to monitor fluctuations in the abundance of AM fungi in fields will give growers a good sense of the health of their soil and the effects of various treatments and practices, even if the exact taxa of AM fungi are not known.

The abundance of AM fungi in the soil is typically measured using fatty acid assays. Although currently the best technique available, this method is tedious and requires specialized equipment not found in most laboratories.

While fatty acids have been successfully used to determine the mycorrhizal potential of soils (Plenchette *et al.*, 2007), living biomass cannot be assessed as the lipids in question persist in the soil following the death of the fungi themselves. Accuracy is also an issue, as the diagnostic fatty acids vary in their abundance among different AM fungal species (Graham *et al.*, 1995) and also occur in some primitive organisms and Gram-negative soil bacteria.

Several studies have attempted to use real-time PCR to quantify fungal DNA in soil. The first (Filion *et al.*, 2003b) used specific primers targeting a fragment of the small subunit (SSU) rRNA gene region in *G. intraradices*. Results indicated that the technique was effective in comparing root colonization and spore numbers between soil samples of similar composition in a controlled experiment (Filion *et al.*, 2003a), but that absolute quantification was problematic due to a non-linear relationship between spore concentration and amplified genomic DNA. It was suggested that this discrepancy might have been due to the amount of polymorphism inherent in the ribosomal sequences.

A more recent study was done in which real-time PCR was used to quantify AM fungal DNA in soil using both actin and 18S ribosomal genes (Gamper *et al.*, 2008). Although the actin gene assays were less sensitive than expected, the authors found that, using the 18S gene, this method was sensitive and reliable in quantifying a specific taxon, even at low template concentrations. In fact, this study showed a good correlation between amplified genetic material and spore numbers. There was difficulty, however, in finding ribosomal gene regions which could amplify entire families or even genera of AM fungi. The authors stated that the limited availability of sequence information made using non-ribosomal genes impractical, and that future development of the method would rest largely on the expansion of sequence databases. This study highlights again the need for further research into appropriate AM fungal marker genes.

Furthermore, a major issue with this manner of quantification is that the spatial heterogeneity of nuclei within the hyphae makes the amount of DNA a poor predictor of the prevalence of hyphae within the soil. Gamper *et al.* (2008) tested a combination of nuclear and vital stains and found that living hyphal sections do not necessarily contain nuclei. Thus, test samples containing mostly hyphae or root fragments may underestimate the amount of AM fungal DNA, and conversely, samples with spores may lead to overestimation of AM fungal biomass. Our current inability to amplify entire groups of AM fungal species and the inapplicability of this method to taxa which sporulate infrequently make real-time PCR, though promising, ineffective for quantification in an agricultural setting.

## 11.5 How Management Practices Will Change and What Needs to Be Done

The development of a fast, cheap means of quantifying and identifying AM fungi in agricultural fields will lead to a better understanding of how the agroecosystem functions as a whole. At such time as efficient mycorrhizal profiling becomes possible, producers will have a new level of control over their fields. New tillage or cropping regimes can be evaluated for their effects on AM fungi; fungal species and strains endemic to an area can be noted and, if sparse or ineffective, supplemented with a commercial inoculum. If they are aware of the level of AM fungal colonization in their crops, producers can adjust their fertilizer applications accordingly, saving money while encouraging the further development of the fungal community. The ability to correlate fungal abundance with crop phosphate uptake over the course of the growing season will enable more accurate phosphorous requirement forecasts to be made, accounting not only for the amount of AM fungi in a field, but for seasonal variations in their functionality. In the face of growing evidence of the importance of both host species and soil type on the mycorrhizal efficiency of a given fungus, easy, accurate field testing could lead to the development of 'designer' inoculum, produced for use under specific conditions. The practical experience of farmers in such endeavours will, in turn, lead to more advanced research on the life cycles and functioning of these organisms in the field.

Before these goals can be realized, much more needs to be learned about the extent and arrangement of variation in AM fungi. Appropriate marker genes need to be characterized which, ideally, would allow for some distinction even between intraspecific isolates. Sequence libraries will need to be built up so that even less common variants can be identified.

One exciting possibility for future research is the potential to go a step beyond identifying and quantifying fungi, and actually assay for their physiological activity in the soil. With different gene products produced in the presymbiotic and symbiotic stages, it could be possible to break down the proportions of the fungi in an area according to their stage in the life cycle, and to actually measure their nutrient-uptake efficiency once they have colonized the host. One recent study used real-time quantitative reverse transcriptase PCR to measure the expression of a *G. intraradices* phosphate dehydrogenase involved in phosphorous metabolism (Stewart *et al.*, 2006). While this experiment was conducted under laboratory conditions with a single known species, it may be possible one day to use the technique with more diverse field-collected samples. Other emerging techniques, such as untargeted expressed sequence tags (EST)-sequencing and microarrays are beginning to see use in profiling mycorrhiza-related genes in both plants and fungi (Küster *et al.*, 2007). In the future, these methods may elucidate many currently unknown pathways, bringing much more precision to enzyme-based assays for fungal activity in the soil.

## 11.6 Conclusion

AM fungi, with their many peculiarities, present us with both great possibilities for improving the health and stress tolerance of our crops, and with a number of unusual obstacles to our understanding of them. These fungi defy many of the rules we have come to think of as applying to all multicellular organisms, as well as our usual means of classification. Heterokaryosis in AM fungi requires us to develop a more fluid concept of identification than is allowed for in the label of ‘species’. The populations of these organisms at work in agricultural fields and elsewhere are more a large collection of interacting genotypes than a small collection of interacting species. Filing various specimens as similar aggregates of gene variants rather than under hard-and-fast phylogenetic labels will allow for a more flexible and dynamic understanding of the true nature of AM fungi, and may aid in our thinking as we develop the techniques necessary to fully comprehend their function. With today’s myriad environmental crises and the need for meaningful change, AM fungi will come to represent a cornerstone in agricultural sustainability.

## References


Barea, J.-M., Azcón, R. and Azcón-Aguilar, C. (2002) Mycorrhizosphere interactions to improve plant fitness and soil quality. *Antonie van Leeuwenhoek* 81, 343–351.

Beauchemin, S. and Simard, R.R. (1999) Soil phosphorus saturation degree: review of some indices and their suitability for P management in Quebec, Canada. *Canadian Journal of Soil Science* 79, 615–625.

Bever, J.D. (2002) Negative feedback within a mutualism: host-specific growth of mycorrhizal fungi reduces plant benefit. *Proceedings of the Royal Society, London* 269, 2595–2601.

Bever, J.D. and Morton, J. (1999) Heritable variation and mechanisms of inheritance of spore shape within a population of *Scutellospora pellucida*, an arbuscular mycorrhizal fungus. *American Journal of Botany* 86, 1209–1216.

Bever, J.D., Morton, J.B., Antonovics, J. and Schultz, P.A. (1996) Host-dependent sporulation and species diversity of arbuscular mycorrhizal fungi in a mown grassland. *Journal of Ecology* 84, 71–82.

Carlyle, W.J. (2004) The rise of specialty crops in Saskatchewan, 1981–2001. *The Canadian Geographer* 48, 137–151.

Caron, M., Fortin, J.A. and Richard, C. (1986) Effect of inoculation sequence on the interaction between *Glomus intraradices* and *Fusarium oxysporum* f. sp. *radicis-lycopersici* in tomatoes. *Canadian Journal of Plant Pathology* 8, 12–16.

Clapp, J.P., Fitter, A.H. and Young, J.P.W. (1999) Ribosomal small subunit sequence variation within spores of an arbuscular mycorrhizal fungus, *Scutellospora* sp. *Molecular Ecology* 8, 915–921.

Corradi, N., Hijri, M., Fumagalli, L. and Sanders, I.R. (2004a) Arbuscular mycorrhizal fungi (*Glomeromycota*) harbour ancient fungal tubulin genes that resemble those of the chytrids (*Chytridiomycota*). *Fungal Genetics and Biology* 41, 1037–1045.

Corradi, N., Kuhn, G. and Sanders, I.R. (2004b) Monophyly of *β-tubulin* and *$H^+$-ATPase* gene variants in *Glomus intraradices*: consequences for molecular evolutionary studies of AM fungal genes. *Fungal Genetics and Biology* 41, 262–273.

Corradi, N., Croll, D., Colard, A., Kuhn, G., Ehinger, M. and Sanders, I.R. (2007) Gene copy number polymorphisms in an arbuscular mycorrhizal fungal population. *Applied and Environmental Microbiology* 73, 366–369.

Croll, D., Wille, L., Gamper, H.A., Mathimaran, N., Lammers, P.J., Corradi, N. and Sanders, I.R. (2008) Genetic diversity and host plant preferences revealed by simple sequence repeat and mitochondrial markers in a population of the arbuscular mycorrhizal fungus *Glomus intraradices*. *New Phytologist* 178(3), 672–687.

de Souza, F.A., Kowalchuk, G.A., Leeflang, P., van Veen, J.A. and Smit, E. (2004) PCR-denaturing gradient gel electrophoresis profiling of inter- and intraspecies 18S rRNA gene sequence heterogeneity is an accurate and sensitive method to assess species diversity of arbuscular mycorrhizal fungi of the genus *Gigaspora*. *Applied and Environmental Microbiology* 70, 1413–1424.

Dickie, I.A. and FitzJohn, R.G. (2007) Using terminal restriction fragment length polymorphism (T-RFLP) to identify mycorrhizal fungi: a methods review. *Mycorrhiza* 17, 259–270.

Filion, M., St-Arnaud, M. and Jabaji-Hare, S.H. (2003a) Direct quantification of fungal DNA from soil substrate using real-time PCR. *Journal of Microbiological Methods* 53, 67–76.

Filion, M., St-Arnaud, M. and Jabaji-Hare, S.H. (2003b) Quantification of *Fusarium solani* f. sp. *phaseoli* in mycorrhizal bean plants and surrounding mycorrhizosphere soil using real-time polymerase chain reaction and direct isolations on selective media. *Phytopathology* 93, 229–235.

Gamper, H.A., Young, J.P.W., Jones, D.L. and Hodge, A. (2008) Real-time PCR and microscopy: are the two methods measuring the same unit of arbuscular mycorrhizal fungal abundance? *Fungal Genetics and Biology* 45, 581–596.

Gavito, M.E. (2007) Mycorrhizae and crop production in a world of rapid climate change: a warning call. In: Hamel, C. and Plenchette, C. (eds) *Mycorrhizae in Crop Production.* Haworth Food and Agricultural Products Press, Binghampton, New York, pp. 293–310.

Giovannetti, M., Azzolini, D. and Citernesi, A.S. (1999) Anastomosis formation and nuclear and protoplasmic exchange in arbuscular mycorrhizal fungi. *Applied and Environmental Microbiology* 65, 5571–5575.

Graham, J.H., Hodge, N.C. and Morton, J.B. (1995) Fatty acid methyl ester profiles for characterization of Glomalean fungi and their endomycorrhizae. *Applied and Environmental Microbiology* 61, 58–64.

Hamel, C. (2004) Impact of arbuscular mycorrhizal fungi on N and P cycling in the root zone. *Canadian Journal of Soil Science* 84, 383–395.

Hawkins, H.-J., Johansen, A. and George, E. (2000) Uptake and transport of organic and inorganic nitrogen by arbuscular mycorrhizal fungi. *Plant and Soil* 226, 275–285.

Helgason, T. and Fitter, A. (2005) The ecology and evolution of the arbuscular mycorrhizal fungi. *Mycologist* 19, 96–101.

Helgason, T., Watson, I.J. and Young, J.P.W. (2003) Phylogeny of the Glomerales and Diversisporales (Fungi: Glomeromycota) from actin and elongation factor 1-alpha sequences. *FEMS Microbiology Letters* 229, 127–132.

Hepper, C.M., Sen, R., Azconaguilar, C. and Grace, C. (1988) Variation in certain isozymes amongst different geographical isolates of the vesicular arbuscular mycorrhizal fungi *Glomus clarum*, *Glomus monosporum* and *Glomus mosseae*. *Soil Biology and Biochemistry* 20, 51–59.

Hijri, I., Sykorova, Z., Oehl, F., Ineichen, K., Mäder, P., Wiemken, A. and Redecker, D. (2006) Communities of arbuscular mycorrhizal fungi in arable soils are not necessarily low in diversity. *Molecular Ecology* 15, 2277–2289.

Hijri, M., Hosny, M., van Tuinen, D. and Dulieu, H. (1999) Intraspecific ITS polymorphism in *Scutellospora castanea* (Glomales, Zygomycota) is structured within multinucleate spores. *Fungal Genetics and Biology* 26, 141–151.

Hosny, M., Hijri, M., Passerieux, E. and Dulieu, H. (1999) rDNA units are highly polymorphic in *Scutellospora castanea* (Glomales, Zygomycetes). *Gene* 226, 61–71.

James, T.Y., Kauff, F., Schoch, C.L., Matheny, P.B. and Hofstetter, V. (2006) Reconstructing the early evolution of fungi using a six-gene phylogeny. *Nature* 443, 818–822.

Jansa, J., Smith, F.A. and Smith, S.E. (2008) Are there benefits of simultaneous root colonization by different arbuscular mycorrhizal fungi? *New Phytologist* 177, 779–789.

Kistner, C. and Parniske, M. (2002) Evolution of signal transduction in intracellular symbiosis. *Trends in Plant Science* 7, 511–518.

Koch, A.M., Kuhn, G., Fontanillas, P., Fumagalli, L., Goudet, J. and Sanders, I.R. (2004) High genetic variability and low local diversity in a population of arbuscular mycorrhizal fungi. *Proceedings of the National Academy of Sciences, USA* 101, 2369–2374.

Koch, A.M., Croll, D. and Sanders, I.R. (2006) Genetic variability in a population of arbuscular mycorrhizal fungi causes variation in plant growth. *Ecology Letters* 9, 103–110.

Kuhn, G., Hijri, M. and Sanders, I.R. (2001) Evidence for the evolution of multiple genomes in arbuscular mycorrhizal fungi. *Nature* 414, 745–748.

Küster, H., Becker, A., Firnhaber, C., Honjec, N., Manthey, K., Perlick, A.M., Bekel, T. and Dondrup, M. (2007) Development of bioinformatic tools to support EST-sequencing, *in silico-* and microarray-based transcriptome profiling in mycorrhizal symbioses. *Phytochemistry* 68, 19–32.

Li, H., Smith, S.E., Holloway, R.E., Zhu, Y. and Smith, F.A. (2006) Arbuscular mycorrhizal fungi contribute to phosphorus uptake by wheat grown in a phosphorus-fixing soil even in the absence of positive growth responses. *New Phytologist* 172, 536–543.

Liu, A., Hamel, C., Hamilton, R.I., Ma, B.L. and Smith, D.L. (2000) Acquisition of Cu, Zn, Mn and Fe by mycorrhizal maize (*Zea mays* L.) grown in soil at different P and micronutrient levels. *Mycorrhiza* 9, 331–336.

Lloyd-MacGilp, S.A., Chambers, S.M., Dodd, J.C., Fitter, A.H., Walker, C. and Young, J.P.W. (1996) Diversity of the ribosomal internal transcribed spacers within and among isolates of *Glomus mosseae* and related mycorrhizal fungi. *New Phytologist* 133, 103–111.

McKenzie, R.H. and Bremer, E. (2003) Relationship of soil phosphorus fractions to phosphorus soil tests and fertilizer response. *Canadian Journal of Soil Science* 83, 443–449.

Pirozynski, K.A. and Malloch, D.W. (1975) The origin of land plants: a matter of mycotropism. *Biosystems* 6, 153–164.

Plenchette, C., Fraser, T. and Hamel, C. (2007) Mycorrhizal soil infectivity: a new approach for its determination. In: *Proceedings of the Annual Meeting of the Canadian Society of Agronomy, Plant Canada 2007.* Canadian Society of Agronomy, Saskatoon, Saskatchewan, p. 164.

Pozo, M.J. and Azcón-Aguilar, C. (2007) Unraveling mycorrhiza-induced resistance. *Current Opinion in Plant Biology* 10, 393–398.

Redecker, D., Kodner, R. and Graham, L.E. (2000) Glomalean fungi from the Ordovician. *Science* 289, 1920–1921.

Remy, W., Taylor, T.N., Hass, H. and Kerp, H. (1994) Four hundred-million-year-old vesicular arbuscular mycorrhizae. *Proceedings of the National Academy of Sciences, USA* 91, 11841–11843.

Rillig, M.C. and Mummey, D.L. (2006) Mycorrhizas and soil structure. *New Phytologist* 171, 41–53.

Rivera, R., Fernández, F., Fernández, K., Ruiz, L., Sánchez, C. and Riera, M. (2007) Advances in the management of effective arbuscular mycorrhizal symbiosis in tropical ecosystems. In: Hamel, C. and Plenchette, C. (eds) *Mycorrhizae in Crop Production.* Haworth Food and Agricultural Products Press, Binghampton, New York, pp. 151–195.

Rodriguez, A., Clapp, J.P. and Dodd, J.C. (2004) Ribosomal RNA gene sequence diversity in arbuscular mycorrhizal fungi (Glomeromycota). *Journal of Ecology* 92, 986–989.

Rodriguez, A., Clapp, J.P., Robinson, L. and Dodd, J.C. (2005) Studies on the diversity of the distinct phylogenetic lineage encompassing *Glomus claroideum* and *Glomus etunicatum. Mycorrhiza* 15, 33–46.

Rosendahl, S. (2008) Communitites, populations and individuals of arbuscular mycorrhizal fungi. *New Phytologist* 178(2), 253–266.

Rosendahl, S. and Sen, R. (1992) Isozyme analysis of mycorrhizal fungi and their mycorrhiza. *Methods in Microbiology* 24, 169–194.

Ruiz-Lozano, J.M., Collados, C., Barea, J.-M. and Azcón, R. (2001) Arbuscular mycorrhizal symbiosis can alleviate drought-induced nodule senescence in soybean plants. *New Phytologist* 151, 493–502.

Sanders, I.R. (2002) Ecology and evolution of multigenomic arbuscular mycorrhizal fungi. *The American Naturalist* 160, S128–S141.

Sanders, I.R., Alt, M., Groppe, K., Boller, T. and Wiemken, A. (1995) Identification of ribosomal DNA polymorphisms among and within spores of the Glomales: application to studies on the genetic diversity of arbuscular mycorrhizal fungal communities. *New Phytologist* 130, 419–427.

Schüßler, A., Schwarzott, D. and Walker, C. (2001) A new fungal phylum, the *Glomeromycota*: phylogeny and evolution. *Mycological Research* 105, 1413–1421.

Sharma, S., Pant, D., Singh, S., Sinha, R.R. and Adholeya, A. (2007) Mycorrhizae in Indian agriculture. In: Hamel, C. and Plenchette, C. (eds) *Mycorrhizae in Crop Production*. Haworth Food and Agricultural Products Press, Binghampton, New York, pp. 239–291.

Simon, L., Bousquet, J., Lévesque, R.C. and Lalonde, M. (1993) Origin and diversification of endomycorrhizal fungi and coincidence with vascular land plants. *Nature* 363, 67–69.

Smith, S.E. and Read, D.J. (1997) *Mycorrhizal Symbiosis.* Academic Press, London.

St-Arnaud, M. and Vujanovic, V. (2007) Effects of the arbuscular mycorrhizal symbiosis on plant diseases and pests. In: Hamel, C. and Plenchette, C. (eds) *Mycorrhizae in Crop Production.* Haworth Food and Agricultural Products Press, Binghampton, New York, pp. 67–122.

St-Arnaud, M., Hamel, C., Caron, M. and Fortin, J.A. (1994) Inhibition of *Pythium ultimum* in roots and growth substrate of mycorrhizal *Tagetes patula* colonized with *Glomus intraradices. Canadian Journal of Plant Pathology* 16, 187–194.

Stewart, L.I., Jabaji-Hare, S.H. and Driscoll, B.T. (2006) Effects of external phosphate concentration on glucose-6-phosphate dehydrogenase gene expression in the arbuscular mycorrhizal fungus *Glomus intraradices. Canadian Journal of Microbiology* 52, 823–830.

Subramanian, K.S. and Charest, C. (1999) Acquisition of N by external hyphae of an arbuscular mycorrhizal fungus and its impact on physiological responses in maize under drought-stressed and well-watered conditions. *Mycorrhiza* 9, 69–75.

Tanaka, Y. and Yano, K. (2005) Nitrogen delivery to maize via mycorrhizal hyphae depends on the form of N supplied. *Plant, Cell and Environment* 28, 1247–1254.

Toro, M., Azcón, R. and Barea, J.-M. (1997) Improvement of arbuscular mycorrhiza development by inoculation of soil with phosphate-solubilizing rhizobacteria to improve rock phosphate bioavailability ($^{32}P$) and nutrient cycling. *Applied and Environmental Microbiology* 63, 4408–4412.

Toussaint, J.-P., St-Arnaud, M. and Charest, C. (2004) Nitrogen transfer and assimilation between the arbuscular mycorrhizal fungus *Glomus intraradices* Schenck and Smith and Ri T-DNA roots of *Daucus carota* L. in an *in vitro* compartmented system. *Canadian Journal of Microbiology* 50, 251–260.

van der Heijden, M.G.A., Klironomos, J.N., Ursic, M., Moutoglis, P., Streitwolf-Engel, R., Boller, T., Wiemken, A. and Sanders, I.R. (1998) Mycorrhizal fungal diversity determines plant biodiversity, ecosystem variability and productivity. *Nature* 396, 69–72.

# 12 Microbial Traits Associated with Actinobacteria Interacting with Plants

Anne-Marie Simao-Beaunoir, Sébastien Roy and Carole Beaulieu

*Université de Sherbrooke, Sherbrooke, Québec, Canada*

## Abstract

Actinobacteria (syn. Actinomycetes) are a large group that includes diverse genera of Gram-positive bacteria. Several of these genera are characterized by vegetative growth with aerial hyphae resembling those of fungi, and by a complex morphological differentiation leading to the formation of asexual spores. Actinobacteria are widely distributed in terrestrial environments and some, like the nitrogen-fixing *Frankia* bacterium, are known to form associations with plants through symbiotic relationships. Several strains, present in the rhizosphere, can also be beneficial to plant growth and act as plant growth-promoting rhizobacteria (PGPR) or as biocontrol agents suppressing soil-borne plant pathogens. Though most actinobacteria are saprophytes, some of them are phytopathogenic. They may cause common scab of potato and other symptoms like galls, stunting and wilts. In this chapter, we examine the mechanisms involved in the interactions between actinobacteria and plants. We also review the physiological traits and bacterial genes important to plant colonization, as well as the production of secondary metabolites, extracellular enzymes and soluble compounds such as plant hormones.

## 12.1 Introduction

The class of *Actinobacteria* is a group of Gram-positive bacteria characterized by a genome with high G+C ratio (Stackebrandt *et al.*, 1997). Several actinobacteria form branching filaments and share common features with fungi; they have a mycelial growth and some species produce external spores. Most members of the actinobacteria are aerobic, but a few can grow under anaerobic conditions. Actinobacteria are present in a wide variety of environments and although they are best known as soil-borne bacteria, they can also be found in marine and freshwater aquatic environments (Zaitlin *et al.*, 2003; Jensen *et al.*, 2005). Actinobacteria play an important role in decomposition of organic materials. Their ability to produce a wide variety of extracellular enzymes significantly contributes to degradation of complex plant biopolymers such as cellulose, pectin and lignin (Williams *et al.*, 1984). They are also well known as secondary metabolite producers and several secondary metabolites produced by actinobacteria are exploited by the pharmacological and agricultural industries.

## 12.2 Interactions Between Actinobacteria and Plants

### Plant pathogenic actinobacteria

Although actinobacteria are essentially saprophytes, some species develop associations with plants through pathogenic or symbiotic relationships. Several actinobacteria are opportunistic pathogens that cause disease only following fortuitous entry through wound sites but only a few actinobacteria species are considered as typical phytopathogens. The actinomycetal phytopathogens that have been most studied attack agronomical (potato, beans, sugarcane, tomato) or ornamental (lilies, poinsettia, bermudagrass) plants (Table 12.1). Table 12.1 presents the diseases caused by the most prevalent phytopathogenic actinobacteria, the plants affected, and some genetic characteristics of these microorganisms.

Bacteria belonging to the *Clavibacter*, *Curtobacterium* and *Leifsonia* genera of the family Microbacteriaceae, and *Rhodococcus fascians*, included in the Nocardiaceae family, have been shown to induce various symptoms on plants including galls, fasciation, gummosis, stunting and wilt (Table 12.1). Phytopathogenic *Streptomyces* species may also cause diseases, including sweet potato soil rot, netted and russet scab of potato (Harrison, 1962; Faucher *et al.*, 1993; Bouchek-Mechiche *et al.*, 2000; Agbessi *et al.*, 2003). Common scab of potato is, however, the best-known disease caused by streptomycetes. Among *Streptomyces* species causing common scab, *Streptomyces scabies*, *Streptomyces turgidiscabies* and *Streptomyces acidiscabies* are the most studied.

While some phytopathogenic actinobacteria have a broad host range, others cause disease on a limited number of plant species. For example, several subspecies of *Clavibacter michiganensis*, *Leifsonia xyli* and many *Curtobacterium flaccumfaciens* pathovars can be differentiated on the basis of plant host specificity. Generally, these high-specificity pathogens do not produce symptoms on non-host plants. *Leifsonia xyli* subsp. *cynodontis* that causes stunt disease on bermudagrass can colonize other crop plants such as maize, rice and sugarcane but will not cause disease (Haapalainen *et al.*, 2000). Furthermore, strains of *C. flaccumfaciens* have been isolated as endophytes from many crops and some strains can reduce symptom severity induced by *Xylella fastidiosa* when inoculated in the non-host *Catharanthus roseus* (Sturz *et al.*, 1998; Lacava *et al.*, 2007).

**Table 12.1.** Main actinobacteria plant pathogens.

| Name | Diseases | Major host | Sequenced genome strain (Centre) | Genome size | Genome and gene characteristics | References |
|---|---|---|---|---|---|---|
| *Clavibacter michiganensis* subsp. *michiganensis* | Bacterial canker | Tomato and pepper | Strain NCPPB 382 (Centre for Biotechnology, Bielefeld University, Germany) | Chromosome (3.3 Mbp), pCM1 (0.0274Mbp), pCM2 (0.07 Mbp) | *celA* (pCM1), *pat-1* (pCM2), *chp* and *tomA* (chromosome) | Rothwell (1968), Vidaver (1982), Meletzus *et al.* (1993), Bermpohl *et al.* (1996), Dreier *et al.* (1997), Jahr *et al.* (1999, 2000), Burger *et al.* (2005), Kaup *et al.* (2005), Gartemann *et al.* (2008) |
| *C. michiganensis* subsp. *sepedonicus* | Bacterial ring rot | Potato | Strain ATCC 33113 (Welcome Trust Sanger Institute) | Chromosome (3.3 Mbp), pCSL1 (0.095 Mbp), pCS1 (0.05 Mbp) | *celA* (pCS1), *pat-1* (pCSL1), homologues of *pat-1* (chromosome) | Manzer and Genereux (1981), Vidaver (1982), Laine *et al.* (1996, 2000), Metzler *et al.* (1997), Jahr *et al.* (1999), Nissinen *et al.* (2001), Marques *et al.* (2003), Romanenko *et al.* (2003), Shafikova *et al.* (2003), Bentley *et al.* (2008), Holtsmark *et al.* (2008) |
| *C. michiganensis* subsp. *tessellarius* | Leaf spot | Wheat | | | | Carlson *et al.* (1982) |
| *C. michiganensis* subsp. *nebraskensis* | Goss's bacterial wilt, blight | Maize | | | | Smidt and Vidaver (1987), Jahr *et al.* (1999) |
| *C. michiganensis* subsp. *insidosus* | Bacterial wilt | Lucerne | | | | McCulloch (1925), Cormack and Moffatt (1956), Paschke and Van Alfen (1993), Jahr *et al.* (1999) |
| *Curtobacterium flaccumfaciens* pv. *flaccumfaciens* | Bacterial wilt | Beans | | | | Hedges (1922), Collins and Jones (1983), Harding *et al.* (2007) |
| *C. flaccumfaciens* pv. *oortii* | Yellow pock | Tulip bulb | | | | Saaltink and Maas Geesteranus (1969), Collins and Jones (1983) |
| *C. flaccumfaciens* pv. *poinsettiae* | Bacterial canker | Common poinsettia | | | | Collins and Jones (1983), Starr and Pirone (1942) |

| | | | | | | |
|---|---|---|---|---|---|---|
| *C. flaccumfaciens* pv. *basellae* | Bacterial leaf spot | Malabar spinach | | | | Chen *et al.* (2000) |
| *C. flaccumfaciens* pv. *betae* | Silvering disease | Red beet | | | | Keyworth *et al.* (1956), Collins and Jones (1983) |
| *C. flaccumfaciens* pv. *beticola* | Bacterial leaf spot | Sugarbeet | | | | Chen *et al.* (2007) |
| *Leifsonia xyli* subsp. *cynodontis* | Stunt disease | Bermudagrass | | Chromosome (2.7 Mbp), pCXC100 (51 kbp) | | Davis *et al.* (1984), Taylor *et al.* (1993), Evtushenko *et al.* (2000), Haapalainen *et al.* (2000), Li *et al.* (2004) |
| *Leifsonia xyli* subsp. *xyli* | Ratoon stunting disease | Sugarcane | Strain CTCB07 (Sao Paulo state (Brazil) Consortium) | Chromosome (2.58 Mbp) | *celA* and *pat-1* | Davis *et al.* (1980, 1984), Evtushenko *et al.* (2000), Monteiro-Vitorello *et al.* (2004) |
| *Streptomyces scabies* | Common scab | Potato | Strain 87.22 (Welcome Trust Sanger Institute) | Chromosome (10.15 Mbp) | *txt* operon, *nos*, *tomA*, *nec1* genes | King *et al.* (1992), Babcock *et al.* (1993), Manulis *et al.* (1994), Natsume *et al.* (1996), Bukhalid and Loria (1997), Goyer and Beaulieu (1997), Loria *et al.* (1997, 2006), Beauséjour *et al.* (1999), Healy *et al.* (2000), Bukhalid *et al.* (2002), Wanner (2004), Kers *et al.* (2005), Simao-Beaunoir and Beaulieu (2006), Johnson *et al.* (2007), Joshi *et al.* (2007b) |
| *Streptomyces turgidiscabies* | Common scab | Potato | | | *txt* operon, *nos*, *tomA*, *nec1* and *fas* genes | Loria *et al.* (1997, 2006), Healy *et al.* (2000), Bukhalid *et al.* (2002), Kers *et al.* (2005), Johnson *et al.* (2007), Joshi and Loria (2007), Joshi *et al.* (2007a), Wach *et al.* (2007) |

*Continued*

**Table 12.1** – *Continued.*

| Name | Diseases | Major host | Sequenced genome strain (Centre) | Genome size | Genome and gene characteristics | References |
|---|---|---|---|---|---|---|
| *Streptomyces acidiscabies* | Common scab | Potato | | | *txt* operon, *nos*, *tomA*, *nec1* genes | Loria *et al.* (1997, 2006), Healy *et al.* (2000, 2002), Bukhalid *et al.* (2002), Wanner (2004), Kers *et al.* (2005), Johnson *et al.* (2007), Wach *et al.* (2007) |
| *Streptomyces ipomoeae* | Pox | Sweet potato | | | | Person and Martin (1940), Clark and Matthews (1987), King *et al.* (1994), Loria *et al.* (1997), Healy *et al.* (2000), Grau *et al.* (2006) |
| *Rhodococcus fascians* | Fasciation, leaf and crown gall | Various monocot and dicot plants | | | *fas*, *att* and *hyp* (conjugative linear Fi plasmid), *vic* (chromosome) | Crespi *et al.* (1992, 1994), Eason *et al.* (1996), Cornelis *et al.* (2001, 2002), Goethals *et al.* (2001), Maes *et al.* (2001), Vereecke *et al.* (2002, 2003), de O. Manes *et al.* (2004), Vandeputte *et al.* (2005) |

In contrast, and although well adapted to potato tubers, the common scab-inducing pathogens are neither tissue- nor host-specific but the pathogens can only infect plant tissue that is actively growing. However, common scab-inducing pathogens can easily induce necrosis in roots of most dicot and monocot plants (Leiner *et al.*, 1996; Goyer and Beaulieu, 1997; Joshi *et al.*, 2007b).

## Symbiotic actinobacteria

Symbiotic actinobacteria belonging to the genus *Frankia* are filamentous soil-borne microorganisms capable of nitrogen fixation. Frankiae were first isolated relatively recently, in 1978, and are characterized by fastidious *in vitro* culture (Lechevalier and Lechevalier, 1990). Culture media are complex and doubling times are long (hours to days), thus slowing research efforts. Isolation attempts have mainly been directed at isolating them from root nodules, and when these attempts are successful, strains typically require months before showing signs of proliferation in liquid media. These observations are in stark contrast to the fitness of frankiae in natural environments. Different characteristics influence their success as soil saprophytes and as root (nodule-forming) endosymbionts in their host plants, which are termed actinorhizal plants. The actinorhizal symbiosis is indeed found on all continents except Antarctica, and actinorhizal plants colonize a large range of environments which tend to be harsh and with poor soils (Baker and Schwintzer, 1990). Actinorhizal plants are thus often early successional species that play a key role in ecosystems disturbed by natural events (e.g. fires, landslides), by conditioning soil and rendering it capable of supporting other plants species (Roy *et al.*, 2007). These Gram-positive bacteria are among the few microorganisms that have evolved mechanisms enabling them to fix nitrogen in their free-living life cycle, as well as within symbiosis. The contribution of *Frankia* and actinorhizal plants to annual global nitrogen fixation in terrestrial ecosystems may be as high as 25% (Dawson, 2008).

Morphological structures in frankiae include: hyphae, sporangia, spores and vesicles – the site of nitrogen fixation. There have also been reports of cell differentiation leading to the formation of ‘reproductive torulose hyphae’, enlarged spore-like segments of hyphae that may play an important role in *Frankia* survival (Diem and Dommergues, 1985). Hyphae and spores are typically 0.5–2 μm, and 1 μm in diameter, respectively (Lechevalier and Lechevalier, 1990) and *Frankia* does not form aerial mycelium. Sporulation is variable from strain to strain and this trait is shared in the saprophyte and *in planta* life cycle. Typical to actinobacteria, frankiae also produce pigments, as secondary metabolites, that are widely varied in colour (e.g. red, green, orange, yellow, brown). The vesicle wall is composed of multiple lipid layers (hopanoids), the number of which increases along with oxygen concentration in the local environment, acting as a barrier to ambient oxygen levels, which would inhibit nitrogenase activity. This adaptation is at the centre of the capability of these organisms to fix nitrogen in the free-living form, and distinguishes them from *Rhizobia*, that may only fix nitrogen within the oxygen-protected environment of legume roots.

Not all *Frankia* strains will be capable of establishing a root nodule symbiosis with all actinorhizal plants, and there is a consensus among researchers regarding the lines of host plant symbiotic compatibility. *Frankia* strains are known to nodulate 200 actinorhizal plants (that belong to 24 genera). They are divided into the following host-infection groups: *Alnus* and *Myrica* (Group 1), *Casuarina* and *Myrica* (Group 2), *Myrica* and *Elaegnus* (Group 3), and those capable of nodulating members of the *Elaegnaceae* (*Elaegnus*, *Hippophae*, *Sherpherdia*) (Group 4) (Pawlowski and Sirrenberg, 2003; Roy *et al.*, 2007).

## Other actinobacteria beneficial to plant development

Several saprophytic actinobacteria are beneficial to plant development by acting as plant growth-promoting rhizobacteria (PGPR) or as biocontrol agents of diseases caused by soil-borne phytopathogens (Doumbou *et al.*, 2001; El-Tarabily and Sivasithamparam, 2006, Schrey and Tarkka, 2008). Plant growth promotion is usually associated with the biosynthesis of hormones such as auxins, gibberellins and cytokinins by actinobacteria strains (Persello-Cartieaux *et al.*, 2003). Hormones secreted by actinobacteria might also be linked to the biocontrol efficiency of some strains (Doumbou *et al.*, 2001; El-Tarabily, 2008). Most studies about biological control involved *Streptomyces* species although other genera from the order *Actinomycetales* have been reported as potential biocontrol agents (El-Tarabily and Sivasithamparam, 2006). Some studies even suggest that plant growth promotion induced by *Frankia* spp. is not strictly dependent on nitrogen fixation but involved other mechanisms such as plant disease suppression (Gopinathan, 1995).

Such beneficial interactions are at least partly due to the ability of actinobacteria to both colonize plant surfaces and compete with plant pathogens. Actinobacteria can suppress the growth of a wide variety of plant pathogens by parasiting these organisms (Yuan and Crawford, 1995) or by impeding their growth via production of antibiotics (Toussaint *et al.*, 1997; Agbessi *et al.*, 2003) or lytic enzymes (Toussaint *et al.*, 1997; El-Tarabily, 2006). Beneficial actinobacteria could also suppress plant diseases by degrading molecules involved in pathogenicity quorum-sensing (QS) signal molecules (Uroz *et al.*, 2003), or by inducing plant resistance mechanisms (Shimizu *et al.*, 2005; Conn *et al.*, 2008).

## 12.3 Genetic Traits Associated with Actinobacteria Interacting with Plants

Forty-nine genomes of different strains of actinobacteria are already completely sequenced and sequencing of genomes from 65 other actinobacteria are in progress (National Center for Biotechnology Information, 1988). The genomes of numerous phytopathogens are already available, such as that of *S. scabies* strain 87.22, *C. michiganensis* subsp. *michiganensis* NCPPB 382, *C. michiganensis* subsp. *sepedonicus* ATCC 33113, and that of *L. xyli* subsp. *xyli* CTCB07, which is closely related to the genus *Clavibacter* in which it was formerly included (Evtushenko *et al.*, 2000). These genomes range in size from 2.58 Mbp to more than 10 Mbp (Table 12.1). The sequencing of the circular genome of three *Frankia* strains, CcI3, ACN14a and Ean1pec, has also recently been achieved and their genome size is 5.4, 7.5 and 9 Mbp, respectively. It is known that genome size in bacteria is related to their life cycle and ecological niche. Bacteria with smaller genomes generally colonize specific environments that have conditions that limit microflora development (Andersson and Kurland, 1998). Examples of such bacteria are *Leifsonia* and *Clavibacter*, which are xylem-limited phytopathogens. Bacteria with larger genomes, such as those from the genera *Frankia* and *Streptomyces*, are capable of adapting to a larger spectrum of environmental conditions using strategies based, for example, on morphological differentiation and elaborate secondary metabolisms.

Strains belonging to the genus *Frankia* are currently not attributed a species name because of issues pertaining to the basis on which to attribute them. However, molecular biology approaches have been used to study frankiae biodiversity as well as functional traits in strains that are still studied through conventional microbiology and biochemistry. Molecular biology has been useful to identify *Frankia* metabolism genes, notably for nitrogen metabolism (*nifD*, *nifH*, *nifK* genes) (Roy *et al.*, 2007). In parallel, the recent sequencing of the genome of the three *Frankia* strains mentioned earlier (ACN14a, Ean1pec and CcI3) is opening new avenues of investigation and suggests the divergent evolution into a wide spectrum of strains ranging from obligate symbionts to facultative soil saprophytes (Normand *et al.*, 2007). Further study of these genomes will undoubtedly provide more understanding of the ecology and adaptability of frankiae, as is demonstrated by a recent study by Sen *et al.* (2008) that revealed the genes most expressed in the three sequenced *Frankia* strains. This study correlated the frequency of expression of genes of different classes with the hypothesis of divergent evolution (and specialization) of strains (Sen *et al.*, 2008).

Similar to *Frankia*, *C. michiganensis* subsp. *michiganensis* and subsp. *sepedonicus*, and *L. xyli* subsp. *xyli* and subsp. *cynodontis* also present a long generation time when cultivated *in vitro*. However, they appear to be restricted to an endophytic lifestyle in nature. This may be the result of adaptation to the plant host accompanied by gene loss. The genome of *L. xyli* subsp. *xyli* indeed contains a high proportion of pseudogenes (13%). This genome shrinkage can have led to the inability for the bacterium to grow in the absence of plant material (Monteiro-Vitorello *et al.*, 2004). An example of this is the enhanced *in vitro* growth observed in cultures of *L. xyli* subsp. *cynodontis* (a species closely related to *L. xyli* subsp. *xyli*) that are supplemented with maize xylem fluid components (Haapalainen *et al.*, 2000).

Genomic analyses are also useful to understand mechanisms linked to pathogenicity of actinobacteria. For *R. fascians*, some pathogenicity factors are chromosomic, such as the locus (*vic*) encoding for a malate synthase that contributes to virulence, but other loci such as *fas* responsible for plant fasciation, and *att* and *hyp*, which affect the degree of virulence of the bacteria were found in the 200 kb linear Fi plasmid (Crespi *et al.*, 1992, 1994; Maes *et al.*, 2001; Cornelis *et al.*, 2002; Vereecke *et al.*, 2002). As for the main pathogenicity genes of *R. fascians*, *C. michiganensis* subsp. *michiganensis* and *C. michiganensis* subsp. *sepedonicus*, the cellulase-encoding *celA* gene and the serine protease encoding *pat-1* gene are both carried on plasmids (Meletzus *et al.*, 1993; Dreier *et al.*, 1997; Laine *et al.*, 2000; Gartemann *et al.*, 2003). Actinobacteria often contain plasmids that vary in size and copy numbers (Piret and Demain, 1988) and that can play a role in genomic rearrangements and acquisition of new traits. Indeed, analysis of the *L. xyli* subsp. *xyli* sequenced genome revealed chromosomic regions that may be the result of plasmid integration events, since this region harbours genes required for the horizontal transfer of plasmids (Monteiro-Vitorello *et al.*, 2004). Moreover, pathogenicity genes are often clustered in a genomic region. These clusters may contain mobile genetic elements. The presence of genes encoding transposases and resolvases, of pseudogenes and of genes with a different G+C content suggests that the pathogenicity cluster may have been acquired by a horizontal gene transfer from other taxa. Such gene clusters are termed PAthogenicity Islands (PAIs). In phytopathogenic bacteria, PAIs carry virulence genes encoding for example for phytotoxic compounds and phytohormones (Arnold *et al.*, 2003).

A large PAI of 325–600 kb has been found in the chromosome of *S. turgidiscabies* (Kers *et al.*, 2005; Loria *et al.*, 2006). Thirty genes were described from the *S. turgidiscabies* PAI which includes genes involved in the biosynthesis of phytotoxic compounds called thaxtomins (*txtA*, *txtB*, *txtC* and *nos*) and tomatinase (*tomA*), genes encoding factors responsible for plant necrosis (*nec1*) and plant fasciation (*fas*), as well as several genes possibly encoding for transposases and resolvases (Kers *et al.*, 2005). Preliminary analysis of *S. scabies* strain 87.22 genome revealed that all pathogenicity genes do not cluster in a continuous region of the bacterial chromosome as was observed in *S. turgidiscabies*. In *S. scabies*, genes responsible for toxin biosynthesis (*txtA*, *txtB*, *txtC* and *nos*) are separated from the *nec1* and the tomatinase genes by approximately 5000 kbp compared to approximately 18 kbp in *S. turgidiscabies* (Simao-Beaunoir and Beaulieu, 2006). Absence of the *fas* gene in the sequenced genome of *S. scabies* is another genetic difference between both species. Hybridization studies revealed the presence of the *txt*, *nec1* and *tomA* genes in *S. acidiscabies* but not the presence of a *fas* gene. Although, *Streptomyces ipomoeae* produces thaxtomins, it is unknown whether *S. ipomoeae* also possesses a PAI similar to the common scab-inducing strain of *Streptomyces*.

The genome sequence of *C. michiganensis* subsp. *michiganensis* also revealed a putative PAI on the chromosome although *celA*, the main pathogenicity gene, is plasmid-borne. The chromosomic island presents a significantly lower G+C content, an unusual codon usage and the pathogenicity factor-bearing loci *chp* (for chromosomal homologue of *pat-1*) and *tomA* (Gartemann *et al.*, 2008). The *L. xyli* subsp. *xyli* genome also contains regions that are described as putative genomic islands and harbours *celA* and *pat-1* that have previously been shown to confer the wilt-inducing phenotype of *C. michiganensis* subsp. *michiganensis* (Monteiro-Vitorello *et al.*, 2004).

## 12.4 Physiological Traits Associated with Actinobacteria Interacting with Plants

### Plant colonization

Plant-associated bacteria interact with plant tissue surfaces during pathogenesis, symbiosis or commensal relationships. The presence of actinobacteria in rhizospheric soils and on plant tissues is well documented and pathogenic as well as non-pathogenic endophytic actinobacteria have also been isolated and characterized (Barakate *et al.*, 2002; Merzaeva and Shirokikh, 2006; Franco *et al.*, 2007). Biofilm production appears to be an important factor for the colonization of plant tissues. Vascular pathogens inhabit the xylem or phloem of the plant host where they often form biofilms. Scanning electron microscopy revealed that *C. michiganensis* subsp. *sepedonicus* formed large bacterial, matrix-encased aggregates attached to the xylem vessels (Marques *et al.*, 2003). Formation of vascular clogging and seed contamination involved biofilm production by *C. flaccumfaciens* pv. *flaccumfaciens*, the causal agent of bean vascular wilt (Harding *et al.*, 2007).

By participating in plant attachment, exopolysaccharides (EPSs) play a role in pathogenicity of Gram-negative phytopathogens, and an inability to produce EPS often correlates with loss of virulence (Coplin and Cook, 1990; Denny, 1995). Involvement of EPS in pathogenicity mechanisms of actinobacteria has been mostly studied on *Clavibacter*. EPSs produced by *C. michiganensis* subsp. *michiganensis* are apparently involved in determination of host specificity. Variations in the sugar composition of EPSs were shown to impede tomato colonization (Bermpohl *et al.*, 1996). Furthermore, Shafikova *et al.* (2003) showed that EPSs produced by *C. michiganensis* subsp. *sepedonicus*, the causal agent of bacterial ring rot of potato, bind to receptor proteins present on the plasma membrane of potato cells. These EPS receptors are present in larger amounts in the cell walls of susceptible cultivars than in the cell walls of resistant ones. The receptors may facilitate binding of bacteria to the plant cells, formation of bacterial aggregates, and consequently disease development (Shafikova *et al.*, 2003). *Clavibacter michiganensis* subsp.

*sepedonicus* EPSs were also shown to considerably increase peroxidase activity, an early plant defence response, in cells of a resistant potato cultivar suggesting a role for EPSs as elicitors of the plant defence mechanism (Romanenko *et al.*, 2003). Nevertheless, no study established a direct correlation between EPSs and virulence of *C. michiganensis* subsp. *michiganensis* or subsp. *sepedonicus.* A mutant completely deficient in EPS biosynthesis has not yet been obtained. Bermpohl *et al.* (1996) produced a *C. michiganensis* subsp. *michiganensis* mutant that produces 90% less EPS than the wild strain. This mutant retains the ability to colonize plants and remains pathogenic (Bermpohl *et al.*, 1996). Similar results have been previously obtained with *C. michiganensis* subsp. *insidosus* (Paschke and Van Alfen, 1993). If EPSs play an important role in pathogenicity, it seems that this property does not strictly depend on the amount of EPS produced or the number of genes involved in EPS production. Hypothetically, the *C. michiganensis* subsp. *sepedonicus* genome contains twice as many genes involved in EPS biosynthesis as the *C. michiganensis* subsp. *michiganensis* genome (Bentley *et al.*, 2008).

In animal pathogens, teichoic acids of Gram-positive bacteria are involved in host recognition, adhesion during biofilm formation and virulence (Neuhaus and Baddiley, 2003). The chemical structure of teichoic acids and other anionic polymers in the cell wall of common scab-inducing streptomycete strains inducing potato scab disease has been determined. Anionic polymers associated with the cell wall were shown to contain a polymer of 3-deoxy-D-glycero-D-galacto-non-2-ulopyranosonic acid (Kdn-polymer) which is presumably essential for attachment of bacterial pathogens to host plant cells. Indeed, an acidic polysaccharide belonging to the same family of the Kdn-polymer has been shown to be involved in *Agrobacterium tumefaciens* attachment to carrot cells (Reuhs *et al.*, 1997; Shashkov *et al.*, 2002; Tul'skaya *et al.*, 2007).

## Quorum sensing (QS)

Many bacteria depend on QS-regulated gene systems to establish symbiotic or pathogenic interactions with hosts (von Bodman *et al.*, 2003; Ramey *et al.*, 2004). Bacteria use QS to coordinate their gene expression according to the local density of their population. QS-regulated genes include those involved in the biosynthesis of EPSs, extracellular hydrolytic enzymes, siderophores, antibiotics, pigments, hypersensitive reaction and pathogenicity (HRP) proteins, biofilm, as well as those involved in conjugation and epiphytic fitness (von Bodman *et al.*, 2003). In actinobacteria, QS is dependent on the production of autoregulatory factors of different chemical classes, mainly the butyrolactone class and the nucleotide-like B-factor class. The first class includes the A-factor of *Streptomyces griseus* (Yamada and Nihira, 1998) and similar compounds found in other actinobacteria (Yamada *et al.*, 1987; Chater and Bibb, 1997; Kawabuchi *et al.*, 1997). The nucleotide-like B-factor class has been isolated from *Amycolatopsis* (*Nocardia*) *mediterranei* (Kawaguchi *et al.*, 1988). In actinobacteria, research efforts on QS have mostly focused on morphological differentiation and secondary metabolism. The role of QS during the interaction between actinobacteria and plants has not been examined in detail.

## Extracellular enzymes degrading plant polymers

Extracellular hydrolytic enzymes are commonly produced by saprophytic actinobacteria for the decomposition of plant residues (Peczynska-Czoch and Mordarski, 1988). These enzymes might however play an important role in the invasion of host plants. As with any actinobacteria, we may expect *Frankia* species to possess the enzymatic capability to degrade natural polymers such as cellulose, lignin, chitosan and pectin. The presence in frankiae of both pectinase and cellulase enzymes have been confirmed (Séguin and Lalonde, 1989; Igual *et al.*, 2001). Such catabolic capabilities may have significance to the actinorhizal symbiosis itself since the formation of root nodules may involve local hydrolysis of plant cell walls. However, the ability of *Frankia* to produce extracellular enzymes appears to be weak when compared to other actinobacteria. There are five times fewer genes coding for extracellular polysaccharide-degrading enzymes than in *Streptomyces coelicolor* and *Streptomyces avermitilis*. However, examination of the secretome of three *Frankia* strains by Mastronunzio *et al.* (2008) revealed that a third of the proteins secreted by *Frankia* appear to be specific to this genus. Mastronunzio *et al.* (2008) suggested that this weak production of extracellular enzymes is a strategy to avoid eliciting host defence responses. A less aggressive attack on plant cell walls may also be an advantage for bacteria establishing a symbiotic and coordinated interaction with plants.

Involvement of extracellular hydrolytic enzymes in pathogenicity varies considerably depending on phytopathogenic actinobacteria species. Histological and genetic studies of *C. michiganensis* suggest that plant cell wall-degrading enzymes strongly contribute to symptom development (Benhamou, 1991; Meletzus *et al.*, 1993; Jahr *et al.*, 1999). Cellulase production appears to be necessary for virulence of the xylem-limited pathogens *C. michiganensis* subsp. *michiganensis* and *C. michiganensis* subsp. *sepedonicus*. An endo-β-1,4-glucanase of the A1 cellulase gene family (*celA*) is found on the plasmid pCM1 belonging to *C. michiganensis* subsp. *michiganensis*, and an orthologue gene is present on plasmid pCS1 from subsp. *sepedonicus* (Meletzus *et al.*, 1993; Laine *et al.*, 1996). The sequence of *celA* revealed a leader sequence for secretion and two cellulase domains (Laine *et al.*, 2000; Bentley *et al.*, 2008). Holtsmark *et al.* (2008) showed that *celA* expression increased in subsp. *sepedonicus* during infection of detached potato leaves. When cured of pCM1 or pCS1, both *C. michiganensis* subspecies became deficient in cellulase production. They also became non-pathogenic or their virulence was drastically reduced (Meletzus *et al.*, 1993; Laine *et al.*, 2000; Jahr *et al.*, 2000; Nissinen *et al.*, 2001; Gartemann *et al.*, 2003).

Genome analysis of *L. xyli* subsp. *xyli* revealed the presence of a gene encoding a protein exhibiting 64% amino acid similarity with CelA of *C. michiganensis* subsp. *sepedonicus*. This gene is present in one of the four putative genomic islands of the bacterium. These genomic islands also harbour genes encoding polygalacturonases, enzymes that degrade the plant cell wall pectin (Monteiro-Vitorello *et al.*, 2004). These polygalacturonases certainly provide nutriments for the bacteria but their specific function in pathogenicity has still to be determined. *Clavibacter michiganensis* subsp. *sepedonicus* was also found to produce an amylase that affects virulence but its role in pathogenicity is minor compared to that of CelA (Metzler *et al.*, 1997).

As most non-pathogenic streptomycetes, common scab-inducing strains produce enzymes involved in plant tissue degradation. However, the ability to produce pectinases and cellulases did not correlate with pathogenicity (Spooner and Hammerschmidt, 1989). There are indications however that esterase production by *S. scabies* may be involved in pathogenicity. McQueen and Schottel (1987), as well as Beauséjour *et al.* (1999), have shown that *S. scabies* produces esterases involved in potato suberin degradation. Suberin is a waxy polymer that covers potato tubers. One could therefore hypothesize that production of esterases would facilitate pathogen entry in plant tissues by depolymerizing suberin.

Pox symptoms, caused by *S. ipomoeae*, differ from those of common scab and the disease is characterized by substantial yield reductions (Clark and Matthews, 1987). The ability of *S. ipomoeae* to cause extensive fibrous root rot should be related to the production of an arsenal of plant cell degrading enzymes.

No cell wall degrading enzymes have been associated with gall and bacterial fasciation. *Rhodococcus fascians* resides primarily on the plant exterior surface. Bacteria grow epiphytically on aerial plant surfaces, protected by a bacterial slime layer. The bacteria cause a collapse of some epidermal cells, and thereby penetrate the plant. Bacteria are then found in the intercellular spaces of gall tissues and sometimes inside plant cells (Cornelis *et al.*, 2001).

## Enzymes interfering with plant defence mechanisms

Some plant pathogens produce extracellular proteins that interfere with plant defence responses. *Clavibacter michiganensis* subsp. *sepedonicus* and subsp. *michiganensis* produce the pathogenicity factor Pat-1. This factor is a putative extracellular serine protease from family S1 (chymotrypsin) and is essential to *C. michiganensis* subsp. *michiganensis* pathogenicity (Dreier *et al.*, 1997; Burger *et al.*, 2005). Proteases interfere with defence responses by promoting or repressing induction of the hypersensitive reaction (HR) in several host–microbe interactions (de Jong *et al.*, 2000; Mudgett, 2005). Homologues of the *pat-1* gene are also found in *L. xyli* subsp. *xyli* strain CTCB07 genome and in several Gram-negative phytopathogenic bacteria, but not in saprophytes (Monteiro-Vitorello *et al.*, 2004).

The major antimicrobial plant saponin, α-tomatine, produced in tomato, can be hydrolysed by tomatinase and consequently lead to the suppression of induced plant defence responses (Bouarab *et al.*, 2002). Tomatinases are well-characterized enzymes produced by phytopathogenic fungi like *Fusarium oxysporum* f. sp. *lycopersici* or *Septoria lycopersici* (Sandrock *et al.*, 1995; Lairini *et al.*, 1996). A tomatinase gene was found in *S. acidiscabies*, *S. scabies* and *S. turgidiscabies*, but not in the non-pathogen *S. coelicolor* (Kers *et al.*, 2005). Nevertheless, function of this tomatinase during plant–microbe interaction has yet to be elucidated. *Clavibacter michiganensis* subsp. *michiganensis* possesses a homologue of the *S. turgidiscabies tomA* gene (Joshi *et al.*, 2007b). The fact that growth inhibition of *C. michiganensis* subsp. *michiganensis* induced by α-tomatine was stronger in *tomA* mutants than in wild-type strains suggests that this enzyme might be of importance during *in planta* growth (Kaup *et al.*, 2005). However, hybridization experiments between the *tomA* probe from *C. michiganensis* subsp. *michiganensis*, and total DNA of *C. michiganensis* subsp. *sepedonicus*, *nebraskensis*, *tesselarius* and *insidosus* and *C. flaccumfaciens* pv. *oortii*, have revealed that this gene is absent in these phytopathogen actinobacteria (Kaup *et al.*, 2005).

## Extracellular enzyme in biological control of plant disease

Extracellular enzyme production is also a trait that has been associated with biocontrol agents. Indeed, biocontrol ability has been linked to production of lytic enzymes attacking the cell wall of phytopathogenic fungi and oomycetes. Most of the studies conducted to find new potential biological control agents consist in the *in vitro* selection of actinobacteria producing high levels of extracellular chitinases (Mahadevan and Crawford, 1997; El-Tarabily, 2003; Hoster *et al.*, 2005; Sousa *et al.*, 2008) or glucanases (Valois *et al.*, 1996; Fayad *et al.*, 2001; El-Tarabily, 2006). For example, Valois *et al.* (1996) screened a collection of non-pathogenic actinobacteria for the ability to produced β-1,3, β-1,4 and β-1,6 glucanases in the presence of *Phytophthora* mycelia. Several of these isolates when inoculated on raspberry protected the plant against root rot induced by oomycetes. El-Tarabily (2003) showed that chitinase production of an *Actinoplanes missouriensis* isolate, antagonist to *Plectosporium tabacinum*, plays an important role in the suppression of root rot of lupin. A mutant strain of *A. missouriensis* producing a low level of chitinase failed to lyse hyphae of *P. tabacinum in vitro* and to reduce root rot of lupin in glasshouse experiments, although it colonized the lupin root as well as the wild type. Many studies about mechanisms involved in biological control of plant disease showed that biocontrol agents not only produced extracellular enzymes but also antimicrobial antibiotics, providing a synergistic or additive antagonistic effect against pathogens (Yuan and Crawford, 1995; Toussaint *et al.*, 1997; Trejo-Estrada *et al.*, 1998).

## Antibiosis

Actinobacteria are the most prolific microbial producers of various antimicrobial compounds like antibiotics and bacteriocins. The genus *Streptomyces* has extensively been studied as a source of new antibiotics and other actinomycetal genera also appear to be of interest (Ouhdouch *et al.*, 2001; Lamari *et al.*, 2002; Boudjella *et al.*, 2006; El-Tarabily and Sivasithamparam, 2006).

The role of antibiotic production during actinobacteria–plant interactions has mostly been studied in non-pathogenic actinobacteria exhibiting antimicrobial activity against plant pathogenic agents. Numerous biocontrol agents have been selected for their ability to produce antibiotic compounds inhibiting the growth of pathogens *in vitro* and to reduce the disease symptoms *in planta* (Tanaka and Omura, 1993; Eckwall and Schottel, 1997; Hwang *et al.*, 2001; Park *et al.*, 2008). Tan *et al.* (2006) showed also that the proportion of actinobacteria with antibacterial activity against the pathogen *Ralstonia solanacearum* was higher in healthy tomato plants than in wilting plants. Although most of the actinomycetal biocontrol agents that have been identified to date belong to the genus *Streptomyces*, some members of the *Actinoplanes*, *Micromonospora*, *Microbispora*, *Streptosporangium*, *Intrasporangium* and *Nonomuraea* genera demonstrated interesting and promising biocontrol properties (El-Tarabily and Sivasithamparam, 2006; Joshi *et al.*, 2007b; Okudoh and Wallis, 2007).

Most of the antibiotics are directly used as pesticides in agriculture (Tanaka and Omura, 1993; Kim and Hwang, 2007) but commercial biopesticides containing, as the active ingredient, an actinobacteria strain are also available. For example, Mycostop®, a biofungicide used for the control of *Fusarium* wilt of carnation and root rot disease of cucumber, contains living *Streptomyces griseoviridis* cells (Lahdenpera *et al.*, 1991). Although antibiosis is considered an important trait in biocontrol agents, few studies have established a direct link between biocontrol and antibiotic production *in planta*. Nevertheless, Agbessi *et al.* (2003) showed that a *Streptomyces melanosporofaciens* mutant, deficient in geldanamycin biosynthesis, lost the ability to protect potato against common scab. Indirect evidence for the role of the antibiotic produced by the antagonist *Streptomyces violaceusniger* strain G10 on *F. oxysporum* f. sp. *cubense*, the causative agent of wilt disease of banana, has also been shown (Getha and Vikineswary, 2002). *In vitro* effects of extracellular metabolites produced by *S. violaceusniger* strain G10 that were observed on mycelial development and spore germination of *F. oxysporum* were similar to those induced when fungal spores were incubated with the antagonist strain in soil. Nevertheless, most scientists interested in biological control estimate that biocontrol efficiency depends not only on antibiosis or production of lytic enzymes but also on other traits such as competition for nutrients, plant tissue colonization and adaptation to environmental stress.

Non-pathogenic bacteria may influence the aggressiveness of pathogens not only by affecting their growth, but also by modulating the production of virulence factors or by degrading these compounds. Doumbou *et al.* (1998) have, for example, identified non-pathogenic streptomycetes able to protect potato tubers and to degrade thaxtomin, a toxin produced by common scab-inducing *Streptomyces* species. An interspecific molecular signalling mechanism may also affect the outcome of a pathogenic interaction. For example, the coinoculation in soil of *S. scabies* and some non-pathogenic strains such as

*S. melanosporofaciens* strain FP-60 increases the incidence of common scab on potato tubers (Agbessi *et al.*, 2003). The increased virulence of *S. scabies* appears to be due to the production by FP-60 of small extracellular molecules that act as signals to increase thaxtomin A production. In contrast, some other non-pathogenic streptomycete strains have been recently shown to produce extracellular molecules that inhibit thaxtomin biosynthesis in *S. scabies* (C. Beaulieu, 2008, unpublished results).

## Production of compounds toxic to plants

Bacterial phytotoxic compounds are secondary metabolites produced during plant–microbe interactions. Thaxtomins are the best-characterized actinomycetal toxins. They are produced by common scab-inducing *Streptomyces* (*S. scabies*, *S. acidiscabies*, *S. turgidiscabies*) and by *S. ipomoeae*, the causal agent of sweet potato pox (Loria *et al.*, 1997). Thaxtomins have been described as unique 4-nitroindol-3-yl containing 2,5-dioxopiperazines (King *et al.*, 1992). Over ten thaxtomin analogues are produced by *S. scabies*. They differ in the presence of specific methyl and hydroxyl residues (King *et al.*, 2003). Thaxtomins A and B are the principal toxins associated with common scab of potato (King *et al.*, 1992) and thaxthomin C, produced by *S. ipomoeae*, is associated with pox disease of the sweet potato (King *et al.*, 1994). Thaxtomins can mimic common scab symptoms when applied to potato tubers but these toxins also cause necrosis when applied to other potato tissues or even to other plant species. Inoculating most seedlings with thaxtomin-producing streptomycetes leads to root swelling and necrosis as well as root and shoot stunting (Loria *et al.*, 2008). The cellular target of thaxtomins in plant cells has not been clearly identified yet. However, thaxtomins are responsible for cellulose synthesis inhibition in expanding plant tissues (Loria *et al.*, 1997). Although thaxtomins were shown to be essential to *S. scabies* pathogenicity, some studies suggest that not all common scab-inducing strains produce these plant toxins (Park *et al.*, 2003; Wanner, 2004).

Synthesis of thaxtomins A and B is dependent on the presence of *txtA*, *txtB*, *txtC* and *nos*. The genes *txtAB* encode the peptide synthetase which links tryptophan and phenylalanine to form the cyclic dipeptide structure of thaxtomins. The gene *txtC* encodes a cytochrome P450 monooxygenase for the hydroxylation of phenylalanine and a nitric oxide synthase, encoded by *nos*, allows the nitration of tryptophan (Healy *et al.*, 2000, 2002; Kers *et al.*, 2004). The thaxtomin C biosynthetic gene cluster of *S. ipomoeae* has been found in a cosmid clone and was sequenced (Grau *et al.*, 2006). *Streptomyces ipomoeae* has homologous sequences for *txtA*, *txtB* and *nos* and the genes are organized as found in *S. turgidiscabies* except for the *txtC* homologue which is not located downstream of *txtB* (Grau *et al.*, 2006). Since *S. ipomoeae* produces mostly thaxtomin C, an analogue of thaxtomin A lacking two hydroxyl and one methyl group, the authors have suggested that the *txtC* gene might not be required for thaxtomin C synthesis (King *et al.*, 1994; Bukhalid and Loria, 1997).

*Streptomyces scabies* does not synthesize thaxtomins in minimal media, but the presence of plant extracts in a culture medium activated thaxtomin production. Biosynthesis of thaxtomins by *S. scabies* strains has been demonstrated in oatmeal and oat bran media (Babcock *et al.*, 1993; Goyer *et al.*, 1998) and in medium supplemented with potato flesh, potato peels, radish and sweet potato peels (Beauséjour *et al.*, 1999). Minimal media supplemented with cellulose, pectate and starch did not allow toxin biosynthesis, while the addition of suberin in culture medium induced thaxtomin A production in *S. scabies* strains (Beauséjour *et al.*, 1999). In *S. acidiscabies*, cell-wall components, including cellobiose and cereal xylans, have been implicated in stimulating thaxtomin biosynthesis (Wach *et al.*, 2007). Cellobiose as a sole carbon source did not allow sufficient growth of *S. scabies* strain 87.22, *S. acidiscabies* strain 84.104 and *S. turgidiscabies* strain Car8. A minimal medium with sorbose as the source of carbon, with and without cellobiose supplement, was used to study the induction of thaxtomin production. For all three strains, thaxtomin production was positively correlated with cellobiose amendment of minimal media, and cellobiose induction of thaxtomin was also shown to be dose-dependent (Johnson *et al.*, 2007). Cellobiose alone however was not a strong inducer of dipeptide synthesis in several other *S. scabies* strains. The presence of this disaccharide, however, could improve thaxtomin A production in a medium containing suberin (S. Lerat, N. Beaudoin and C. Beaulieu, 2008, unpublished results).

In addition to thaxtomins, *S. scabies* also produces another type of phytotoxic compound called concanamycin, a plecomacrolide belonging to the class of macrolactone antibiotics (Natsume *et al.*, 1996). The phytotoxic activity of the concanamycins is due to the inhibition of V-type H+-ATPase (Dröse *et al.*, 1993). Despite the fact that toxicity of concanamycins on plant tissues has been demonstrated, the role of these toxins in common scab disease is still unknown. Other phytopathogenic streptomycetes also produce plant toxins. This is the case for a *Streptomyces* sp. strain named *Streptomyces chelonium*, a causal agent of russet scab in Japan. The strain did not seem to produce thaxtomin A nor concanamycins A and B, but synthesized 18-membered macrolides that are analogous to concanamycin A and are capable of inducing necrosis in potato slices (Natsume *et al.*, 2005).

In addition to small molecules such as thaxtomins and concanamycins, *S. scabies* also produces a small protein encoded by *nec1* which has no homologue in sequence databases and lacks characterized motifs. The necrosis gene, *nec1*, includes a sequence encoding an N-terminal secretion signal. When *nec1* is introduced in the non-pathogenic *Streptomyces lividans*, the transformants gain the ability to colonize and cause necrosis on potato slices. The *nec1* gene was not required for pathogenicity in *S. turgidiscabies* but the Δnec1 mutant had a reduced aggressiveness on *Arabidopsis thaliana* seedlings compared to wild type and the mutant failed to colonize expanding cells proximal to the root meristem of radish roots (Loria *et al.*, 2006; Joshi *et al.*, 2007a). The Nec1 protein is not required for thaxtomin production and therefore represents an independent virulence factor (Kers *et al.*, 2005). Further, *S. ipomoeae* strains do not carry the necrogenic factor *nec1*, which is found in strains of *S. scabies*,

*S. acidiscabies* and *S. turgidiscabies* (King *et al.*, 1994; Bukhalid and Loria, 1997; Bukhalid, *et al.*, 2002) and a recently isolated *Streptomyces* sp. strain causing common scab in potato appears more virulent than *S. scabies* whereas the bacteria lacks *nec1* (Wanner, 2007).

## Bacterial phytohormones

Actinobacteria grown in pure culture are known to produce several of the known phytohormones: auxins, gibberellins and cytokinins. The role of bacterial phytohormones in pathogenicity has been most studied in *R. fascians,* the causal agent of plant fasciation and galls. Symptoms induced by *R. fascians* are similar to those induced by direct application of cytokinins. Nevertheless, even if *R. fascians* produces multiple forms of cytokinins *in vitro*, the quantities of the phytohormones produced seem insufficient to induce leafy gall symptoms (Eason *et al.*, 1996; Vereecke *et al.*, 2003). It is hypothesized that *in planta*, the close proximity of cytokinins produced by the bacteria has a greater impact on the plant (Eason *et al.*, 1996). In *R. fascians*, cytokinin production is under the control of the plant fasciation (*fas*) operon, identified by sequence homology to other cytokinin biosynthetic pathways. The *fas* locus contains six genes that encode an isopentenyltransferase (IPT) and other enzymes involved in the modification of the IPT product. In *R. fascians*, these genes are located on a linear plasmid that is required for leafy gall formation in plants (Crespi *et al.*, 1992; Goethals *et al.*, 2001). The production of cytokinin via the *fas* operon is tightly controlled in *R. fascians* and requires induction by extracts of infected plant tissue; no induction occurs with healthy plant tissue extract (Crespi *et al.*, 1992). The chromosomal PAI from *S. turgidiscabies* also contains a plant fasciation (*fas*) operon homologous to and nearly colinear with the *fas* operon present in the phytopathogen *R. fascians*. *Streptomyces turgidiscabies* produces all of the cytokinin-dependent symptoms produced by *R. fascians*, including leafy galls on tobacco and *Arabidopsis* (Goethals *et al.*, 2001; de O. Manes *et al.*, 2004; Joshi and Loria, 2007). There is also evidence for cytokinin production by *Frankia* strains (Stevens and Berry, 1988) but it is unknown if this factor contributes to symbiosis mechanisms. Interestingly, if bacterial-produced cytokinins can influence the growth of plants, the phytohormone has also been shown to be a signalling molecule for secondary metabolite production in *Streptomyces* species (Yang *et al.*, 2006).

In addition to cytokinin, *R. fascians* also produces auxin. The production of indole-3-acetic acid (IAA), the most important member of the auxin family, by a plasmid-free strain of *R. fascians* shows that the biosynthetic genes are located on the bacterial chromosome, although plasmid-encoded genes contribute to the kinetics and regulation of IAA biosynthesis. Phenotypic analysis of a leafy gall suggests that auxin may play an important role in the development of the symptoms but no evidence has been shown to confirm it. Nevertheless, as for the production of cytokinins, auxin production has been shown to be highly induced by infected-plant extracts (Vandeputte *et al.*, 2005).

Biosynthesis of IAA in *S. scabies* has been demonstrated (Manulis *et al.*,

1994), but involvement of this phytohormone in pathogenesis has not been extensively studied yet. The genome carries genes with homology to *iaaM* and *iaaH* of *S. avermitilis*. In the latter, these genes encode a tryptophan monooxygenase and an indole acetamide hydrolase, respectively. Addition of tryptophan, the precursor of both auxin and thaxtomin biosynthetic pathways, in *S. scabies* culture media promoted auxin biosynthesis but had a negative effect on thaxtomin biosynthesis. In environmental conditions where thaxtomin biosynthesis is inhibited and auxin production is stimulated, *S. scabies* induces no symptoms but rather promotes plant growth (G. Legault, S. Lerat and C. Beaulieu, 2008, unpublished results).

## Nodulation and nitrogen fixation

The actinorhizal symbiosis and the legume symbiosis have similarities and differences. These have been reviewed extensively by Pawlowski and Bisseling (1996) and recently by Pawlowski and Sprent (2008), and this comparison will only be touched upon here. When compatible frankiae and host plants are in contact and the infection process of the actinorhizal symbiosis is initiated, frankiae cells will either proceed by root hair deformation, or penetrate the root epidermis and cortex. Although these pathways are similar to those in the legume symbiosis (being inter- and intracellular routes), the resulting nodule will be markedly different in its internal structure (Benson and Silvester, 1993; Pawlowski and Sprent, 2008; Wall and Berry, 2008). The lateral roots of actinorhizal plants are modified by the infection process, and become individual lobes. Multiple lobes make up each root nodule. It is within each lobe that the structural differences between the actinorhizal nodule and the legume nodule become apparent. The prominent difference is the proximity of infected cells (thus *Frankia*) to the outer periphery of the nodule lobe (Baker and Schwintzer, 1990). In legume nodules, layers of vascular tissue surround the infected cells, likely reducing the exposure of nitrogenase in *Rhizobia* bacteroids to ambient oxygen. *Frankia* has its own oxygen protection mechanism in vesicle wall composition and thickness. In actinorhizal nodules, extracellular *Frankia* cells are surrounded by plasmalemma, induce cortical cell division and infect some of these cells (Benson and Silvester, 1993; Pawlowski and Sprent, 2008). It is within these cells that *Frankia* develops vesicles and fixes nitrogen. The mediation by Nod factors of the actinorhizal symbiosis, in contrast to that in legumes, has not been established (Pawlowski and Sprent, 2008), and Nod factors have not been shown to induce root deformation in actinorhizal plants (Wall and Berry, 2008). However, a root hair deformation factor (Had factor) has been isolated in fractions of pure *Frankia* cultures and N-acetyl-glucosamine (common to Nod factors) was also identified in these same culture fractions (Wall and Berry, 2008). There are also preliminary reports of compounds isolated from alder seeds and root exudates that are similar to flavonoids and these may influence the establishment of nodulation by *Frankia*. This would be coherent with observations in other symbioses, including those from *Rhizobia* where flavonoids induce microbial genes (Wall and Berry, 2008).

Following root nodule formation and the onset of nitrogen fixation by *Frankia*, the microsymbiont derives its carbon source from plant photosynthates while the host plant benefits from the ammonium and also auxins produced by the endosymbiont (Wall and Berry, 2008). The nitrogen fixed by *Frankia* in root nodules will supply 70–100% of the host plant's nitrogen requirement (Nickel *et al.*, 2001; Myrold and Huss-Danell, 2003). Interestingly, nodules may harbour little or no nitrogen fixation. Nitrogen fixation rates are influenced by diurnal variations in the plant, however it is also known that *Frankia* strains may have very different nitrogen fixation capabilities *in vitro* versus *in planta*. Microorganism genotype, compatibility with the host plant, and nodule age may play important roles in the level of nodule nitrogenase activity (Verghese and Misra, 2000). Finally, environmental conditions may affect the actinorhizal symbiosis. If plant nitrogen requirements are met by soil nutrients, nodulation will be inhibited. It is reported that total soil nitrogen concentrations of the order of 10–30 mg/kg will have this effect (Benoit and Berry, 1990). Nodulation itself may also inhibit, and regulate further nodule development. Although precise mechanisms and compounds are not yet identified, it is known that the root region most responsive for nodule formation (the region near the young root tip) will be less susceptible to colonization by *Frankia* within days of prior exposure to the microorganism, even before nitrogen fixation has started in the existing pre-nodules (Wall and Berry, 2008). Feedback inhibition of nodule development, through nitrogen fixation in existing nodules, is systemic.

Actinorhizal plants can enter tripartite and tetrapartite symbioses that involve frankiae, and ectomycorrhizal and/or arbuscular fungi (Gardner and Barrueco, 1995; Sprent and Parsons, 2000). These multiple partner interactions are complex and have not been explored to their full extent. Photosynthate production is finite and their consumption by microsymbionts is likely to be a key factor in the development of the individual symbioses. There are conflicting reports in regard to the increase in nodule biomass per plant, nitrogen fixation activity and plant development in the combined presence of frankiae, vesicular arbuscular mycorrhizae and ectomycorrhizae (Chatarpaul *et al.*, 1989; Gardner and Barrueco, 1995; Diem, 1996; Sprent and Parsons, 2000; Yamanaka *et al.*, 2003; Orfanoudakis *et al.*, 2004). Further research in this field is warranted.

## 12.5 Conclusion

Actinobacteria have evolved in a wide spectrum of ecological niches. In soil, they are intimately associated with the degradation of naturally occurring, sometimes recalcitrant, polymers. Our increasing knowledge of their genomes is today helping us to decipher their life cycles which often include direct and indirect interactions with plants, in addition to that with other microbes. The 49 genomes sequenced to date reflect the diversity of this class of bacteria, their highly developed physiological and morphological traits, as well as their distinct niches in the environment. Actinobacteria are indeed characterized by their capability to thrive in very specific niches (as endosymbionts, or in the rhizosphere of particular plants) and in wider niches (as free-living saprophytes). In their interactions with plants, they may be commensal, pathogenic, symbiotic (*Frankia* sp.) of PGPR. Their secretome is often highly developed, allowing interactions with other microorganisms that are specific in nature (QS and antibiosis) or less specific (lytic enzymes). Many species of actinobacteria are capable of sporulation, thus allowing efficient dispersal, as well as adaptation to environmental stress. These characteristics explain why actinobacteria have evolved to become an important and versatile class of microorganisms in our environment and also highlight their potential in biotechnology.

## References

Agbessi, S., Beauséjour, J., Déry, C. and Beaulieu, C. (2003) Antagonistic properties of two recombinant strains of *Streptomyces melanosporofaciens* obtained by intraspecific protoplast fusion. *Applied Microbiology and Biotechnology* 62, 233–238.

Andersson, S.G.E. and Kurland, C.G. (1998) Reductive evolution of resident genomes. *Trends in Microbiology* 6, 263–268.

Arnold, D.L., Pitman, A. and Jackson, R.W. (2003) Pathogenicity and other genomic islands in plant pathogenic bacteria. *Molecular Plant Pathology* 4, 407–420.

Babcock, M.J., Eckwall, E.C. and Schottel, J.L. (1993) Production and regulation of potato-scab-inducing phytotoxins by *Streptomyces scabies*. *Journal of General Microbiology* 139, 1579–1586.

Baker, D.D. and Schwintzer, C.R. (1990) Introduction. In: Schwintzer, C.R. and Tjepkema, J.D. (eds) *The Biology of Frankia and Actinorhizal Plants.* Academic Press, Toronto, Canada, pp. 1–13.

Barakate, M., Ouhdouch, Y., Oufdou, K.H. and Beaulieu, C. (2002) Characterization of rhizospheric soil strepomycetes from Moroccan habitat and their antimicrobial activities. *World Journal of Microbiology and Biotechnology* 18, 49–54.

Beauséjour, J., Goyer, C., Vachon, J. and Beaulieu, C. (1999) Production of thaxtomin A by *Streptomyces scabies* strains in plant extract containing media. *Canadian Journal of Microbiology* 45, 764–768.

Benhamou, N. (1991) Cell surface interactions between tomato and *Clavibacter michiganense* subsp. *michiganense*: localization of some polysaccharides and hydroxyproline-rich glycoproteins in infected host leaf tissues. *Physiological and Molecular Plant Pathology* 38, 15–28.

Benoit, L.F. and Berry, A.M. (1990) Methods for production and use of actinorhizal plants in forestry, low maintenance landscapes, and revegetation. In: Schwintzer, C.R. and Tjepkema, J.D. (eds) *The Biology of Frankia and Actinorhizal Plants.* Academic Press, Toronto, Canada, pp. 281–297.

Benson, D.R. and Silvester, W.B. (1993) Biology of *Frankia* strains, actinomycete symbionts of actinorhizal plants. *Microbiological Reviews* 57, 293–319.

Bentley, S.D., Corton, C., Brown, S.E., Barron, A., Clark, L., Doggett, J., Harris, B., Ormond, D., Quail, M.A., May, G., Francis, D., Knudson, D., Parkhill, J. and Ishimaru, C.A. (2008) Genome of the actinomycete plant pathogen *Clavibacter michiganensis* subsp. *sepedonicus* suggests recent niche adaptation. *Journal of Bacteriology* 190, 2150–2160.

Bermpohl, A., Dreier, J., Bahro, R. and Eichenlaub, R. (1996) Exopolysaccharides in the pathogenic interaction of *Clavibacter michiganensis* subsp. *michiganensis* with tomato plants. *Microbiological Research* 151, 391–399.

Bouarab, K., Melton, R., Peart, J., Baulcombe, D. and Osbourn, A. (2002) A saponin-detoxifying enzyme mediates suppression of plant defences. *Nature* 418, 889–392.

Bouchek-Mechiche, K., Pasco, C., Andrivon, D. and Jouan, B. (2000) Differences in host range, pathogenicity to potato cultivars and response to soil temperature among *Streptomyces* species causing common and netted scab in France. *Plant Pathology* 49, 3–10.

Boudjella, H., Bouti, K., Zitouni, A., Mathieu, F., Lebrihi, A. and Sabaou, N. (2006) Taxonomy and chemical characterization of antibiotics of *Streptosporangium* Sg 10 isolated from a Saharan soil. *Microbiological Research* 161, 288–298.

Bukhalid, R.B. and Loria, R. (1997) Cloning and expression of a gene from *Streptomyces scabies* encoding a putative pathogenicity factor. *Journal of Bacteriology* 179, 7776–7783.

Bukhalid, R.A., Takeuchi, T., Labeda, D. and Loria, R. (2002) Horizontal transfer of the plant virulence gene, *nec1*, and flanking sequences among genetically distinct *Streptomyces* strains in the diastatochromogenes cluster. *Applied and Environmental Microbiology* 68, 738–744.

Burger, A., Gräfen, I., Engemann, J., Niermann, E., Pieper, M., Kirchner, O., Gartemann, K.-H. and Eichenlaub, R. (2005) Identification of homologues to the pathogenicity factor Pat-1, a putative serine protease of *Clavibacter michiganensis* subsp. *michiganensis*. *Microbiological Research* 160, 417–427.

Carlson, R.R., Randall, R. and Vidaver, A.K. (1982) Bacterial mosaic, a new corynebacterial disease of wheat (*Triticum aestivum*). *Plant Disease* 66, 76–79.

Chatarpaul, L., Chakravarty, P. and Subramaniam, P. (1989) Studies in tetrapartite symbioses. *Plant and Soil* 118, 145–150.

Chater, K. and Bibb, M. (1997) Regulation of bacterial antibiotic production. In: Kleinkauf, H. and von Döhren, H. (eds) *Biotechnology: Products of Secondary Metabolism.* Vol. 7. VCH Verlag, Weinheim, Germany, pp. 59–105.

Chen, Y.-F., Guo, J.-H. and Fang, Z.D. (2000) A new pathovar of *Curtobacterium flaccumfaciens* on malabar spinach. *Acta Phytopathologica Sinica* 30, 171–175.

Chen, Y.-F., Yin, Y.-N., Zhang, X.-M. and Guo, J.-H. (2007) *Curtobacterium flaccumfaciens* pv. *beticola*, a new pathovar of pathogens in sugar beet. *Plant Disease* 91, 677–684.

Clark, C.A. and Matthews, S.W. (1987) Histopathology of sweet potato root infection by *Streptomyces ipomoea. Phytopathology* 77, 1418–1423.

Collins, M.D. and Jones, D. (1983) Reclassification of *Corynebacterium flaccumfaciens, Corynebacterium betae, Corynebacterium oortii* and *Corynebacterium poinsettiae* in the genus *Curtobacterium*, as *Curtobacterium flaccumfaciens* comb. nov. *Journal of General Microbiology* 129, 3545–3548.

Conn V.M., Walker, A.R. and Franco, C.M.M. (2008) Endophytic actinobacteria induce defense pathways in *Arabidopsis thaliana. Molecular Plant–Microbe Interactions* 21, 208–218.

Coplin, D.L. and Cook, D. (1990) Molecular genetics of extracellular polysaccharide biosynthesis in vascular phytopathogenic bacteria. *Molecular Plant–Microbe Interactions* 3, 271–279.

Cormack, M.W. and Moffatt, J.E. (1956) Occurrence of the bacterial wilt organism in alfalfa seed. *Phytopathology* 46, 407–409.

Cornelis, K., Ritsema, T., Nijsse, J., Holsters, M., Goethals, K. and El Jaziri, M. (2001) The plant pathogen *Rhodococcus fascians* colonizes the exterior and interior of the aerial parts of plants. *Molecular Plant–Microbe Interactions* 14, 599–608.

Cornelis, K., Maes, T., Jaziri, M., Holsters, M. and Goethals, K. (2002) Virulence genes of the phytopathogen *Rhodococcus fascians* show specific spatial and temporal expression patterns during plant infection. *Molecular Plant–Microbe Interactions* 15, 398–403.

Crespi, M., Messens, E., Caplan, A.B., Van Montagu, M. and Desomer, J. (1992) Fasciation induction by the phytopathogen *Rhodococcus fascians* depends upon a linear plasmid encoding a cytokinin synthase gene. *EMBO Journal* 11, 795–804.

Crespi, M., Vereecke, D., Temmerman, W., Van Montagu, M. and Desomer, J. (1994) The *fas* operon of *Rhodococcus fascians* encodes new genes required for efficient fasciation of host plants. *Journal of Bacteriology* 176, 2492–2501.

Davis, M.J., Gillaspie, A.G., Jr, Harris, R.W. and Lawson, R.H. (1980) Ratoon stunting disease of sugarcane: isolation of the causal bacterium. *Science* 210, 1365–1367.

Davis, M.J., Gillaspie, A.G., Jr, Vidaver, A.K. and Harris, R.W. (1984) *Clavibacter*, a new genus containing some phytopathogenic coryneform bacteria *Clavibacter xyli* subsp. *xyli* sp. nov., subsp. nov. and *Clavibacter xyli* subsp. *cynodontis* subsp. nov., pathogens that cause ratoon stunting disease of sugarcane and Bermudagrass stunting disease. *International Journal of Systematic Bacteriology* 34, 107–117.

Dawson, J.O. (2008) Ecology of actinorhizal plants. In: Pawlowski, K. and Newton, W.E. (eds) *Nitrogen-fixing Actinorhizal Symbioses*. Vol. 6. Springer Netherlands, New York, pp. 199–234.

de Jong, A.J., Hoeberichts, F.A., Yakimova, E.T., Maximova, E. and Woltering, E.J. (2000) Chemical-induced apoptotic cell death in tomato cells: involvement of caspase-like proteases. *Planta* 211, 656–662.

de O. Manes, C.L., Beeckman, T., Ritsema, T., Van Montagu, M., Goethals, K. and Holsters, M. (2004) Phenotypic alterations in *Arabidopsis thaliana* plants caused by *Rhodococcus fascians* infection. *Journal of Plant Research* 117, 139–145.

Denny, T.P. (1995) Involvement of bacterial polysaccharides in plant pathogenesis. *Annual Review of Phytopathology* 33, 173–197.

Diem, H.G. (1996) Mycorrhizae of actinorhizal plants. *Acta Botanica Gallica* 143, 581–592.

Diem, H.G. and Dommergues, Y.R. (1985) *In vitro* production of specialized reproductive torulose hyphae by *Frankia* strain ORS 021001 isolated from *Casuarina junghuhniania* root nodules. *Plant and Soil* 87, 17–29.

Doumbou, C.L., Akimov, V. and Beaulieu, C. (1998) Selection and characterization of microorganisms utilizing thaxtomin A, a phytotoxin produced by *Streptomyces scabies*. *Applied and Environmental Microbiology* 64, 4313–4316.

Doumbou, C.L., Salove, M.K.H., Crawford, D.L. and Beaulieu, C. (2001) Actinomycetes, promising tools to control plant diseases and to promote plant growth. *Phytoprotection* 82, 85–102.

Dreier, J., Meletzus, D. and Eichenlaub, R. (1997) Characterization of the plasmid encoded virulence region *pat-1* of phytopathogenic *Clavibacter michiganensis* subsp. *michiganensis*. *Molecular Plant–Microbe Interactions* 10, 195–206.

Dröse, S., Bindseil, K.U., Bowman, E.J., Siebers, A., Zeeck, A. and Altendorf, K. (1993) Inhibitory effect of modified bafilomycins and concanamycins on P- and V-type adenosinetriphosphatases. *Biochemistry* 32, 3902–3906.

Eason, J.R., Morris, R.O. and Jameson, P.E. (1996) The relationship between virulence and cytokinin production by *Rhodococcus fascians* (Tilford 1936) Goodfellow 1984. *Plant Pathology* 45, 323–331.

Eckwall, E.C. and Schottel, J.L. (1997) Isolation and characterization of an antibiotic produced by the scab disease-suppressive *Streptomyces diastatochromogenes* strain PonSSII. *Journal of Industrial Microbiology and Biotechnology* 19, 220–225.

El-Tarabily, K.A. (2003) An endophytic chitinase-producing isolate of *Actinoplanes missouriensis*, with potential for biological control of root rot of lupine caused by *Plectosporium tabacinum*. *Australian Journal of Botany* 51, 257–266.

El-Tarabily, K.A. (2006) Rhizosphere-competent isolates of streptomycete and non-streptomycete actinomycetes capable of producing cell-wall degrading enzymes to control *Pythium aphanidermatum* damping-off disease of cucumber. *Canadian Journal of Botany* 84, 211–222.

El-Tarabily, K.A. (2008) Promotion of tomato (*Lycopersicon esculentum* Mill.) plant growth by rhizosphere competent 1-aminocyclopropane-1-carboxylic acid deaminase-producing streptomycete actinomycetes. *Plant and Soil* 308, 161–174.

El-Tarabily, K.A. and Sivasithamparam, K. (2006) Non-streptomycete actinomycetes as biocontrol agents of soil-borne fungal plant pathogens and as plant growth promoters. *Soil Biology and Biochemistry* 38, 1505–1520.

Evtushenko, L.I., Dorofeeva, L.V., Subbotin, S.A., Cole, J.R. and Tiedje, J.M. (2000) *Leifsonia poae* gen. nov., sp. nov., isolated from nematode galls on *Poa annua*, and reclassification of '*Corynebacterium aquaticum*' Leifson 1962 as *Leifsonia aquatica* (ex Leifson 1962) gen. nov., nom. rev., comb.nov. and *Clavibacter xyli* Davis *et al.* 1984 with two subspecies as *Leifsonia xyli* (Davis *et al.* 1984) gen. nov., comb. nov. *International Journal of Systematic and Evolutionary Microbiology* 50, 371–380.

Faucher, E., Otrysko, B., Paradis, É., Hodge, N.C., Stall, R.E. and Beaulieu, C. (1993) Characterization of streptomycetes causing russet scab in Québec. *Plant Disease* 77, 1217–1220.

Fayad, K.P., Simao-Beaunoir, A.-M., Gauthier, A., Leclerc, C., Mamady, H., Beaulieu, C. and Brzezinski, R. (2001) Purification and properties of a β-1,6-glucanase from *Streptomyces* sp. EF-14, an actinomycete antagonistic to *Phytophthora* spp. *Applied Microbiology and Biotechnology* 57, 117–123.

Franco, C., Michelsen, P., Percy, N., Conn, V., Listiana, E., Moll, S., Loria, R. and Coombs, J. (2007) Actinobacterial endophytes for improved crop performance. *Australasian Plant Pathology* 36, 524.

Gardner, I.C. and Barrueco, C.R. (1995) Mycorrhizal and actinorhizal biotechnology-problems and prospects. In: Varma, A.K. and Hock, B. (eds) *Mycorrhiza: Structure, Function, Molecular Biology and Biotechnology*, 2nd edn. Springer, New York, pp. 469–495.

Gartemann, K.-H., Kirchner, O., Engemann, J., Grafen, I., Eichenlaub, R. and Burger, A. (2003) *Clavibacter michiganensis* subsp. *michiganensis*: first steps in the understanding of virulence of a Gram-positive phytopathogenic bacterium. *Journal of Biotechnology* 106, 179–191.

Gartemann, K.-H., Abt, B., Bekel, T., Burger, A., Engemann, J., Flügel, M., Gaigalat, L., Goesmann, A., Gräfen, I., Kalinowski, J., Kaup, O., Kirchner, O., Krause, L., Linke, B., McHardy, A., Meyer, F., Pohle, S., Rückert, C., Schneiker, S., Zellermann, E.-M., Pühler, A., Eichenlaub, R., Kaiser, O. and Bartels, D. (2008) The genome sequence of the tomato-pathogenic actinomycete *Clavibacter michiganensis* subsp. *michiganensis* NCPPB382 reveals a large island involved in pathogenicity. *Journal of Bacteriology* 190, 2138–2149.

Getha, K. and Vikineswary, S. (2002) Antagonistic effects of *Streptomyces violaceusniger* strain G10 on *Fusarium oxysporum* f. sp. *cubense* race 4: indirect evidence for the role of antibiosis in the antagonistic process. *Journal of Industrial Microbiology and Biotechnology* 28, 303–310.

Goethals, K., Vereecke, D., Jaziri, M., Van Montagu, M. and Holsters, M. (2001) Leafy gall formation by *Rhodococcus fascians*. *Annual Review of Phytopathology* 39, 27.

Gopinathan, S. (1995) Biological control of *Rhizoctonia* sp. root rot of *Casuarina equisetifolia* seedlings by *Frankia* spp. strains. *Biology and Fertility of Soils* 20, 221–225.

Goyer, C. and Beaulieu, C. (1997) Host range of streptomycete strains causing common scab. *Plant Disease* 81, 901–904.

Goyer, C., Vachon, J. and Beaulieu, C. (1998) Pathogenicity of *Streptomyces scabies* mutants altered in thaxtomin a production. *Phytopathology* 88, 442–445.

Grau, B.L., Guan, D., Loria, R. and Pettis, G.S. (2006) Cloning and characterization of the thaxtomin C biosynthetic gene cluster of the bacterial sweet potato pathogen *Streptomyces ipomoeae*. *Phytopathology* 96, S42.

Haapalainen, M., Mattinen, J. and Metzler, M.C. (2000) The growth of a plant-parasitic bacterium, *Clavibacter xyli* subsp. *cynodontis*, is enhanced by xylem fluid components. *Physiological and Molecular Plant Pathology* 56, 147–155.

Harding, M.W., Marques, L.L.R., Howard, R.J. and Olson, M.E. (2007) Biofilms are associated with vascular clogging and seed contamination by *Curtobacterium flaccumfaciens* pv. *flaccumfaciens* in bean vascular wilt. *Canadian Journal of Plant Pathology* 29, 209.

Harrison, M.D. (1962) Potato russet scab, its cause and factors affecting its development. *American Potato Journal* 39, 368–381.

Healy, F.G., Wach, M., Krasnoff, S.B., Gibson, D.M. and Loria, R. (2000) The txtAB genes of the plant pathogen *Streptomyces acidiscabies* encode a peptide synthetase required for phytotoxin thaxtomin A production and pathogenicity. *Molecular Microbiology* 38, 794–804.

Healy, F.G., Krasnoff, S.B., Wach, M., Gibson, D.M. and Loria, R. (2002) Involvement of a cytochrome P450 monooxygenase in thaxtomin A biosynthesis by *Streptomyces acidiscabies*. *Journal of Bacteriology* 184, 2019–2029.

Hedges, F. (1922) A bacterial wilt of the bean caused by *Bacterium flaccumfaciens* nov. sp. *Science* 55, 433–434.

Holtsmark, I., Takle, G.W. and Brurberg, M.B. (2008) Expression of putative virulence factors in the potato pathogen *Clavibacter michiganensis* subsp. *sepedonicus* during infection. *Archives of Microbiology* 189, 131–139.

Hoster, F., Schmitz, J.E. and Daniel, R. (2005) Enrichment of chitinolytic microorganisms: isolation and characterization of a chitinase exhibiting antifungal activity against phytopathogenic fungi from a novel *Streptomyces* strain. *Applied Microbiology and Biotechnology* 66, 434–442.

Hwang, B.K., Lim, S.W., Kim, B.S., Lee, J.Y. and Moon, S.S. (2001) Isolation and *in vivo* and *in vitro* antifungal activity of phenylacetic acid and sodium phenylacetate from *Streptomyces humidus*. *Applied and Environmental Microbiology* 67, 3739–3745.

Igual, J.M., Velazquez, E., Mateos, P.F., Rodriguez-Barrueco, C., Cervantes, E. and Martinez-Molina, E. (2001) Cellulase isoenzyme profiles in *Frankia* strains belonging to different cross-inoculation groups. *Plant and Soil* 229, 35–39.

Jahr, H., Bahro, R., Burger, A., Ahlemeyer, J. and Eichenlaub, R. (1999) Interactions between *Clavibacter michiganensis* and its host plants. *Environmental Microbiology* 1, 113–118.

Jahr, H., Dreier, J., Meletzus, D., Bahro, R. and Eichenlaub, R. (2000) The endo-β-1,4-glucanase CelA of *Clavibacter michiganensis* subsp. *michiganensis* is a pathogenicity determinant required for induction of bacterial wilt of tomato. *Molecular Plant–Microbe Interactions* 13, 703–714.

Jensen, P.R., Mincer, T.J., Williams, P.G. and Fenical, W. (2005) Marine actinomycete diversity and natural product discovery. *Antonie Van Leeuwenhoek* 87, 43–48.

Johnson, E.G., Joshi, M.V., Gibson, D.M. and Loria, R. (2007) Cello-oligosaccharides released from host plants induce pathogenicity in scab-causing *Streptomyces* species. *Physiological and Molecular Plant Pathology* 71, 18–25.

Joshi, M.V. and Loria, R. (2007) *Streptomyces turgidiscabies* possesses a functional cytokinin biosynthetic pathway and produces leafy galls. *Molecular Plant–Microbe Interactions* 20, 751–758.

Joshi, M.V., Bignell, D.R.D., Johnson, E.G., Sparks, J.P., Gibson, D.M. and Loria, R. (2007a) The AraC/XylS regulator TxtR modulates thaxtomin biosynthesis and virulence in *Streptomyces scabies*. *Molecular Microbiology* 66, 633–642.

Joshi, M.V., Rong, X., Moll, S., Kers, J., Franco, C. and Loria, R. (2007b) *Streptomyces turgidiscabies* secretes a novel virulence protein, Nec1, which facilitates infection. *Molecular Plant–Microbe Interactions* 20, 599–608.

Kaup, O., Gräfen, I., Zellermann, E.-M., Eichenlaub, R. and Gartemann, K.-H. (2005) Identification of a tomatinase in the tomato-pathogenic actinomycete *Clavibacter michiganensis* subsp. *michiganensis* NCPPB382. *Molecular Plant–Microbe Interactions* 18, 1090–1098.

Kawabuchi, M., Hara, Y., Nihira, T. and Yamada, Y. (1997) Production of butyrolactone autoregulators by *Streptomyces coelicolor* A3(2). *FEMS Microbiology Letters* 157, 81–85.

Kawaguchi, T., Azuma, M., Horinouchi, S. and Beppu, T. (1988) Effect of B-factor and its analogues on rifamycin biosynthesis in *Nocardia* sp. *Journal of Antibiotics* 41, 360–365.

Kers, J.A., Wach, M.J., Krasnoff, S.B., Widom, J., Cameron, K.D., Bukhalid, R.A., Gibson, D.M., Crane, B.R. and Loria, R. (2004) Nitration of a peptide phytotoxin by bacterial nitric oxide synthase. *Nature* 429, 79–82.

Kers, J.A., Cameron, K.D., Joshi, M.V., Bukhalid, R.A., Morello, J.E., Wach, M.J., Gibson, D.M. and Loria, R. (2005) A large, mobile pathogenicity island confers plant pathogenicity on *Streptomyces* species. *Molecular Microbiology* 55, 1025–1033.

Keyworth, W.G., Howell, J.S. and Dowson, W.J. (1956) *Corynebacterium betae* (sp. nov.) the causal organism of silvering disease of red beet. *Plant Pathology* 5, 88–90.

Kim, B.S. and Hwang, B.K. (2007) Microbial fungicides in the control of plant diseases. *Journal of Phytopathology* 155, 641–653.

King, R.R., Lawrence, C.H. and Calhoun, L.A. (1992) Chemistry of phytotoxins associated with *Streptomyces scabies*, the causal organism of potato common scab. *Journal of Agricultural and Food Chemistry* 40, 834–837.

King, R.R., Lawrence, C.H., Calhoun, L.A. and Ristaino, J.B. (1994) Isolation and characterization of thaxtomin-type phytotoxins associated with *Streptomyces ipomoeae*. *Journal of Agricultural and Food Chemistry* 42, 1791–1794.

King, R.R., Lawrence, C.H., Embleton, J. and Calhoun, L.A. (2003) More chemistry of the thaxtomin phytotoxins. *Phytochemistry* 64, 1091–1096.

Lacava, P.T., Li, W., Araujo, W.L., Azevedo, J.L. and Hartung, J.S. (2007) The endophyte *Curtobacterium flaccumfaciens* reduces symptoms caused by *Xylella fastidiosa* in *Catharanthus roseus*. *Journal of Microbiology* 45, 388–393.

Lahdenpera, M.L., Simon, E. and Uoti, J. (1991) Mycostop – a novel biofungicide based on *Streptomyces* bacteria. In: Beemster, A.B.R., Bollen, G.J., Gerlagh, M., Ruissen, M.A., Schippers, B. and Tempel, A. (eds) *Biotic Interactions and Soil-Borne Diseases*. Elsevier, Amsterdam, The Netherlands, pp. 258–263.

Laine, M.J., Nakhei, H., Dreier, J., Lehtila, K., Meletzus, D., Eichenlaub, R. and Metzler, M.C. (1996) Stable transformation of the Gram-positive phytopathogenic bacterium *Clavibacter michiganensis* subsp. *sepedonicus* with several cloning vectors. *Applied and Environmental Microbiology* 62, 1500–1506.

Laine, M.J., Haapalainen, M., Wahlroos, T., Kankare, K., Nissinen, R., Kassuwi, S. and Metzler, M.C. (2000) The cellulase encoded by the native plasmid of *Clavibacter michiganensis* ssp. *sepedonicus* plays a role in virulence and contains an expansin-like domain. *Physiological and Molecular Plant Pathology* 57, 221–233.

Lairini, K., Perez-Espinosa, A., Pineda, M. and Ruiz-Rubio, M. (1996) Purification and characterization of tomatinase from *Fusarium oxysporum* f. sp. *lycopersici*. *Applied and Environmental Microbiology* 62, 1604–1609.

Lamari, L., Zitouni, A., Boudjella, H., Badji, B., Sabaou, N., Lebrihi, A., Lefebvre, G., Seguin, E. and Tillequin, F. (2002) New dithiolopyrrolone antibiotics from *Saccharothrix* sp. sa 233.1. taxonomy, fermentation, isolation and biological activities. *Journal of Antibiotics* 55, 696–701.

Lechevalier, M.P. and Lechevalier, H.A. (1990) Systematics, isolation, and culture of *Frankia*. In: Schwintzer, C.R. and Tjepkema, J.D. (eds) *The Biology of Frankia and Actinorhizal Plants*. Academic Press, Toronto, Canada, pp. 35–60.

Leiner, R.H., Fry, B.A., Carling, D.E. and Loria, R. (1996) Probable involvement of thaxtomin A in pathogenicity of *Streptomyces scabies* on seedlings. *Phytopathology* 86, 709–713.

Li, T.Y., Yin, P., Zhou, Y., Zhang, Y., Zhang, Y.Y. and Chen, T.A. (2004) Characterization of the replicon of a 51-kb native plasmid from the Gram-positive bacterium *Leifsonia xyli* subsp. *Cynodontis. FEMS Microbiology Letters* 236, 33–39.

Loria, R., Bukhalid, R.A., Fry, B.A. and King, R.R. (1997) Plant pathogenicity in the genus *Streptomyces. Plant Disease* 81, 836–846.

Loria, R., Kers, J. and Joshi, M. (2006) Evolution of plant pathogenicity in *Streptomyces. Annual Review of Phytopathology* 44, 469–487.

Loria, R., Bignell, D.R.D., Moll, S., Huguet-Tapia, J.C., Joshi, M.V., Johnson, E.G., Seipke, R.F. and Gibson, D.M. (2008) Thaxtomin biosynthesis: the path to plant pathogenicity in the genus *Streptomyces. Antonie Van Leeuwenhoek* 94, 3–10.

Maes, T., Vereecke, D., Ritsema, T., Cornelis, K., Ngo Thi Thu, H., Van Montagu, M., Holsters, M. and Goethals, K. (2001) The *att* locus of *Rhodococcus fascians* strain D188 is essential for full virulence on tobacco through the production of an autoregulatory compound. *Molecular Microbiology* 42, 13–28.

Mahadevan, B. and Crawford, D.L. (1997) Properties of the chitinase of the antifungal biocontrol agent *Streptomyces lydicus* WYEC108. *Enzyme and Microbial Technology* 20, 489–493.

Manulis, S., Shafrir, H., Epstein, E., Lichter, A. and Barash, I. (1994) Biosynthesis of indole-3-acetic-acid via the indole-3-acetamide pathway in *Streptomyces* spp. *Microbiology* 140, 1045–1050.

Manzer, F. and Genereux, H. (1981) Ring rot. In: Hooker, W.J. (ed.) *Compendium of Potato Diseases.* American Phytopathological Society (APS) Press, St Paul, Minnesota, pp. 31–32.

Marques L.L.R., De Boer, S.H., Ceri, H. and Olson, M.E. (2003) Evaluation of biofilms formed by *Clavibacter michiganensis* subsp. *sepedonicus. Phytopathology* 93, S57.

Mastronunzio, J.E., Tisa, L.S., Normand, P. and Benson, D.R. (2008) Comparative secretome analysis suggests low plant cell wall degrading capacity in *Frankia* symbionts. *BMC Genomics* 9, 47–63.

McCulloch, L. (1925) *Aplanobacter insidiosum* n. sp., the cause of an alfalfa disease. *Phytopathology* 15, 496–497.

McQueen, D.A. and Schottel, J.L. (1987) Purification and characterization of a novel extracellular esterase from pathogenic *Streptomyces scabies* that is inducible by zinc. *Journal of Bacteriology* 169, 1967–1971.

Meletzus, D., Bermpohl, A., Dreier, J. and Eichenlaub, R. (1993) Evidence for plasmid-encoded virulence factors in the phytopathogenic bacterium *Clavibacter michiganensis* subsp. *michiganensis* NCPPB382. *Journal of Bacteriology* 175, 2131–2136.

Merzaeva, O.V. and Shirokikh, I.G. (2006) Colonization of plant rhizosphere by actinomycetes of different genera. *Microbiology* 75, 226–230.

Metzler, M.C., Laine, M.J. and De Boer, S.H. (1997) The status of molecular biological research on the plant pathogenic genus *Clavibacter. FEMS Microbiology Letters* 150, 1–8.

Monteiro-Vitorello, C.B., Camargo, L.E.A., Van Sluys, M.A., Kitajima, J.P., Truffi, D., do Amaral, A.M., Harakava, R., de Oliveira, J.C.F., Wood, D., de Oliveira, M.C., Miyaki, C., Takita, M.A., da Silva, A.C.R., Furlan, L.R., Carraro, D.M., Camarotte, G., Almeida, N.F., Jr, Carrer, H., Coutinho, L.L., El-Dorry, H.A., Ferro, M.I.T., Gagliardi, P.R., Giglioti, E., Goldman, M.H.S., Goldman, G.H., Kimura, E.T., Ferro, E.S., Kuramae, E.E., Lemos, E.G.M., Lemos, M.V.F., Mauro, S.M.Z., Machado, M.A., Marino, C.L., Menck, C.F., Nunes, L.R., Oliveira, R.C., Pereira, G.G., Siqueira, W., de Souza, A.A., Tsai, S.M., Zanca, A.S., Simpson, A.J.G., Brumbley, S.M. and Setúbal, J.C. (2004) The genome sequence of the

Gram-positive sugarcane pathogen *Leifsonia xyli* subsp. *xyli*. *Molecular Plant–Microbe Interactions* 17, 827–836.

Mudgett, M.B. (2005) New insights to the function of phytopathogenic bacterial type III effectors in plants. *Annual Review of Plant Biology* 56, 509–531.

Myrold, D.D. and Huss-Danell, K. (2003) Alder and lupine enhance nitrogen cycling in a degraded forest soil in Northern Sweden. *Plant and Soil* 254, 47–56.

National Center for Biotechnology Information (1988) Available at: http://www.ncbi.nlm.nih.gov/genomes/lproks.cgi (accessed 4 August 2008).

Natsume, M., Ryu, R. and Abe, H. (1996) Production of phytotoxins, concanamycins A and B by *Streptomyces* spp. causing potato [*Solanum tuberosum*] scab. *Annals of the Phytopathological Society of Japan*, 62, 411–413.

Natsume, M., Komiya, M., Koyanagi, F., Tashiro, N., Kawaide, H. and Abe, H. (2005) Phytotoxin produced by *Streptomyces* sp. causing potato russet scab in Japan. *Journal of General Plant Pathology* 71, 364–369.

Neuhaus, F.C. and Baddiley, J. (2003) A continuum of anionic charge: structures and functions of D-alanyl-teichoic acids in Gram-positive bacteria. *Microbiology and Molecular Biology Reviews* 67, 686–723.

Nickel, A., Pelz, O., Hahn, D., Saurer, M., Siegwolf, R. and Zeyer, J. (2001) Effect of inoculation and leaf litter amendment on establishment of nodule-forming *Frankia* populations in soil. *Applied and Environmental Microbiology* 67, 2603–2609.

Nissinen, R., Kassuwi, S., Peltola, R. and Metzler, M.C. (2001) *In planta* – complementation of *Clavibacter michiganensis* subsp. *sepedonicus* strains deficient in cellulase production or HR induction restores virulence. *European Journal of Plant Pathology* 107, 175–182.

Normand, P., Lapierre, P., Tisa, L.S., Gogarten, J.P., Alloisio, N., Bagnarol, E., Bassi, C.A., Berry, A.M., Bickhart, D.M., Choisne, N., Couloux, A., Cournoyer, B., Cruveiller, S., Daubin, V., Demange, N., Francino, M.P., Goltsman, E., Huang, Y., Kopp, O.R., Labarre, L., Lapidus, A., Lavire, C., Marechal, J., Martinez, M., Mastronunzio, J.E., Mullin, B.C., Niemann, J., Pujic, P., Rawnsley, T., Rouy, Z., Schenowitz, C., Sellstedt, A., Tavares, F., Tomkins, J.P., Vallenet, D., Valverde, C., Wall, L.G., Wang, Y., Medigue, C. and Benson, D.R. (2007) Genome characteristics of facultatively symbiotic *Frankia* sp. strains reflect host range and host plant biogeography. *Genome Research* 17, 7–15.

Okudoh, V.I. and Wallis, F.M. (2007) Antimicrobial activity of rare actinomycetes isolated from natural habitats in KwaZulu-natal, South Africa. *South African Journal of Science* 103, 216–222.

Orfanoudakis, M.Z., Hooker, J.E. and Wheeler, C.T. (2004) Early interactions between arbuscular mycorrhizal fungi and *Frankia* during colonization and root nodulation of *Alnus glutinosa*. *Symbiosis* 36, 69–82.

Ouhdouch, Y., Barakate, M. and Finance, C. (2001) Actinomycetes of Moroccan habitats: isolation and screening for antifungal activities. *European Journal of Soil Biology* 39, 69–74.

Park, C.N., Lee, J.M., Lee, D. and Kim, B.S. (2008) Antifungal activity of valinomycin, a peptide antibiotic produced by *Streptomyces* sp. strain M10 antagonistic to *Botrytis cinerea*. *Journal of Microbiology and Biotechnology* 18, 880–884.

Park, D.H., Kim, J.S., Kwon, S.W., Wilson, C., Hur, J.H. and Lim, C.K. (2003) *Streptomyces luridiscabiei* sp. nov., *Streptomyces puniciscabiei* sp. nov. and *Streptomyces niveiscabiei* sp. nov., which cause potato common scab disease in Korea. *International Journal of Systematic and Evolutionary Microbiology* 53, 2049–2054.

Paschke, M.C. and Van Alfen, N.K. (1993) Extracellular polysaccharide impaired mutants of *Clavibacter michiganense* subsp. *insidiosum*. *Physiological and Molecular Plant Pathology* 42, 309–319.

Pawlowski, K. and Bisseling, T. (1996) Rhizobial and actinorhizal symbioses: what are the shared features? *Plant Cell* 8, 1899–1913.

Pawlowski, K. and Sirrenberg, A. (2003) Symbiosis between *Frankia* and actinorhizal plants: root nodules of non-legumes. *Indian Journal of Experimental Biology* 41, 1165–1183.

Pawlowski, K. and Sprent, J.I. (2008) Comparison between actinorhizal and legume symbiosis. In: Pawlowski, K. and Newton, W.E. (eds) *Nitrogen-fixing Actinorhizal Symbioses.* Vol. 6. Springer Netherlands, New York, pp. 261–288.

Peczynska-Czoch, W. and Mordarski, M. (1988) Actinomycete enzymes. In: Goodfellow, M., Williams, S.T. and Mordarski, M. (eds) *Actinomycetes in Biotechnology.* Academic Press, London, pp. 219–283.

Persello-Cartieaux, F., Nussaume, L. and Robaglia, C. (2003) Tales from the underground: molecular plant–rhizobacteria interactions. *Plant, Cell and Environment* 26, 189–199.

Person, L.H. and Martin, W.J. (1940) Soil rot of sweet potatoes in Louisiana. *Phytopathology* 30, 913–926.

Piret, J.M. and Demain, A.L. (1988) Actinomycetes in biotechnology: an overview. In: Goodfellow, M., Williams, S.T. and Mordarski, M. (eds) *Actinomycetes in Biotechnology.* Academic Press, London, pp. 461–482.

Ramey, B.E., Koutsoudis, M., von Bodman, S.B and Fuqua, C. (2004) Biofilm formation in plant–microbe associations. *Current Opinion in Microbiology* 7, 602–609.

Reuhs, B.L., Kim, J.S. and Matthysse, A.G. (1997) Attachment of *Agrobacterium tumefaciens* to carrot cells and *Arabidopsis* wound sites is correlated with the presence of a cell-associated, acidic polysaccharide. *Journal of Bacteriology* 179, 5372–5379.

Romanenko, A.S., Lomovatskaya, L.A., Shafikova, T.N., Borovskii, G.B. and Krivolapova, N.V. (2003) Potato cell plasma membrane receptors to ring rot pathogen extracellular polysaccharides. *Journal of Phytopathology* 151, 1–6.

Rothwell, A. (1968) Bacterial canker of tomato caused by *Corynebacterium michiganense* (Smith) Jensen. *Rhodesia Agricultural Journal* 1871, 75–78.

Roy, S., Khasa, D.P. and Greer, C.W. (2007) Review: combining alders, frankiae, and mycorrhizae for soil remediation and revegetation. *Canadian Journal of Botany* 85, 237–251.

Saaltink, G.J. and Maas Geesteranus, H.P. (1969) A new disease in tulip caused by *Corynebacterium oortii* nov. spec. *Netherlands Journal of Plant Pathology* 75, 123–128.

Sandrock, R.W., DellaPenna, D. and VanEtten, H.D. (1995) Purification and characterization of beta 2-tomatinase, an enzyme involved in the degradation of alpha-tomatine and isolation of the gene encoding beta 2-tomatinase from *Septoria lycopersici. Molecular Plant–Microbe Interactions* 8, 960–970.

Schrey, S.D. and Tarkka, M.T. (2008) Friends and foes: streptomycetes as modulators of plant disease and symbiosis. *Antonie Van Leeuwenhoek* 94, 11–19.

Séguin, A. and Lalonde, M. (1989) Detection of pectolytic activity and *pel* homologous sequences in *Frankia. Plant and Soil* 118, 221–229.

Sen, A., Sur, S., Bothra, A.K., Benson, D.R., Normand, P. and Tisa, L.S. (2008) The implication of life style on codon usage patterns and predicted highly expressed genes for three *Frankia* genomes. *Antonie Van Leeuwenhoek* 93, 335.

Shafikova, T.N., Romanenko, A.S. and Borovskii, G.B. (2003) Plasma membrane receptors for exopolysaccharides of the ring rot causal agent in potato cells. *Russian Journal of Plant Physiology* 50, 220–223.

Shashkov, A.S., Kosmachevskaya, L.N., Streshinskaya, G.M., Evtushenko, L.I., Bueva, O.V., Denisenko, V.A., Naumova, I.B. and Stackebrandt, E. (2002) A polymer with a backbone of 3-deoxy-D-glycero-D-galacto-non-2-ulopyranosonic acid, a teichuronic acid, and a β-glucosylated ribitol teichoic acid in the cell wall of plant pathogenic *Streptomyces* sp. VKM Ac-2124. *European Journal of Biochemistry* 269, 6020–6025.

Shimizu, M., Suzuki, T., Mogami, O. and Kunoh, H. (2005) Disease resistance of plants induced by endophytic actinomycetes. In: Tsuyumu, S., Leach, J.E., Shiraishi, T. and Wolpert, T. (eds) *Genomic and Genetic Analysis of Plant Parasitism and Defense.* American Phytopathological Society (APS) Press, St Paul, Minnesota, pp. 292–293.

Simao-Beaunoir, A.-M. and Beaulieu, C. (2006) Caractérisation des îlots de pouvoir pathogène chez *Streptomyces scabies. Phytoprotection* 87, 93.

Smidt, M.L. and Vidaver, A.K. (1987) Variation among strains of *Clavibacter michiganense* subsp. *nebraskense* isolated from a single popcorn field. *Phytopathology* 77, 388–392.

Sousa, C.S., Soares, A.C.F. and Garrido, M.S. (2008) Characterization of streptomycetes with potential to promote plant growth and biocontrol. *Scientia Agricola* 65, 50–55.

Spooner, F.R. Jr and Hammerschmidt, R. (1989) Characterization of extracellular pectic enzymes produced by *Streptomyces* species. *Phytopathology* 79, 1190.

Sprent, J.I. and Parsons, R. (2000) Nitrogen fixation in legume and non-legume trees. *Field Crops Research* 65, 183–196.

Stackebrandt, E., Rainey, F.A. and Ward-Rainey, N.L. (1997) Proposal for a new hierarchic classification system, *Actinobacteria* classis nov. *International Journal of Systematic and Evolutionary Microbiology* 47, 479–491.

Starr, M.P. and Pirone, P.P. (1942) *Phytomonas poinsettiae* n. sp., the cause of a bacterial disease of poinsettia. *Phytopathology* 32, 1076–1082.

Stevens, G.A. and Berry, A.M. (1988) Cytokinin secretion by *Frankia* sp. HFP ArI3 in defined medium. *Plant Physiology* 87, 15–16.

Sturz, A.V., Christie, B.R. and Matheson, B.G. (1998) Associations of bacterial endophyte populations from red clover and potato crops with potential for beneficial allelopathy. *Canadian Journal of Microbiology* 44, 162–167.

Tan, H.M., Cao, L.X., He, Z.F., Su, G.J., Lin, B. and Zhou, S.N. (2006) Isolation of endophytic actinomycetes from different cultivars of tomato and their activities against *Ralstonia solanacearum. World Journal of Microbiology and Biotechnology* 22, 1275–1280.

Tanaka, Y.T. and Omura, S. (1993) Agroactive compounds of microbial origin. *Annual Review of Microbiology* 47, 57–87.

Taylor, J., Stearman, R.S. and Uratani, B.B. (1993) Development of a native plasmid as a cloning vector in *Clavibacter xyli* subsp. *cynodontis. Plasmid* 29, 241–244.

Toussaint, V., Valois, D., Dodier, M., Faucher, E., Déry, C., Brzezinski, R., Ruest, L. and Beaulieu, C. (1997) Characterization of actinomycetes antagonistic to *Phytophthora fragariae* var. *rubi*, the causal agent of rasberry root rot. *Phytoprotection* 78, 43–51.

Trejo-Estrada, S.R., Paszczynski, A. and Crawford, D.L. (1998) Antibiotics and enzymes produced by the biocontrol agent *Streptomyces violaceusniger* YCED-9. *Journal of Industrial Microbiology and Biotechnology* 21, 81–90.

Tul'skaya, E.M., Shashkov, A.S., Senchenkova, S.N., Akimov, V.N., Bueva, O.V., Stupar, O.S. and Evtushenko, L.I. (2007) Anionic polymers of the cell wall of *Streptomyces* sp. VKM Ac-2534. *Russian Journal of Bioorganic Chemistry* 33, 251–257.

Uroz, S., D'Angelo-Picard, C., Carlier, A., Elasri, M., Sicot, C., Petit, A., Oger, P., Faure, D. and Dessaux, Y. (2003) Novel bacteria degrading N-acylhomoserine lactones and their use as quenchers of quorum-sensing-regulated functions of plant-pathogenic bacteria. *Microbiology* 149, 1981–1989.

Valois, D., Fayad, K., Barasubiye, T., Garon, M., Déry, C., Brzezinski, R. and Beaulieu, C. (1996) Glucanolytic actinomycetes antagonistic to *Phytophthora fragariae* var. *rubi*, the causal agent of raspberry root rot. *Applied and Environmental Microbiology* 62, 1630–1635.

Vandeputte, O., Öden, S., Mol, A., Vereecke, D., Goethals, K., El Jaziri, M. and Prinsen, E. (2005) Biosynthesis of auxin by the Gram-positive phytopathogen *Rhodococcus fascians* is controlled by compounds specific to infected plant tissues. *Applied and Environmental Microbiology* 71, 1169–1177.

Vereecke, D., Cornelius, K., Temmerman, W., Jaziri, M., Van Montagu, M., Holsters, M. and Goethals, K. (2002) Chromosomal locus that affects pathogenicity of *Rhodococcus fascians*. *Journal of Bacteriology* 184, 1112–1120.

Vereecke, D., Temmerman, W., Jaziri, M., Holsters, M. and Goethals, K. (2003) Toward an understanding of the *Rhodococcus fascians*–plant interaction. In: Stacy, G. and Kean, N. (eds) *Plant–Microbe Interactions.* Vol. 6. American Phytopathological Society (APS) Press, St Paul, Minnesota, pp. 53–79.

Verghese, S.K. and Misra, A.K. (2000) PCR-RFLP based screening of *Frankia* in alder nodules having different levels of nitrogenase activity. *Symbiosis* 28, 337–350.

Vidaver, A.K. (1982) The plant pathogenic *Corynebacteria*. *Annual Review of Microbiology* 36, 495–517.

von Bodman, S.B., Dietz Bauer, W. and Coplin, D.L. (2003) Quorum-sensing in plant-pathogenic bacteria. *Annual Review of Phytopathology* 41, 455–482.

Wach, M.J., Krasno, S.B., Loria, R. and Gibson, D.M. (2007) Effect of carbohydrates on the production of thaxtomin A by *Streptomyces acidiscabies*. *Archives of Microbiology* 188, 81–88.

Wall, L.G. and Berry, A.M. (2008) Early interactions, infection and nodulation in actinorhizal symbiosis. In: Pawlowski, K. and Newton, W.E. (eds) *Nitrogen-fixing Actinorhizal Symbioses.* Vol. 6. Springer, New York, pp. 147–166.

Wanner, L.A. (2004) Field isolates of *Streptomyces* differ in pathogenicity and virulence on radish. *Plant Disease* 88, 785–796.

Wanner, L.A. (2007) A new strain of *Streptomyces* causing common scab in potato. *Plant Disease* 91, 352–359.

Williams, S.T., Lanning, S. and Wellington, E.M.H. (1984) Ecology of actinomycetes. In: Goodfellow, M., Mordarski, M. and Williams, S.T. (eds) *The Biology of the Actinomycetes.* Academic Press, London, pp. 481–528.

Yamada, Y. and Nihira, T. (1998) Microbial hormones and microbial chemical ecology. In: Barton, D.H.R. and Nakanishi, K. (eds) *Comprehensive Natural Products Chemistry.* Vol. 8. Elsevier, Oxford, pp. 377–413.

Yamada, Y., Sugamura, K., Kondo, K., Yanagimoto, M. and Okada, H. (1987) The structure of inducing factors for virginiamycin production in *Streptomyces virginias*. *Journal of Antibiotics* 40, 496–504.

Yamanaka, T., Li, C.Y., Bormann, B.T. and Okabe, H. (2003) Tripartite associations in an alder: effects of *Frankia* and *Alpova diplophloeus* on the growth, nitrogen fixation and mineral acquisition of *Alnus tenuifolia*. *Plant and Soil* 254, 179–186.

Yang, Y.Y., Zhao, X.Q., Jin, Y.Y., Huh, J.H., Cheng, J.H., Singh, D., Kwon, H.J. and Suh, J.W. (2006) Novel function of cytokinin: signaling molecule for promotion of antibiotic production in streptomycetes. *Journal of Microbiology and Biotechnology* 16, 896–900.

Yuan, W.M. and Crawford, D.L. (1995) Characterization of *Streptomyces lydicus* WYEC108 as a potential biocontrol agent against fungal root and seed rots. *Applied and Environmental Microbiology* 61, 3119–3128.

Zaitlin, B., Watson, S.B., Dixon, J. and Steel, D. (2003) Actinomycetes in the Elbow River Basin, Alberta, Canada. *Water Quality Research Journal of Canada* 38, 115–125.

# 13 Insight into *Fusarium*–Cereal Pathogenesis

RAJAGOPAL SUBRAMANIAM, CHARLES G. NASMITH, LINDA J. HARRIS AND THÉRÈSE OUELLET

*Agriculture and Agri-Food Canada, Ottawa, Ontario, Canada*

## Abstract

*Fusarium* is one of the most important genera of plant pathogenic fungi, causing diseases including blights, wilts, root and stem rots on a wide variety of economically important cereal crops. The ability to grow both as a saprophyte and a pathogen provides an important strategy for *Fusarium* to become a formidable foe. A sustainable solution to *Fusarium* infection of cereals is the development of resistant cultivars. Breeding has provided improved levels of resistance to Fusarium head blight (FHB) in wheat and gibberella ear rot in maize, but the molecular mechanism underlying this resistance is unknown. Until recently, knowledge of how cereal species respond to *Fusarium* infection was almost non-existent, and biological tools to address this question were very limited. With the availability of barley, maize and wheat Affymetrix GeneChip or other microarray platforms, we are getting insight into the world of *Fusarium* pathogenesis. This review will be a comprehensive analysis between barley, wheat and maize infection to *Fusarium*.

## 13.1 Introduction

The genus *Fusarium* is remarkable, comprising over 1000 species that cause diseases on agricultural crops with serious economic consequences worldwide. The distribution of this genus ranges from tropical and temperate areas to zones with extreme climatic conditions, such as desert, alpine and arctic (Stoner, 1981). In addition to occupying various habitats, *Fusarium* phytopathogenic species also have a broad host range. They have been identified on all cereal crops in Western Europe (Cassini, 1981) and North America (Cook, 1981a), on wheat, cotton and barley in China, (Cook, 1981b), on rice in Japan, Taiwan and Thailand (Sun and Snyder, 1981), on all economically important crops of the tropics (Stoner, 1981), and on timber trees in forest nurseries worldwide (Bloomberg, 1981). Finally, success of this genus can be attributed to the variety of ways it compromises and acquires nutrition from hosts. Fusarium diseases can affect crops in the pre-harvest (vascular wilts, stem and root rots, and infection of reproductive structures) and postharvest stages (tuber, bulb, and seed rots) (Parry, 1990; Agrios, 1997).

*Fusarium* species, besides being responsible for a wide variety of diseases, also produce a plethora of toxins that contaminate cereal grains and other plant-based foods, thereby impairing human and animal health (Desjardins, 2006). Major classes of mycotoxins include the fumonisins, trichothecenes and zearalenones; other minor mycotoxins, such as beauvericin, fusaproiliferin, fusarins and moniliformin are also found (Desjardins, 2006). Mechanisms of action of these toxins are as varied as their chemical composition. For example, fumonisins inhibit sphingolipid biosynthesis; zearalenones mimic oestrogens, while trichothecenes compromise their host by binding to ribosomes and inhibiting protein synthesis (Desjardins, 2006). Some of these mycotoxins have been shown to increase the virulence of the pathogen and increase disease severity (Desjardins, 2006; Glenn *et al.*, 2008).

Fusarium head blight (FHB) of wheat, barley and oats has reached epidemic proportions in the world, especially in areas experiencing moderately high temperatures and high humidity during the heading and blossoming period. FHB has been the most significant disease of barley in parts of western Canada (Tekauz *et al.*, 2000). Severe FHB outbreaks have occurred between 1993 and 1998. Shifts from primary plant hosts, from wheat to barley may have resulted from changes in pathogen population and environmental conditions (Tekauz *et al.*, 2000). This review will focus on the species that has a major impact on cereal crops worldwide, *Fusarium graminearum*. Currently, *F. graminearum* has been split into at least nine phylogenetic species (Starkey *et al.*, 2007). Species are often limited to particular regions of the world, although the origin of many of these species, including *F. graminearum sensu stricto* remains uncertain. This indicates that while newer taxonomic tools will undoubtedly resolve some questions, there remains a challenge to achieve taxonomic stability for both taxonomists and plant pathologists.

## 13.2 *Fusarium graminearum* Infection Process

The cycle begins with the source of the inoculum arising from the infected plant debris on which the fungus over-winters as a saprophyte. As the weather conditions improve in the spring, perithecia develop to produce ascospores (Markell and Francl, 2003). The ascospores from mature perithecia (Trail *et al.*, 2002) are discharged and dispersed by wind, rain or insects to flowering host plants (Sutton, 1982; Parry *et al.*, 1995). A comprehensive review of the FHB infection process has been published recently (Bushnell *et al.*, 2003).

Deposition of spores on or inside spike tissue of cereals initiates the infection process. Spore germination occurs on wheat within 6 h after infection (hai) while on barley, germination has not been observed until 24 hai (Pritsch *et al.*, 2000; Boddu *et al.*, 2006). A brief biotrophic phase of the fungus is initiated by the development of fungal hyphae on the exterior surfaces of florets and glumes, allowing it to grow towards stomata and other susceptible sites within the inflorescence (Bushnell *et al.*, 2003). This phase is unlike other fungal infection where specific structures such as appressoria develop to directly penetrate the epidermis. However, lobed structures resembling appressorium have been observed between the cuticle and epidermal cell wall on the surface of inoculated glumes (Pritsch *et al.*, 2000). Such subcuticular growth on the glume, lemma and palea is thought to serve as a mechanism for fungal spread and, probably leads to direct penetration of epidermal cells (Bushnell *et al.*, 2003). Other avenues for direct entry include stomata and underlying parenchyma, partially or fully exposed anthers, openings between the lemma and palea of the spikelet/floret during dehiscence (Lewandowski and Bushnell, 2001; Bushnell *et al.*, 2003) and through the base of the wheat glumes where the epidermis and parenchyma are thin walled. Colonization of the vascular bundles significantly contributes to the spread of *F. graminearum* in wheat reproductive and vegetative tissues (Ribichich *et al.*, 2000; Guenther and Trail, 2005). Under wet conditions, mycelia can spread over the exterior surfaces of the glume, lemma and palea in both wheat and barley (Bushnell *et al.*, 2003).

The infection of maize ears by *F. graminearum* occurs via the silk channel or directly into the ear when facilitated by bird, insect or extreme weather damage (Sutton, 1982). The presence of maize pollen on the silks stimulates macroconidia germination (Naik and Busch, 1978). Macroconidia or ascospores germinate on exposed silks, initiating localized mycelial growth followed by directed hyphal growth within and on the surface of the silk towards the developing kernels (Miller *et al.*, 2007). The mycelia accesses the developing kernels through the silk attachment point and can subsequently also infect the cob through the kernel pedicel. As the silk travels over the surface of other developing kernels, the mycelia can also colonize the inter-kernel spaces and enter the cob tissue. In some instances, *F. graminearum* may also colonize plant tissues asymptomatically, such as stalks of corn (Bushnell *et al.*, 2003), and can be isolated from non-symptomatic wild grass hosts (Farr *et al.*, 1989; Inch and Gilbert, 2003).

## 13.3 Host Responses to *F. graminearum* Infection

### Types of resistance

In cereals, two prevalent types of resistance against FHB are observed: a generalized resistance to infection termed type I, where the pathogen is unable to infect; and a type II resistance in which the pathogen is able to infect but is unable to spread beyond the infection site (Goswami and Kistler, 2004). In barley, FHB symptoms do not seem to spread internally beyond the initially infected spikelet, such that even susceptible cultivars exhibit a natural type II resistance. Genetic resistance to FHB in wheat and barley is partial and quantitatively controlled by many loci. Similarly, mapping of resistance in wheat has revealed multiple quantitative trait loci (QTL) that confer partial resistance. So in neither barley nor wheat has loci that control resistance (R) in a gene-for-gene manner been identified (Bai and Shaner, 1994). Resistance to FHB in some wheat cultivars is derived from the Chinese cv. Sumai 3 and its derivatives. Sumai 3-derived resistance to FHB is a complex, quantitative trait which confers type II resistance (Bai and Shaner, 1994).

## Gene expression profiling

Many of the earlier comparative studies between resistant and susceptible cultivars of wheat infected with *F. graminearum* revealed differences in the accumulation of several classes of biotic stress-related transcripts such as pathogenesis-related genes *PR1*, β-glucanases, chitinases and thaumatin-like proteins; genes involved in oxidative bursts such as superoxide dismutase, catalase and peroxidases (Kang and Buchenauer, 2000; Kruger *et al.*, 2002). Some of these genes, especially those involved in cell-wall fortification showed upregulation as early as 6 h after infection – the earliest time point for *F. graminearum* conidia to germinate on wheat (Kruger *et al.*, 2002).

In addition, many genes involved in abiotic stress were also abundantly expressed, especially those involved in cold and drought acclimation (Kruger *et al.*, 2002; Kong *et al.*, 2007). Other types of stress-related genes that are expressed include enzymes of the phenylpropanoid pathway, lipid transfer proteins and the cellular protectant glutathione *S*-transferases. A positive correlation between level of expression and resistance has been shown for some genes, including two cytochrome P450s, a multi-drug resistance-like protein and a disease resistance-like protein (Kong *et al.*, 2007). Although these studies alluded that many of these may be involved in curtailing FHB severity, a direct relationship between type II resistance and stress-related gene expression has not been established.

Boddu *et al.* (2006) utilized the Barley1 Affymetrix GeneChip representing ~22,000 genes to understand the *F. graminearum*–barley interaction in a susceptible cultivar (Boddu, *et al.*, 2006). Transcriptional profiling revealed that in addition to > 250 genes that were quantitatively different, there were also > 200 genes that were qualitatively upregulated. Genes with qualitative changes were only detected in the pathogen's presence, while the quantitative set represents genes that display greater transcript accumulation after pathogen inoculation compared with mock inoculation. The latter set of genes is likely to be part of the plant's basal resistance and it is activated regardless of the type of pathogen infection (Boddu, *et al.*, 2006).

Genome-wide transcriptional analyses in barley also revealed three distinct stages of the infection process (Boddu, *et al.*, 2006). In the first stage (within 48 h after inoculation), *F. graminearum* behaves like a biotroph and does not necessarily penetrate the host cell. During this period, however, a limited number (~100–150) of host genes localized to the infected site are expressed.

An important note is that during this period trichothecenes are not detectable, suggesting that this first phase is strictly a pathogen recognition period. The majority of the response in barley occurs during the second or intermediate stage. In this stage, the pathogen is recognized; trichothecenes are synthesized and begin to accumulate in the host (Boddu, *et al.*, 2006, 2007). A full-blown induction of the defence response is activated. It is also during this phase, when necrosis is apparent, that the pathogen switches from a biotroph to a necrotroph. The third and final stage is the necrosis stage resulting in large-scale accumulation of trichothecenes and a concomitant reduction in the number of transcripts (Boddu, *et al.*, 2006, 2007).

The Barley1 GeneChip has also been used to compare gene expression profiles between samples inoculated with either a wild-type *F. graminearum* strain or a DON non-producing strain, to separate the host response to the fungus from the response to the mycotoxin DON (Boddu *et al.*, 2007). Some genes were found to be induced only or mainly in the presence of DON, including genes with predicted detoxification and transport activities, and genes coding for proteins associated with ubiquitination and cell death (Boddu *et al.*, 2007). Therefore, in the compatible interaction, DON not only induces genes that attempt to directly detoxify trichothecenes but also induces genes that increase disease progression through accelerated cell death. Here again, additional work is needed to confirm the role of the genes identified.

## 13.4 Basal Resistance – the First Stage of Defence Response

Unlike their animal counterparts, plants lack a somatic adaptive immune system and therefore rely on their innate immunity to thwart attack by any potential pathogens. Foremost in the activation of defence mechanisms is a quick and efficient detection of invading microbes. This occurs in plants regardless of the final outcome, be it disease or resistance (Jones and Dangl, 2006). Evidence gathered in other pathosystems suggests that pathogens are perceived by evolutionary conserved features called PAMPs (pathogen-associated molecular patterns) (Zipfel, 2008). These PAMPs are present in both pathogenic and non-pathogenic microbes and are molecular structures that are unique to microorganisms. They are recognized either directly or indirectly by pattern recognition receptors (PRRs) and appear to be a prerequisite for the induction of defence responses. Pattern recognition is unusual in that each host receptor has broad host specificity and can potentially bind a large number of molecules that have a common structural motif or pattern (Medzhitov, 2007).

In oomycetes and fungal plant pathogens, molecules identified as PAMPs include structurally diverse ergosterols, fungal-specific glycosylated proteins, and the wall components of chitin and β-glucan (Kaku *et al.*, 2006; Kamoun, 2007). The best-studied examples include, a Pep13-domain of the cell-wall transglutaminase which activates resistance responses in *Solanaceae* and a heptaglucoside found in the cell wall of the oomycete *Phytophtora sojae* that functions as an elicitor activating the defence response in soybean (Altenbach and Robatzek, 2007). In addition, enzymes such as xylanase isolated from the commercially available enzyme cocktail cellulysin, isolated from the fungus *Trichoderma viride*, act as PAMPs in a variety of species (Ron and Avni, 2004). The recently identified protein Nep1 (necrosis and ethylene-inducing peptide-1) derived from the pathogen *Hyaloperonospora parasitica*, triggers responses similar to flg22 in *Arabidopsis* (Qutob *et al.*, 2006). Nep-like proteins or NLPs have been identified in *Fusarium* species; however, their role as a PAMP has not been documented (Bae *et al.*, 2006). It is important to note that while the above-mentioned PAMPs trigger responses in a wide variety of plant species; recognized PAMPs such as the Pep13-domain of transglutaminase and fungal cell-wall components such as chitin and bacterial cold shock protein (CSP) are restricted to specific hosts (Zipfel, 2008). This underscores the fact that individual plant species recognize only a subset of potential PAMPs and additionally plants respond variably to different PAMPs (Felix *et al.*, 1999; Kunze *et al.*, 2004). Host genes that are expressed during the first phase of *Fusarium* infection are likely part of the PAMP recognition.

A potential source of PAMPs in *F. graminearum* could be found in its exoproteome (Phalip *et al.*, 2005). There are over 24 different classes of enzymes that are secreted in response to various carbon sources. Most of these proteins are enzymes associated with cell-wall degradation, including cellulases, chitinases, pectin esterases and xylanases. Thirty-two *F. graminearum* genes encoding cell wall-degrading enzymes are induced during infection of susceptible barley and not on other nutritional medium (Güldener *et al.*, 2006; Cuomo *et al.*, 2007). It would be important to characterize these PAMPs in relation to their host specificity, given the fact that *Fusarium* species infect a broad host range encompassing both monocot and dicot plants. In a proteomic study of *F. graminearum* under mycotoxin-inducing conditions *in vitro*, FG04741 was observed to be significantly upregulated both by 2D-PAGE and by a non-gel-based proteomic method (Taylor *et al.*, 2008). FG04741 exhibits 44% amino acid identity with a 22 kDa glycoprotein secreted by *Ophiostoma ulmi*, the causal agent of Dutch elm disease. The *Ophiostoma* glycoprotein is an elicitor of mansonones (elm sesquiterpene quinone phytoalexins) (Yang *et al.*, 1989).

## Pathogen perception: PRRs

PRRs act as sentinels and perceive microbial patterns – PAMPs. Generally, these receptors have been divided into two groups. Those that are found on the surface include receptor-like kinases (RLKs) and receptor-like proteins (RLPs) (Altenbach and Robatzek, 2007). The second group includes intracellular receptors possessing a nucleotide-binding site (NBS) and a leucine-rich repeat (LRR). The intracellular receptors are further subdivided into two types harbouring either the coiled-coil (CC) or the TOLL and IL-1 (TIR) domain at their N terminus. Astonishingly, in *Arabidopsis*, there are over 650 proteins belonging to the RLPs and RLKs families that have been identified (Zipfel, 2008). In comparison, there are only ten proteins structurally related to RLPs found in mammals. To date, very few of these proteins have been functionally characterized in plants and even fewer have been implicated in defence responses. These differences of PAMP receptors could be attributed to the fact that plants lack an adaptive immune system and thus rely on the PRRs with broad specificities against both conserved and invariant features of microbes to activate defence (Medzhitov, 2007).

PRR that binds to the heptaglucoside from oomycetes was the first PRR identified in plants (Fleigmann *et al.*, 2004). It binds to glucan with high affinity and possesses endo-β-glucansae activity (Fleigmann *et al.*, 2004). As has been reported for other PAMPs, this high affinity binding protein and elicitor response is limited to a few species of the *Leguminosae* (Mithofer *et al.*, 2000). The structure of this receptor is not yet characterized. Two high affinity receptors for xylanase have been characterized from tomato (Ron and Avni, 2004). The two proteins LeEIX1 and LeEIX2, from tomato are capable of binding xylanase and they belong to the RLP class of PRRs with extracellular LRRs and a short cytoplasmic tail. A PRR that binds to chitin, a major structural component of fungal cell walls, has also been characterized (Kaku *et al.*, 2006). This CEBiP (Chitin oligosaccharide elicitor-binding protein) is also a member of the RLP family with two extracellular lysM (lysine motif) and a short cytoplasmic tail at its C terminus (Kaku *et al.*, 2006). LysM domains first identified in bacterial lysins and muramidases, are enzymes that degrade cell-wall peptidoglycans. More interesting is the discovery of RLKs with extracellular LysM domains, namely NFR1 and NFR5 from *Lotus japonicus* (Mulder *et al.*, 2006). These RLKs are involved in the recognition of Nod-factors (bacterial chitin-like molecules) during the nitrogen-fixing legume–*Rhizobium* symbiosis.

Detailed study with Nep-like protein or NLP, the sole PRR identified from *H. parasitica* revealed activation of at least 35 LRR-RLKs. Included in these is the wall-associated kinase RFO1 (Qutob *et al.*, 2006). This RLK member has recently been shown to be the largest factor contributing to resistance to *Fusarium oxysporum* (Diener and Ausubel, 2005). Furthermore, *RFO1* has been shown recently to be essential for quantitative resistance to *Verticillium longisporum*, a fungus with lifestyle and infection strategies similar to *F. oxysporum* (Berrocal-Lobo and Molina, 2007; Johansson *et al.*, 2007). The role of other RLKs is currently unknown, but 18 of the RLKs activated by the NLP in *Arabidopsis* exhibit similar expression patterns as when activated by flg22 (Kunze *et al.*, 2004). The speculation is that these RLKs have other roles in addition to perception of the PAMPs.

There are up to 149 potential RLKs and RLPs, and up to 98 potential NBS-LRR proteins represented on the wheat Affymetrix GeneChip (Table 13.1). In our RNA profiling experiments with *F. graminearum*-infected wheat heads, only a small number of those are differentially regulated during the first 4 days of infection by *F. graminearum* and their expression change was more robust in susceptible than in resistant cultivars (T. Ouellet, 2008, unpublished data). However, the basal expression levels of a few RLK and NBS-LRR genes were significantly higher in resistant than in susceptible plants as shown in Table 13.1, and this could contribute to the resistance phenotype.

It would be important to demonstrate if any of these potential PRRs are the target of proteins secreted by *F. graminearum*. It is likely then that a

**Table 13.1.** Overview of potential PRR genes represented on the wheat Affymetrix GeneChip.

| | Number of probe sets | |
|---|---|---|
| Protein type[a] | On wheat chip | Affected by *F. graminearum* |
| LRR RLK | 9 | 1↑ |
| Other RLK | 132 | 17↑, 17↓, 4 with basal level higher in resistant plants |
| RLP | 8 | 1↓ |
| LysM domain | 6 | 2↑ |
| NBS-LRR | 66 | 4↑, 4↓, 2 with basal level higher in resistant plants |
| NBS-LRR, RGA type | 29 | 0 |
| CC-NBS-LRR | 3 | 1↑ |

[a] CC, coiled-coil (domain); LRR, leucine-rich repeat; LysM, lysine motif; NBS, nucleotide-binding site; RGA, resistance gene analogue type; RLK, receptor-like kinases; RLP, receptor-like proteins.

function of the different types of PRRs activated following pathogen perception is to mutually reinforce the defence response. This obviates a necessity for a full-blown response which may compromise the fitness of the plant (Van Hulten *et al.*, 2006). Typically, following PRR activation, there is an activation of local genes and this priming of plant defence is a prerequisite for a sustained response when pathogen-specific virulence factors or effectors are detected by the intracellular receptors (NBS-LRR) (Jones and Dangl, 2006). This double layer of recognition differentiates those microbes that cause disease from those that are beneficial or harmless to the plants. Only when this second layer of resistance is breached does a plant become susceptible and eventually diseased. However, in a majority of cases, plants detect these effectors and enact various counter-effective measures, collectively termed induced responses.

## 13.5 Induced Responses

### Recognition of virulence factors

Induced responses are the result of the failure of basal defence and the recognition of pathogen specific virulence factors injected into the host plant. Our knowledge of fungal and oomycete pathogenicity has been mainly restricted to specialized infection structures, secretion of hydrolytic enzymes and production of host specific toxins (Phalip *et al.*, 2005). However, new findings have broadened our knowledge of both the variety of effectors that these microorganisms secrete into the plant, as well as the methods they employ to target and eventually colonize their hosts (Kamoun, 2007). Effectors that are secreted from the fungus generally act inside the plant cells and interfere with host defence responses (Schmelzer, 2002; Hückelhoven, 2007). However, their effect is also felt on the extracellular matrix of the plant cell. For example, AVR2 from the fungal pathogen *Cladosporium fulvum* binds and inhibits the function of the tomato extracellular cysteine protease RCR3 (Rivas and Thomas, 2005). This interaction between the fungal effector and plant receptor activates the RLP, Cf2, leading to the induced response in tomato. Similarly, other RLPs such as Cf9, Cf4 and Cf5 respond specifically to extracellular effectors secreted by the same pathogen (Rivas and Thomas, 2005).

In a study of the extracellular proteome of maize cell suspension cultures, treatment by *Fusarium verticillioides* elicitor prompted the accumulation of xylanase inhibitor protein, GAPDH, heat shock proteins and the rapid dephosphorylation of peroxidases (Chivasa *et al.*, 2005). This suggests that the extracellular matrix plays a major role in the *Fusarium*–host interaction. Interestingly, the only effector protein studied so far from *Fusarium* species, *F. oxysporum* Six1, encodes a small cysteine-rich protein (Martijn *et al.*, 2005). This gene is essential for virulence in *F. oxysporum* and the resistance is mediated by the resistance gene *I-3* in tomato, a NB-LRR family member (Martijn *et al.*, 2005). Up to now, trichothecenes and a secreted lipase have been identified as *F. graminearum* virulence factors (Proctor *et al.*, 1995; Voigt *et al.*, 2005). However, no host receptor has been identified so far. For a more extensive list of effectors from filamentous pathogens, please refer to a recent review by Sophien Kamoun (2007).

## Cellular responses

The cellular response is initiated at the site of plant–microbe interaction. Failures of fungal ingress are frequently associated with the formation of papillae (Schmelzer, 2002). These are cell-wall appositions and commonly contain callose and phenolics of a highly divergent nature. These depositions are thought to provide a physical barrier to halt growth and contain the penetrating pathogen. Callose depositions have been observed in the non-host *Arabidopsis* infected with various pathogens including *F. graminearum* (R. Subramaniam, 2008, unpublished data). In resistant and susceptible wheat plants where infection was inhibited, the formation of vascular occlusions containing pectin was observed (Ribichich *et al.*, 2000; Guenther and Trail, 2005). Other constituents of papillae can include lignin, cellulose, suberin, chitin and proteins such as hydroxyl-rich glycoproteins (HRGPs) and peroxidases (Takemoto *et al.*, 2003). Another common feature is the rearrangements of the Golgi network and endoplasmic reticulum, which become polarized towards the site of infection (Takemoto *et al.*, 2003; Hückelhoven, 2007). Mutations influencing the structure of the supporting cytoskeleton invariably compromise this initial stage of pathogen ingress (Bhat *et al.*, 2005; Stein *et al.*, 2006). For example, mutation of *AtPEN2*, which encodes for syntaxin, an integral component of cytoskeleton inhibits non-host response to the powdery mildew infection in *Arabidopsis* (Bhat *et al.*, 2005; Stein *et al.*, 2006). Microarray analysis of *F. graminearum*-infected wheat revealed that genes involved in cell-wall fortification such as HRGPs, peroxidases and the syntaxin-like ROR2 are upregulated by the fungus suggesting a role for the cytoskeleton in *Fusarium* defence (Boddu *et al.*, 2006).

The synthesis, deposition and assembly of the papillae constituents require the action of reactive oxygen species (ROS) which includes $H_2O_2$. Necrotrophic fungi such as Fusaria are able to produce hydrolytic enzymes and induce plant ROS to promote cell death which facilitates access to nutrients (Berrocal-Lobo and Molina, 2007). ROS production depends on the type of effector that a plant encounters. For example, Nep1, a PAMP identified initially in *H. parastica*, and conserved in many fungi including *Fusarium* species, induces the expression of the host gene *AtrbohD*, which encodes a NADPH oxidase to produce $H_2O_2$ (He *et al.*, 2001; Bae *et al.*, 2006). However, in *Arabidopsis* cell suspension cells, peroxidases seem to be the source of ROS upon *F. oxysporum* infection (Davies *et al.*, 2006). Our initial experiments with the *Arabidopsis* mutant *AtrbohD* showed no increased susceptibility to *F. graminearum* (R. Subramaniam, 2008, unpublished data). So it is possible that ROS production following *F. graminearum* infection occurs via peroxidases. Mohammadi and Kazemi (2002) found that, after *F. graminearum* inoculation, peroxidase enzymatic activity was significantly higher in resistant wheat cultivars relative to susceptible cultivars at the dough stage. Furthermore, microarray analyses in wheat–, barley– and maize–*F.graminearum* interactions showed upregulation of peroxidases along with other oxidative burst-type genes such as oxalate oxidases (Boddu *et al.*, 2006). Class III peroxidases (POXs) have been implicated in the final polymerization of phenolic derivatives into lignin and other cell-wall fortifications as well as wound healing and enhanced production of ROS and phytoalexins (Hiraga *et al.*, 2001; Mohammadi and Kazemi, 2002). The PeroxiBase database (http://peroxidase.isb-sib.ch) currently lists 113 *Zea mays*, 98 *Triticum aestivum* and 104 *Hordeum vulgare* class III peroxidase genes. It is still uncertain whether expression of any of these offers protection to these plants.

ABC and multi-drug and toxic compound extrusion (MATE) transporters function in cross-membrane metabolite transport (Hückelhoven, 2007). Traditionally, the focus has been on how the plant metabolites are transported to the site of pathogen ingress. However, in the context of *Fusarium*–host interaction, the transporters need to not only transport plant metabolites to combat the ingress but they also have to efficiently pump out the mycotoxins that are produced by the pathogens. *Fusarium graminearum* produces various kinds of mycotoxins, including trichothecenes, a potent inhibitor of host protein synthesis (Desjardins, 2006). An ABC transporter NpPDR1 from tobacco has been shown to be essential for transport of antifungal diterpene sclareol upon infection with the necrotrophic pathogen *Botrytis cinerea* (Stukkens *et al.*, 2005). The expression of this gene is augmented after infection with this pathogen. More recently, an *Arabidopsis* ABC transporter mutant *Atpen3/Atpdr8* was rendered susceptible to and allowed early penetration of the non-host pathogen *B. graminis* from barley (Stein *et al.*, 2006). In wheat, the predicted homologue of *Pen3* is robustly expressed in response to *F. graminearum* infection. Additionally, many other genes predicted to be ABC and MATE transporters were induced by the infection (Table 13.2) and concomitantly, a subset of the MATE and ABC genes was also repressed by *F. graminearum*; however, the downregulation was observed only in the susceptible varieties of wheat (Table 13.2). In barley and maize, genes encoding for both ABC transporters and the MATE transporters were induced by infection (Boddu *et al.*, 2006; T. Ouellet, 2008, unpublished data), thus offering a scenario of reducing the concentration of mycotoxins by shuttling it from the cytoplasm.

## Intracellular signalling

Activation of NB-LRR results in a complex network of responses in part to differentiate the nature of attack. In the model systems, when the pathogen is a biotroph, then the response favours the pathway associated with the signals obtained from the accumulation of salicylic acid (SA) (Jones and Dangl, 2006). However, a combination of the jasmonic acid (JA) and ethylene pathways are favoured for the host that is attacked by necrotrophs. These straightforward responses cannot be applied to pathogens that have a mixed lifestyle such as *F. graminearum*.

*Arabidopsis* mutants defective in either SA biosynthesis, such as *sid2*, or responsiveness, such as *npr1*, are compromised both in basal defence and in systemic acquired resistance (SAR). Overexpression of AtNPR1 in transgenic *Arabidopsis* provides enhanced resistance to bacterial and oomycete pathogens (Cao *et al.*, 1998; Friedrich *et al.*, 2001). The enhanced disease resistance in *Arabidopsis* conferred by the overexpression of *AtNPR1* was associated with the faster response of these plants to SA (Cao *et al.*, 1998; Friedrich *et al.*, 2001). Expression of wheat *PR* genes is induced in response to *F. graminearum* inoculation and SA application (Pritsch *et al.*, 2000; Kruger *et al.*, 2002; Anand *et al.*, 2003) and more importantly, resistance to FHB is correlated with an increase in the endogenous level of SA (Anand *et al.*, 2003). Furthermore, constitutive expression of bacterial *NahG* in *Arabidopsis* which catabolizes SA into catechol, renders those plants susceptible to *F. graminearum* infection (C. Nasmith, 2008, unpublished data). A positive role for SA is also supported by microarray analyses in barley, wheat and maize. In barley, the SA regulator *NPR* and *PR1*, a marker gene for SA signalling, are upregulated (Boddu *et al.*, 2006). *PR1* genes are also strongly upregulated in wheat and maize (T. Ouellet, 2008, unpublished data).

JA signalling mediates diverse cellular responses including those originating from pathogen attack. JA biosynthetic genes such as *cis*-oxophytodienoic acid

**Table 13.2.** Predicted MATE and ABC transporters on the wheat Affymetrix GeneChip and their response to *F. graminearum* infection in spike tissues.

| Transporter type | Total[a] | Upregulated | Downregulated |
|---|---|---|---|
| MATE | 7 | 3 | 2 |
| ABC | 46 | 26 | 11 |

[a] Total values include transporters that are upregulated, those that are downregulated and those that remain unchanged (e.g. of the seven MATE transporters, three are upregulated, two are downregulated and the others remain unchanged).

(OPDA) reductase, lipoxygenase 2 and allene oxide synthase (AOS) are upregulated upon attack by necrotrophs (Lorenzo and Solano, 2005). Large-scale transcriptional profiling in wheat and barley following *F. graminearum* infection did not reveal upregulated expression of these genes but this may be the result of limited sampling times. Inferences from limited studies should be viewed with caution, since a number of genes related to either the lipoxygenases or precursors to JA biosynthesis such as 12-OPDA have been shown to be upregulated in maize during its interaction with *F. verticillioides* (Nemchenko *et al.*, 2006; Gao *et al.*, 2007). Transcriptional profiling also revealed that in the early part of the resistance interaction, enzymes that are involved in ethylene biosynthesis are upregulated. 1-Aminocyclopropane-1-carboxylic acid (ACC) synthase and ACC oxidase, part of the biosynthetic pathway of ethylene, are activated early and are sustained for a considerable period of time in maize, barley and wheat (L. Harris, 2008, unpublished data). Molecular markers such as PR4, indicative of ethylene responsiveness, are highly upregulated in this interaction (Guo and Ecker, 2004; Van Loon *et al.*, 2006). Interestingly, an ethylene insensitive mutant 2 (EIN2) homologue, an essential component of ethylene signal transduction, is downregulated in the susceptible varieties of wheat compared to the resistant ones (T. Ouellet, 2008, unpublished data). Recently, an EIN2 has been shown to be involved in multi-hormone interaction (O'Donnell *et al.*, 2003). Mutation in this gene renders the plant hypersensitive to abscisic acid (ABA). Thus, the ethylene response pathway seems to intersect with ABA and other hormone pathways in a complex regulatory network that awaits further analyses. Therefore, early and sustained activation of ethylene and SA-responsive genes in the infection process may be important for *F. graminearum* resistance. Moreover, any responses that augment the SA signalling will be favoured in the resistance interaction.

## 13.6 Perspectives

With the availability of DNA-microarrays such as the barley and wheat Affymetrix GeneChip, we have begun to understand the mechanism of *F. graminearum* infection. Comparative analyses between resistant varieties of wheat and a susceptible one from barley infected with *F. graminearum* reveal that there is a substantial overlap in their transcriptional profiles. The common element between the two pathosystems is that both exhibit type II resistance, in which the pathogen is able to infect but is unable to spread beyond the infection site (Goswami and Kistler, 2004). One of the studies utilized mutant strains of *F. graminearum* compromised in mycotoxin biosynthesis to further differentiate genes necessary for toxin detoxification and cell death (Boddu *et al.*, 2007). Therefore, similar studies with other mutant strains of *F. graminearum* that are compromised in various aspects of virulence will identify genes that are necessary for defence against this pathogen.

Overall, there is a recognition that defence against *F. graminearum* is multifaceted and complex. Both branches of the plant immune system, the basal response and the induced response seem to be involved. The basal response occurs through transmembrane PRRs that respond to slowly evolving microbial patterns (PAMPs). This is supported by microarray data in which we see differential expression of many PRRs, which is likely to be a result of recognition of multiple PAMPs. The induced response is intracellular in nature. The induced response against *F. graminearum* resembles many aspects of non-host resistance (Mysore and Choong-Min, 2004). Non-host resistance is a composite of overlapping and multi-component forms of resistance. While *Arabidopsis* offers an optimal system to decipher this type of resistance, the recently sequenced grass species *Brachypodium distachyon* provides a plant genetic resource that is evolutionarily more relevant to cereal crops such as wheat and maize (Draper *et al.*, 2001). As such, development of this pathosystem could be valuable to understand *Fusarium* pathogenesis. Since non-host resistance is generally thought to be broad spectrum, it is expected to be more durable under field conditions.

In conclusion, while *Fusarium* taxonomy, biodiversity, host specialization, economic threats and ever-increasing numbers of species, have presented challenges for both taxonomists and plant pathologists, these challenges may now be viewed as new opportunities for insights into *Fusarium* research. No other fungal plant system has the urgency for answers that *Fusarium* has with respect to global crop protection, mycotoxin contamination and genetics. There is now an opportunity to address these questions with the resources currently available.

## References

Agrios, G.N. (1997) *Plant Pathology*, 4th edn. Academic Press, Berkeley, California.

Altenbach, D. and Robatzek, S. (2007) Pattern recognition receptors: from the cell surface to intracellular dynamics. *Molecular Plant–Microbe Intearctions* 20, 1031–1039.

Anand, A., Zhou, T., Trick, H.N., Gill, B.S., Bockus, W.W. and Muthukrishnan, S. (2003) Greenhouse and field testing of transgenic wheat plants stably expressing genes for thaumatin-like protein, chitinase and glucanase against *Fusarium graminearum. Journal of Experimental Botany* 54, 1101–1111.

Bae, H., Kim, M.S., Sichr, R.C., Bae, H.-J. and Bailey, B.A. (2006) Necrosis- and ethylene-inducing peptide from *Fusarium oxysporum* induces a complex cascade of transcripts associated with signal transduction and cell death in *Arabidopsis. Plant Physiology* 141, 1056–1067.

Bai, G.-H. and Shaner, G. (1994) Scab of wheat: prospects for control. *Plant Disease* 78, 760–770.

Berrocal-Lobo, M. and Molina, A. (2007) *Arabidopsis* defense response against *Fusarium oxysporum. Trends in Plant Science* 13, 145–150.

Bhat, R.A., Miklis, M., Schmelzer, E., Schulze-Lefert, P. and Panstruga, R. (2005) Recruitment and interaction dynamics of plant penetration resistance components in a plasma membrane microdomain. *Proceedings of National Academy of Sciences, USA* 102, 3135–3140.

Bloomberg, W.J. (1981) Diseases caused by *Fusarium* in forest nurseries. In: Nelson, P.E., Toussoun, T.A. and Cook, R.J. (eds) *Fusarium: Diseases, Biology, and Taxonomy.* Pennsylvania State University Press, University Park, Pennsylvania, pp. 178–187.

Boddu, J., Cho, S., Kruger, W.M. and Muehlbauer, G.J. (2006) Transcriptome analysis of the barley–*Fusarium graminearum* interaction. *Molecular Plant–Microbe Interactions* 19, 407–417.

Boddu, J., Cho, S. and Muehlbauer, G.J. (2007) Transcriptome analysis of trichothecene-induced gene expression in barley. *Molecular Plant–Microbe Interactions* 20, 1364–1375.

Bushnell, W.R., Hazen, B.E. and Pritsch, C. (2003) Histology and physiology of Fusarium head blight of wheat and barley. In: Leonard, K.J. and Bushnell, W.R (eds) *Fusarium Head Blight of Wheat and Barley.* American Phytopathological Society (APS) Press, St Paul, Minnesota, pp. 44–83.

Cao, H., Li, X. and Dong, X. (1998) Generation of broad-spectrum disease resistance by overexpression of an essential regulatory gene in systemic acquired resistance. *Proceedings of National Academy of Sciences, USA* 95, 6531–6536.

Cassini, R. (1981) *Fusarium* diseases of wheat and corn in Western Europe. In: Nelson, P.E., Toussoun, T.A. and Cook, R.J. (eds) *Fusarium: Diseases, Biology, and Taxonomy.* Pennsylvania State University Press, University Park, Pennsylvania, pp. 56–65.

Chivasa, S., Simon, W.J., Yu, X.-L., Yalpani, N. and Slabas, A.R. (2005) Pathogen elicitor-induced changes in the maize extracellular matrix proteome. *Proteomics* 5, 4894–4904.

Cook, R.J. (1981a) *Fusarium* diseases in the People's Republic of China. In: Nelson, P.E., Toussoun, T.A. and Cook, R.J. (eds) *Fusarium: Diseases, Biology, and Taxonomy.* Pennsylvania State University Press, University Park, Pennsylvania, pp. 39–52.

Cook, R.J. (1981b) *Fusarium* diseases of wheat and other small grains in North America. In: Nelson, P.E., Toussoun, T.A. and Cook, R.J. (eds) *Fusarium: Diseases, Biology, and Taxonomy.* Pennsylvania State University Press, University Park, Pennsylvania, pp. 53–55.

Cuomo, C.A., Güldener, U., Xu, J.-R., Trail, F., Turgeon, B.G., Di Pietro, A., Walton, J.D., Ma, J.-L., Baker, S.E., Rep, M., Adam, G., Antoniw, J., Baldwin, T., Calvo, S., Chang, Y.-L., DeCaprio, D., Gale, L.R., Gnerre, S., Goswami, R.S., Hammond-Kosack, K., Harris, L.J., Hilburn, K., Kennell, J.C., Kroken, S., Magnuson, J.K., Mannhaupt, G., Mauceli, E., Mewes, H.-W., Mitterbauer, R., Muehlbauer, G., Münsterkötter, M., Nelson, D., O'Donnell, K., Ouellet, T., Qi, W., Quesneville, H., Roncero, I.G., Seong, K.-Y., Tetko, I.V., Urban, M., Waalwijk, C., Ward, T.J., Yao, J., Birren, B.W. and Kistler, H.C. (2007) The *Fusarium graminearum* genome reveals a link between localized polymorphism and pathogen specialization. *Science* 317, 1400–1402.

Davies, D.R., Bindschedler, L.V., Strickland, T.S. and Bolwell, P.G. (2006) Production of reactive oxygen species in *Arabidopsis thaliana* cell suspension cultures in response to an elicitor from *Fusarium oxysporum*: implications for basal resistance. *Journal of Experimental Botany* 57, 1817–1827.

Desjardins, A.E. (2006) *Fusarium Mycotoxins Chemistry, Genetics and Biology.* The American Phytopathological Society (APS) Press, St Paul, Minnesota.

Diener, A.C. and Ausubel, F.M. (2005) *RESISTANCE TO FUSARIUM OXYSPORUM 1*, a dominant *Arabidopsis* disease-resistance gene, is not race specific. *Genetics* 171, 305–321.

Draper, J., Mur, A.J.L., Jenkins, G., Ghouk-Biswas, G.C., Bablak, P., Hasterok, R. and Routledge, A.P.M. (2001) *Brachypodium distachyon.* A new model system for functional genomics in grasses. *Plant Physiology* 127, 1539–1555.

Farr, D.F., Bills, G.F., Chamuris, G.P. and Rossman, A.Y. (1989) *Fungi on Plants and Plant Products in the United States.* American Phytopathological Society (APS) Press, St Paul, Minnesota.

Felix, G., Duran, J.D., Volko, S. and Boller, T. (1999) Plants have a sensitive perception system for the most conserved domain of bacterial flagellin. *The Plant Journal* 18, 265–276.

Fleigmann, J., Mithofer, A., Wanner, G. and Ebel, J. (2004) An ancient enzyme domain hidden in the putative β-glucan elicitor receptor of soybean may play an active part in the perception of pathogen-associated molecular patterns during broad host resistance. *Journal of Biological Chemistry* 279, 1132–1140.

Friedrich, L., Lawton, K., Dietrich, R., Willits, M., Cade, R. and Ryals, J. (2001) *NIM1* overexpression in *Arabidopsis* potentiates plant disease resistance and results in enhanced effectiveness of fungicides. *Molecular Plant–Microbe Interactions* 14, 1114–1124.

Gao, X., Won-Bo, S., Gobel, C., Kunze, S., Feussner, I., Meeley, R., Balint-Kurti, P. and Kolomiets, M. (2007) Disruption of a maize 9-lipoxygenase results in increased resistance to fungal pathogens and reduced levels of contamination with mycotoxin fumonisin. *Molecular Plant–Microbe Interactions* 20, 922–933.

Glenn, A.E., Zitomar, N.C., Zimeri, A.M., Williams, L.D., Riley, R.T. and Proctor, R.H. (2008) Transformation-mediated complementation of a FUM gene cluster deletion in *Fusarium verticillioides* restores both fumonisin production and pathogenicity on maize seedlings. *Molecular Plant–Microbe Interactions* 21, 87–97.

Goswami, R.S. and Kistler, C.H. (2004) Heading for disaster: *Fusarium graminearum* on cereal crops. *Molecular Plant Pathology* 5, 515–525.

Guenther, J.C. and Trail, F. (2005) The development and differentiation of *Gibberella zeae* (anamorph: *Fusarium graminearum*) during colonization of wheat. *Mycologia* 97, 229–237.

Güldener, U., Seong, K.-Y., Boddu, J., Cho, S., Trail, F., Xu, J.-R., Adam, G., Mewes, H.-W., Muehlbauer, G.J. and Kistler, C. (2006) Development of a *Fusarium graminearum* Affymetrix GeneChip for profiling fungal gene expression *in vitro* and *in planta. Fungal Genetics and Biology* 43, 316–325.

Guo, H. and Ecker, J.R. (2004) The ethylene signaling pathway: new insights. *Current Opinion in Plant Biology* 7, 40–49.

He, C., Hsiang, T. and Wolyn, D.J. (2001) Activation of defense responses to *Fusarium* infection in *Asparagus densiflorus. European Journal of Plant Pathology* 107, 473–483.

Hiraga, S., Sasaki, K., Ito, H., Ohashi, Y. and Matsui, H. (2001) A large family of Class III plant peroxidases. *Plant and Cell Physiology* 42, 462–468.

Hückelhoven, R. (2007) Transport and secretion in plant–microbe interactions. *Current Opinion in Plant Biology* 10, 573–579.

Inch, S.A. and Gilbert, J. (2003) Survival of *Gibberella zeae* in *Fusarium*-damaged wheat kernels. *Plant Disease* 87, 282–287.

Johansson, A., Staal, J. and Dixelius, C. (2007) Early responses in the *Arabidopsis-Verticillium longisporum* pathosystem are dependent on NDR1, JA- and ET-associated signals via cytosolic NPR1 and RFO1. *Molecular Plant–Microbe Interactions* 19, 958–969.

Jones, J.D.G. and Dangl, J.L. (2006) The plant immune system. *Nature* 444, 323–329.

Kaku, H., Nishizawa, Y., Ishii-Minami, N., Akimoto-Tomiyama, C., Dohmae, N., Takio, K., Minami, E. and Shibuya, N. (2006) Plant cells recognize chitin fragments for defense signaling through a plasma membrane receptor. *Proceedings of National Academy of Sciences, USA* 103, 11086–11091.

Kamoun, S. (2007) Groovy times: filamentous pathogen effectors revealed. *Current Opinion in Plant Biology* 10, 358–365.

Kang, Z. and Buchenauer, H. (2000) Ultrastructural and immunocytochemical investigation of pathogen development and host responses in resistant and susceptible wheat spikes infected by *Fusarium culmorum. Physiological and Molecular Plant Pathology* 57, 255–268.

Kong, L., Ohm, H.W. and Anderson, J.M. (2007) Expression analysis of defense-related genes in wheat in response to infection by *Fusarium graminearum. Genome* 50, 1038–1048.

Kruger, W.M., Pritsch, C., Chao, S. and Muehlbauer, G.J. (2002) Functional and comparative bioinformatic analysis of expressed genes from wheat spikes infected with *Fusarium graminearum*. *Molecular Plant–Microbe Interactions* 15, 445.

Kunze, G., Zipfel, C., Robatzek, S., Niehaus, K., Boller, T. and Felix, G. (2004) The N terminus of bacterial elongation factor Tu elicits innate immunity in *Arabidopsis* plants. *The Plant Cell* 16, 3496–3507.

Lewandowski, S. and Bushnell, W.R. (2001) Development of *Fusarium graminearum* on floret surfaces of field-grown barley. In: *Proceedings of 2001 National Fusarium Head Blight Forum*. 2001 National Head Blight Forum Proceedings, Erlanger, Kentucky, p. 128.

Lorenzo, O. and Solano, R. (2005) Molecular players regulating the jasmonate signalling network. *Current Opinion in Plant Biology* 8, 532–540.

Markell, S.G. and Francl, L.J. (2003) Fusarium head blight inoculum: species prevalence and *Gibberella zeae* spore type. *Plant Disease* 87, 814–820.

Martijn, R., Meijer, M., Houterman, P.M., Van der Does, H.C. and Corneilssen, B.J.C. (2005) *Fusarium oxysporum* evades I-3-mediated resistance without altering the matching avirulence gene. *Molecular Plant–Microbe Interactions* 18, 15–23.

Medzhitov, R. (2007) Recognition of microorganisms and activation of the immune response. *Nature* 449, 819–826.

Miller, S.S., Reid, L.M. and Harris, L.J. (2007) Colonization of maize silks by *Fusarium graminearum*, the causative organism of gibberella ear rot. *Canadian Journal of Botany* 85, 369–376.

Mithofer, A., Fliegmann, J., Neuhaus-Url, G., Schwarz, H. and Ebel, J. (2000) The hepta-geta-glucoside elicitor-binding proteins from legumes represent a putative family. *Biological Chemistry*. 381, 705–713.

Mohammadi, M. and Kazemi, H. (2002) Changes in peroxidase and polyphenol oxidase activities in susceptible and resistant wheat heads inoculated with *Fusarium graminearum* and induced resistance. *Plant Science* 162, 491–498.

Mulder, L., Lefebvre, B., Cullimore, J. and Imberty, A. (2006) LysM domains of *Medicago truncatula* NFP protein involved in Nod factor perception. Glycosylation state, molecular modeling and docking of chitooligosaccharides and Nod factors. *Glycobiology* 16, 801–809.

Mysore, K.S. and Choong-Min, R. (2004) Nonhost resistance: how much do we know? *Trends in Plant Science* 9, 97–104.

Naik, D.M. and Busch, L.V. (1978) Stimulation of *Fusarium graminearum* by maize pollen. *Canadian Journal of Botany* 56, 1113–1117.

Nemchenko, G., Kunze, S., Feussner, I. and Kolomiets, M. (2006) Duplicate maize 13-lipoxygenase genes are differentially regulated by circadian rhythm, cold stress, wounding, pathogen infection, and hormonal treatments. *Journal of Experimental Botany* 57, 3767–3779.

O'Donnell, P.J., Schelmz, E., Block, A., Miersch, O., Wasternack, C., Jones, J.B. and Klee, H.J. (2003) Multiple hormones act sequentially to mediate a susceptible tomato pathogen defense response. *Plant Physiology* 133, 1181–1189.

Parry, D.W. (1990) The incidence of *Fusarium* spp. in stem bases of selected crops of winter wheat in the Midlands. *Plant Pathology* 39, 619–622.

Parry, D.W., Jenkinson, P. and McLeod, L. (1995) Fusarium ear blight (scab) in small grain cereals – a review. *Plant Pathology* 44, 207–238.

Phalip, V., Delalande, F., Carapito, C., Goubet, F., Hatsch, D., Leize-Wagner, E., Dupree, P., Dorsselaer, A.V. and Jeltsch, J.M. (2005) Diversity of the exoproteome of *Fusarium graminearum* grown on plant cell wall. *Current Genetics* 48, 366–379.

Pritsch, C., Muehlbauer, G.J., Bushnell, W.R., Somers, D.A. and Vance, C.P. (2000) Fungal development and induction of defense response genes during early infection of wheat spikes by *Fusarium graminearum*. *Molecular Plant–Microbe Interactions* 13, 159–169.

Proctor, R.H., Hohn, T.M. and McCormick, S.P. (1995) Reduced virulence of *Gibberella zeae* caused by disruption of a trichothecene toxin biosynthetic gene. *Molecular Plant–Microbe Interactions* 8, 593.

Qutob, D., Kemmerling, B., Brunner, F., Kuffner, I., Engelhardt, S., Gust, A.A., Leberaki, B., Seitz, H.U., Stahl, D., Rauhut, T., Glawischnig, E., Scheen, G., Lacombe, B., Watanabe, N., Lam, E., Schlichting, R., Scheel, D., Nau, K., Dodt, G., Hubert, D., Gijzen, M. and Nurnberger, T. (2006) Phytotoxicity and innate immune responses induced by Nep1-like proteins. *The Plant Cell* 18, 3721–3744.

Ribichich, A., Lopez, S.E. and Vegetti, A.C. (2000) Histopathological spikelet changes produced by *Fusarium graminearum* in susceptible and resistant wheat cultivars. *Plant Disease* 84, 794–802.

Rivas, S. and Thomas, C.M. (2005) Molecular interactions between tomato and the leaf mold pathogen *Cladosporium fulvum*. *Annual Review of Phytopathology* 43, 395–436.

Ron, M. and Avni, A. (2004) The receptor for the fungal elicitor ethylene-inducing xylanase is a member of a resistance-like gene family in tomato. *The Plant Cell* 16, 1604–1615.

Schmelzer, E. (2002) Cell polarization, a crucial process in fungal defence. *Trends in Plant Science* 7, 411–415.

Starkey, D.E., Ward, T.J., Aoki, T., Gale, L.R., Kistler, H.C., Geiser, D.M., Suga, H., Tóth, B., Varga, J. and O'Donnell, K. (2007) Global molecular surveillance reveals novel Fusarium head blight species and trichothecene toxin diversity. *Fungal Genetics and Biology* 44, 1191–1204.

Stein, M., Dittgen, J., Sanchez-Rodriguez, C., Hou, B.H., Molina, A., Schulze-Lefert, P., Lipka, V. and Somerville, S. (2006) *Arabidopsis* PEN3/PDR8, an ATP binding cassette transporter, contributes to nonhost resistance to inappropriate pathogens that enter by direct penetration. *The Plant Cell* 18, 731–746.

Stoner, M.F. (1981) Ecology of *Fusarium* in non-cultivated soil. In: Nelson, P.E., Toussoun, T.A. and Cook, R.J. (eds) *Fusarium: Diseases, Biology, and Taxonomy.* Pennsylvania State University Press, University Park, Pennsylvania, pp. 276–286.

Stukkens, Y., Bultreys, A., Grec, S., Trombik, T., Vanham, D. and Boutry, M. (2005) NpPDR1, a pleiotropic drug resistance-type ATP-binding cassette transporter from *Nicotiana plumbaginifolia*, plays a major role in plant pathogen defense. *Plant Physiology* 139, 341–352.

Sun, S.-K. and Snyder, W.C. (1981) The bakanae disease of the rice plant. In: Nelson, P.E., Toussoun, T.A. and Cook, R.J. (eds) *Fusarium: Diseases, Biology, and Taxonomy.* Pennsylvania State University Press, University Park, Pennsylvania, pp. 104–113.

Sutton, J.C. (1982) Epidemiology of wheat head blight and maize ear rot caused by *Fusarium graminearum*. *Canadian Journal of Plant Pathology* 4, 195–209.

Takemoto, D., Jones, D.A. and Hardham, A.R. (2003) GFP-tagging of cell components reveals the dynamics of subcellular re-organization in response to infection of *Arabidopsis* by oomycete pathogens. *The Plant Journal* 33, 775–792.

Taylor, R.D., Saparno, A., Blackwell, B., Anoop, V., Gleddie, S., Tinker, N.A. and Harris, L.J. (2008) Proteomic analyses of *Fusarium graminearum* grown under mycotoxin-inducing conditions. *Proteomics* 8, 2256–2265.

Tekauz, A., McCallum, B. and Gilbert, J. (2000) Review: Fusarium head blight of barley in western Canada. *Canadian Journal of Plant Pathology* 22, 9–16.

Trail, F., Xu, H., Loranger, R. and Gadoury, D. (2002) Physiological and environmental aspects of ascospore discharge in *Gibberella zeae* (anamorph *Fusarium graminearum*). *Mycologia* 94, 181–189.

Van Hulten, M., Pelser, M., Van Loon, L.C., Pieterse, M.J. and Ton, J. (2006) Costs and benefits of priming for defense in *Arabidopsis*. *Proceedings of National Academy of Sciences, USA* 103, 5602–5607.

Van Loon, C.L., Geraats, B.P.J. and Linthorst, H.J.M. (2006) Ethylene as a modulator of disease resistance in plants. *Trends in Plant Science* 11, 184–191.

Voigt, C.A., Schäfer, W. and Salomon, S. (2005) A secreted lipase of *Fusarium graminearum* is a virulence factor required for infection of cereals. *The Plant Journal* 42, 364–375.

Yang, D., Jeng, R.S. and Hubbes, M. (1989) Mansonone accumulation in elm callus induced by elicitors of *Ophiostoma ulmi*, and general properties of elicitors. *Canadian Journal of Botany* 67, 3490–3497.

Zipfel, C. (2008) Pattern-recognition receptors in plant innate immunity. *Current Opinion in Immunology* 20, 10–16.

# Index